POWER ELECTRONICS

Dr. J. S. Chitode

M.E. (Electronics), Ph.D.

Formerly Professor & Head,

Department of Electronics Engineering

Bharati Vidyapeeth Deemed University

College of Engineering, Pune

TECHNICAL PUBLICATIONS

SINCE 1993

An Up-Thrust for Knowledge

POWER ELECTRONICS

First Edition : December 2020

Published by :

TECHNICAL PUBLICATIONS
SINCE 1993 An Up-Thrust for Knowledge

Amit Residency, Office No.1, 412, Shaniwar Peth, Pune - 411030, M.S. INDIA
Ph.: +91-020-24495496/97, Email : sales@technicalpublications.org
Website : www.technicalpublications.org

ISBN 978-93-332-2401-7

9789333224017

PREFACE

The importance of **Power Electronics** is well known in various engineering fields. Overwhelming response to my books on various subjects inspired me to write this book. The book is structured to cover the key aspects of the subject **Power Electronics.**

The book uses plain, lucid language to explain fundamentals of this subject. The book provides logical method of explaining various complicated concepts and stepwise methods to explain the important topics. Each chapter is well supported with necessary illustrations, practical examples and solved problems. All the chapters in the book are arranged in a proper sequence that permits each topic to build upon earlier studies. All care has been taken to make students comfortable in understanding the basic concepts of the subject.

Representative questions have been added at the end of each section to help the students in picking important points from that section.

The book not only covers the entire scope of the subject but explains the philosophy of the subject. This makes the understanding of this subject more clear and makes it more interesting. The book will be very useful not only to the students but also to the subject teachers. The students have to omit nothing and possibly have to cover nothing more.

I wish to express my profound thanks to all those who helped in making this book a reality. Much needed moral support and encouragement is provided on numerous occasions by my whole family. I wish to thank the **Publisher** and the entire team of **Technical Publications** who have taken immense pain to get this book in time with quality printing.

Any suggestion for the improvement of the book will be acknowledged and well appreciated.

Author
Dr. J. S. Chitode

Dedicated at Lotus feet of Lord Ranganatha

TABLE OF CONTENTS

Chapter - 3 AC Voltage Controllers and Cyclo Converters (3 - 1) to (3 - 58)

ABBREVIATIONS AND SYMBOLS

A	Anode
AC	AC supply current or voltage
B	Base or B-phase voltage
C	Collector or capacitor
D	Diode
D_{FW}	Freewheeling diode
E	Back emf of the motor, Emitter
E_g , e_g	Back emf of the motor
f	Frequency
f_{ripple}	Ripple frequency
G	Gate
HF	Harmonic factor
I_s , i_s	Supply current
I_m	Peak or maximum current
I_c, i_c	Capacitor current or charging current
I_o, i_o	Output current
i_L	Inductance current
I_g , i_g	Gate current
i_B	Base current
I_{AK}	Anode to cathode current
I_T	Average current rating (SCR)
I_{TR}	RMS current rating (SCR)
i_{ripple}	Ripple current
i_{FW}	Freewheeling current
$I_{(av)}$	Average current
I_{rms}	RMS current
i_{cir}	Circulating current
K	Cathode
L, L_r	Inductance, or current limiting reactor
PF	Power factor
P_o, P_i	Output, input power
R	Resistance or R-phase voltage
T	Period of the waveform, transistor or SCR
T_{on}	On or conduction period of switch
t_{on}	Turn-on time of switch
t_c , t_{off}	Circuit turn-off time
t_q	Turn-off time of switch
T_{off}	Off period of switch
t_r	Rise time
t_p	Spread time
t_d	Delay time
t_{rr}	Reverse recovery time
t_{gr}	Gate recovery time
V_s , v_s	Supply voltage
V_m	Peak value of supply phase voltage
V_{BO}	Forward break-over voltage
V_{BR}	Reverse break-down voltage
V_{dc}	Supply or load DC voltage
V_o, v_o	Output voltage
V_{rms}	RMS voltage
$V_{(av)}$	Average voltage
v_g	Gate voltage
v_B	Base voltage
v_c	Capacitor voltage
V_p , v_p	Peak voltage in UJT or supply
V_{BB} , V	Bias voltage or UJT supply voltage
v_{ripple}	Ripple voltage
v_R , v_Y , v_B	3 ϕ supply voltages
V_{AK}	Anode to cathode voltage
η	Intrinsic standoff ratio
$\frac{dv}{dt}$	Rate of change of voltage
$\frac{di}{dt}$	Rate of change of current
δ	Duty cycle
π	Pi or half cycle period of supply
α	Triggering angle or firing angle
ω	Angular frequency
β	Extinction angle

1 Power Devices and Commutation Circuits

Objectives

- *Principle of operation of power converter . Necessity, input, output and controls.*
- *Applications of power electonics. Advantages and disadvantages.*
- *Classification of power converters depending upon input and output.*
- *Introduction to power semiconductor devices, their important ratings.*
- *Characteristics of power devices.*
- *Basics of triggering circuits for power devices.*
- *Requirements of triggering circuits.*
- *Triggering circuits for SCR, BJT, MOSFET and IGBTs.*
- *Isolation in triggering circuits.*
- *Protection of power devices from* dv/dt *and over voltages by snubber circuit.*
- *Protection of power devices against hot spots and* di/dt *.*
- *Metal oxide varistors (MOVs) for the protection of power devices.*
- *Semiconductor fuses for protection against over currents.*
- *Concept of thermal resistance and heat sink design.*
- *Mounting of devices for power cooling.*
- *Derating of devices due to heating.*
- *Cooling techniques such as forced air cooling, liquid cooling and vapour phase cooling.*

1.1 Introduction

- Power electronics is one of the important branch of electronics and electrical engineering. It deals with conversion and control of electric energy. We know that AC voltage and current of fixed frequency is available from mains. This supply cannot be used always directly. For example computer needs SMPS (Switched Mode Power Supply) for its working.

- Fig. 1.1.1 shows the basic functioning of power electronic system. The electric energy in one form is given at the input. The power electronic system converts the electric energy in the other form. For example, the input may be AC and the output can be DC. We know that such conversion is performed by rectifier. Thus rectifier is a power electronic system.

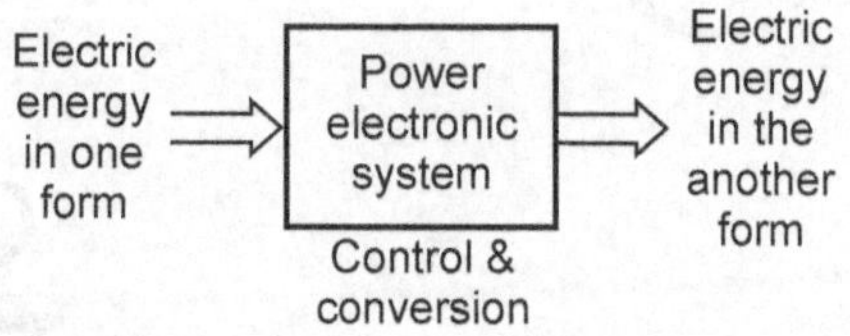

Fig. 1.1.1 Basic inputs/outputs of power electronic systems

- The power electronic system thus performs *conversion* of electric energy. It also *controls* the amount of electric energy to be given to the output.
- The word *power* means high amplitudes of current and voltages.

1.1.1 Applications of Power Electronics

We will briefly present the various applications where power electronic systems are used. There are numerous applications coming up every day in power electronics. Some of them are mentioned below :

1. Uninterruptible power supplies and stand by power supplies (emergency power supplies) for critical loads such as computers, medical equipments etc.
2. Power control in resistance welding induction heating, electrolysis, process industry etc.
3. Power conversion for HVDC and HVAC transmission systems.
4. Speed control of motors which are used in traction drives, textile mills, rolling mills, cranes, lifts, compressors, pumps etc.
5. Solid state power compensators, static contactors, transformer tap changers etc.
6. High voltage supplies for electrostatic precipitators, and x-ray generators, etc.
7. Power supplies for communication systems, telephone exchanges, satellite systems etc.

These are some of the important applications of power electronics.

1.1.2 Advantages and Disadvantages of Power Electronic Controllers

The control and conversion of electric power is performed with the help of power electronic controllers. Thus power electronic systems consists of *controllers*. The power electronic *controllers* are also called as power electronic *converters*.

The power electronic controllers have following advantages :

1. Fast dynamic response due to static devices.
2. High efficiency of conversion due to low losses in electronic devices.
3. Compact size and light weight of the controllers due to electronic devices.
4. Increased operating life and reduced maintenance since there are no moving parts.
5. Power electronic controllers use digital or microprocessor based control. Hence their operation is highly flexible.
6. Since solid state devices are used, the electromagnetic interference and a acoustic noise is reduced.

The power electronic controllers have some drawbacks also. They are as follows :

1. The power electronic controllers generate harmonics. These harmonics affect the performance of other loads.
2. The power factor of some power electronic controllers is very low. Hence power factor correction is necessary to reduce reactive power.
3. For very simple conversion requirements, power electronic converters may be costly.

Here note that the advantages outweigh the disadvantages. Hence power electronic controllers are used in large number of applications.

1.1.3 Block Diagram of Power Electronic Controller

Fig. 1.1.2 shows the block diagram of the system which uses power controller. Here note that the power controller or converter can use thyristor (SCR), GTO, MOSFET, BJT or IGBT as a switch.

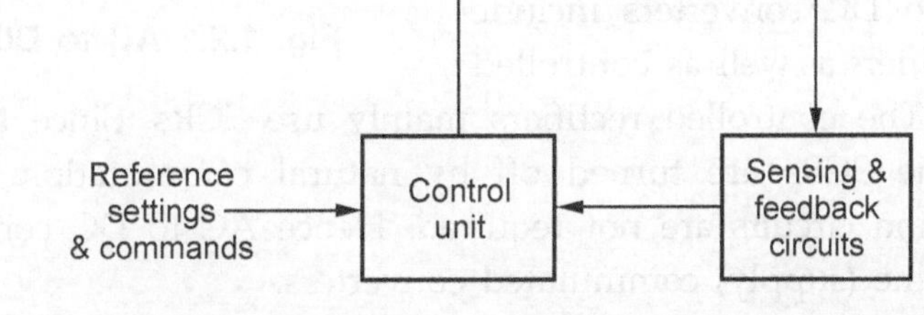

Fig. 1.1.2 Block diagram of the system having power controller

- The power source can be AC mains, generator or batteries. The power controller converts the input power which is suitable for the load. For example, if the load is DC motor, then it needs dc voltage and current. Then the power controller converts the input AC into controlled DC which is suitable for the motor.
- The sensing and feedback circuits monitor the load conditions. For example, speed and output current are sensed in case of motor loads. This feedback is given to the control unit.
- The control unit consists of drive circuits of the power controller. The drives of the switches are adjusted according to feedback and the reference settings. In case of motor drives, the reference settings can be required speeds at different times.
- The control unit adjusts the drives whenever there is difference between feedback (actual) speed and reference speed. The control unit also accepts commands form the user. These commands are given for the proper functioning of the power electronic system and the load.

Our major study of power electronics is centered around the study of various power controllers (converters), their performance for different loads, control and drive circuits, design of feedback and sensing circuits etc.

1.2 Types of Power Electronic Converters

The power electronic circuits are classified into five broad categories depending upon the input, output and job they perform. These power electronic circuits are discussed briefly as follows :

1.2.1 AC to DC Converters

- Fig. 1.2.1 shows the inputs and outputs of AC to DC converters. The input is single phase or three phase AC supply normally available from the mains. The output is the controlled DC voltage and current.

Fig. 1.2.1 AC to DC converters

- The AC to DC converters include diode rectifiers as well as controlled rectifiers. The controlled rectifiers mainly use SCRs. Since the input is AC supply, the SCRs are turned off by natural commutation. Hence external commutation circuits are not required. Hence AC to DC converters are also called as line (supply) commutated converters.
- These converters are used for DC drives, UPS and HVDC systems.

1.2.2 DC to AC Converters (Inverters)

- Fig. 1.2.2 shows the inputs and outputs of DC to AC converters. These converters are commonly called inverters. The input to the inverters is fixed DC voltage. Normally this DC voltage is obtained from the batteries. The output of the inverter is the fixed or variable frequency AC voltage. The AC voltage magnitude is also variable.

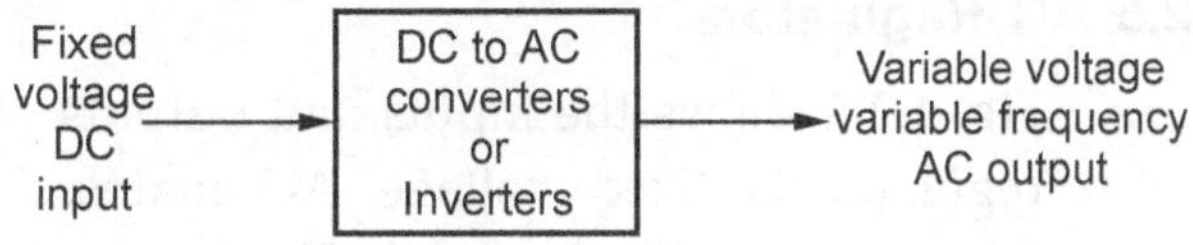

Fig. 1.2.2 DC to AC converters or inverters

- Inverters are mainly used whenever mains is not available. For example UPS use inverters inside to generate AC output from batteries. Inverters are also used for speed control of induction motors. The voltage, frequency or both are varied by inverter to control the speed of induction motors. Inverters are also used in standby and emergency power supplies.

1.2.3 DC to DC Converters (Choppers)

- Fig. 1.2.3 shows the inputs and outputs of DC to DC converters or choppers. The choppers take input from fixed voltage DC supply such as battery or output of uncontrolled rectifier. The output of the chopper is fixed or variable DC voltage.
- The choppers are normally used in DC drives. The speed of the motor can be controlled in forward and reverse directions. The choppers are also used in switched mode power supplies (SMPS).

Fig. 1.2.3 Inputs and outputs of the DC to DC converter or chopper

1.2.4 AC to AC Converters (Cycloconverters)

- Fig. 1.2.4 shows the inputs and outputs of cycloconverters. The input to the cycloconverters is normally 1ϕ or 3ϕ AC mains supply. It is fixed voltage and fixed frequency. The cycloconverters provide the output

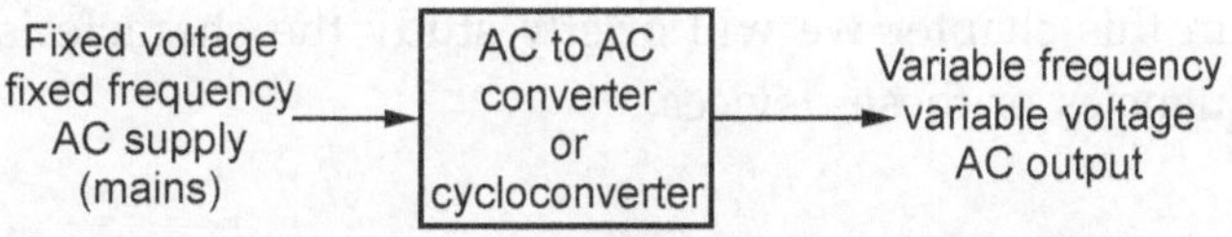

Fig. 1.2.4 Inputs and outputs of cycloconverters

which has variable voltage and variable frequency. The output frequency is lower than the input frequency.

- The cycloconverters are used mainly for AC traction drives.

1.2.5 AC Regulators

- Fig. 1.2.5 shows the inputs and outputs of AC regulator. The input to the AC regulator is fixed voltage AC mains. The output is variable AC voltage which is suitable for load. Here note that the output frequency is same as input frequency. Thus AC regulators does not change the frequency. Whereas cycloconverters change the frequency also. This is the difference between AC regulators and cycloconverters.

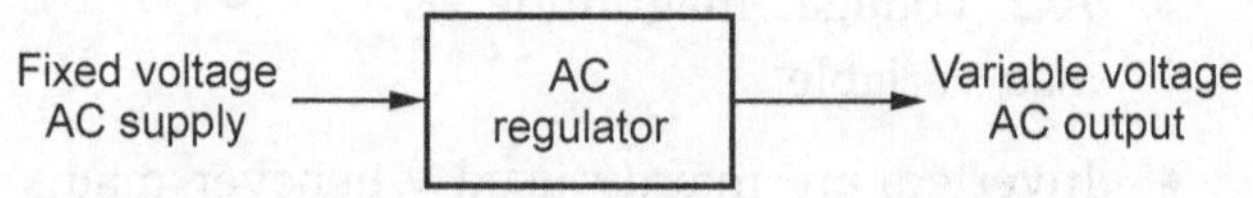

Fig. 1.2.5 Inputs and outputs of AC regulators

- The AC regulators are used for the speed control of large fans and pumps.

1.3 Power Semiconductor Devices

The power semiconductor devices are used as on/off *switches* in power control circuit. These devices are classified as follows :

Power Devices

Diodes
1. General purpose
2. High speed
3. Schottky

Thyristors
1. SCRs
2. GTO
3. RCT
4. SITH
5. GATT
6. LASCR
7. MCT
8. TRIAC

Transistors
1. BJT
2. MOSFET
3. IGBT
4. SIT

In this chapter we will briefly study the characteristics, power rating and operating frequencies of these devices.

1.4 Power Diodes

Power diodes are required in most of the power converters. Power diode is uncontrolled device. Fig. 1.4.1 shows the symbol of power diode.

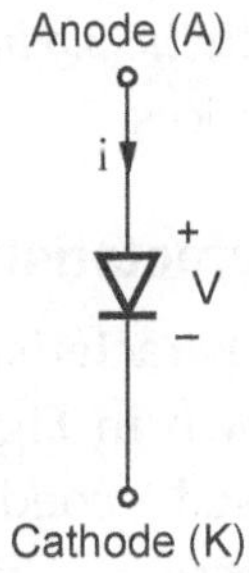

Fig. 1.4.1 Symbol of power diode

When anode (A) is positive with respect to cathode (K), diode starts conducting. Normally a forward bias of 1 volts is sufficient to push the diode in conduction. Current flows from anode to cathode. The diode does not conduct when anode to cathode voltage is negative. The diode is said to be reverse biased.

1.4.1 Structure of Power Diode

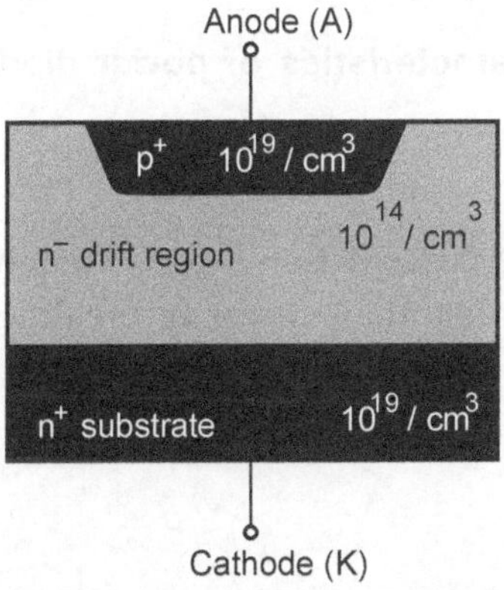

Fig. 1.4.2 Structure of power diode

The structure of the power diode is little different than the small signal diodes. Fig. 1.4.2 shows the structure of power diode. In this figure observe that there is heavily doped n^+ substrate with doping level of $10^{19}/cm^3$. This substrate forms a cathode of the diode. On n^+ substrate, lightly doped n^- epitaxial layer is grown.

This layer is also called drift region. The doping level of n^- layer is about $10^{14}/cm^3$. Then the pn junction is formed by diffusing a heavily doped p^+ region. This p^+ region forms anode of the diode. The doping level of p^+ region is $10^{19}/cm^3$. The thickness of p^+ region is 10 μm. The thickness of n^+ substrate is 250 μm. The thickness of n^- drift layer depends upon the breakdown voltage of the diode. For higher breakdown voltages, the drift region is wide. The n^- drift region is absent in low power signal diodes. The drift region absorbs depletion layer of the reverse biased p^+n^- junction.

Conductivity modulation of the drift layer

Now let us see what will happen when the power diode is forward biased. Under this condition, the holes will be injected from the p^+ region into the drift region. Some of the holes combine with the electrons in the drift region. Since injected holes are large, they attract electrons from the n^+layer. Thus holes and electrons are injected in the drift region simultaneously. Hence resistance of the drift region reduces significantly. Thus diode current goes on increasing, but drift region resistance remains almost constant. Hence on-state losses in the diode are reduced. This phenomena is called *conductivity modulation* of drift region. This phenomena is present in almost all the power devices.

1.4.2 I-V Characteristics

The I-V characteristics of power diode are shown in Fig. 1.4.3. In the forward biased condition, anode current increases linearly with voltage. In lower power diodes, current increases exponentially. The linear rise takes place because of ohmic resistance in n^- layer. The n^- drift region is lightly doped. Hence, it appears as low value internal resistance of the diode. Therefore current is linearly proportional to voltage. A forward bias of 1 V is sufficient to trigger diode into conduction.

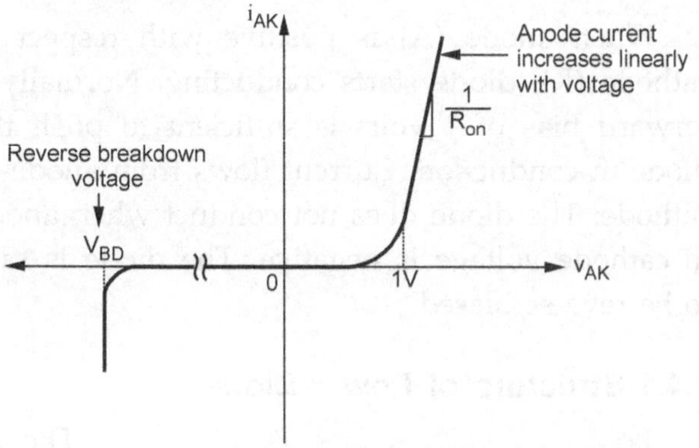

Fig. 1.4.3 v-i characteristics of power diode

When the diode is reverse biased, a very small anode current flows. This current is called leakage current. When the reverse bias is greater than reverse breakdown voltage, anode current starts rising rapidly. Hence, large power dissipation takes place in the diode and it is damaged.

1.4.3 Switching Characteristics of Diodes

Power diodes are mainly used in commutation and freewheeling circuits. The $\frac{di}{dt}$ through the diode is controlled by these circuits. The voltage across the diode changes according to current through it. Fig. 1.4.4 shows the current and voltage through the diode during switching periods. These waveforms are also called switching characteristics.

As shown in the figure, the diode is reverse biased $(-V_R)$ initially. Hence space charge is stored in the depletion region. When diode current starts increasing during period t_1, the voltage also increases. At the end of t_1, the voltage across diode becomes at peak. This voltage overshoot occurs, because there is no conductivity modulation during t_1. Hence, ohmic resistance of the drift region and inductance offered by silicon wafer is high. Therefore there is large voltage drop in the diode. There is large amount of carrier injection in the drift region. Hence, space charge in the drift region is discharged to its thermal equilibrium. By this time diode current reaches to its steady state value. Then during time t_2, conductivity modulation begins and resistance of drift region starts reducing. Hence, voltage across diode reduces and comes down to minimum value. At the end of time t_2, diode voltage drop becomes minimum and remains steady.

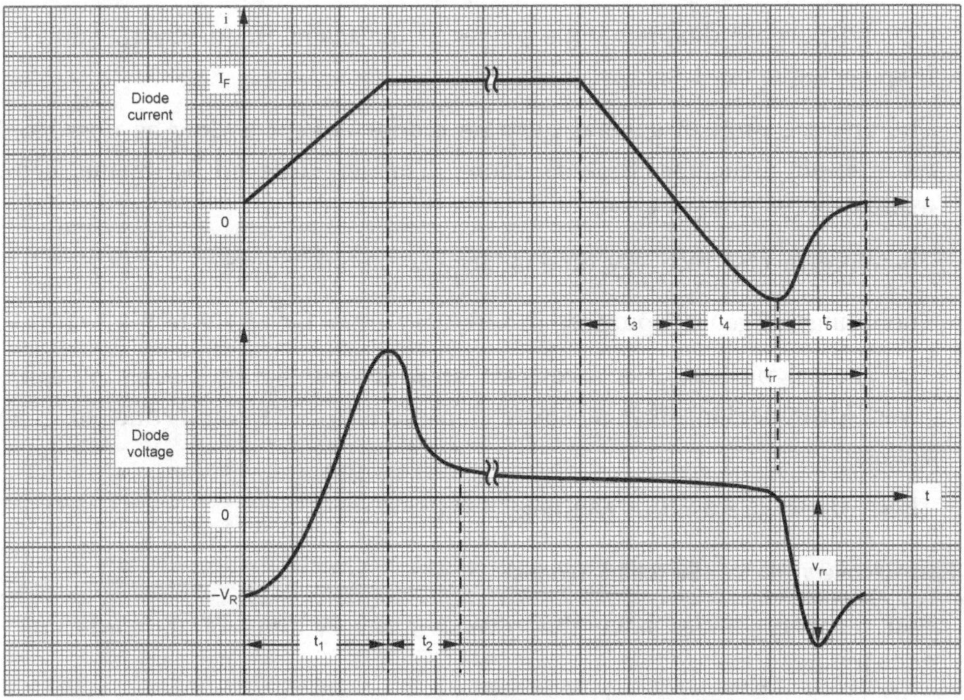

Fig. 1.4.4 Switching characteristics of power diode

Now let us consider the case when $\frac{di}{dt}$ is negative. That is when diode current reduces. As shown in Fig. 1.4.4, the voltage also reduces by small value. During t_3 and t_4, diode current is reducing. But voltage drops by a small value. During this period there are excess carriers in the drift region. Hence, even if diode current is negative in t_4, the voltage drops by a small value. At the end of t_4, all the excess carriers in drift region are removed. Hence, the junctions p^+n^- and n^+n^- are reverse biased. The negative current continues to flow in t_5. Hence, depletion regions are created in drift layer. Therefore diode voltage becomes negative. The negative diode current goes to zero at the end of t_5 since there are no carriers. Hence, the diode voltage falls to negative bias voltage $-V_R$ as shown in Fig. 1.4.4.

During t_4 and t_5, excess carriers are removed from the diode and reverse voltage is buildup. This period is called *reverse recovery period* of the diode. And the portion of characteristics (Fig. 1.4.4) from t_3 to t_5 is called reverse recovery characteristics.

1.4.4 Types of Diodes

There are three types of diodes as discussed next :

1.4.4.1 General Purpose Diodes

- General purpose diodes handle power at the mains frequency (i.e. 50 Hz) or frequencies below 1 kHz.
- The on-state drop of general purpose diodes is kept as low as possible, but its trun-off switching time is more. This is achieved with the help of increasing the doping level.
- Therefore reverse recovery time is large i.e. about 25 μsec.
- The ratings of general purpose diodes vary from 1 A/50 V to 5 kA/10 kV.
- The general purpose diode consists of additional lightly doped n^- epitaxial layer (drift region) that absorbs depletion layer formed due to reverse voltage.
- These diodes are used in rectifiers and converters.

1.4.4.2 Fast Recovery Diodes (Switching Diodes)

- Fast recovery diodes have the reverse recovery time of the order of 25 – 100 ns.
- Fast recovery diode constists of a highly doped p-type and n^+ type layers sandwitched by a lightly doped n^- layer.
- The thickness of n^- layer is increased to increase the voltage blocking capacity.
- The reverse recovery time is reduced by reducing the stored charge in the neutral region and minimizing carrier life time. This is achieved by forming recombination centers in neutral region.
- For very high frequency applications p-i-n diodes are used.
- Fast recovery diodes have ratings from 1 A / 50 V to 1 kA / 3 kV with reverse recovery time less than 5 μs.

Applications

Fast recovery diodes are used in inverters, choppers, SMPS, UPS, dc-dc converters, resonant converters etc.

1.4.4.3 Schottky Diodes

- In schottky diodes, the pn junction is eliminated. A thin film of metal is placed directly on the semiconductor as shown in Fig. 1.4.5. Normally aluminium is deposited on n-type semiconductor. The metal is anode and semiconductor is cathode.

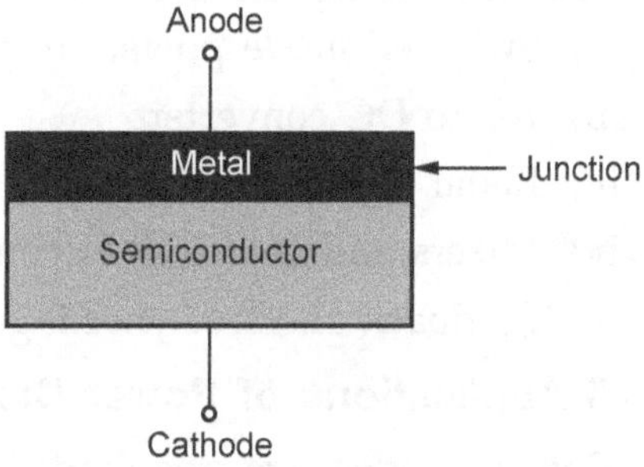

Fig. 1.4.5 A schottky diode

- Since, there is no pn junction, the storage time is absent. Hence, turn-off time is very small. Hence schottky diodes have high switching frequencies.
- The drift layer is absent. Hence on-state losses are very low. But schottky diodes have large reverse leakage currents.

I-V characteristics of schottky diodes

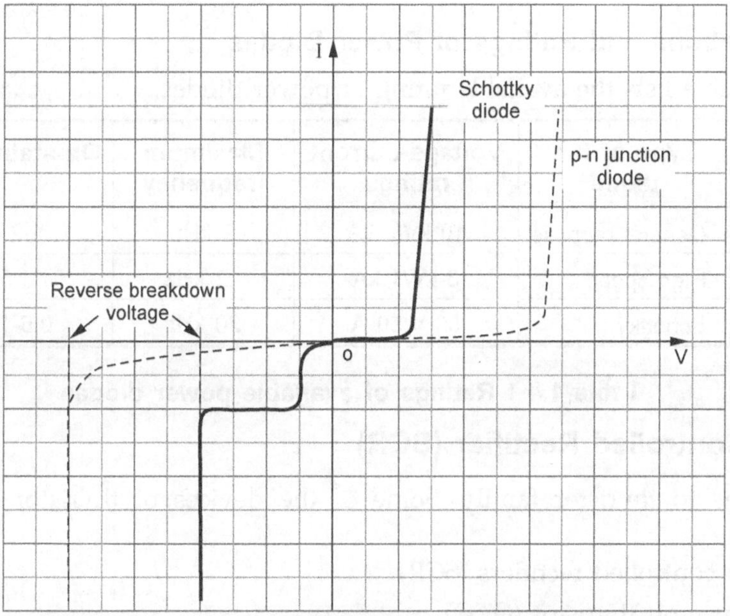

Fig. 1.4.6 I-V characteristics of schottky diodes

- The schottky diodes have similar I-V characteristics to that of pn junction diodes.
- Schottky diodes have very low cut-in voltage (0.2 V).
- Reverse breakdown voltage of the schottky diode is about 50 V compared to 150 V of p-n junction diodes.

Applications of schottky diodes

Schottky diodes are used in

i) Switched mode power supplies.

ii) AC to DC converters

iii) Radar systems

iv) Mixers and detectors in communication circuits.

v) Feedback and freewheeling operations of power converters.

1.4.5 Applications of Power Diodes

Power diodes are required in almost all the power converters. Some of the applications are mentioned below :

1. Power diodes are used in uncontrolled rectifiers.
2. Feedback and freewheeling operations in choppers, inverters and controlled converters use power diodes.
3. Almost all the commutating circuits for SCRs use power diodes.
4. Half controlled converters and half bridge inverters use power diodes.

1.4.6 Specifications and Ratings of Power Diodes

Following table lists the available rating of power diodes.

Serial No.	Type of the diode	Voltage-current ratings	Maximum frequency	On-state drop
1.	General purpose	10 kV/5 kA	2 kHz	1-2 V
2.	High speed	3 kV/1 kA	12 kHz	1-1.5 V
3.	Schottky	50 V/50 A	20 kHz	0.5-1 V

Table 1.4.1 Ratings of available power diodes

1.5 Silicon Controlled Rectifier (SCR)

SCR belongs to thyristor family. Some of the devices of thyristor family are as follows :

1. Silicon controlled rectifiers (SCR)
2. Gate turn-off thyristor (GTO)
3. Reverse conducting thyristor (RCT)
4. Static induction thyristor (SITH)
5. Gate assisted turn-off thyristor (GATT)
6. Light activated silicon controlled rectifier (LASCR)
7. MOS-controlled thyristor (MCT)
8. TRIAC

Out of all these devices, SCR is the most commonly used thyristor.

1.5.1 Construction of SCR

- We know that SCR is a four layer device. Fig. 1.5.1 (a) shows the symbol of the SCR. It has three terminals : Anode (A), Cathode (K) and Gate (G). A small positive voltage between gate and cathode turns on the SCR.
- Fig. 1.5.1(b) shows the detailed structure. The p^+ layer is doped at 1019 / cm^3. The player is doped at $10^{17}/cm^3$. The p and p^+ layers from anode (A) of the SCR. The thickness of the p-layer is 30 to 50 μm. The n^- layer is lightly doped. The doping level of this layer is $10^{14}/cm^3$. The width of n^- layer is 50 to 1000 μm. This layer absorbs depletion layer of the junction J_2.

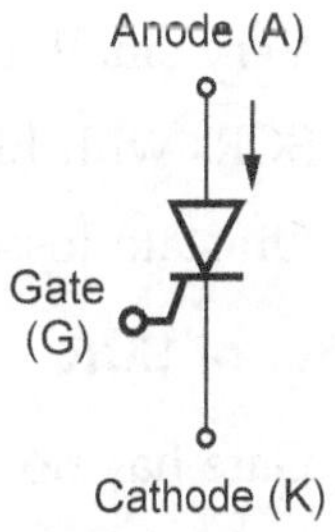

Fig. 1.5.1 (a) Symbol

- When SCR is forward biased (V_{AK} positive), junction J_2 is reverse biased. And J_1 and J_3 are forward biased. The depletion layer of J_2 is absorbed by n^- layer when SCR is forward biased. The width of n^- layer decides forward blocking capability of the SCR. The next p-layer, having doping level of $10^{17}/cm^3$ forms the gate of SCR. The width of this layer is 30 to 100 μm. The next, i.e. n^+ layer (doping level of $10^{19}/cm^3$) forms the cathode of SCR.

Fig. 1.5.1(c) shows the simplified structure of SCR. The gate - cathode junction is J_3. When this junction is forward biased, (i.e. gate signal applied) SCR can be turn-on. Due to gate signal, current starts flowing across J_3. Some carriers flow across J_2 also. Hence, internal regeneration starts and SCR turns on. This process is explained in detail with the help of two transistor analogy in next section.

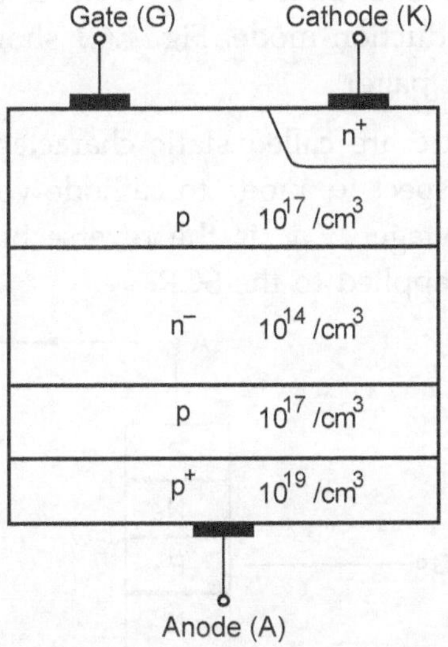

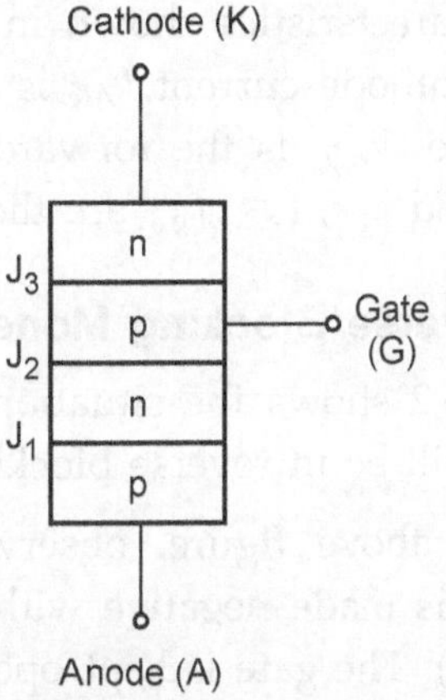

(b) Structure of SCR Fig. 1.5.1 (c) Simplified structure of SCR

1.5.2 Merits, Demerits and Applications of SCR

Merits of SCR

i. Very small amount of gate drive is required since SCR is a regenerative device.

ii. SCRs with high voltage and current ratings are available.

iii. On-state losses in SCRs are reduced.

Demerits of SCR

i. Gate has no control, once the SCR is turned on.

ii. External circuits are required to turn-off the SCR.

iii. Operating frequencies are very low.

iv. Snubbers (RC circuits) are required for dv/dt protection.

Applications of SCR

i. SCRs are best suitable for controlled rectifiers.

ii. AC regulators, lighting and heating applications.

iii. DC motor drives, large power supplies and electronic circuit breakers.

Review Question

1. *With the help of characteristics, explain the modes of operation of the thyristor.*

1.6 SCR Characteristics and Modes of Operation

The working of the SCR can be discussed into three modes : Reverse blocking mode, forward blocking mode and forward conduction mode. Fig. 1.6.1 shows the V-I characteristics of the SCR. (See Fig. 1.6.1 on next page)

The characteristics shown in the above figure are called static characteristics. The anode to cathode current I_{AK} is plotted with respect to anode to cathode voltage V_{AK}. The voltage 'V_{BO}' is the forward break-over voltage. 'V_{BR}' is the reverse break down voltage. And I_{g1}, I_{g2} ,I_{g3} are the gate currents applied to the SCR.

1.6.1 Reverse Blocking Mode

Fig. 1.6.2 shows the situation when the thyristor will be in reverse blocking mode.

In the above figure, observe that the anode (A) is made negative with respect to cathode (K). The gate is kept open. There

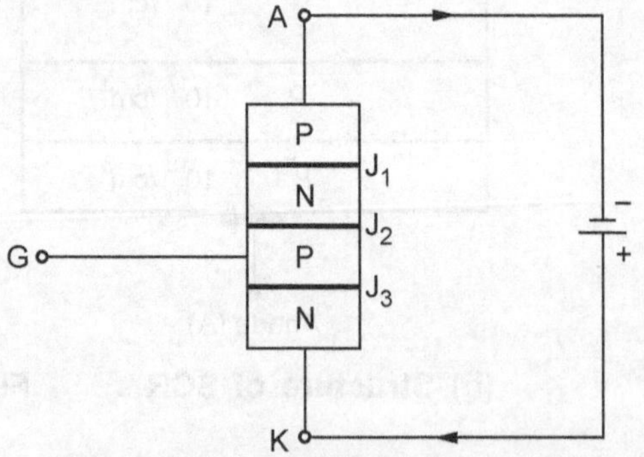

Fig. 1.6.2 A reverse biased SCR

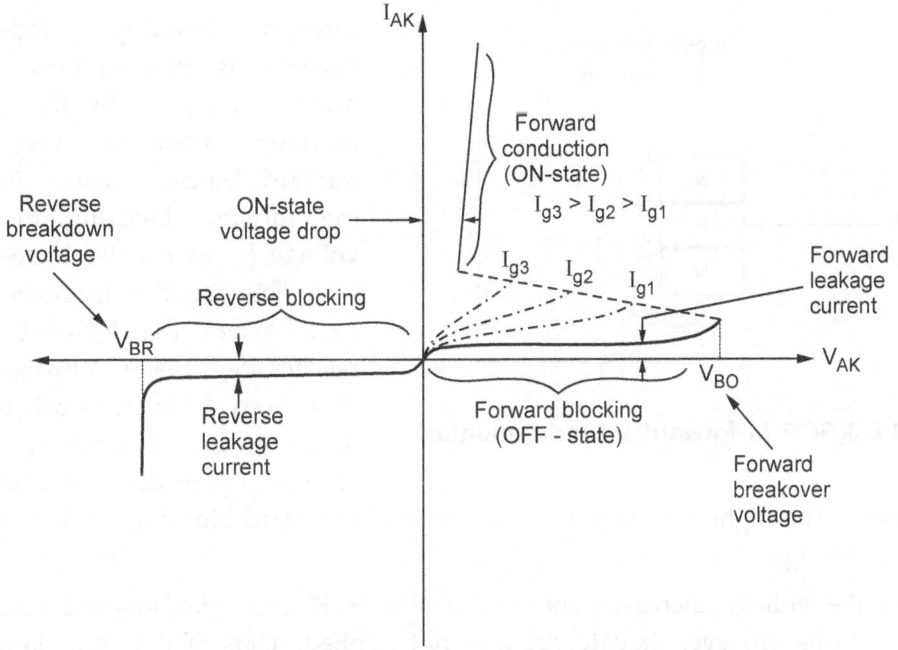

Fig. 1.6.1 Static V-I characteristics of a SCR

are three PN junctions in the SCR : J_1, J_2 and J_3. Due to this reverse bias, junctions J_1 and J_3 are also reverse biased. And junction J_2 is forward biased. The SCR doesnot conduct due to this reverse bias. A very small current flows from cathode to anode. This current is called *reverse leakage current* of the SCR. This mode is called reverse blocking mode. Fig. 1.6.1 shows the characteristic of SCR in reverse blocking mode. Observe that reverse voltage increases but very small current flows. At reverse breakdown voltage (V_{BR}), the reverse current increases rapidly. At the time of reverse breakdown, the high voltage is present across the SCR and heavy current flows through it. Hence large power dissipation takes place in the thyristor. Due to this dissipation, the junction temperature exceeds the permissible value and the SCR is damaged. Hence a reverse voltage across the SCR should never exceed V_{BO}.

During the reverse blocking mode, the positive gate signal should not be applied. If the positive signal is applied between gate and cathode, junction J_3 is forward biased. Hence current starts flowing through it. This current adds to reverse leakage current of the SCR. Hence dissipation is also increased.

1.6.2 Forward Blocking Mode

The SCR is said to be forward biased when anode is made positive with respect to cathode as shown in Fig. 1.6.3. Due to this forward bias the junction J_1 and J_3 is forward biased and J_2 is reverse biased. Hence the forward voltage is to be hold by junction J_2. A very small current flows from anode to cathode. This current is called forward leakage current. This current is of the order of few milliamperes. In the

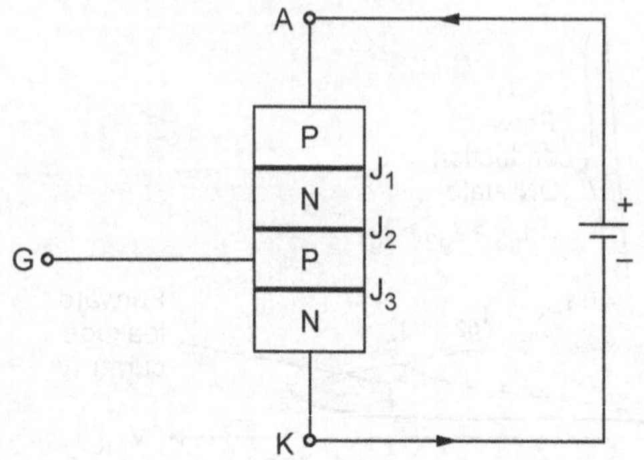

Fig. 1.6.3 SCR in forward biased condition

forward blocking mode, the thyristor is forward biased but it doesnot turn-on. In the forward blocking mode a very small forward leakage current flows. In the forward blocking mode the voltage (V_{AK}) can be increased till V_{BO}. This situation is shown in Fig. 1.6.1. When the forward voltage reaches V_{BO}, the SCR turns on. The SCR goes from forward blocking mode to forward conduction mode. Normally gate drive is applied for this purpose. The highest voltage to be sustained in forward blocking mode is forward break-over voltage, V_{BO}.

When the voltage increases above V_{BO}, the SCR goes into forward conduction mode (i.e. turns on) even if gate drive is not applied. Thus SCR is not damaged if voltage $V_{AK} > V_{BO}$, rather it is turned on.

1.6.3 Forward Conduction Mode

When the SCR is forward biased, then it can go into forward conduction by following techniques :

i) When $V_{AK} > V_{BO}$

ii) When gate drive is applied

iii) When $\frac{dv}{dt}$ exceeds permissible value

iv) When gate cathode junction is exposed to light

Here note that the SCR can go in the forward conduction mode only if it is in the forward blocking mode earlier.

(i) When $V_{AK} > V_{BO}$

The SCR is driven into forward conduction mode when anode to cathode voltage (V_{AK}) exceeds the forward break-over voltage (V_{BO}). The SCR is said to have *turned-on* when it operates in forward conduction mode. When $V_{AK} > V_{BO}$, the SCR is driven in forward conduction even if gate is open. From Fig. 1.6.3, it is clear that junction J_2 is reverse biased during forward blocking mode ($V_{AK} < V_{BO}$). When V_{AK} exceeds V_{BO}, the avalanche break down of junction J_2 takes place even if gate drive is not applied. Hence heavy current starts flowing through the SCR and anode to cathode voltage falls to very small value. This is shown in Fig. 1.6.1. The dotted line (...........) indicates

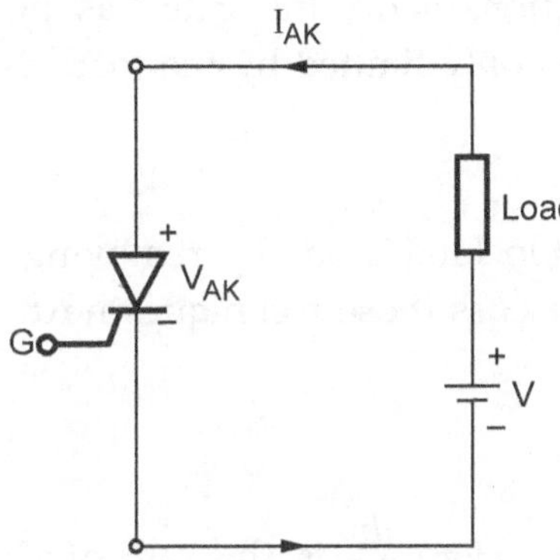

Fig. 1.6.4 Use of SCR in forward conduction

switching of SCR from forward blocking state (i.e. OFF) to forward conduction state (i.e. ON). The anode to cathode current of the SCR is only limited by the load. Fig. 1.6.4 shows such situation :

When the SCR conducts in the forward conduction mode, it is said to have turned 'ON'. The anode to cathode voltage is less than 2 volts. This voltage is normally neglected in calculations. Then the current through the load and SCR will be,

$$I_{AK} = \frac{V}{Load} \quad \text{... (1.6.1)}$$

Thus the SCR current is only limited by the load, once the SCR turns 'on'.

(ii) When gate drive is applied

A positive gate to cathode signal is applied whenever the SCR is to be driven into forward conduction mode (ON state). This is also called *gate triggering* of the SCR. Such situation is shown by the typical circuit of Fig. 1.6.5. The SCR is in forward blocking mode when gate drive is not applied. When the positive gate to cathode voltage is applied, current flows from gate to cathode. This current adds to the forward leakage current. Hence avalanche breakdown of junction J_2 takes place at lower anode to cathode voltage also. Thus SCR is driven into forward conduction mode (ON state) even if $V_{AK} < V_{BO}$. Fig. 1.6.1 shows the characteristic by center (–·–·–·–) lines when gate drive is applied. Observe that, as the gate current is increased, the SCR turns on at lower and lower values of anode to cathode voltages. All these anode to cathode voltages are less than V_{BO}. Thus gate triggering is the most convenient way of triggering the SCR.

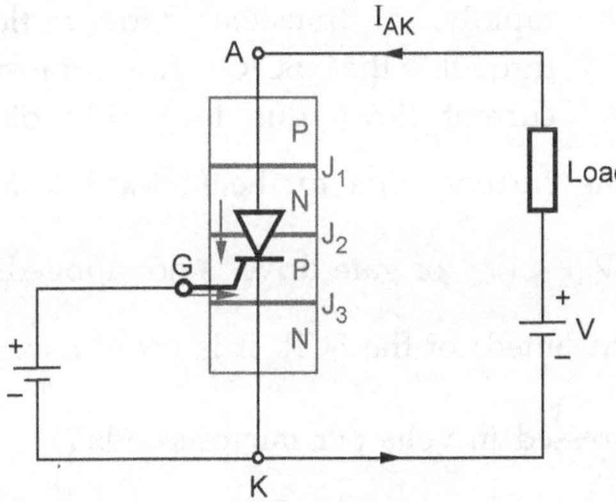

Fig. 1.6.5 Gate triggering is used to turn-on the SCR

Once the thyristor goes into forward conduction mode, the gate has no control over the conduction of thyristor. The current I_{AK} is only limited by the load, i.e.,

$$I_{AK} = \frac{V}{Load}$$

The SCR cannot be driven back into forward blocking mode by removing the gate drive. There are some other techniques. We will discuss those techniques next.

(iii) When $\frac{dv}{dt}$ exceeds permissible value

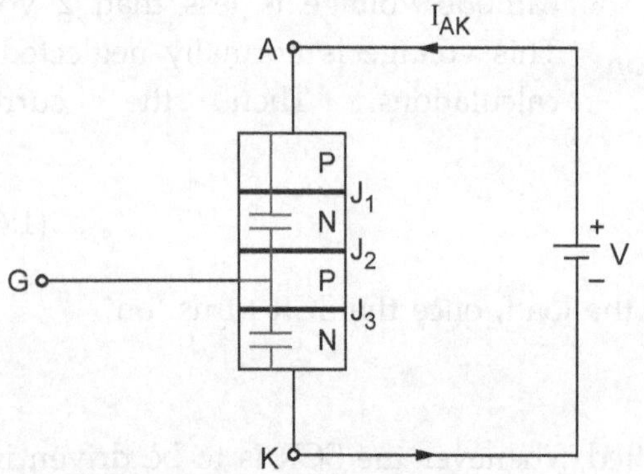

Fig. 1.6.6 SCR turns on by $\frac{dv}{dt}$ due to current flow in equivalent internal capacitor

Here $\frac{dv}{dt}$ is the rate of change of anode to cathode voltage with respect to time. Whenever the SCR is in forward blocking state, only forward leakage current flows through the SCR. In such state an equivalent internal capacitor is formed inside the SCR from anode to gate and gate to cathode. Fig. 1.6.6 shows such internal circuit. Whenever the voltage applied across the SCR changes rapidly, a transient current flows through the SCR. This transient current flows due to rapid voltage variations $\left(\frac{dv}{dt}\right)$ and internal capacitance. This current adds to the forward leakage current. And hence the SCR turns on even if $V_{AK} < V_{BO}$ or gate drive is not applied.

The $\frac{dv}{dt}$ turn-on makes false triggering (unwanted) of the SCR. It is never used for triggering. Every SCR has $\frac{dv}{dt}$ rating. It is expressed in volts per microseconds ($V / \mu s$). The voltage variations across the SCR must be kept less than permissible value of $\frac{dv}{dt}$ to avoid false triggering. Normally a small resistance is connected between gate and cathode to avoid false triggering of SCR due to $\frac{dv}{dt}$. This resistance acts as a external path for leakage current generated by the internal capacitor.

(iv) When a gate cathode junction is exposed to light

When the gate cathode junction is exposed to a beam of light, the current flows in the junction due to photons of light. This current acts as a gate drive to the SCR and it is driven into conduction. This type of triggering is normally used in light activated SCRs (LASCR).

1.6.4 Latching and Holding Currents

Now let us briefly discuss the two important currents which flow through the SCR. These currents are : latching current and the holding current.

1.6.4.1 Latching Current (I_L)

Consider that the SCR is in forward blocking state. Then the SCR can be turned on by applying a gate drive. Then the SCR goes into forward conduction mode (ON state). For the SCR to remain in the 'ON'-state, the anode to cathode current (I_{AK}) must be greater than *latching current*. i.e.,

$I_{AK} \geq I_L$; to remain in ON-state after triggering. Fig. 1.6.7 shows the V-I characteristics of the SCR showing latching current.

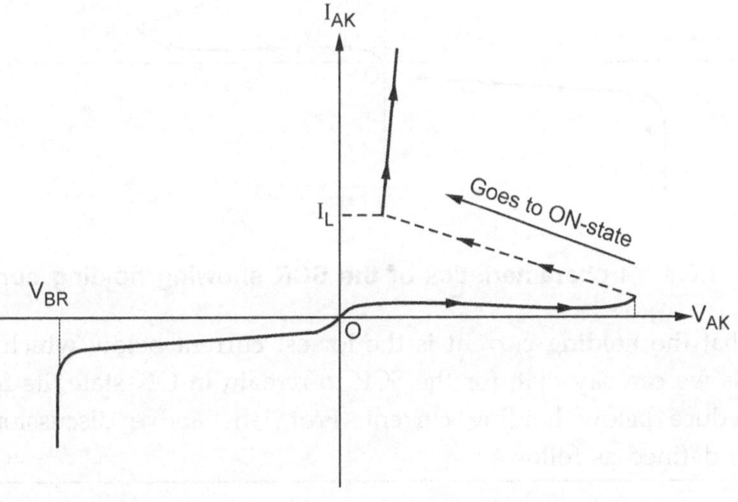

Fig. 1.6.7 V-I characteristics of the SCR showing latching current

Observe that latching current is the lowest current which flows through the SCR to remain in forward conduction (ON-state) after triggering. If the current through the SCR is less than latching current, then the SCR goes back into forward blocking state as soon as gate drive is removed. This is said to be SCR is not *latched* (i.e. not turned on). From the above discussion, the latching current can be defined as follows :

Latching current is the minimum forward current that flows through the SCR to keep it in forward conduction mode (i.e. ON-state) at the time of triggering. If forward current is less than latching current, SCR doesnot turn-on.

The latching current is of the order of 10 to 15 milliamperes.

1.6.4.2 Holding Current (I_H)

Consider that the SCR is in forward conduction state (i.e. ON-state). The SCR goes into forward blocking state when current through it falls below *holding current* (I_H). i.e., if $I_{AK} < I_H$; SCR turns-off. Fig. 1.6.8 shows the V-I characteristics of the SCR showing holding current.

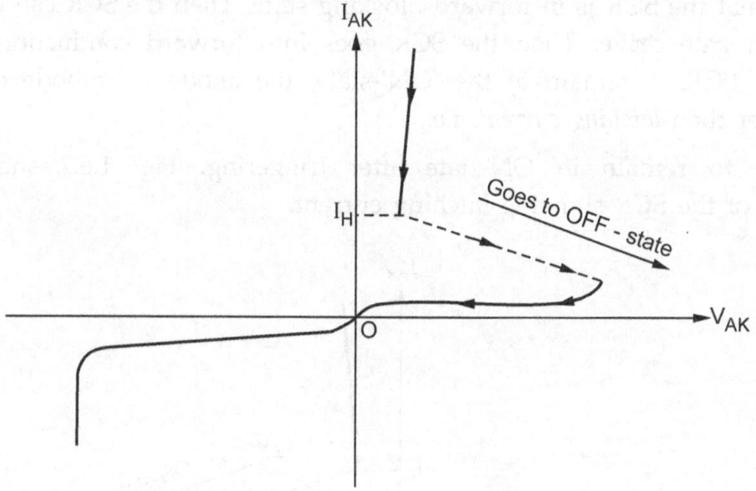

Fig. 1.6.8 V-I characteristics of the SCR showing holding current

Observe that the holding current is the lowest current below which SCR turns-off. In other words we can say that, for the SCR to remain in ON-state, its forward current should not reduce below holding current. From the above discussion, the holding current can be defined as follows :

Holding current is the minimum forward current that flows through the SCR to keep it in forward conduction mode. When forward current reduces below holding current, SCR turns off.

The holding current of the SCRs is of the order of 8 to 10 milliamperes.

1.6.4.3 Comparison (Difference) between Holding and Latching Currents

The definitions of holding current and latching current appear similar but they are totally different. The differences are mentioned below :

1. Latching current is effective at the time of turning-ON, whereas holding current is effective at the time of turning-OFF the SCR.
2. Latching current is the minimum current that should flow at the time of triggering to turn-ON the SCR. Whereas once the SCR is already in ON-state, its current should not reduce below holding current otherwise it turns-OFF.
3. Latching current is greater than holding current even though their magnitudes are much related.

Example 1.6.1 : *The SCR shown in Fig. 1.6.9 has the latching current of 20 mA and is fired by the pulse of width 50 μsec. Determine whether the SCR triggers or not.*

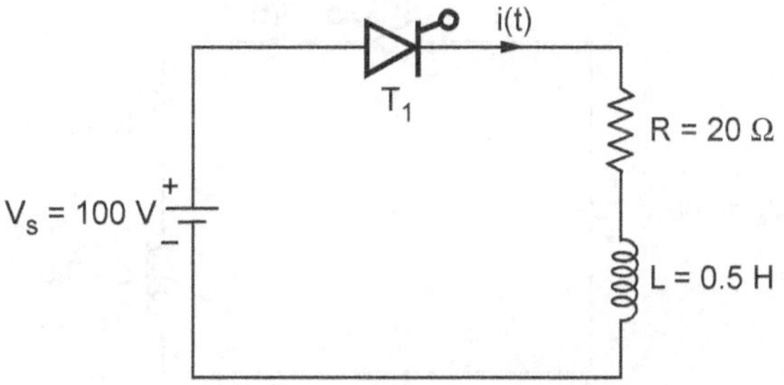

Fig. 1.6.9 Circuit of example 1.6.1

Solution : A step of voltage is applied to the R-L load when SCR turns on. The current through the RL circuit for step input is given as,

$$i(t) = \frac{V_s}{R}\left(1 - e^{-t\frac{R}{L}}\right) \quad \text{... (1.6.2)}$$

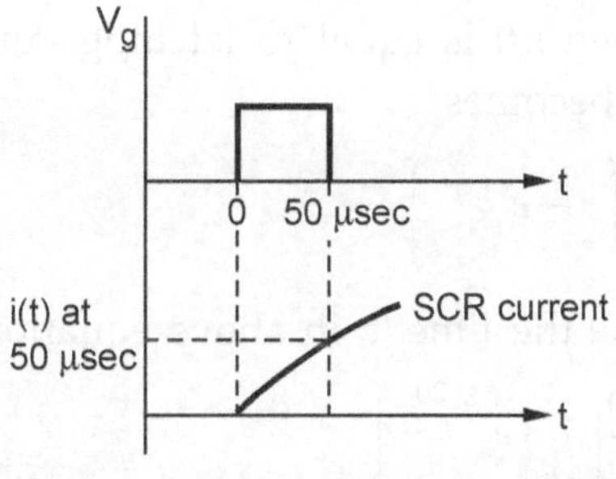

Fig. 1.6.10 After 50 μsec, $i(t) > I_L$ to trigger the SCR properly (triggered)

Fig. 1.6.10 shows the gate pulse and current waveform. Here observe that the SCR will be latched (triggered) if $i(t)$ is greater than latching current when gate triggering pulse is removed after 50 μsec. Hence let us calculate current $i(t)$ through the SCR at 50 μsec,

$$i(t) = \frac{100}{20}\left(1 - e^{-50\times10^{-6}\times\frac{20}{0.5}}\right) = 10 \text{ mA}$$

Here note that current through the SCR is 10 mA. It is not reached to the latching current level and trigger pulse is removed at 50μ*sec.* Hence the SCR will not be triggered.

Example 1.6.2 : *A SCR is connected in series with a 0.5 H inductor and 20 Ω resistance. A 100 V DC voltage is applied to this circuit. If the latching current of the SCR is 4 mA, find the minimum width of the gate trigger pulse required to properly turn-on the SCR.*

Solution : Fig. 1.6.11 shows the circuit diagram.

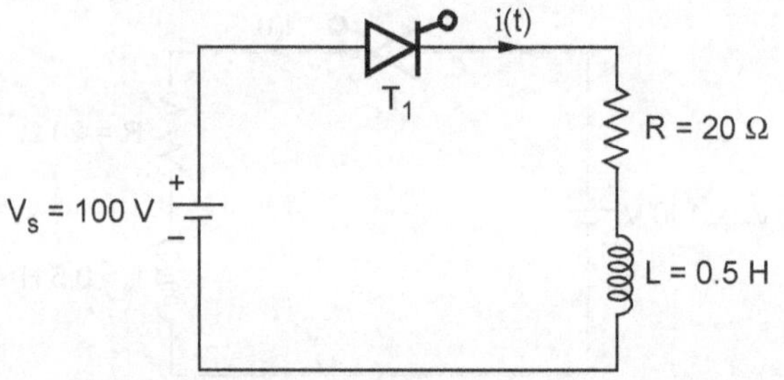

Fig. 1.6.11 Circuit of example 1.6.2

Latching current, $I_L = 4\,mA$ (Given). The current through the RL circuit is given by equation 1.6.2 as,

$$i(t) = \frac{V_s}{R}\left(1 - e^{-t\frac{R}{L}}\right)$$

In the above equation when *i(t)* is equal to latching current, SCR turns on. Hence with $i(t) = I_L$, above equation becomes,

$$I_L = \frac{V_s}{R}\left(1 - e^{-t\frac{R}{L}}\right)$$

Now we have to determine the time 't' in above equation. Putting other values,

$$4 \times 10^{-3} = \frac{100}{20}\left(1 - e^{-t\frac{20}{0.5}}\right)$$

Solving above equation for t,

$$t = 20\ \mu\text{ sec}$$

Thus the width of the gate trigger pulse must be 20 μ sec to turn-on the SCR.

Example 1.6.3 : *The latching current of an SCR used in a phase controlled circuit, comprising an inductive load of R = 10 Ω and L = 0.1 H is 15 mA. The input voltage is 325 sin 314 t. Obtain the minimum gate pulse width required for reliable triggering of the SCR if gated at $\frac{\pi}{3}$ angle in every positive half cycle.*

Solution : Fig. 1.6.12 shows the circuit diagram of this example.

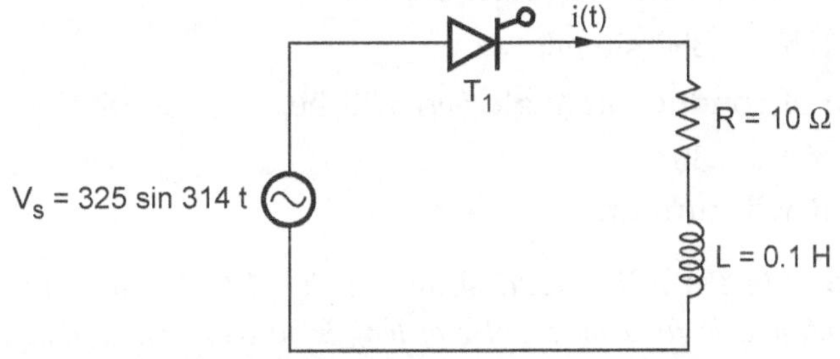

Fig. 1.6.12 Circuit of example 1.6.3

Thus SCR is triggered at $\frac{\pi}{3}$. Hence applied voltage at this angle will be,

$$V_s = 325 \sin \frac{\pi}{3} = 281.458 \text{ volts}$$

Thus 281.458 volts is applied at the time when SCR is triggered. For short duration (till SCR turns on) this voltage can be considered constant. The current through load is then given by equation 1.6.2 as,

$$i(t) = \frac{V_s}{R}\left(1 - e^{-t\frac{R}{L}}\right)$$

In this equation we have to determine the pulse width when SCR triggers successfully. SCR will be triggered successfully when $i(t) = I_L$ = 15 mA. Putting other values in above equation.

$$15 \times 10^{-3} = \frac{281.458}{10}\left(1 - e^{-t\frac{10}{0.1}}\right)$$

Solving the above equation,

$$t = 5.33 \ \mu\text{sec}$$

Thus, the minimum gate pulse should be 5.33 μsec to reliably turn-on the SCR

Example 1.6.4 : *A SCR has a forward breakover voltage of 175 volts when a gate pulse of 2 mA is made to flow. Find the conduction angle if a sinusoidal voltage of 350 V peak is applied.*

Solution : When the gate pulse is applied, the SCR turns on at 175 volts. The applied voltage is,

$$v_S = 350 \sin \omega t$$

when v_S reaches to 175 SCR will turn on. i.e.

$$175 = 350 \sin \omega t$$

Hence the value of conduction angle (ωt) will be,

$$\therefore \quad \omega t = 30°$$

Thus, at 30°, SCR will turn-on.

Example 1.6.5 : *In the SCR circuit shown in Fig. 1.6.13 below, the SCR has a latching current of 50 mA and is fired by a pulse of length 50 μsec. Show that without resistance R, the SCR will fail to remian on, when the firing pulse ends and then find the maximum value of R to ensure firing.*

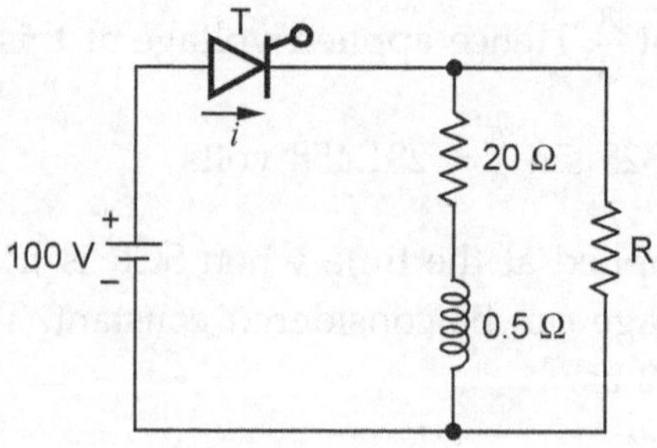

Fig. 1.6.13 SCR circuit of example 1.6.5

Solution : To show that SCR does not latch

Here first consider the SCR circuit without resistance R. This circuit is shown below in Fig. 1.6.14.

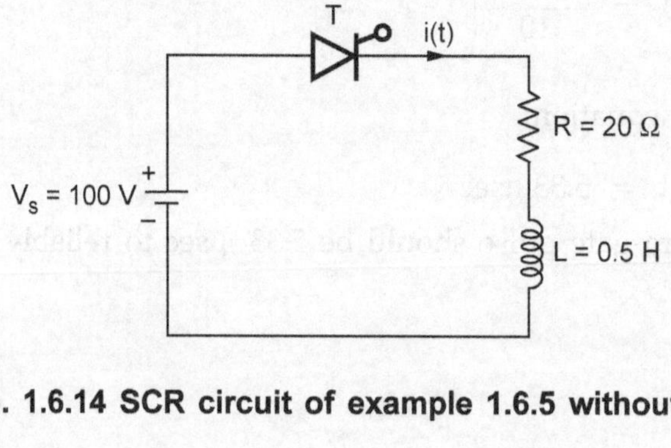

Fig. 1.6.14 SCR circuit of example 1.6.5 without R

For this circuit the given data is,

$$V_S = 100$$

$$R = 20\ \Omega$$

$$L = 0.5\ \text{H}$$

$$I_L = 50\ \text{mA}$$

$$\text{Pulse width} = 50\ \mu s$$

Now let us check whether the SCR current rises above latching current in the firing pulse duration of 50 μs. The current in the RL circuit is given by equation 1.6.2 as,

$$i(t) = \frac{V_S}{R}\left(1 - e^{-t\frac{R}{L}}\right)$$

Putting the value of R, L and V_S

$$i(t) = \frac{100}{20}\left(1 - e^{-t\frac{20}{0.5}}\right)$$

$$= 5(1 - e^{-40t})$$

The current after t = 50 μs will be,

$$i(t = 50\mu s) = 5(1 - e^{-40 \times 50 \times 10^{-6}})$$

$$= 9.99 \times 10^{-3} \approx 10\ \text{mA}$$

Thus during the firing pulse width of 50 μs, the SCR current rises upto 10 mA. Since this current is less than latching current of 50 mA. SCR will fail to remain on when firing pulse ends.

To determine value of R

The additional resistance connected in parallel with RL circuit increases the current through SCR. SCR takes 10 mA current when firing pulse of width 50 μs ends. To latch the SCR, 50 mA current should be passed through it. Hence additional 40 mA current can be passed through 'R' as shown in Fig. 1.6.15.

If we neglect the voltage drop in the SCR, full V_S will appear across R. Hence,

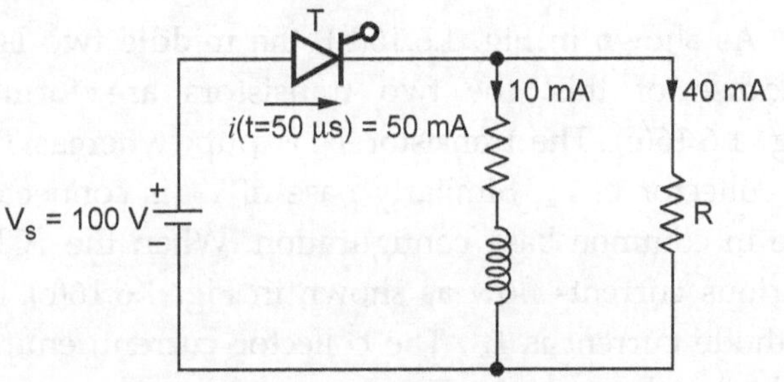

Fig. 1.6.15 Currents at t = 50 μs

$$V_s = 40\,mA \times R$$

$$\therefore \quad R = \frac{V_s}{40\,mA}$$

$$= \frac{100}{40 \times 10^{-3}}$$

$$= 2500\ \Omega$$

Thus a maximum R = 2.5 kΩ will ensure firing of the SCR.

1.6.5 Two Transistor Model of SCR

The operation of the SCR can be explained with the help of two transistor model. Fig. 1.6.16 shows how the two transistor model of the SCR is formed.

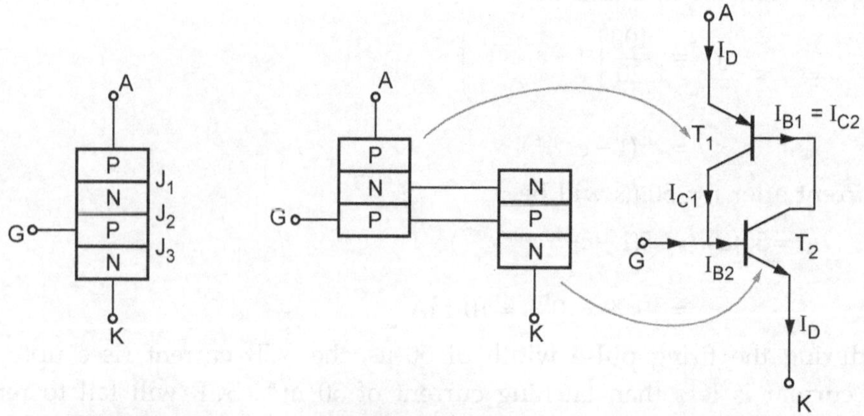

Fig. 1.6.16 A two transistor model of the SCR

As shown in Fig. 1.6.16(b), the middle two layers are split into two separate parts. Because of this, the two transistors are formed. These transistors are shown in Fig. 1.6.16(c). The transistor T_1 is pnp, whereas T_2 is npn. The base of T_1 is connected to collector of T_2. Similarly base of T_2 is connected to collector of T_1. These transistors are in common base configuration. When the SCR is forward biased and gate is open, various currents flow as shown in Fig. 1.6.16(c). As shown in this figure, the anode to cathode current is I_D. The collector current, emitter current and leakage currents of T_1 are related as,

$$I_{C1} = \alpha_1 I_{E1} + I_{CO1} \quad \text{... (1.6.3)}$$

Here $I_{E1} = I_D$ and I_{CO1} is leakage current of T_1. Similarly for T_2,

$$I_{C2} = \alpha_2 I_{E2} + I_{CO2} \quad \text{... (1.6.4)}$$

Here $I_{E2} = I_D$ and I_{CO2} is leakage current of T_2.

Therefore equation 1.6.3 and 1.6.4 can be written as,

$$\left.\begin{aligned} I_{C1} &= \alpha_1 I_D + I_{CO1} \\ I_{C2} &= \alpha_2 I_D + I_{CO2} \end{aligned}\right\} \quad \text{... (1.6.5)}$$

In Fig. 1.6.16(c), observe that the current I_D flows through the collectors of T_1 and T_2. Hence we can write,

$$I_D = I_{C1} + I_{C2}$$

Putting the values from equation 1.6.5 in above equation,

$$I_D = \alpha_1 I_D + I_{CO1} + \alpha_2 I_D + I_{CO2}$$

$$\therefore \quad I_D = (\alpha_1 + \alpha_2) I_D + I_{CO1} + I_{CO2}$$

$$\therefore \quad I_D = \frac{I_{CO1} + I_{CO2}}{1-(\alpha_1 + \alpha_2)} \quad \text{... (1.6.6)}$$

$I_{CO1} + I_{CO2}$ can be considered as total reverse leakage current of junction J_2. This current can be denoted by the I_{CO}. Then above equation can be written as,

$$\boxed{I_D = \frac{I_{CO}}{1-(\alpha_1 + \alpha_2)}} \quad \text{... (1.6.7)}$$

Here I_{CO} is the reverse leakage current of the reverse biased junction J_2. And α_1 is the common base current gain of T_1 and α_2 is common base current gain of T_2. Initially when forward voltage is small, $(\alpha_1 + \alpha_2)$ is very small and less than 1. Hence forward blocking current as given by equation 1.6.7 is also small. As forward voltage applied across the SCR increases, the values of α_1 and α_2 also increase. When $(\alpha_1 + \alpha_2)$ tends unity, then $I_D$approaches infinity as given by equation 1.6.7. At this instant, internal regeneration starts and the SCR goes into forward conduction (ON-state) mode. The current through the SCR is only limited by the external load.

Once the SCR goes into conduction, the two transistor model is no more applicable. Here note that the internal regeneration takes place in the SCR due to avalanche breakdown of reverse biased junction J_2. It does not take place when SCR is reverse biased. When the current through the SCR falls below holding current, the forward blocking state is regained. Then α_1 and α_2 of transistors are also reduced to small values.

When the gate current I_g is applied, then equation 1.6.7 will be written as,

$$I_D = \frac{I_{CO} + I_g}{1-(\alpha_1 + \alpha_2)} \quad ...\ (1.6.8)$$

Thus the forward leakage current (I_D) is increased due to gate drive (I_g). This leakage current flows through junction J_2 and its avalanche break-down occurs at lower forward voltage. Thus with the gate drive, the SCR is turned on at voltages less than V_{BO}. Hence gate becomes convenient way of triggering the SCR. Once the SCR is turned-on, the gate has no control over its conduction.

Review Questions

1. *Explain the terms latching current and holding current and compare them.*
2. *Explain the operation of the SCR with the help of two transistor analogy.*

1.7 SCR Turn-on and Turn-off

1.7.1 Different Ways to Turn-on the SCR

We know that SCR can be turned-on if the anode current is above latching current. There is regenerative action in the SCR. SCR can be turned-on by following ways:

1. **Gate drive**

   SCR can be turned on by applying positive gate-cathode voltage. Injected gate carriers increase the anode current and regenerative action starts. As shown in equation 1.6.7, $(\alpha_1 + \alpha_2)$ approaches unit and anode current (I_D) becomes large. It is limited only by external load. Once the SCR is turned-on, there is no need of gate drive. Hence it can be removed. Normally pulsed gate drive is applied to reduce losses in the SCR gate.

2. **High forward voltage**

   SCR turns on when its anode-cathode voltage exceeds forward break over voltage, i.e. $V_{Ak} > V_{BO}$. This is shown in Fig. 1.6.1. At these voltages, the leakage current is so high, that internal regenerative starts in the device.

3. $\frac{dv}{dt}$

   SCR can be thought of as a capacitor in the forward biased state. When the anode-cathode voltage changes rapidly, leakage current thought the device increases due to internal capacitor. The leads to turn-on of the SCR.

4. **Light**

   SCR can be turned on by light, when it falls on gate cathode junction of the SCR light induces electronic hole pairs and it helps to increases leakage current.

5. **High temperature**

   SCR turns on due to increased temperature AT higher temperature, there are more electron-hole pairs across junctions. This inverses the leakage current and the SCR turns on.

1.7.2 Turn-on Dynamic Characteristics

Fig. 1.7.1 shows the current and voltage of the SCR during turn-on. The gate pulse is applied at $t = 0$. During the *delay time* (t_d), the anode current rises very slowly and flows only near the narrow region of the gate. Observe that anode to cathode doesnot reduce during t_d. It remains to the forward blocking value. During the *rise time* (t_r), the anode current increases rapidly and anode to cathode voltage falls rapidly. The high voltage and current are present in the SCR. Hence large dissipation takes place in the SCR.

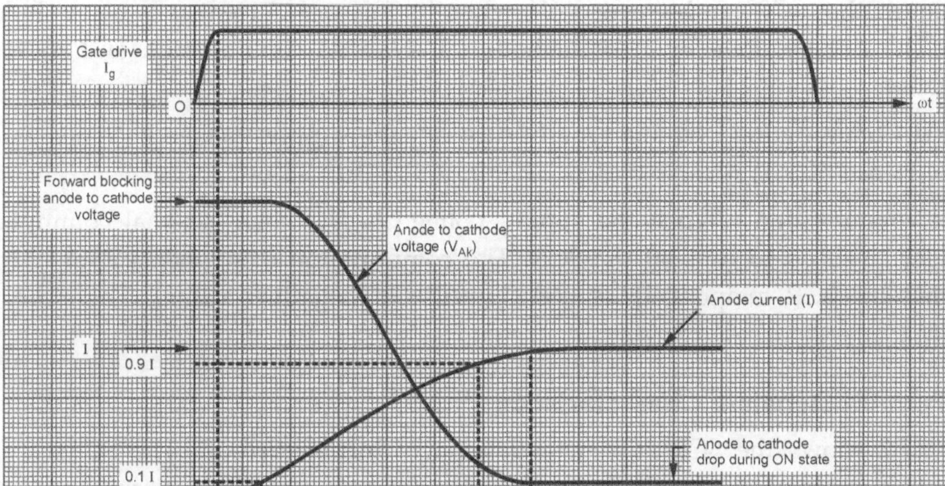

Fig. 1.7.1 Dynamic characteristics of SCR during turn-on

This power dissipation is called switching loss of the SCR. The current starts spreading in the remaining area of the SCR. During the *spread time* (t_p), the conduction spreads over the complete cross-section of the SCR. The anode current reaches to its maximum value. And the anode to cathode voltage falls to lowest value (i.e. less than 2 V). The dissipation in the SCR is also reduced. *The turn-on time* (t_{on}) of the SCR is given as total of t_d, t_r and t_p. Thus,

$$t_{on} = t_d + t_r + t_p$$

The turn-on time can be defined as,

> *The turn-on time of the SCR is defined as the time from initiation of gate drive to the time when anode current reaches to its full value.*

The turn-on time of the SCRs is about 1 to 3 microseconds. The turn-on time can be effectively reduced by applying higher values of gate currents. Because of high gate currents, more electron-holes are injected near junction J_2. Hence avalanche break-down of J_2 takes place fast. Therefore anode current rises fast. Thus effective turn-on time is reduced. To turn-on the SCR, the gate pulse is thus sufficient.

1.7.3 SCR Turn-off

We know that SCR can be turned-off, when its forward current falls below holding current. The can be done by two methods : i) Natural commutation and ii) forced commutation.

i) Natural Commutation : In this type of turn-off, the supply voltage becomes zero or negative, Hence SCR is reverse biased. Therefore it is turned-off.

ii) Force commutation : When the supply voltage is DC, then external commutation component are used to turn-off the SCR. The commutation components apply reverse bias across the SCR temporarily or pass impulse of negative current. Therefore SCR turns-off.

1.7.4 Turn-off Dynamic Characteristics

Fig. 1.7.2 shows the SCR current and voltage during turn-off. The SCRs are not turned off by gate. They need external circuit for turn-off. These circuits are called commutation circuits. These commutation circuits has to hold negative voltage across the SCR during turn-off period. The SCR is said to be turned-off when it regains forward blocking capability after forward conduction. In the above figure observe that anode current falls and then it becomes negative. The negative pulse of current flows through the SCR for short period. During the conducting state, the SCR is flooded with carriers and it acts as short circuit. The negative anode current flows through the SCR till all these carriers are removed. Then junctions J_1 and J_3 achieve their forward blocking state. The time required for this is called *reverse recovery time* (t_{rr}). At the end

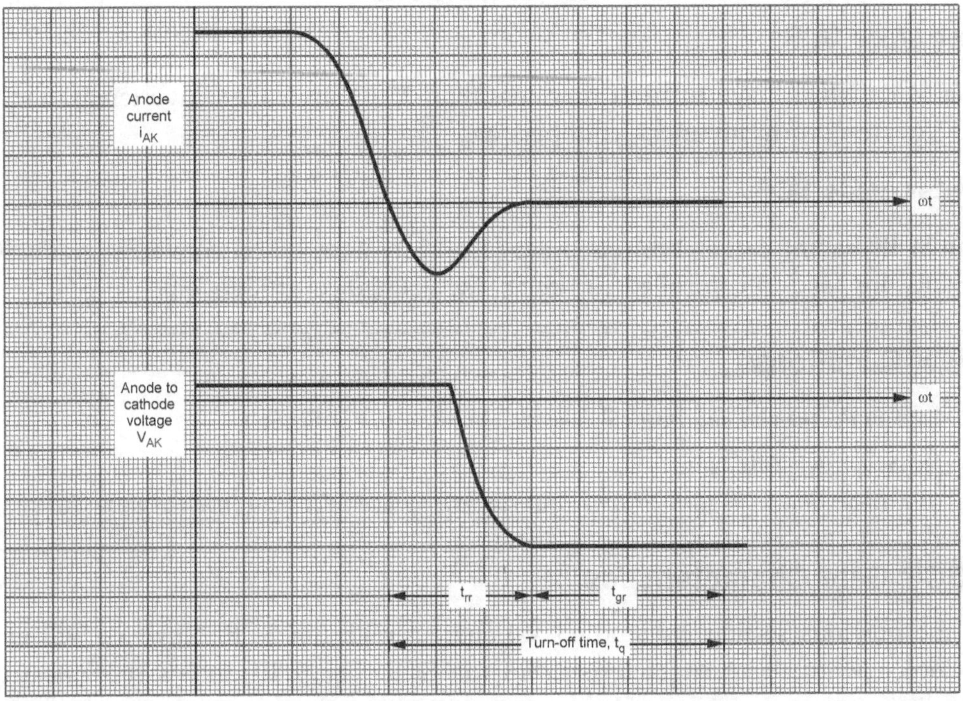

Fig. 1.7.2 Dynamic characteristics of SCR during turn-off

of t_{rr}, reverse voltage appears across the SCR and anode current becomes zero. This is shown in Fig. 1.7.2. But still, the SCR is not turned-on. The commutation circuit has to hold negative voltage across the SCR for *gate recovery time* (t_{gr}). During this time, the excess carriers near junction J_2 are recombined. If negative voltage is removed by commutation circuit before t_{gr}, then SCR may turn-on again due to these excess carrier near junction J_2. Because they act like gate drive to the SCR. Hence the turn-off is complete at the end of gate recovery time. The SCR regains its forward blocking capability. The negative voltage imposed by commutation circuit can be removed at the end of t_{gr}. *The turn-off time* (t_q) of the SCR is the total time required by reverse recovery and gate recovery. i.e.,

$$t_q = t_{rr} + t_{gr}$$

The turn-off time can be defined as follows :

> *The turn-off time of the SCR is the time required to achieve forward blocking capability after commutation is initiated.*

The turn-off time of the SCR varies from 5 to 200 microseconds. The turn-off time of the commutation circuit is called *circuit turn-off time* (t_c). And hence circuit turn-off time must be greater than the turn-off time of the SCR ($t_c > t_q$).

1.7.5 Inverter Grade and Converter Grade SCRs

Inverter grade SCRs

The SCRs which have turn-off time less than 25 μs are called inverter grade SCRs. Such SCRs are used in inverters, choppers etc.

Converter grade SCRs

The SCRs having larger turn-off times $(t_q > 25 \mu s)$ are called converter grade SCRs. Such SCRs are used in controlled rectifiers, AC voltage controllers etc.

Review Questions

1. *Explain the turn-on and turn-off dynamic characteristics of the SCR.*
2. *Define the following :*
   (i) turn-on time (ii) turn-off time (iii) converter grade SCR (iv) Inverter grade SCR

1.8 SCR Gate Characteristics

In the previous section we studied V-I characteristics of SCR. Now we will have a closer look towards gate characteristics of the SCR. Fig. 1.8.1 shows the gate trigger characteristics.

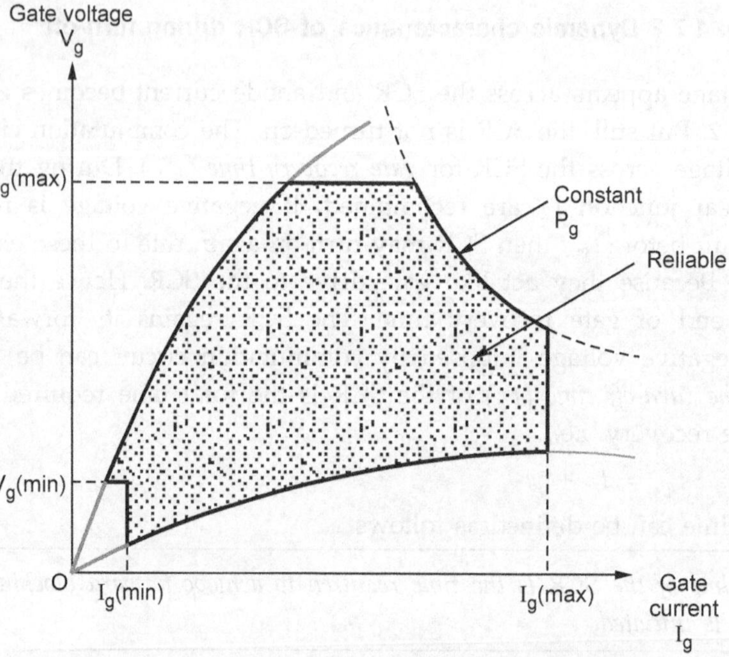

Fig. 1.8.1 Gate trigger characteristics

The gate voltage is plotted with respect to gate current in the above characteristics. $I_{g(\max)}$ is the maximum gate current that can flow through the SCR without damaging it. Similarly $v_{g(max)}$ is the maximum gate voltage to be applied. Similarly $v_{g(min)}$ and $I_{g(min)}$ are minimum gate voltage and current, below which SCR will not be turned-on. Hence to turn-on the SCR successfully the gate current and voltage should be

$$I_{g(min)} < I_g < I_{g(max)}$$

and $$v_{g(min)} < v_g < v_{g(max)}$$

The characteristic of Fig. 1.8.1 also shows the curve for constant gate power (P_g). Thus for reliable turn-on, the (v_g , i_g) point must lie in the shaded area in Fig. 1.8.1. It turns-on SCR successfully. Note that any spurious voltage/current spikes at the gate must be less than $v_{g(min)}$ and $I_{g(min)}$ to avoid false triggering of the SCR. The gate characteristics shown in Fig. 1.8.1 are for DC values of gate voltage and current.

1.8.1 Pulsed Gate Drive

Instead of applying a continuous (DC) gate drive, the pulsed gate drive is used. The gate voltage and current are applied in the form of high frequency pulses. The frequency of these pulses is upto 10 kHz. Hence the width of the pulse can be upto 100 micro seconds. The pulsed gate drive is applied for following reasons (advantages) :

(i) The SCR has small turn-on time i.e. upto 5 microseconds. Hence a pulse of gate drive is sufficient to turn-on the SCR.

(ii) Once SCR turns-on, there is no need of gate drive. Hence gate drive in the form of pulses is suitable.

(iii) The DC gate voltage and current increases losses in the SCR. Pulsed gate drive has reduced losses.

(iv) The pulsed gate drive can be easily passed through isolation transformers to isolate SCR and trigger circuit.

1.8.2 Requirement of Gate Drive

The gate drive has to satisfy the following requirements :

(i) The maximum gate power should not be exceeded by gate drive, otherwise SCR will be damaged.

(ii) The gate voltage and current should be within the limits specified by gate characteristics (Fig. 1.8.1) for successful turn-on.

(iii) The gate drive should be preferably pulsed. In case of pulsed drive the following relation must be satisfied :

(Maximum gate power × pulse width) × (Pulse frequency) ≤ Allowable average gate power

(iv) The width of the pulse should be sufficient to turn-on the SCR successfully.

(v) The gate drive should be isolated electrically from the SCR. This avoids any damage to the trigger circuit if in case SCR is damaged.

(vi) The gate drive should not exceed permissible negative gate to cathode voltage, otherwise the SCR is damaged.

(vii) The gate drive circuit should not sink current out of the SCR after turn-on.

Review Questions

1. *Explain the SCR gate characteristics.*
2. *What are the requirements of gate drive ? What is pulse gate drive ?*

1.9 SCR Ratings

Every SCR is manufactured for particular voltage, current and switching frequencies. If these values are exceeded, then the SCR can be damaged. These are called ratings. The SCRs are to be protected when any of the voltage or current rating tries to exceed. In this section we will discuss these concepts.

1.9.1 Current Ratings

The current flow through the SCR increase the junction temperature. The excess current flow may exceed the permissible junction temperature and damage the device. Hence the current should not exceed the rated value. The various current ratings are discussed next :

(i) Average current rating (I_T)

The average current rating is the maximum repetitive average current that can flow through the SCR. The power loss in the SCR depends upon average current flowing through it. If the SCR is operating at sufficiently high frequency, then switching loss will also be significant. Hence switching losses may be added to losses due to average current.

(ii) R.M.S. current rating (I_{TR})

The RMS current rating is the maximum repetitive rms current that can flow through the SCR. The R.M.S. current rating is same as average current rating for DC current. This rating is required to prevent excessive heating in metallic joints, leads and interfaces of SCRs.

(iii) Surge current rating (I_{TSM})

The surge current rating is the peak amplitude of the surge current that the SCR can withstand only limited number of times in its life cycle. The surge current is normally specified as number of cycles and peak amplitude. The SCR may be damaged when surge current rating and its number of cycles are exceeded.

(iv) $i^2 t$ rating

The i^2t rating is the measure of thermal energy that the device can abosrb for a short period of time. Whenever fault occurs, the fast acting fuse clears such fault. Due to the fault, thermal energy is generated in the device also. The fuse should clear the fault and device should be protected. Hence $i^2 t$ rating is used to determine about how long the device can absorb the thermal energy. The fuse must clear the fault before the device is damaged due to exceeding $i^2 t$ rating.

(v) $\frac{di}{dt}$ rating

The $\frac{di}{dt}$ rating specifies maximum allowable rate of change of current through the device. Due to rapid variations in anode current, the carriers doesnot spread across the junctions at the turn-on time. Hence they are concentrated in a small area of the device, creating local heating. This is called *hot-spot* created due to high current density in the restricted area of the junctions. Because of this, the junction temperature increases and the device may be damaged. The $\frac{di}{dt}$ rating specifies maximum allowable variations in anode current, so that the device will not be damaged. Normally it is specified in Amperes/microseconds and typical values are from 50 A/μs to 800 A / μsec.

1.9.2 Voltage Ratings

The SCR blocks the forward and reverse voltages. The voltage ratings mainly specify the maximum allowable voltages those the device can withstand without damaging the junctions.

(i) Peak repetitive forward blocking voltage (V_{DRM})

This is the maximum voltage that the SCR can block in the forward direction. It is specified with maximum allowable junction temperature and gate open circuited. If this rating is exceeded, the device turns on. Note that device is not damaged.

(ii) Peak repetitive reverse voltage (V_{RRM}) or peak inverse voltage (PIV)

This is the maximum voltage that the device can withstand repetitively in the reverse blocking state. It is also specified at maximum allowable junction temperature. The device is damaged, when this rating is exceeded.

(iii) Non-repetitive peak reverse voltage (V_{RSM})

This the maximum transient voltage that the device can safely withstand in the reverse direction. This transient is not repetitive. The device is damaged if transient is exceeded or it occurs repetitively. This transient voltage can be increased by putting a diode of same current rating in series with the SCR. The total transient voltage capacity becomes due to SCR and diode.

(iv) $\frac{dv}{dt}$ rating

The $\frac{dv}{dt}$ rating specifies maximum allowable rate of change of forward voltage that the device can withstand in forward direction. If the forward voltage variations exceed $\frac{dv}{dt}$ rating, then the device turns on. Such turn-on is false triggering and disturbs the operation of the controller.

The other ratings are : turn-on time (t_{on}), turn-off time (t_q), gate voltage (v_g), gate current (i_g), latching current (I_L) and holding current (I_H). These ratings we have discussed earlier in section 1.6, 1.7 and 1.8.

Review Question

1. *Explain the following ratings :*

   (i) i^2t *rating*

   (ii) $\frac{dv}{dt}$ *rating*

   (iii) $\frac{di}{dt}$ *rating*

1.10 Thyristor Types

Now let us briefly discuss other members of thyristor family. They are gate turn-off thyristor family (GTO), light activated SCR (LASCR) etc.

1.10.1 Gate Turn-off Thyristor (GTO)

1.10.1.1 Structure of GTO

At the beginning of the chapter we discussed structure and working of SCR. The SCR is most commonly used member of thyristor family. But SCR needs external circuits for turn-off. Now we present another thyristor, called GTO. The GTO can be turned-off by gate drive. Thus gate has full control over the operation of GTO. Fig. 1.10.1 shows the structure of GTO.

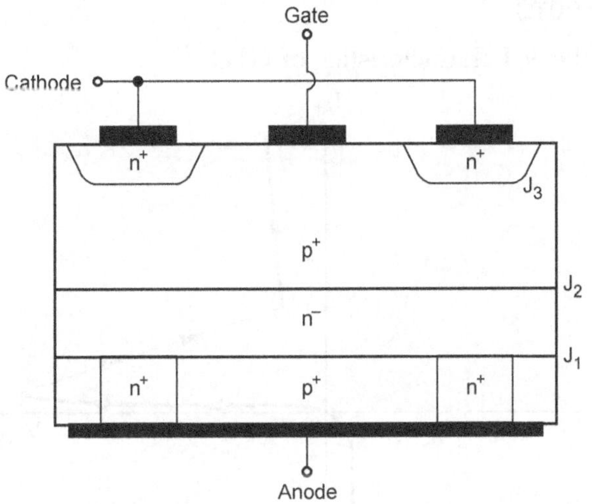

Fig. 1.10.1 Structure of GTO

Observe that the structure of GTO is almost similar to SCR. But there are significant differences that make GTO different than SCR. These differences are :

i. Gate and cathodes are highly interdigited with various geometric forms. This maximizes periphery of the cathode and minimize gate-cathode distance.

ii. There are n^+ regions at regular intervals in the p^+anode layer. This n^+layer makes direct contact with n^- layer. This is called anode short. This speeds up the turn-off mechanism of GTO.

iii. The operation of GTO can be explained with the help of two transistor analogy. The gain of pnp transistor is reduced. This reduces the regenerative action. Hence turn-off of GTO can be achieved by negative current from gate.

Fig. 1.10.2 shows the symbol of GTO.

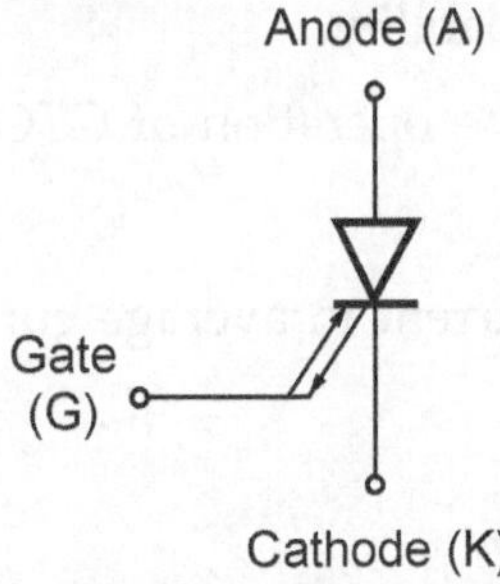

Fig. 1.10.2 Symbol of GTO

Observe that there is double arrow on the gate. This indicates that bidirectional current flows through the gate. The rest of the symbol is similar to SCR.

1.10.1.2 Characteristics of GTO

Fig. 1.10.3 shows the V-I characteristics of GTO.

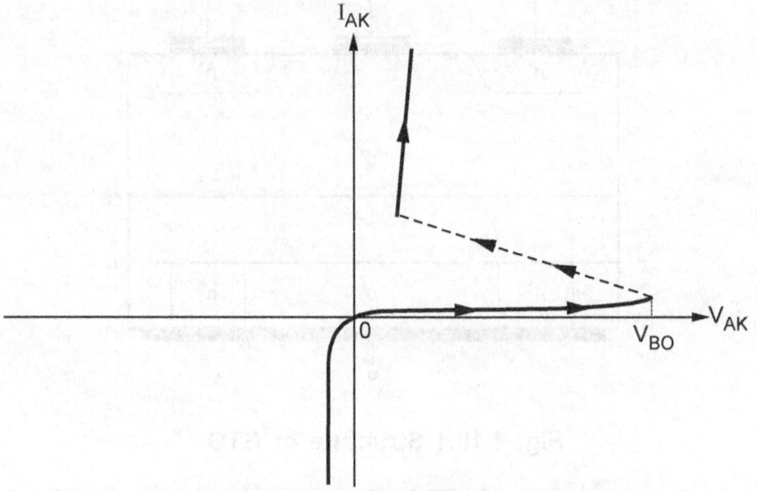

Fig. 1.10.3 V-I characteristics of GTO

In this figure observe that the V-I characteristics of GTO in forward direction are similar to that of SCR. But in reverse direction GTO has virtually no blocking capability. Observe that GTO starts conducting in reverse direction after very small reverse (20 to 30 V) voltage. This is because of the anode short structure.

In Fig. 1.10.1 observe that junction J_3 blocks reverse voltages. But J_3 has very small reverse breakdown voltage. Thus GTO has asymmetric voltage blocking capability.

1.10.1.3 Advantages, Limitations and Applications of GTO

Advantages

i. Higher voltage blocking capability.

ii. Gate has full control over the operation of GTO.

iii. Low on-state loss.

iv. High ratio of peak surge current to average current.

v. High on-state gain.

Limitations

i. GTOs require large negative gate currents for turn-off. Hence they are suitable for low power applications.

ii. Very small reverse voltage blocking capability.

iii. Switching frequencies are very small.

Applications

i. GTOs are suitable mainly for low power applications.

ii. Induction heating and motor drives.

1.10.2 Light Activated SCR (LASCR)

The light activated SCRs can be triggered using a beam of light. Their gate region is photosensitive. Fig. 1.10.4 shows the symbol and V-I characteristics of LASCR.

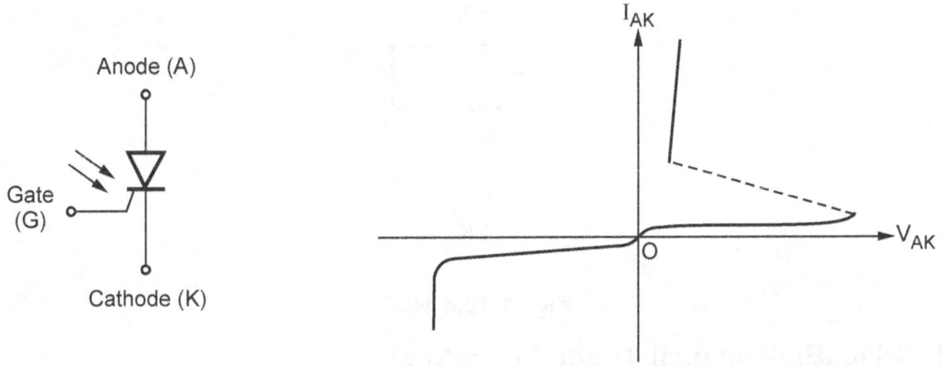

(a) Symbol of LASCR (b) V-I characteristics of LASCR

Fig. 1.10.4

The photons of light induce electrons in the gate-cathode junction. Because of these electrons, current starts flowing across J_3 and SCR turns-on. Once the SCR is turned on, gate has no control over its operation.

Advantages

i. It can be turned-on by a beam of light. Hence isolation is provided between control circuit and SCR.

ii. Because of optical triggering, effects of noise are reduced.

Applications

i. Used in high power applications like HVDC transmission, VAR compensation etc.

ii. Used in noise environments for better triggering control.

1.10.3 Reverse Conducting Thyristor (RCT)

In most of the applications, an antiparallel diode is connected across the thyristor. For example in chopper and inverter circuits, the antiparallel diode improves the turn-off requirement of the circuit. The antiparallel diode is also useful in case of inductive loads to provide the path for feedback currents. A reverse conducting thyristor (RCT) is similar to an SCR with antiparallel diode. Fig. 1.10.5 shows the equivalent circuit of RCT. The RCT is also called as asymmetrical thyristor or ASCR. It conducts in the reverse direction without any control. The conduction in the forward direction is controlled by the gate. The characterictics are similar to SCR in the forward direction. And the characteristics are similar to diode in reverse direction. It has the capability of upto 2000 V/500 A in the forward direction. The reverse blocking voltage is upto 40 V.

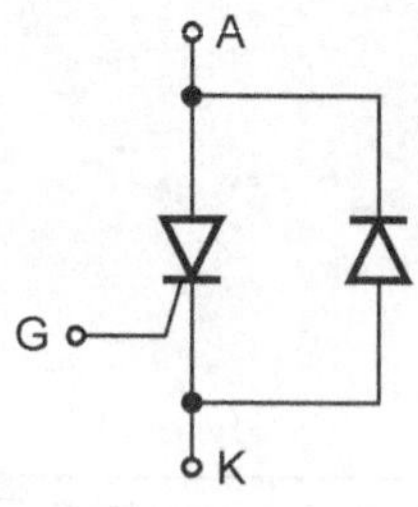

Fig. 1.10.5 RCT

1.10.4 Triac (Bidirectional Triode Thyristors)

The triac is the bidirectional device. It conducts in both the directions. We know that SCR conducts only in one direction. The triac is equivalent to the two antiparallel SCRs as shown in Fig. 1.10.6.

The triac has three terminals : Main Terminal 1 (MT_1), Main Terminal 2 (MT_2) and gate (G). Observe that the symbol also consists of antiparallel devices. The current can flow from MT_1 to MT_2 when MT_1 is forward biased with respect to MT_2. Similarly current flows from MT_2 to MT_1 when MT_2 is forward biased with respect to MT_1. The current flows (i.e. triac is switched 'on') whenever gate drive is applied. The triac is the best device for AC phase control. The input is AC and the load is also AC. The power is to be controlled in positive as well as negative half cycles. The triac is then triggered in every half cycle. The triac turns-off when current falls to zero in every half cycle. Thus the necessity of antiparallel SCRs is eliminated by triac. The word *triac* is abbreviated from *Tri* of triode and *AC*. It is mainly used for AC phase control and similar applications.

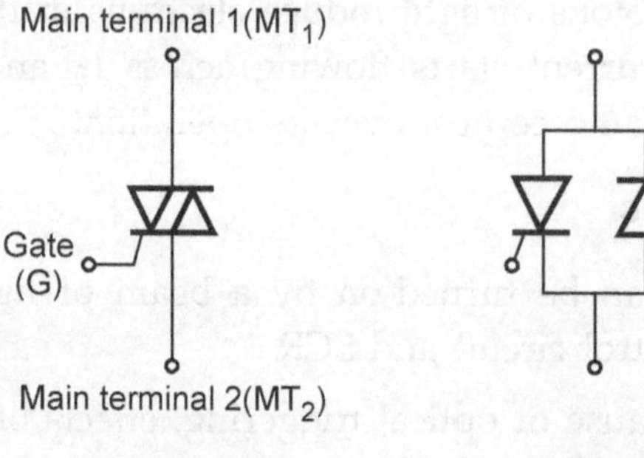

(a) Triac symbol **(b) Triac equivalent circuit**

Fig. 1.10.6 Triac symbol and equivalent circuit

1.10.4.1 V-I Characteristics of Triac

Fig. 1.10.7 shows the V-I characteristics of triac. The current 'I' is plotted with respect to voltage 'V'. Here I_F is the forward current and V_F is the forward voltage. Similarly I_R is the reverse current and V_R is the reverse voltage. V_{BO} is the breakover voltage of the triac. When gate drive is applied, the triac turns 'on' at lower voltage than V_{BO}. Such characteristic is shown by center (–·–·–) line. The dotted line in the characteristic indicates the switching of the triac. Once the triac is turned 'on', the voltage across it drops and current increases rapidly. Observe that the characteristics are exactly similar in the forward and reverse directions.

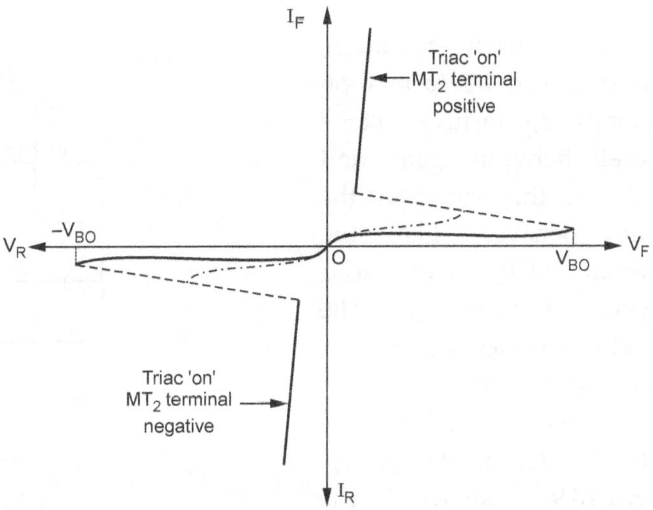

Fig. 1.10.7 V-I characteristics of triac

1.10.4.2 Operating Modes of Triac

Fig. 1.10.8 shows the internal structure of the triac. In this structure observe that the terminal MT_1 is connected to two layers : N_2 and P_2. Similarly Gate (G) is connected to two layers : N_3 and P_2. Also observe that MT_2 is connected to two layers N_4 and P_1.

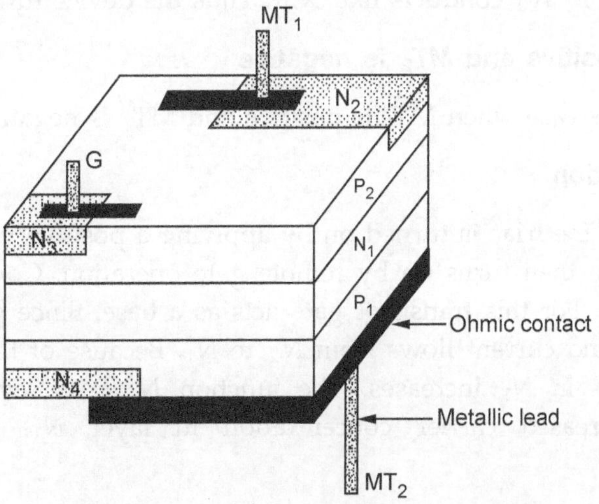

Fig. 1.10.8 Internal structure of triac

The operation of triac can be explained as follows :

(i) When MT_2 is positive and MT_1 is negative

Under such condition, the triac can be turned on by applying positive voltage signal between gate and MT_1. In this situation the triac behaves like the normal SCR, with four layers $P_1 N_1 P_2 N_2$. The anode becomes at P_1 (i.e. MT_2) and cathode becomes at N_2 (i.e. MT_1). The gate will be effectively at P_2. Fig. 1.10.9 shows the effective structure. The positive gate signal at P_2 triggers the device like SCR.

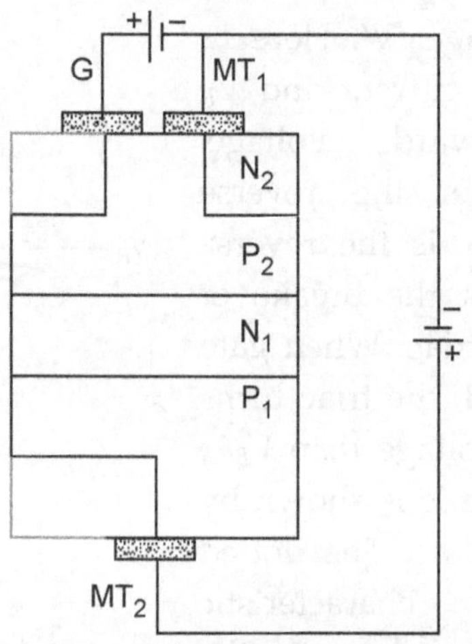

Fig. 1.10.9 Equivalent structure when $MT_2 > MT_1$ and triac triggered by applying a positive voltage at gate

Junction gate operation

When MT_2 is positive and MT_1 is negative, the triac can also be triggered by applying negative signal to gate. This is called junction gate operation. Refer to Fig. 1.10.8. When gate is negative with respect to MT_1, the junction $P_2 N_3$ is forward biased, and the left hand portion $P_1 N_1 P_2 N_3$ conducts like SCR. In this case MT_1 acts like gate initially. The carriers spread in the layer P_2 from left to right, i.e. towards N_2. This carrier flow from P_2 to N_2 is like a conventional gate current and right handside part $P_1 N_1 P_2 N_2$ conducts like SCR. Thus the device turns on.

(ii) When MT_1 is positive and MT_2 is negative

Now consider the case when MT_1 is positive and MT_2 is negative.

Remote gate operation

In this operation the triac is turned on by applying a positive voltage between gate and MT_1. The device then turns on by remote gate operation. Consider the transistor formed by $N_2 P_2 N_1$. For this transistor gate acts as a base. Since gate is positive, this transistor turns on and current flows from N_2 to N_1. Because of this current flow, the carrier concentration in N_1 increases. The junction $N_1 P_1$ is reverse biased. But it breaks due to increased carrier concentration in layer N_1. Once the junction

$N_1 P_1$ breaks, the device is switched 'on' and conducts heavily. The four layers of conduction are $P_2 N_1 P_1 N_4$.

1.10.4.3 Merits, Limitations and Applications of Triac

Merits of Triac

i. Triac is a bidirectional device, i.e. it conducts in both directions.

ii. Triac turns off when voltage is reversed.

iii. Single gate controls conduction is both directions.

iv. Triacs with high voltage and current ratings are available.

Demerits of Triac

i. Triacs are latching devices like SCR. Hence, they are not suitable for DC power applications.

ii. Gate has no control over the conduction once triac is turned on.

iii. Triacs have very small switching frequencies.

Applications of Triac

i. AC power controllers and heater, fan etc. controllers.

ii. Triggering device for SCRs.

Example 1.10.1 : *The gate-cathode characteristic of a triac is given by* $v_g = 2+5I_g$. *A triggering pulse train with an amplitude of 10 V, ON period of 10 μsec is applied to the gate through a* 10 Ω *series register. Calculate :*

(i) Peak gate power.

(ii) Triggering frequency to obtain an average gate power of 0.5 W.

Solution : (i) To obtain peak gate power

Fig. 1.10.10 shows the circuit diagram. From this diagram we can write,

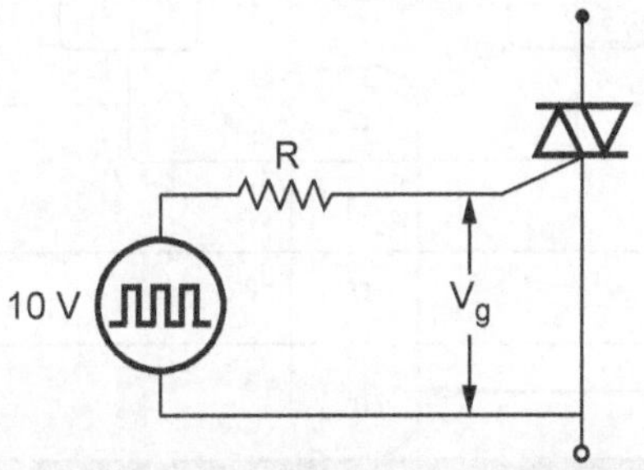

Fig. 1.10.10 Gate triggering circuit

$$10\text{ V} = I_g \times R + V_g$$

$$10\text{ V} = 10 I_g + 2 + 5 I_g$$

$$\therefore \quad I_g = 0.533\text{ A.}$$

Hence gate voltage can be obtained as,

$$V_g = 2 + 5\, I_g = 2 + 5 \times 0.533 = 4.667$$

Here above V_g and I_g are calculated for peak amplitude of 10 V for the pulse train. Hence V_g and I_g indicate peak values.

$$\therefore \quad \text{Peak gate power } P_g \text{ (peak)} = V_g\, I_g = 4.667 \times 0.533 = 2.48\text{ W.}$$

(ii) To obtain triggering frequency

$$\text{Duty cycle} = \frac{P_g(av)}{P_g(peak)} = \frac{0.5}{2.48} = 0.2$$

$$\text{Duty cycle} = \frac{T_{on}}{T} = T_{on} \times f$$

$$\therefore \quad f = \frac{Duty\ cycle}{T_{on}} = \frac{0.2}{10 \times 10^{-6}} = 20\text{ kHz.}$$

1.10.5 MOS Controlled Thyristor (MCT)

The MOS controlled thyristor (MCT) has two built in MOSFETs. One MOSFET is used for turn-on and the other MOSFET is used for turn-off. MCT is similar to GTO except the MOSFETs.

1.10.5.1 Internal Structure of MCT

Fig. 1.10.11 shows the internal structure of MCT.

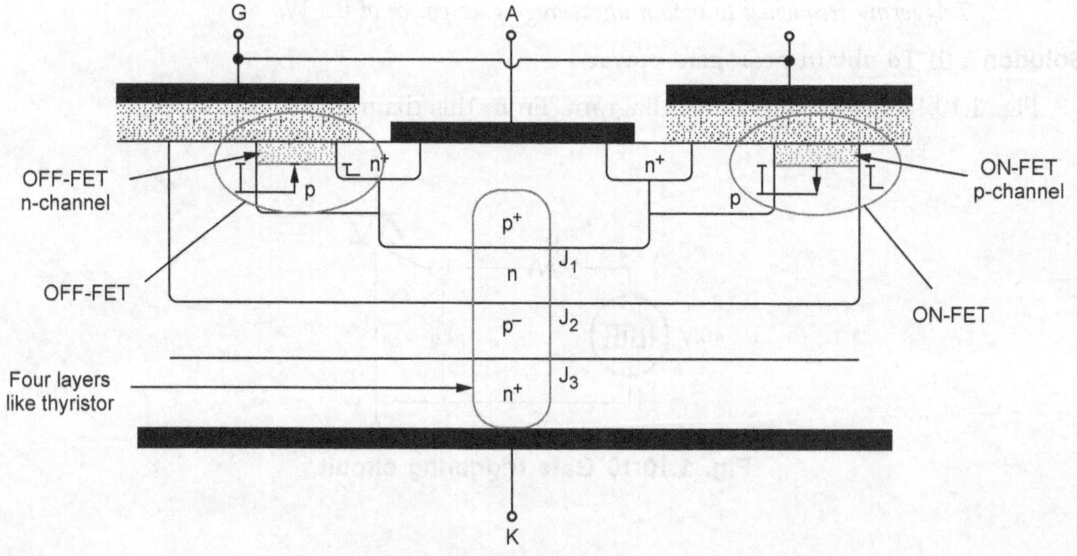

Fig. 1.10.11 Internal structure of p-type MCT

There are two types :

p-MCT and n-MCT. Fig. 1.10.11 shows the internal structure of p-MCT. The p-type of layer is lightly doped. In the forward blocking stage this layer contains the depletion region of junction J_2. The ON-FET is p-channel MOSFET and OFF-FET is n-channel MOSFET. Fig. 1.10.12(a) shows the equivalent circuit of p-MCT and Fig. 1.10.12(b) shows its symbol.

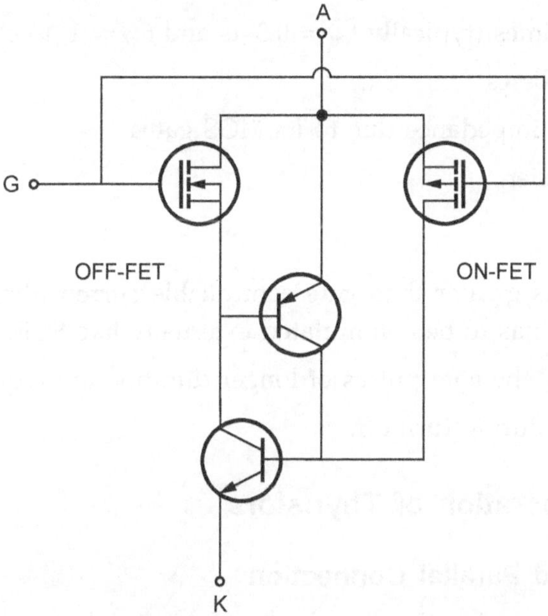

(a) Equivalent circuit

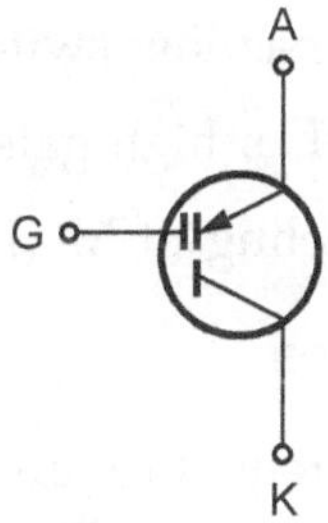

(b) Symbol

Fig. 1.10.12

In the equivalent circuit observe that drains(D) of both the MOSFETs are connected to anode. The source(s) of ON-FET is connected to base region (p^-) of the npn transistor.

1.10.5.2 Turn-on and Turn-off MCT

MCT can be turned-on by applying negative gate-anode voltage. Because of this voltage the p-channel is induced as shown in Fig. 1.10.11. Hence ON-FET starts conducting the current of ON-FET goes to base of the npn transistor and the device turns on by regeneration.

The MCT can be turned-off by applying positive gate-anode voltage. This induces the n-channel in the device as shown in Fig. 1.10.11. The OFF-FET then starts conducting. In Fig. 1.10.12(a) observe that the OFF-FET is connected across emitter base of pnp transistor. Since OFF-FET is turned on, the base-emitter of pnp transistor is shorted. Hence it comes out of saturation and starts turning OFF. Therefore npn transistor also turns-off and the MCT is turned off.

1.10.5.3 Characteristics of MCT

The V-I characteristics of MCT are similar to that of GTO.

1.10.5.4 Advantages and Disadvantages of MCT

Advantages

i. MCT can be turned-on and turned-off by low gate voltages.

ii. MCT has fast switching times (typically t_{on} = 0.3 μs and t_{off} = 1 μsec).

iii. MCT has low switching losses.

iv. MCT has high gate input impedance due to its MOS gates.

v. Paralleling of MCTs is easier.

Disadvantages

i. If current through MCT is greater than gate controllable current then it cannot be turned off. Then MCT has to be commutated externally like SCR.

ii. For larger anode currents, the gate pulses of longer duration are required.

iii. Gate draws peak current during turn-off.

1.11 Series and Parallel Operation of Thyristors

1.11.1 Necessity of Series and Parallel Connection

Series connection of the devices is often required to increase the overall voltage rating. For example the thyristor is to be operated at 1000 volts. But we have thyristors of rating 600 volts. Then the circuit can be implemented by connecting two thyristors in series.

(a) Series connection

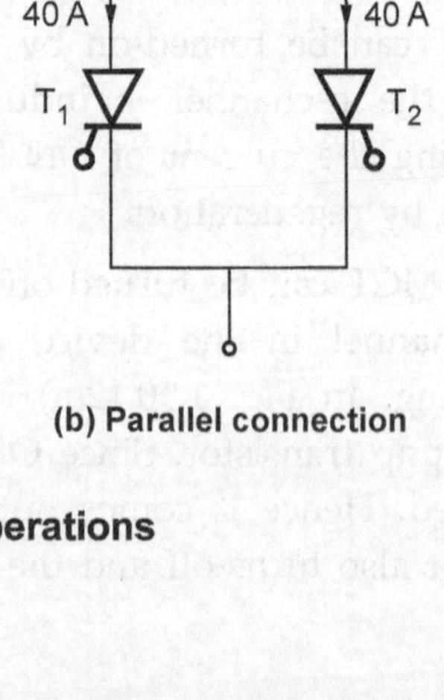

(b) Parallel connection

Fig. 1.11.1 Series and parallel operations

Similarly parallel connection is used to increase current ratings. For example, current in the circuit is 80 A. But we have a thyristor of rating 50 A. Then the problem can be solved by connecting two thyristors in parallel as shown in Fig. 1.11.1 (b). This makes the current sharing among two thyristors and each one carries $\frac{80}{2}$ = 40 A. Thus series and parallel connections are most widely used to cater the need of higher voltage and currents.

1.11.2 Series Connection of Thyristors

1.11.2.1 Problems Encountered in Series Connection

When the thyristors are connected in series, they have small differences in their ratings. For example Fig. 1.11.2 shows the V-I characteristics of two thyristors of same ratings. Observe that there are minor differences between the characteristics. Forward break-over voltages, internal resistance, leakage current etc are not exactly same. The thyristor with highest internal resistance will have minimum leakage current. Hence high voltage will appear across it in off state. This creates voltage imbalance in the series connection. Hence equalization is necessary in the series connection.

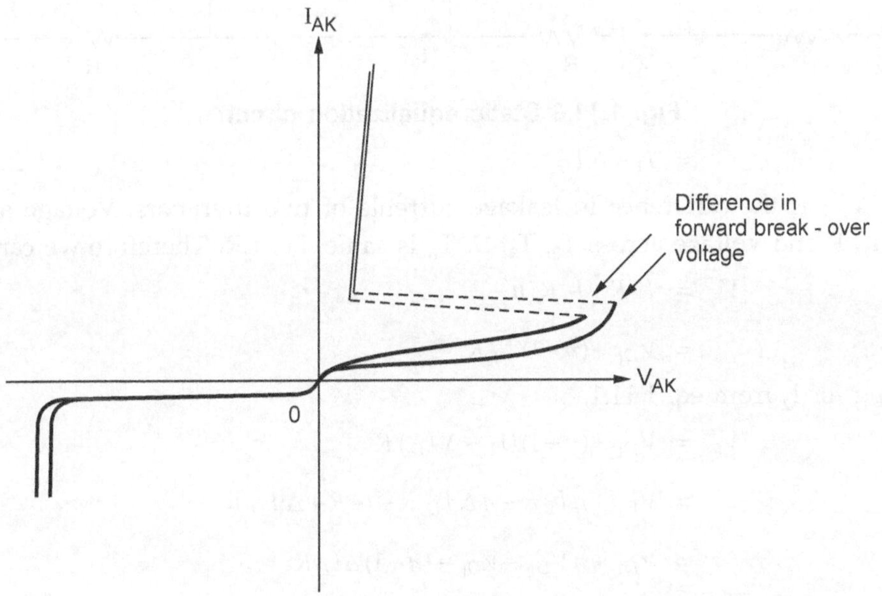

Fig. 1.11.2 Two thyristors of same rating have slightly different characteristics

1.11.2.2 Equalizing Components

Let us consider that 'n' number of thyristors are connected in series. An equalizing resistance 'R' is connected across each thyristor as shown in Fig. 1.11.3. Let us assume that T_1 has maximum internal resistance in off-state. Hence its leakage current I_{D1} is minimum. Let the internal resistance of other devices is same. Hence their leakage current is also same, i.e. I_{D2} current I_1 flows through R. Since other thyristors have same internal resistance, current through their parallel resistors will be same, i.e. I_2. From above circuit we can write,

$$I_{D1} + I_1 = I_{D2} + I_2$$

$$\therefore \quad I_2 = I_1 - I_{D1} - I_{D2}$$

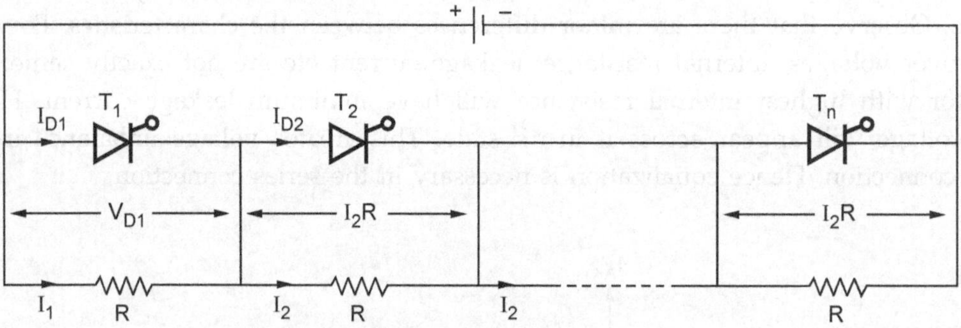

Fig. 1.11.3 Static equalization circuit

$$= I_1 - \Delta I_D \quad \dots (1.11.1)$$

Here ΔI_D is the difference in leakage currents of two thyristors. Voltage across T_1 is $V_{D1} = I_1$, R and voltage across T_2, T_3, ... T_n is same, i.e. I_2R. Therefore we can write,

$$V_s = I_1 R + I_2 R(n-1)$$

$$= V_{D1} + (n-1) I_2 R$$

Putting for I_2 from eq. 1.11.1,

$$V_s = V_{D1} + (n-1)(I_1 - \Delta I_D) R$$

$$= V_{D1} + n I_1 R - n \Delta I_D R - I_1 R + \Delta I_D R$$

$$= V_{D1} + n V_{D1} - V_{D1} - (n-1) \Delta I_D R$$

$$= n V_{D1} - (n-1) \Delta I_D R$$

$$\therefore \quad \boxed{V_{D1} = \frac{V_s + (n-1) \Delta I_D R}{n}} \quad \dots (1.11.2)$$

Maximum value of V_{D1} will occur when ΔI_D is maximum. Value of R can be obtained from above equation as,

$$R = \frac{n V_{D1} - V_s}{(n-1)\Delta I_D} \quad \text{... (1.11.3)}$$

1.11.2.3 Dynamic Equalization Circuit

We studied the equalization when all thyristors are in forward blocking state. There can be some difference in the turn-on and turn-off times of the thyristors. When trigger is given to all the thyristors, they start turning-on. Higher voltage will appear across the thyristor which turns-on late. Hence its maximum voltage rating may be exceeded. This problem can be overcome by dynamic equalizing circuit. Fig. 1.11.4 shows the dynamic equalizing circuit. It is R-C circuit across each thyristor. Static equalization resistors are also shown in this circuit. An RC circuit is placed across each thyristor. C_1 provides the equalization during turn-on and turn-off. Resistance R_1 is used to limit discharge current of C_1.

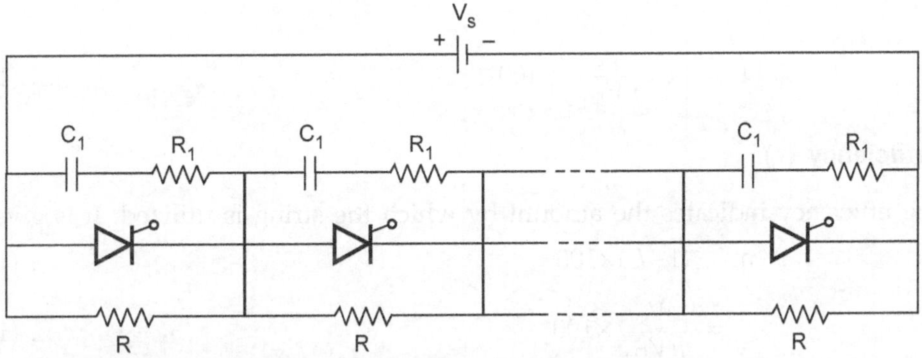

Fig. 1.11.4 Complete equalization circuit

Value of C_1

We know that the relationship between charge on the capacitor, voltage on the capacitor and value of the capacitor is,

$$V = \frac{Q}{C}$$

If V is ΔV, then,

$$\Delta V = \frac{\Delta Q}{C}$$

Consider equation 1.11.2, $V_{D1} = \dfrac{V_s + (n-1)\Delta I_D R}{n}$

If ΔQ indicates the difference in charge stored in C_1 and C_2, then ΔV will indicate the voltage difference across T_1 and T_2. Hence we have $\Delta V = \Delta I_D R$. Therefore above equation becomes,

$$V_{D1} = \frac{V_s + (n-1)\Delta V}{n}$$

Since $\Delta V = \frac{\Delta Q}{C_1}$,

$$V_{D1} = \frac{V_s + (n-1)\Delta Q / C_1}{n} \quad \text{... (1.11.4)}$$

This equation gives the voltage imbalance across T_1. Value of C_1 can be obtained as,

$$\boxed{C_1 = \frac{(n-1)\Delta Q}{nV_{D1} - V_s}} \quad \text{... (1.11.5)}$$

Derating factor of the string

The derating factor (D) indicates the amount by which string is derated. It is given by

$$\% D = \left[1 - \frac{V_s}{nV_{D1}}\right] \times 100 \quad \text{... (1.11.6)}$$

String efficiency (η)

String efficiency indicates the amount by which the string is utilized. It is given as,

$$\% \eta = (1-D) \times 100 \quad \text{... (1.11.7)}$$

$$= \frac{V_s}{n V_{D1}} \times 100 \quad \text{... (1.11.8)}$$

Example 1.11.1 : *How many SCRs are required in a series string to withstand a DC voltage of 3500 volts in steady state, if the SCRS have steady state voltage rating of 1000 V and steady state derating factor of 30% ? Assuming maximum difference in leakage current of SCRs to be 10 mA, calculate the value of voltage sharing resistances to be used. Draw the circuit showing the SCRs and the voltage sharing resistances.*

Solution : Given data

$$V_S = 3500 \text{ V}$$

$$V_{D1} = 1000 \text{ V}$$

$$\% D = 30$$

$$\Delta I_D = 10 \text{ mA}$$

i) To obtain number of SCRs

By equation 1.11.6 we have,

$$\%D = \left[1 - \frac{V_S}{nV_{D1}}\right] \times 100$$

$$\therefore \quad 30 = \left[1 - \frac{3500}{n \times 1000}\right] \times 100$$

$$\therefore \quad n = 5$$

ii) To obtain voltage sharing resistance

By equation 1.11.3 such resistance is given as,

$$R = \frac{nV_{D1} - V_S}{(n-1)\,\Delta I_D} = \frac{5 \times 1000 - 3500}{4 \times 10 \times 10^{-3}}$$

$$= 37.5 \text{ k}\Omega$$

The circuit will be similar to Fig. 1.11.3 having 5 SCRs in series and each resistance of 37.5 kΩ.

Example 1.11.2 : *A 3ϕ converter is used for HVDC transmission system and is operated from 3 ϕ 25 kV supply. Thyristor each of 1600 V/16 A are available. The forward leakage current difference of the device is 35 mA. The string efficiency can be assumed to be 85% and ΔQ_{max} = 25 μC.*

a) Determine the number of devices to be connected in series.

b) Equalizing components.

Solution : Given data

$$V_{line} = 25 \text{ kV}$$

$$\therefore \quad V_s = PIV = \sqrt{2} \times V_{line}$$

$$= \sqrt{2} \times 25 \text{ kV} = 35.35 \text{ kV}$$

$$V_{D1} = 1600 \text{ V}$$

$$\Delta Q_{max} = 25 \ \mu C$$

$$\Delta I_D = 35 \text{ mA}$$

String efficiency, $\eta = 0.85$

a)To obtain number of devices

Form equation 1.11.8,

$$\eta = \frac{V_s}{nV_{D1}}$$

$$\therefore \quad n = \frac{V_s}{\eta V_{D1}} = \frac{35.35\times10^3}{0.85\times1600}$$

$$= 26$$

b) To obtain equalizing components

Value of R is given by equation 1.11.3 as,

$$R = \frac{n V_{D1} - V_s}{(n-1)\Delta I_D}$$

$$= \frac{26\times1600 - 35.35\times10^3}{(26-1)\,35\times10^{-3}}$$

$$= 7.142 \text{ k}\Omega$$

Value of C_1 is given by equation 1.11.5 as,

$$C_1 = \frac{(n-1)\Delta Q}{nV_{D1} - V_s} = \frac{(26-1)\,25\times10^{-6}}{26\times1600 - 35.35\times10^3}$$

$$= 0.1\ \mu F$$

1.11.3 Parallel Connection of Thyristors

We know that parallel connection of thyristors is used to cater higher current demands.

1.11.3.1 Problems Occurred in Parallel Connection

The devices used in parallel connection donot have exactly similar characteristics. The thyristor carrying higher current will have more power dissipation. This will increase its temperature and reduce the internal resistance. Therefore the current further increases. This process continues till thyristor damages.

1.11.3.2 Equalizing Arrangements

Heat sharing can be done by using common heat sinks for all the devices. This will maintain all the devices at same temperature.

Static Current Sharing

Fig. 1.11.5 shows the static current sharing arrangement. Two resistors connected in series with the thyristors.

R_1 and R_2 try to equalize the currents through T_1 and T_2. But power dissipation in R_1 and R_2 is very high.

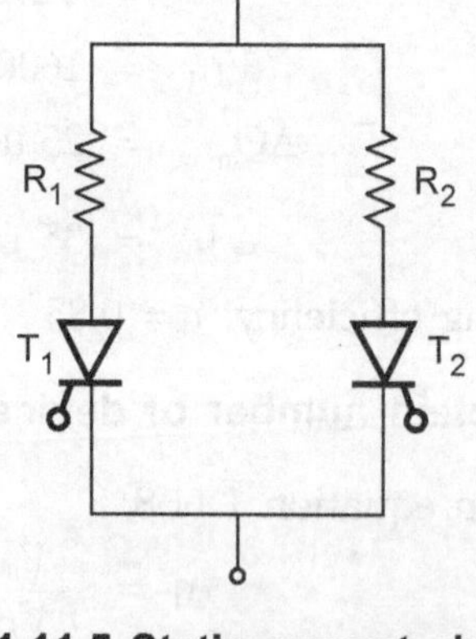

Fig. 1.11.5 Static current sharing

Dynamic Current Sharing

Fig. 1.11.6 shows an arrangement for dynamic current sharing. The inductors are placed in series with thyristors. These inductors are magnetically coupled. But they are connected in opposite direction. Hence if current in T_1 tries to increase, then a voltage of opposite polarity will be induced in L_2. This increases the current in T_2. Thus current balance is maintained.

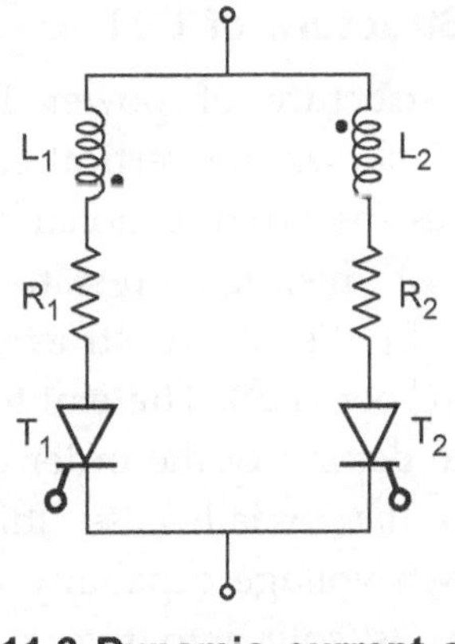

Fig. 1.11.6 Dynamic current sharing

> **Review Questions**
>
> 1. *Derive an equation for components of dynamic equalization circuit of SCRs connected in series.*
> 2. *What is the necessity of series and parallel connection of SCRs ?*

1.12 Power BJT

The power BJT is the bipolar device. Fig. 1.12.1 shows the symbols of BJT. The BJT is also called as power transistors in general. The BJTs are of two types : npn and pnp. BJT has collector (C) base (B) and emitter (E). In the npn BJT, when the base emitter junction is forward biased to saturation, the transistor turns 'on' and current flows from collector to emitter. When the BJT turns 'on', the collector emitter drop becomes negligible. The BJT turns-off as soon as base emitter drive is removed. Similarly in case of pnp BJT, when base emitter junction is forward biased to saturation, the transistor turns 'on' and current flows from emitter to collector. The transistor turns-off as soon as base emitter drive is removed. Thus the drive has full control over the conduction of BJT. No commutating components are required by BJT for turn-off.

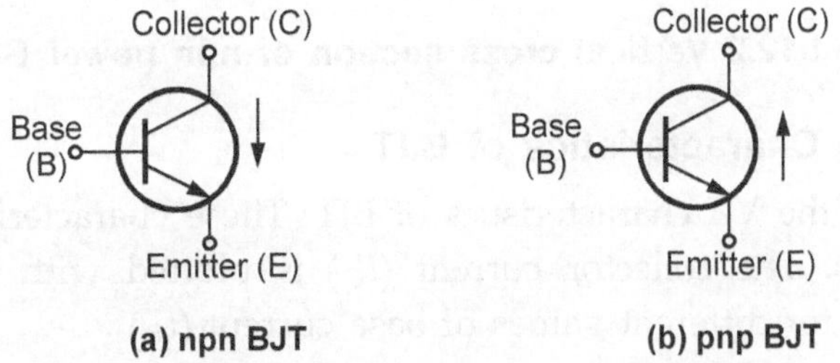

Fig. 1.12.1 Symbols of BJT

1.12.1 Structure of BJT

The structure of power BJT is different compared to small signal transistors. Fig. 1.12.2 shows the vertical cross-section of the npn power BJT. The vertical structure maximizes the cross-sectional area through which the current in the device is flowing. Because of this, the on-state resistance and power dissipation in the transistor is reduced. In the above structure, observe that there is highly doped emitter region (about 10^{19}per cm^3). The emitter region has the thickness of about 10 μm. The base has moderate doping of the order of 10^{16} per cm^3. The thickness of the base can vary from 5 to 20 μm. Small base thickness provides good amplification capabilities. But breakdown voltage capability of the transistor is reduced for small base regions. The collector is split into two regions as shown in Fig. 1.12.2. These two regions are n^-(with 10^{14} per cm^3) and n^+ (with 10^{19} per cm^3). The n^- region has light doping and it is called collector drift region. The thickness of n^- layer determines the breakdown voltage capability of the transistor. The n^+ region has high doping intensity. It's doping is similar to that of emitter. The n^+ region serves as collector contact for external circuits. In the practical power transistors, the emitters and bases are interleaved. This reduces the effects of current crowding and hence possibility of second breakdown.

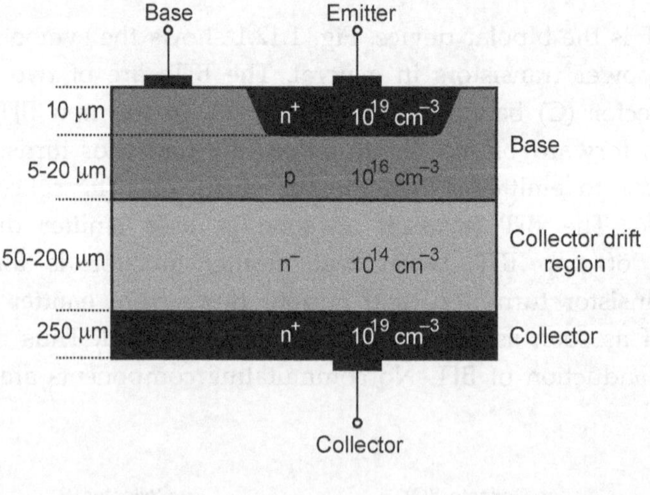

Fig. 1.12.2 Vertical cross section of npn power BJT

1.12.2 Steady State Characteristics of BJT

Fig. 1.12.3 shows the V-I characteristics of BJT. These characteristics are also called output characteristics. The collector current (i_C) is plotted with respect to collector emitter voltage (V_{CE}) for different values of base current (i_B) .

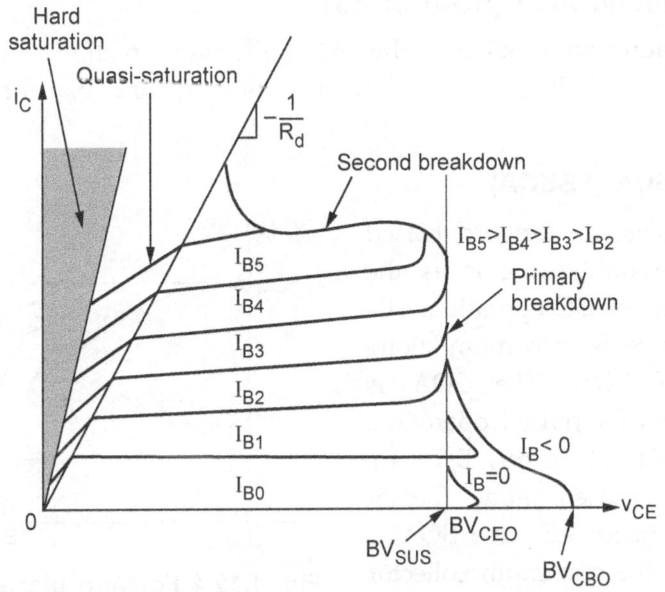

Fig. 1.12.3 V-I characteristics of npn power BJT and different regions of operation

- There are four regions clearly shown : Cut-off region, Active region, quasi-saturation and hard saturation.

Cut-off region : The cut-off region is the area where base current is almost zero. Hence no collector current flows and transistor is 'off'.

Quasi-saturation : In the quasi saturation and hard saturation, the base drive is applied and transistor is said to be 'on'. Hence collector current flows depending upon the load. The BJT is never operated in the active region (i.e. as an amplifier). It is operated in cut-off and saturation. Thus BJT acts as a switch. The 'BV_{SUS}' is the maximum collector to emitter voltage that can be sustained when BJT is carrying substantial collector current. 'BV_{CEO}' is the maximum collector to emitter breakdown voltage that can be sustained when base current is zero (i.e. base open circuitted). And 'BV_{CBO}' is the collector base breakdown voltage when the emitter is open circuitted.

Primary breakdown : The primary breakdown in BJT takes place because of avalanche breakdown of the collector base junction. The large power dissipation normally leads to primary breakdown.

Second breakdown : It is clear from Fig. 1.12.3 that, at the large collector currents, the collector emitter voltage drops. Due to this drop in voltage, the collector current increases. Here there is substantial increase in power dissipation. This power dissipation is not evenly spread across the entire volume of the device. But it is concentrated in the highly localized regions. In these regions the local temperature grows very rapidly and the BJT is damaged.

1.12.3 Safe Operating Area (SOA) of BJT

The safe operating area (SOA) of the BJT is the area on the $i_C - v_{CE}$ plane where BJT should be operated. It gives the combinations of i_C and v_{CE} for which BJT will function safely.

Forward biased SOA (FBSOA)

Fig. 1.12.4 shows the forward biased SOA of BJT. The dotted region is the SOA for continuous or DC operation. The junction temperature is maximum along the boundary of SOA. The SOA is effectively increased for pulsed operation. In power electronic converters, BJTs are normally used in pulsed mode (i.e. at high frequencies) hence effective SOA is larger. The I_{CM} is the maximum collector current even as a pulse. BV_{CEO} is the maximum collector to emitter voltage when base is open circuitted. The SOA is normally supplied by the manufacturers in the specifications.

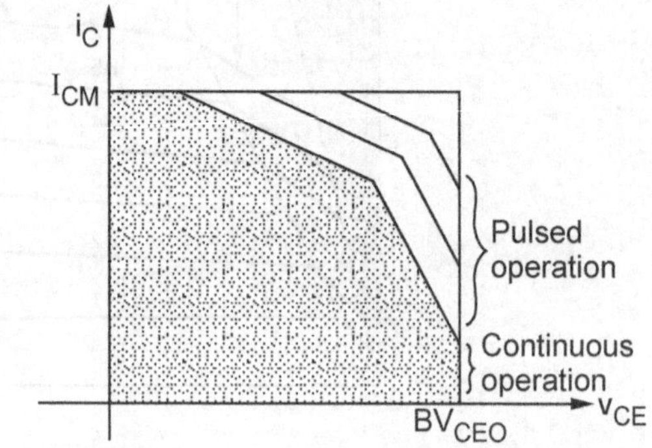

Fig. 1.12.4 Forward biased SOA of BJT

Reverse biased SOA(RBSOA)

Fig. 1.12.5 shows the RBSOA of power BJT. It is pulsed SOA.

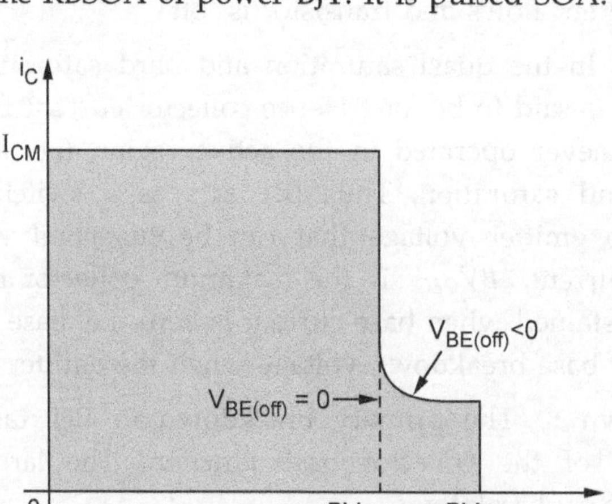

Fig. 1.12.5 The RBSOA of BJT

- This RBSOA is larger than FBSOA because of BV_{CBO} at low collector currents.

1.12.4 Switching Characteristics of BJT

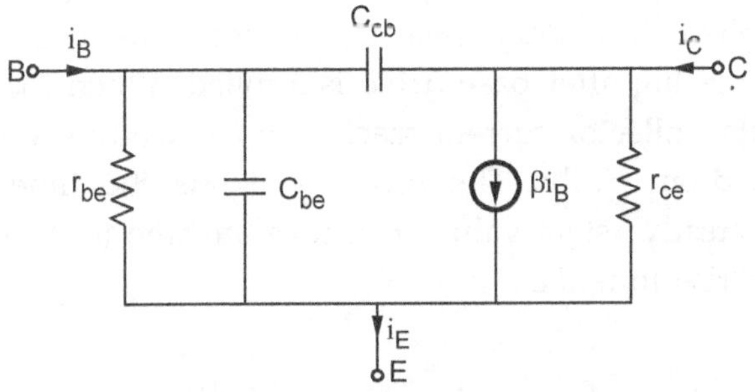

Fig. 1.12.6 Equivalent model of BJT

Consider the model of npn transistor shown in Fig. 1.12.6 The equivalent collector-base junction capacitance (C_{cb}) and base-emitter junction capacitances (C_{be}) play significant role during turn-on and turn-off. Effect of these capacitances can be neglected under steady state conditions. But turn-on and turn-off (i.e. transient or switching conditions) are affected due to internal capacitances of BJT. The values of internal capacitances depend upon junction voltages and physical construction of the BJT.

Fig. 1.12.7 shows the switching waveforms of the BJT. The base-emitter voltage is made positive to turn-on the BJT. This voltage is $v_{BE(on)}$. As the base voltage becomes positive, the base current also starts flowing. The value of base current is $I_{B(on)}$.

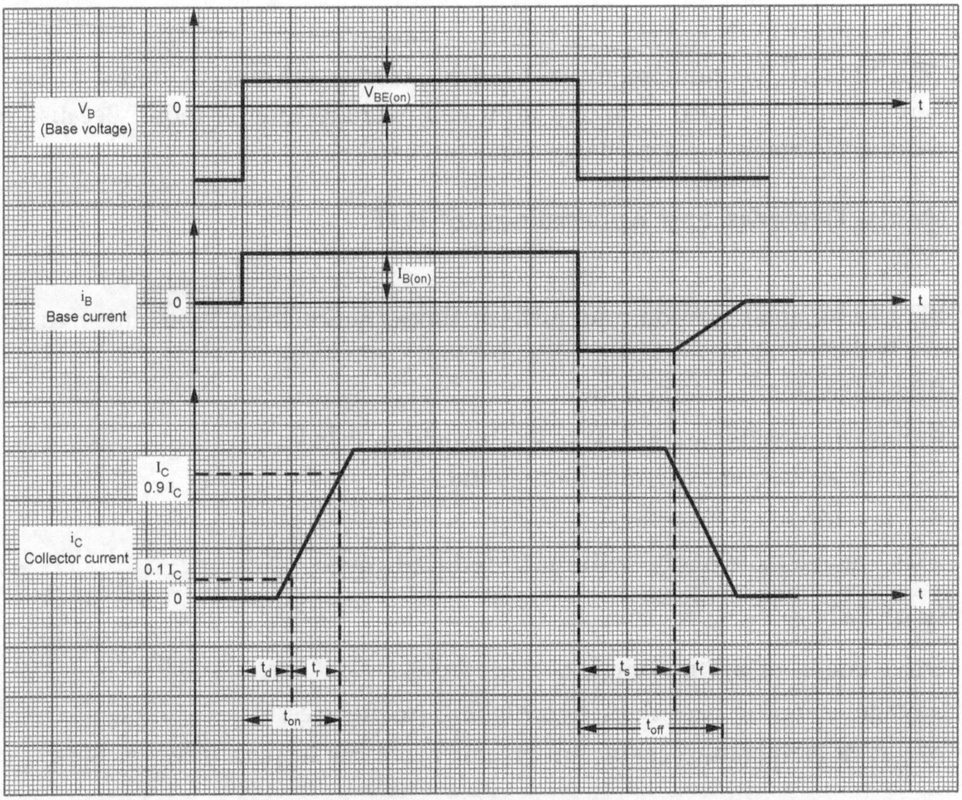

Fig. 1.12.7 Switching waveforms of BJT

In the Fig. 1.12.7 observe that collector current does not start flowing as soon as base drive is applied. This is because, the collector-base junction capacitance (C_{be}) starts charging when base drive is applied. The *delay time* (t_d) is the time delay involved when collector current starts increasing after base drive is applied. When C_{be} charges to forward bias voltage of 0.7 volts, collector current reaches to its steady state value. The BJT is then said to be turned-on fully. The *rise time* (t_r) is the time required to raise collector current to it's steady state value. The turn-on time (t_{on}) of the BJT is equal to sum of delay time and rise time. i.e.,

$$t_{on} = t_d + t_r$$

The turn-off the transistor, base voltage is made negative. Hence, the base current is also negative. But the collector current does not change for time t_s. This is called *storage time*. During this period, the saturating charge is removed from the base. After the stored charges are removed, the base current starts reducing and collector current also starts falling. The stored charge is removed because of the negative base current. Once the stored charge in base is removed, C_{be} is charged to negative base voltage and base current becomes zero. The decay of collector current depends upon the stored charge and hence on C_{be}. The turn-off time of the transistor is equal to sum of storage time and *fall time* (t_f). i.e.,

$$t_{off} = t_s + t_f$$

The fall time is the time required by the collector current to decay to its 10% value. Here note that if base is highly saturated, storage time is more. Because more time is required to remove extra charge from the base.

1.12.5 Merits, Demerits and Applications of BJT

Merits of BJT

i. BJTs have small turn-on and turn-off times, hence their switching frequencies are higher.
ii. BJTs have small turn-on losses.
iii. Base drive has full control over the operation of BJT.
iv. BJTs do not require commutation circuits.
v. BJT is a bipolar device.
vi. BJTs are available easily with much reduced costs.

Demerits of BJT

i. Drive circuit of BJT is complex.
ii. Storage charge in base reduces switching frequencies.
iii. Negative temperature coefficient creates problems in paralleling of BJTs.

Applications of BJT

i. Switched Mode Power Supplies
ii. Bridge inverters
iii. DC to DC converter (choppers)
iv. Power factor correction techniques.

Sr. No.	SCR	BJT
1.	Four layer (PNPN) device.	Three layer (PNP or NPN) device.
2.	Turns-on by regeneration.	No regeneration exists
3.	Gate has no control once SCR is turned on.	Base has full control over the operation of BJT.
4.	External circuits are required to turn-off the SCR	No external circuits are required. BJT turns-off if base drive is removed.
5.	Switching frequencies are low.	Switching frequencies are high.
6.	False triggering takes place if dv/dt is exceeded.	BJT is damaged if dv/dt is exceeded.
7.	Used for controlled rectifiers, AC regulators and DC motor drives.	Used for inverters, UPS, AC motor drives and SMPS.

Table 1.12.1 Comparison of BJT and SCR

Review Question

1. *Explain the switching characteristics of power BJT.*

1.13 Power MOSFET

The metal oxide semiconductor field effect transistors (MOSFET) are majority carrier devices. Fig. 1.13.1 shows the symbols of MOSFETs.

Observe that there are two types of power MOSFETs : n-channel MOSFET and p-channel MOSFET. The MOSFET has three terminals : gate (G), drain (D) and source (S).

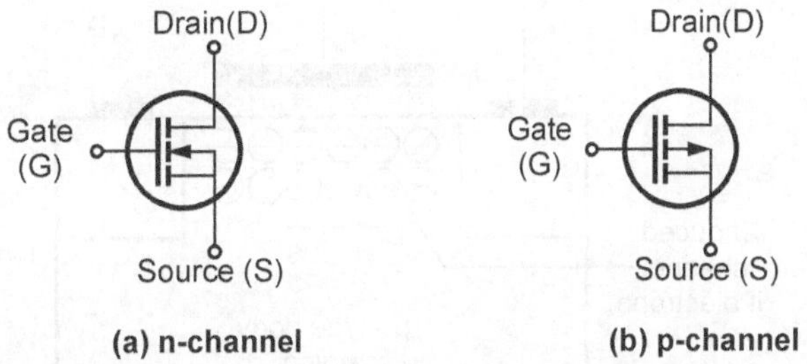

Fig. 1.13.1 Symbols of MOSFETs

When the MOSFET is turned 'on' the current flows from drain to source. The voltage is applied between gate-source to turn 'on' the MOSFET. Very small current flows from gate to source. Only voltage is to be applied to turn on the MOSFET. The MOSFET can be turned-off by removing the gate to source voltage. Thus gate has full control over the conduction of the MOSFET. The turn-on and turn-off times of MOSFETs are very small. Hence they operate at very high frequencies. Hence MOSFETs are preferred in applications such as choppers and inverters. Since only voltage drive (gate-source) is required, the drive circuits of MOSFETs are very simple. The paralleling of MOSFETs is easier due to their positive temperature coefficient (PTC). MOSFETs have high on-state resistance, $R_{DS\,(on)}$. Hence for higher currents, losses in the MOSFETs are substantially increased. Hence MOSFETs are mainly used for low power applications.

1.13.1 Structure of MOSFETs

There are two types of MOSFETs : depletion type MOSFET and enhancement type MOSFET. In both of these types the MOSFETs can be n-channel or p-channel. Fig. 1.13.2 shows the structure of n-channel enhancement type MOSFET. The source and drain are connected n^+ regions. These regions are heavily doped with the intensity of 10^{19} per cm^3. The p-type body region forms the channel between drain and source. The body region has the doping level of 10^{16} per cm^3. The gate is not directly connected to the p-type region. There is insulating oxide (SiO_2) layer between gate metal and p-type layer. When gate is made positive with respect to source an accumulation layer is formed in the channel as shown in Fig. 1.13.3. This accumulation layer is formed because of V_{GS}. The gate terminal (metal) is positive. The other side of oxide layer is p-type of body region. Accumulation layer of electrons is generated in the body region near oxide layer. This is also called induced channel of electrons. Therefore current (i_{DS}) starts flowing through this induced channel. The current flows from drain to source. If V_{GS} = 0, then

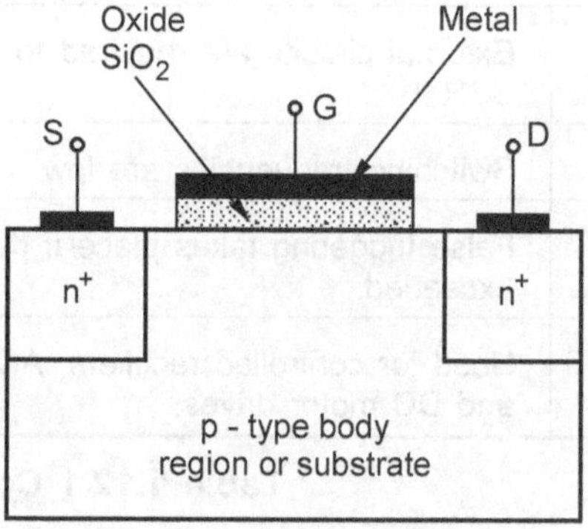

Fig.1.13.2 Structure of n-channel enchancemode MOSFET. (Drift layer is not shown)

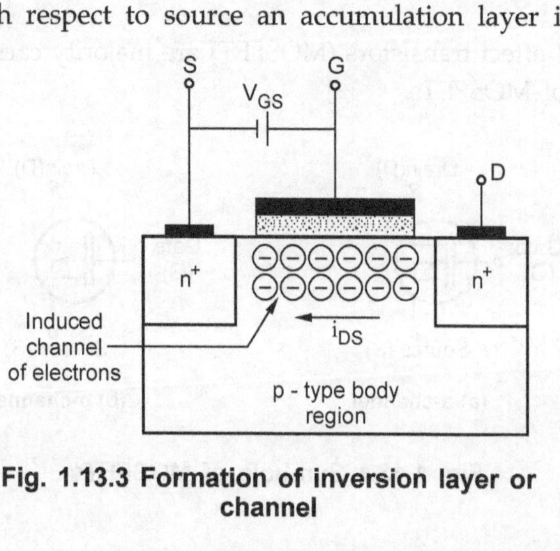

Fig. 1.13.3 Formation of inversion layer or channel

induced channel is absent and no current flows. Since channel is made of electrons, this is called n-channel MOSFET.

Fig. 1.13.4 shows the four layer structure of n-channel enhancement mode MOSFET. This is $n^+pn^-n^+$ structure. A drift region (n^-) is shown in this structure. The drift region is lightly doped (10^{14} per cm^3).

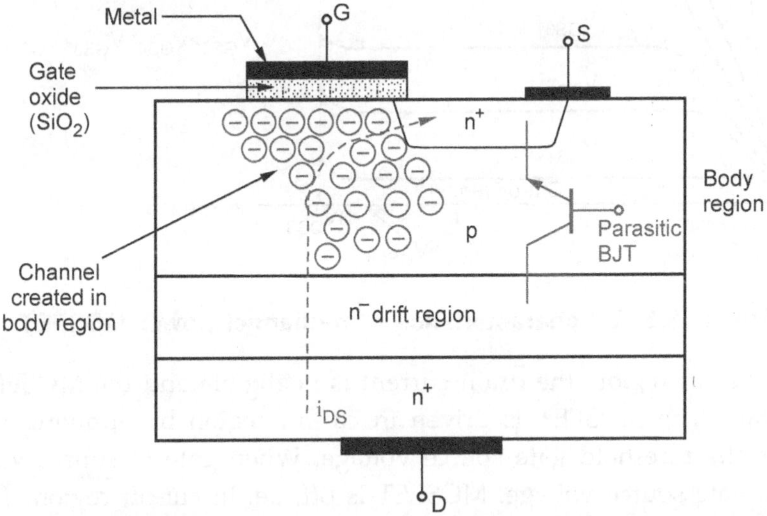

Fig. 1.13.4 Four layer structure of n-channel enhancement mode MOSFET

In the Fig. 1.13.4 structure observe that the source is connected to n^+ region as well as p-type body region. The gate also overlaps p-type region and n^+ region. The gate is isolated from these regions by SiO_2 layer. When V_{GS} is positive, an n-type channel is induced in the body region as shown in Fig. 1.13.4. Hence current (i_{DS}) starts flowing from drain to source as shown. Because of drift region, the on-state drop of MOSFET increases. The thickness of drift region determines breakdown voltage of MOSFET. In Fig. 1.13.4 observe that a parasitic BJT is formed as shown. Base of this parasitic BJT is the p-type body region. Emitter is n^+ region and collector is n^- drift region. The emitter and base of this parasitic BJT are shorted to source. Hence it does not conduct. This is the reason for shorting p-type body region to source.

Advantages of vertical structure

1. On-state resistance of MOSFET is reduced.
2. Width of the gate is maximized. Hence, Gain of the device is increased.

1.13.2 Steady State (V-I) Characteristics of MOSFETs

Fig. 1.13.5 shows the V-I characteristics of n-channel power MOSFET. The drain current i_D is plotted with respect to drain to source voltage v_{DS}. These characteristics are plotted for various values of gate source voltages (V_{GS}). In Fig. 1.13.5 observe that there are three regions in the characteristics : Ohmic, region, active region and cut-off

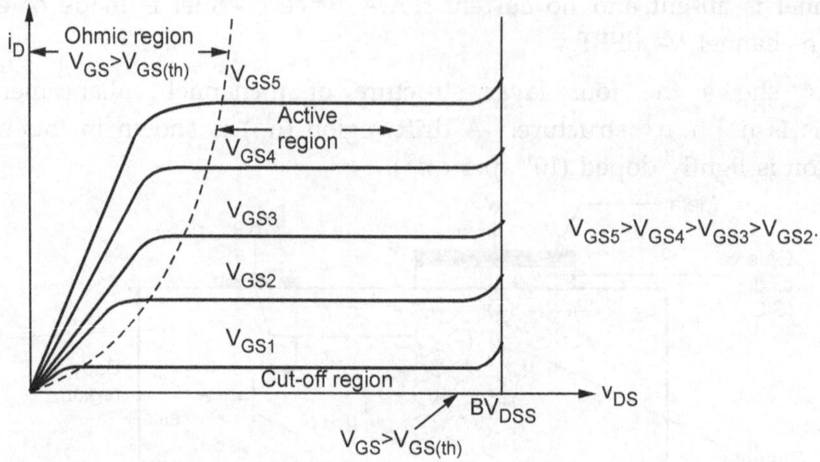

Fig. 1.13.5 V-I characteristics of n-channel power MOSFET

region. In the cut-off region, the drain current is negligible and the MOSFET is said to be in 'OFF' state. The MOSFET is driven in cut-off region by applying $V_{GS} < V_{GS\,(th)}$. Here $V_{GS\,(th)}$ is the threshold gate source voltage. When gate to source voltage is less than threshold gate source voltage, MOSFET is off, i.e. in cut-off region. The MOSFET is driven into ohmic region when $V_{GS} >> V_{GS(th)}$. In the ohmic region, the MOSFET conducts heavily. Hence it is said to be 'on' in the ohmic region. Thus by applying heavy gate to source voltage, MOSFET can be turned on. In the power electronic applications, MOSFET is never operated in the active region. In active region it acts as an amplifier. For switching applications, MOSFET is operated only in ohmic and cut-off regions. The BV_{DSS} is the drain to source breakdown voltage, when the gate is open circuitted. The MOSFET is damaged if drain to source voltage is increased above BV_{DSS}.

1.13.3 Switching Characteristics of MOSFET

The internal capacitances of MOSFET affect the turn-on and turn-off times of MOSFETs. These capacitances have no effect during steady state. Fig. 1.13.6 shows the switching model of MOSFET.

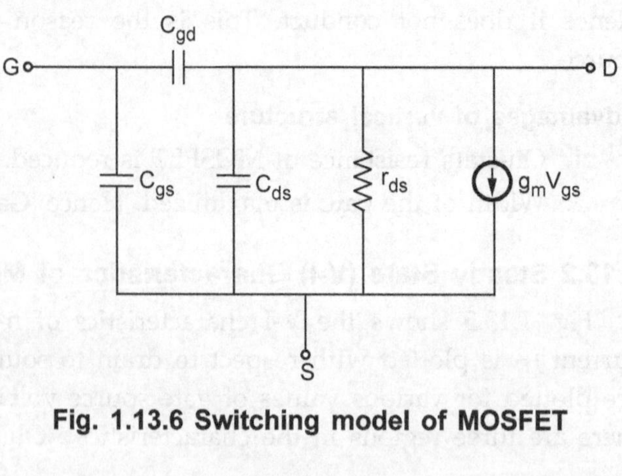

Fig. 1.13.6 Switching model of MOSFET

In the Fig. 1.13.6 C_{gs} is the gate to source parasitic capacitance and C_{gd} is the gate to drain parasitic capacitance. The MOSFET can be turned on by applying positive gate voltage as shown in Fig. 1.13.7.

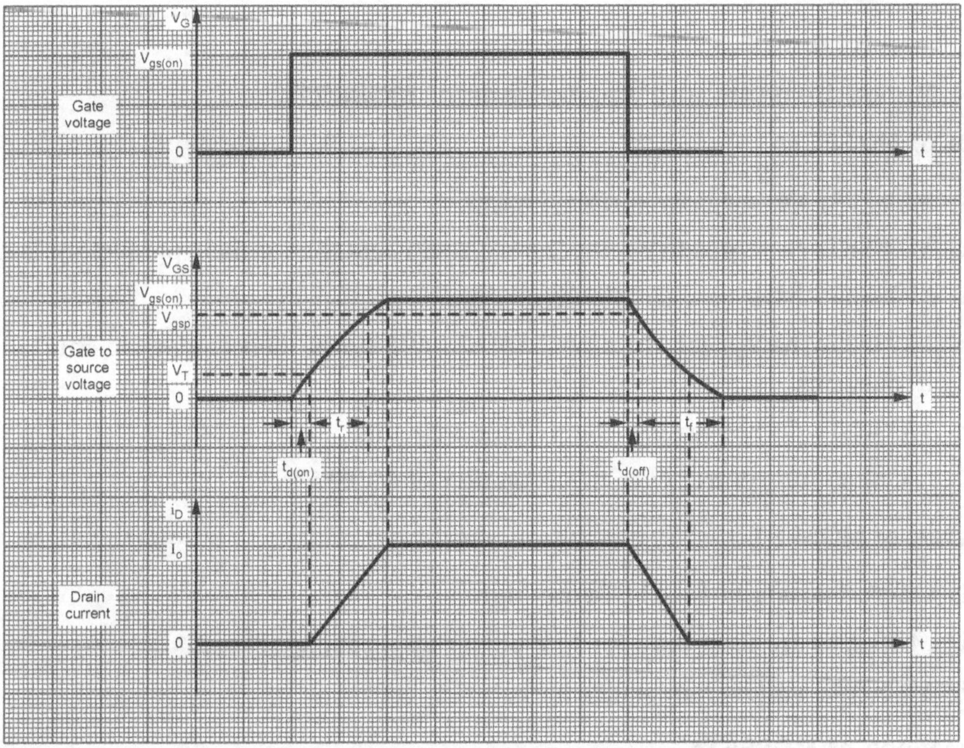

Fig. 1.13.7 Switching characteristics of MOSFET

When the gate voltage is applied, the gate to source capacitance C_{gs} starts charging. The *turn-on delay* ($t_{d(on)}$) is the time required to charge C_{gs} to threshold voltage (V_T). After this voltage, the drain current (i_D) starts rising. The C_{gs} charges from threshold voltage to full gate voltage (v_{gsp}). The time required for this charging is called *rise time* (t_r). Observe that during this period, the drain current rises to its full value, i.e. I_o. The MOSFET is then said to have fully turned on. Thus, the total turn-on time of the MOSFET is,

$$t_{on} = t_{d(on)} + t_r$$

To turn-off the MOSFET, the gate voltage is made negative or zero. The gate to source voltage then reduces from $v_{gs(on)}$ to v_{gsp}. That is, C_{gs} discharges from overdrive to pinch-off region gate voltage. The time required for this discharge is called *turn-off delay* time ($t_{d(ff)}$). The drain current also start reducing. The C_{gs} keeps on discharging and its voltage becomes equal to threshold voltage (V_T). The time required to discharge C_{gs} from v_{gsp} to V_T is called *fall time* (t_f). The drain current becomes zero when $v_{GS} \leq V_T$. The MOSFET is then said to have turned-off. The C_{gs} then discharges to zero voltage. The turn-off time of the MOSFET is equal to sum of turn-off delay time and fall time. i.e.,

$$t_{off} = t_{d(off)} + t_f$$

1.13.4 Merits, Demerits and Applications of MOSFETs

Merits of MOSFETs

i. MOSFETs are majority carrier devices.

ii. MOSFETs have positive temperature coefficient, hence their paralleling is easy.

iii. MOSFETs have very simple drive circuits.

iv. MOSFETs have short turn-on and turn-off times, hence they operate at high frequencies.

v. MOSFETs do not require commutation circuits.

vi. Gate has full control over the operation of MOSFET.

Demerits of MOSFET

i. On-state losses in MOSFETs are high.

ii. MOSFETs are used only for low power applications.

iii. MOSFETs suffer from static charge.

Applications of MOSFETs

i. High frequency and low power inverters.

ii. High frequency SMPS.

iii. High frequency inverters and choppers.

iv. Low power AC and DC drives.

Sr. No.	BJT	MOSFET
1.	This is bipolar device.	This is majority carrier device.
2.	Controlled by base.	Controlled by gate.
3.	Current controlled device.	Voltage controlled device.
4.	Negative temperature coefficient.	Positive temperature coefficient.
5.	Paralleling of BJTs is difficult.	Paralleling of MOSFETs is simple.
6.	Losses are low.	Losses are higher than BJTs.
7.	Drive circuit is complex.	Drive circuit is simple.

8.	Switching frequency is lower than MOSFET.	Switching frequency is high.
9.	BJTs are suitable for high power applications.	MOSFETs are suitable for low power application.
10.	BJTs are available with higher voltage and current ratings.	MOSFETs have less voltage and current ratings.

Table 1.13.1 Comparison of BJT and MOSFET

Review Question

1. *Explain the steady state and switching characteristics of MOSFET.*

1.14 IGBT

The Insulated Gate Bipolar Transistor (IGBT) is the latest device in power electronics. It is obtained by combining the properties of BJT and MOSFET. We know that BJT has lower on-state losses for high values of collector current. But the drive requirement of BJT is little complicated. The drive of MOSFET is very simple (i.e. only voltage is to be applied between gate and source). But MOSFET has high on-state losses. The gate circuit of MOSFET and collector emitter circuits of BJT are combined together to form a new device. This device is called IGBT. Thus IGBT has advantages of both the BJT and MOSFETs. Fig. 1.14.1 shows the symbol of IGBT. Observe that the symbol clearly indicates combination of MOSFET and BJT.

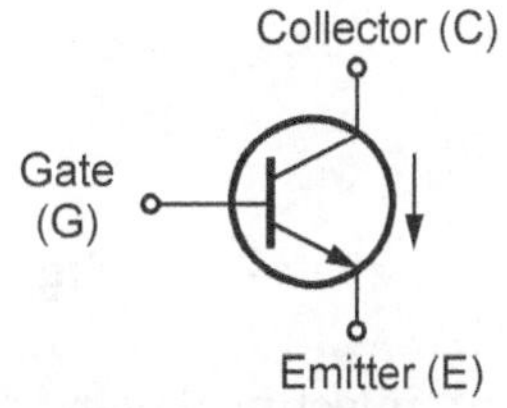

Fig. 1.14.1 Symbol of IGBT

The IGBT has three terminals : Gate (G), collector (C) and emitter (E). Current flows from collector to emitter whenever a voltage between gate and emitter is applied. The IGBT is said to have turned 'on'. When gate emitter voltage is removed, IGBT turns-off. Thus gate has full control over the conduction of IGBT. When the gate to emitter voltage is applied, very small (negligible) current flows. This is similar to the gate circuit of MOSFET. The on-state collector to emitter drop is very small like BJT.

1.14.1 Structure of IGBT

The structure of IGBT is similar to that of MOSFET. Fig. 1.14.2 shows the vertical cross-section of IGBT. In this structure observe that there is additional p^+ layer. This layer is collector (Drain) of IGBT.

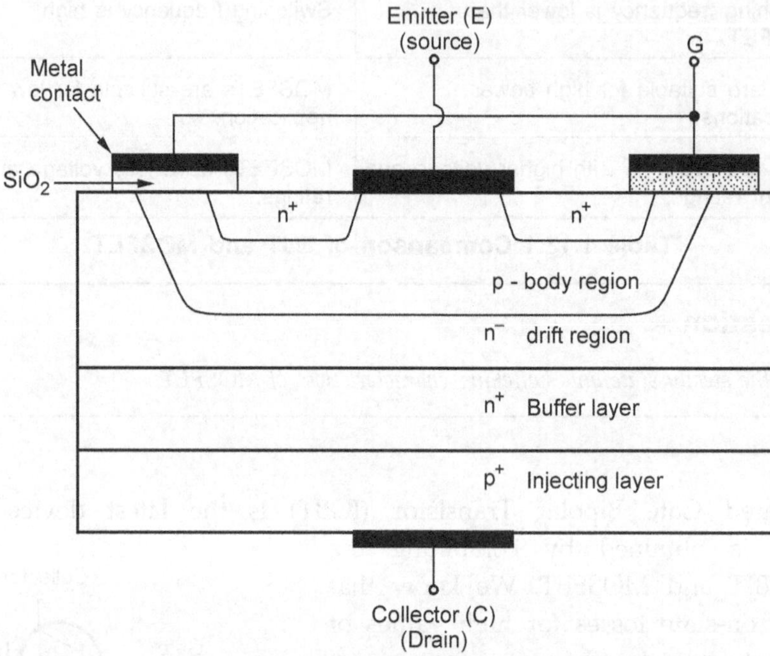

Fig. 1.14.2 Vertical cross section of IGBT

This p^+ injecting layer is heavily doped. It has the doping intensity of 10^{19} per cm^3. The doping of other layers is similar to that of MOSFET. n^+ layers have 10^{19} per cm^3. p-type body region has doping level of 10^{16} per cm^3. The n^- drift region is lightly doped (10^{14} per cm^3).

1.14.1.1 Punch through IGBT

The n^+ buffer layer is not necessary for the operation of IGBT. The IGBTs which have n^+ buffer layer are called punch through IGBTs. Such IGBTs have asymmetric voltage blocking capabilities. Punch through IGBTs have faster turn-off times. Hence they are used for inverter and chopper circuits.

1.14.1.2 Non-punch through IGBT

The IGBTs without n^+ buffer layer are called non-punch through IGBTs. These IGBTs have symmetric voltage blocking capabilities. These IGBTs are used for rectifier type applications.

1.14.1.3 Operation of IGBT

Now let us see how IGBT operates. When $V_{GS} > V_{GS(threshold)}$, then the channel of electrons is formed beneath the gate as shown in Fig. 1.14.3. These electrons attract holes from p^+layer. Hence, holes are injected from p^+ layer into n^- drift region. Thus hole/electron current starts flowing from collector to emitter. When holes enter p-type body region, they attract more electrons from n^+ layer. This action is exactly similar to MOSFET.

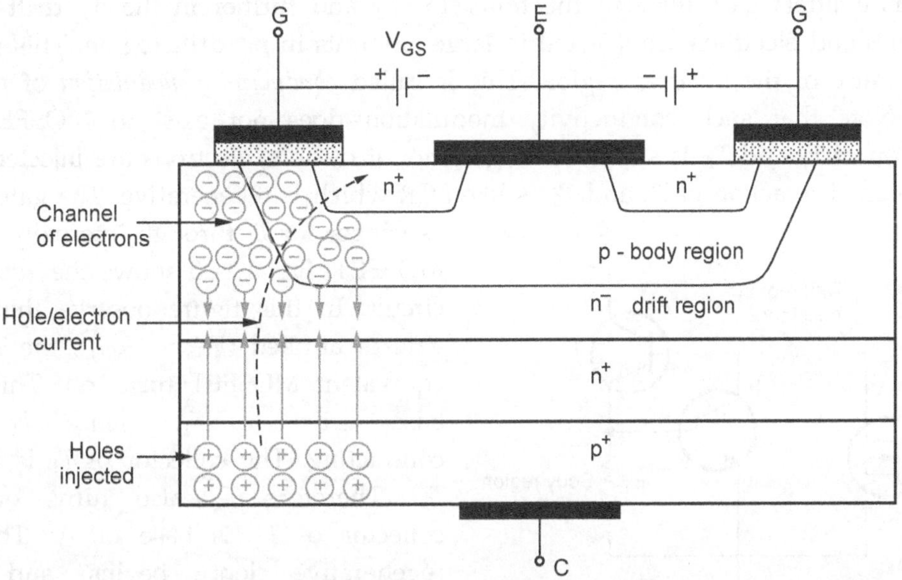

Fig. 1.14.3 A positive gate to source voltage initiates MOSFET action

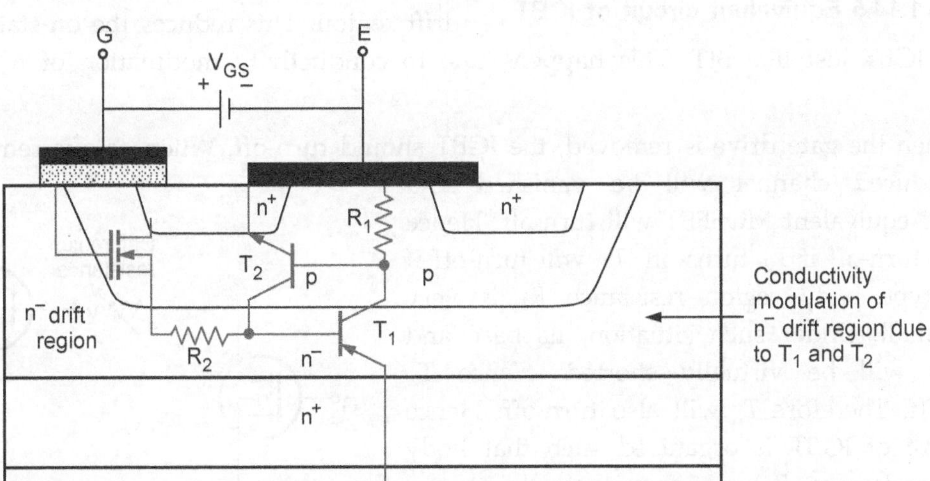

Fig. 1.14.4 Structure of IGBT

Now let us see how p^+ injecting layer makes the operation different than MOSFET. Fig. 1.14.4 shows the structure of IGBT showing how internal MOSFETs and transistors are formed. The MOSFET is formed with input gate, emitter as source and n^- drift region as drain. The two transistors T_1 and T_2 are formed as shown. The holes injected by the p^+ injecting layer go to the n^- drift region. This n^- drift region is base of T_1 and collector of T_2. The holes in the n^- drift region further go to the p-type body region, which is connected to the emitter. The electrons from n^+ region (which is emitter) pass through the transistor T_2 and further in the n^- drift region. Thus holes and electrons are injected in large amounts in n^- drift region. This reduces the resistance of the n^- drift region. This is called *conductivity modulation* of n^- drift region. Note that such conductivity modulation does not exist in MOSFET. The connection of T_1 and T_2 is such that large amount of hole/electrons are injected in n^- drift region. The action of T_1 and T_2 is like SCR which is regenerative. The gate serves as trigger for T_1 through internally formed MOSFET. Fig. 1.14.5 shows the equivalent circuit. In this figure observe that when gate is applied ($V_{GS} > V_{GS(th)}$), the internal equivalent MOSFET turns on. This gives base drive to T_1. Hence T_1 starts conducting. The collector of T_1 is base of T_2. Therefore T_2 also turns on. The collector of T_2 is base of T_1. Thus the regenerative loop begins and large number of carriers are injected in n^- drift region. This reduces the on-state loss of the IGBT just like BJT. This happens due to conductivity modulation of n^- drift region.

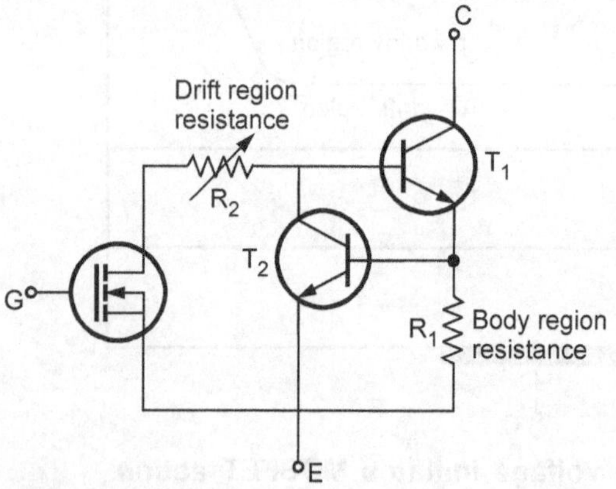

Fig. 1.14.5 Equivalent circuit of IGBT

When the gate drive is removed, the IGBT should turn-off. When gate is removed, the induced channel will be vanished and internal equivalent MOSFET will turn-off. Hence T_1 will turn-off if T_2 turns-off. T_2 will turn-off if the p-type body region resistance R_1 is very very small. Under such situation, its base and emitter will be virtually shorted. Hence T_2 turns-off. Therefore T_1 will also turn-off. Hence structure of IGBT is organized such that body region resistance (R_1) is very very small.

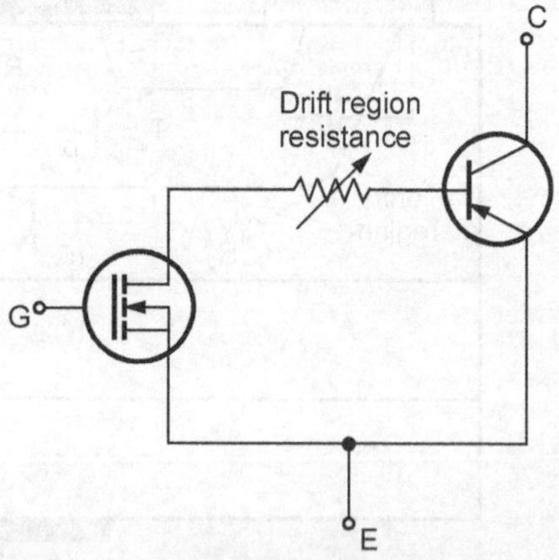

Fig. 1.14.6 Simplified equivalent circuit of IGBT

If R_1 is very very small, then T_2 will never conduct and the equivalent circuit of IGBT will be as shown in Fig. 1.14.6. IGBTs are thus different than MOSFETs because of conduction of current from collector to emitter. For MOSFETs, on state losses are high since resistance of drift region remains same. But in IGBTs, resistance of drift region reduces when gate drive is applied. This resistance reduces because of p^+injecting region. Hence, on-state loss of IGBT is very small.

1.14.1.4 Latchup in IGBT

- Latchup means IGBT remains in ON condition even if gate drive is removed.
- **Causes of latchup :** The current in p-type body region flows vertically as well as laterally. In equivalent circuit of Fig. 1.14.5 observe that the lateral component of current flows through the body region resistance R_1. Therefore the drop across this body region resistance is sufficient to turn-on npn transistor T_2. Therefore pnp transistor T_1 is also turned on and regeneration takes place. Under this situation gate has no control over the current flow. Thus IGBT remains in ON condition even if gate drive is removed.
- **To avoid latchups :**

i) Efforts are made to reduce the body resistance R_1. This can be achieved by using special geometries for the structure of IGBT.

ii) The circuit should be designed in such a way that maximum specified current should not be exceeded.

iii) Latchups can also be avoided by increasing the turn-off time and controlling h_{FE} of the pnp transistor.

1.14.1.5 Body-Source-Short and its Reason

What is body-source-short?

In the structure of IGBT observe that the source (or emitter 'E') is connected to p-type body region as well as n^+ region. This means p-type body region is shorted to source (or emitter 'E').

Reason

In Fig. 1.14.4 observe that various layers of IGBT form internal parasitic thyristor. i.e. $(p^+ - n^+ n^- - p - n^+)$. The body-source short avoids possible turn-on of this parasitic thyristor. The base-emitter (i.e p-type body region and n^+ region) of transistor T_2 are shorted due to body-source-short. Hence it does not conduct. It also helps in avoiding latch up in IGBT.

1.14.2 Safe Operating Area (SOA) of IGBT

Forward biased SOA (FBSOA)

- The FBSOA is the square for turn-on times less than 1 msec.
- For reduced switching times FBSOA is limited by maximum junction temperature.

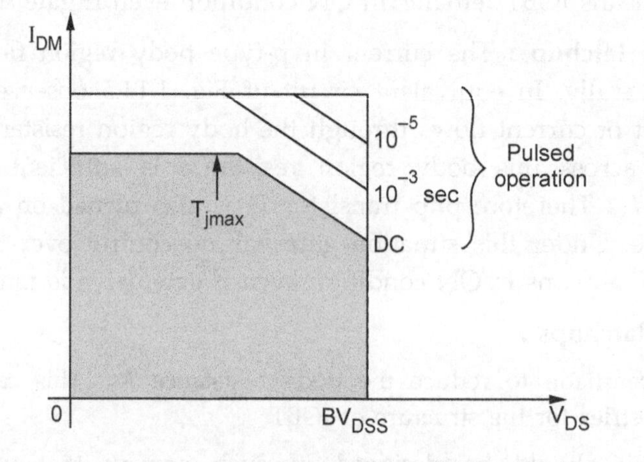

Fig. 1.14.7 FBSOA of IGBT

Reverse biased SOA (RBSOA)

- The RBSOA depends upon the reapplied rate of change of drain-source voltage.
- Observe that the RBSOA reduces as dV_{DS} / dt increases.
- The reduction in RBSOA is necessary to avoid latchups.

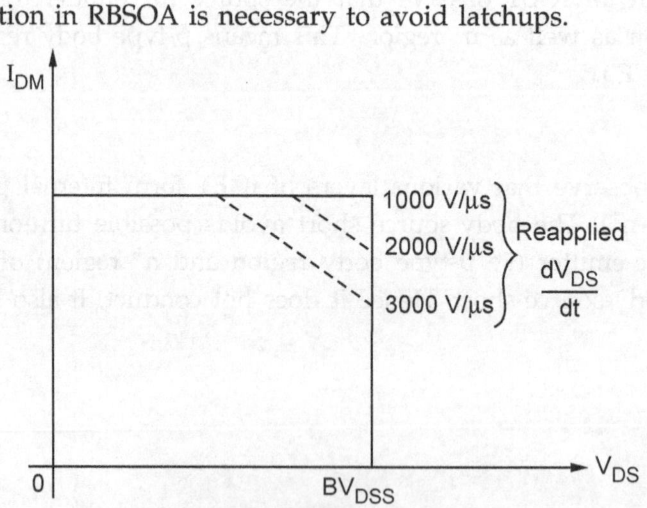

Fig. 1.14.8 RBSOA of IGBT

Superiority of SOA of IGBT

- The SOA of IGBT is square for short switching times.
- The RBSOA of IGBT indicates significantly higher values of reapplied dv/dt.
- The SOA of IGBT is wider compared to that of BJT.
- As switching frequency increases, the SOA of BJT reduces considerably.

1.14.3 Steady State (V-I) Characteristics of IGBT

Fig. 1.14.9 shows the V-I characteristics of n-channel IGBT. Sometime the collector is also called drain and emitter is also called source. The above characteristics are plotted for drain (collector) current i_D with respect to drain source (collector emitter) voltage V_{DS}. The characteristics are plotted for different values of gate to source (V_{GS}) voltages. When the gate to source voltage is greater than the threshold voltage $V_{GS(th)}$, then IGBT turns-on. The IGBT is off when V_{GS} is less than $v_{GS(th)}$. Fig. 1.14.9 shows the 'on' and 'off' regions of IGBT. The BV_{DSS} is the breakdown drain to source voltage when gate is open circuitted. The IGBT is the popular device now-a-days. IGBT has simplest drive circuit and it has low on-state losses.

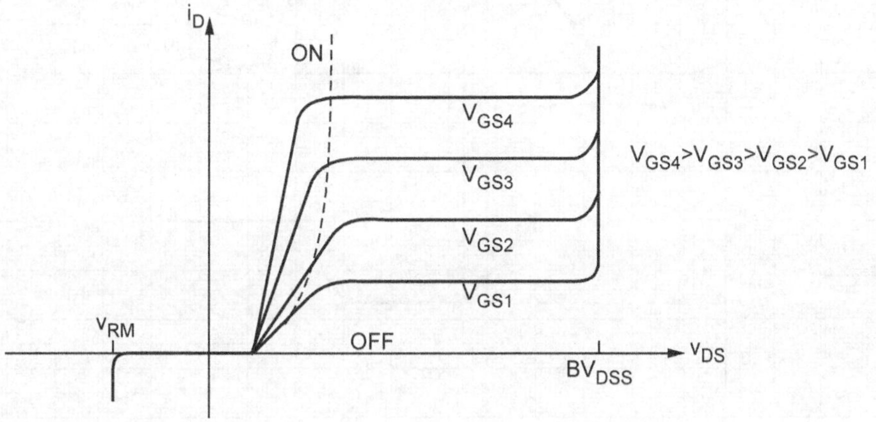

Fig. 1.14.9 V-I characteristics of n-channel IGBT

1.14.4 Switching Characteristics of IGBT

Fig. 1.14.10 shows the switching characteristics of IGBT. The gate to source voltage is normally negative. This voltage is made positive to turn-on the IGBT. When $V_{GS} > V_{GS(th)}$, the collector current starts increasing. Turn-on delay, $t_{d(on)}$ is the delay when gate drive is applied and i_C starts increasing. When i_C increases to its full value, collector emitter voltage starts falling. 't_{ri}' is the rise time of collector and t_{fv} is the fall time of voltage. Thus, turn-on time of IGBT is,

$$t_{on} = t_{d(on)} + t_{ri} + t_{fv}$$

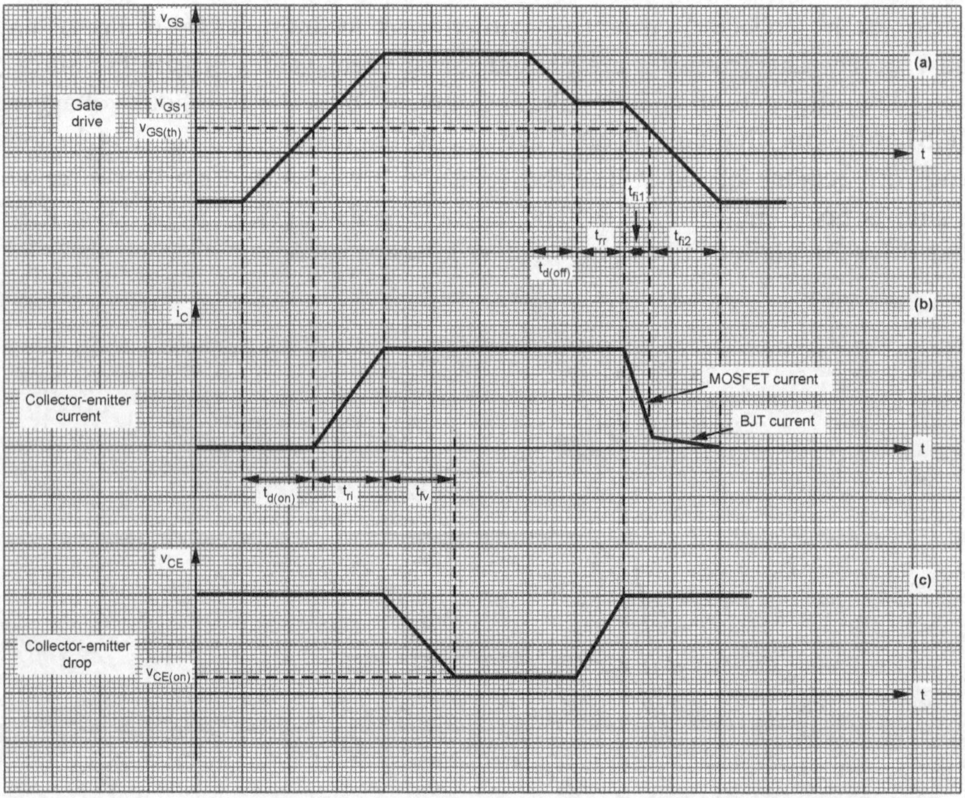

Fig. 1.14.10 Switching characteristics of IGBT

The turn-off of the IGBT is initiated by reducing the gate voltage. When gate voltage falls to the value equal to v_{GS1}, v_{CE} starts rising. v_{GS1} is the voltage where IGBT comes out of saturation. Turn-off delay, $t_{d(off)}$ is the delay time when gate voltage is reduced and v_{CE} starts increasing. When v_{CE} reaches to supply voltage, i_C starts reducing. i_C reduces fast till v_{GS} reaches to $v_{GS(th)}$. This fast decay in i_C is basically due to internal MOSFET. Then v_{GS} goes to zero and becomes negative.

But i_C keeps on flowing for some time. This is internal BJT current. This current flows due to stored carriers in the drift region. Hence, turn-off time of IGBT is higher than IGBT. The turn-off time of IGBT will be,

$$t_{off} = t_{d(off)} + t_{rv} + t_{fi1} + t_{fi2}$$

Here, t_{rv} is voltage rise time

t_{fi1} is MOSFET current fall time.

t_{fi2} is BJT current fall time.

1.14.5 Merits, Demerits and Applications of IGBT

Merits of IGBT

i. Voltage controlled device. Hence drive circuit is very simple.

ii. On-state losses are reduced.

iii. Switching frequencies are higher than thyristors.

iv. No commutation circuits are required.

v. Gate have full control over the operation of IGBT.

vi. IGBTs have approximately flat temperature coefficient.

Demerits of IGBT

i. IGBTs have static charge problems.

ii. IGBTs are costlier than BJTs and MOSFETs.

Applications of IGBTs

i. AC motor drives, i.e. inverters.

ii. DC to DC power supplies, i.e choppers.

iii. UPS systems.

iv. Harmonic compensators.

1.14.6 Comparison of Power Devices

The power devices can be compared on the basis of switching frequency, gate drive circuit, power handling capacity etc. Table 1.14.1 shows the comparison of SCR, BJT, MOSFET and IGBT.

Sr. No.	Parameter	SCR	BJT	MOSFET	IGBT
1	Symbol	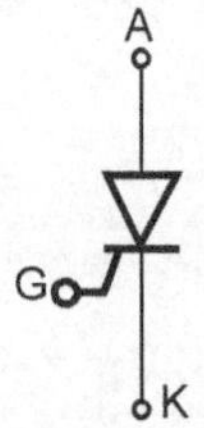	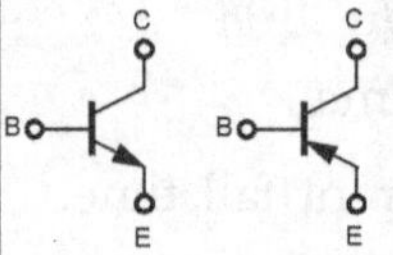	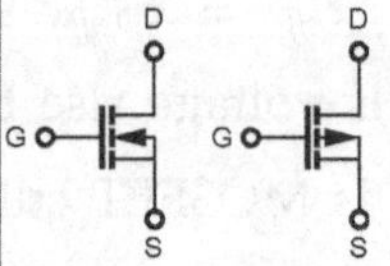	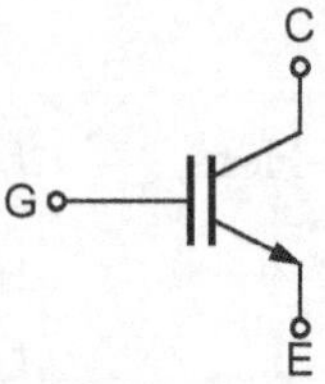
2	Triggered i.e. latching or linear	Triggered or latching device	Linear trigger	Linear trigger	Linear trigger
3	Type of carriers in device	Majority carrier device	Bipolar device	Majority carrier device	Majority carrier device
4	Control of gate or base	Gate has no control once turned on	Base has full control	Gate has full control	Gate has full control
5	On-state drop	< 2 volts	< 2 volts	4-6 volts	3.3 volts
6	Switching frequency	500 Hz	10 kHz	upto 100 kHz	20 kHz
7	Gate drive	Current	Current	Voltage	Voltage
8	Snubber	Unpolarized	Polarized	Not essential	Not essential
9	Temperature coefficient	Negative	Negative	Positive	Approximately flat, but positive at high current
10	Voltage and current ratings	10 kV/ 4 kA	2 kV/1 kA	1kV/50A	1.5 kV/400 A
11	Voltage blocking capability	Symmetric and asymmetric (both)	Asymmetric	Asymmetric	Asymmetric
12	Applications	AC to DC converters, AC voltage controllers, electronic circuit breakers	DC to AC converters, induction motor drives, UPS, SMPS, Choppers	DC choppers, low powers UPS, SMPS, brushless DC motor drives	DC to AC converters, AC motor drives, UPS, choppers, SMPS etc.

Table 1.14.1 Comparison of power devices

Review Questions

1. *Explain the characteristics of IGBT. Explain its operation.*
2. *Explain the characteristics of following devices : i) BJT ii) MOSFET iii) IGBT*
3. *Compare BJT, MOSFET and IGBT.*

1.15 Firing Circuits for the SCR

Thyristor can be turned on by following :

(i) Forward break-over voltage

(ii) $\frac{dv}{dt}$ triggering

(iii) Exceeding internal device temperature

(iv) Focusing light beam on the junction

(v) Gate triggering.

The gate triggering is the most widely used method of turning on the thyristor. In this section we will study various types of gate triggering circuits.

1.15.1 Features of Firing Circuits

The triggering circuits are called firing circuits. We have already discussed the requirements of gate trigger circuits. The following features or requirements must be fulfilled by the firing circuit in addition to those discussed earlier.

(i) The firing circuit should produce the triggering pulses for every thyristor at appropriate instants.

(ii) The triggering pulses generated by the control circuit need to be amplified and passed through the isolation circuit. The triggering pulses generated by the control circuit have very small power. Hence their power is increased by pulse amplifier. Fig. 1.15.1 shows the scheme. The firing circuit operates at low voltage levels (5 to 20 volts). And the thyristor operates at high voltage levels (greater than 250 volts). Hence there must be electrical isolation between firing circuit and thyristor. This isolation is provided by the pulse transformer or optocouplers.

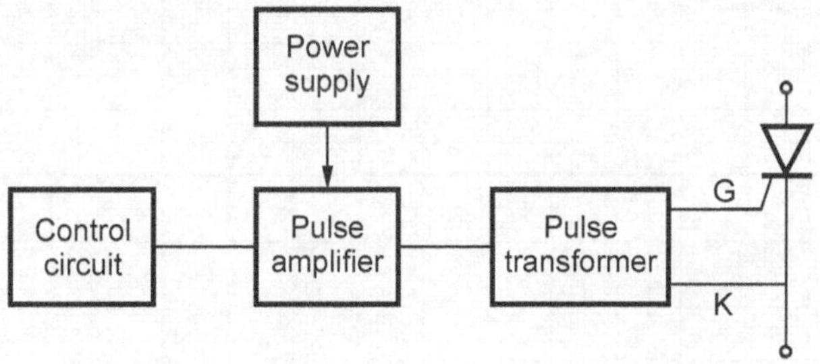

Fig. 1.15.1 Main blocks of firing circuit

1.15.2 R-Firing Circuit

Fig. 1.15.2 shows the simple R-firing circuit.

The resistance R_{min} is used to limit the gate current to its maximum value. If $I_{g(max)}$ is maximum gate current and V_m is the peak supply voltage, then R_{min} will be,

$$R_{min} \geq \frac{V_m}{I_{g(max)}}$$

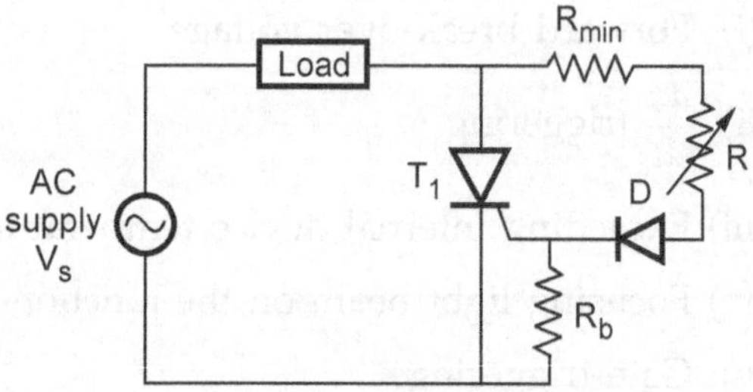

Fig. 1.15.2 R-firing circuit

The resistance R_b is the stabilizing resistance. The voltage across R_b should not exceed minimum gate voltage $(V_{g(min)})$, otherwise thyristor will turn-on directly. Then the variable resistance R is used to trigger the thyristor T_1. When 'R' is zero, the triggering angle is minimum. The triggering angle increases as value of 'R' is increased. Fig. 1.15.3 shows the waveforms of this circuit.

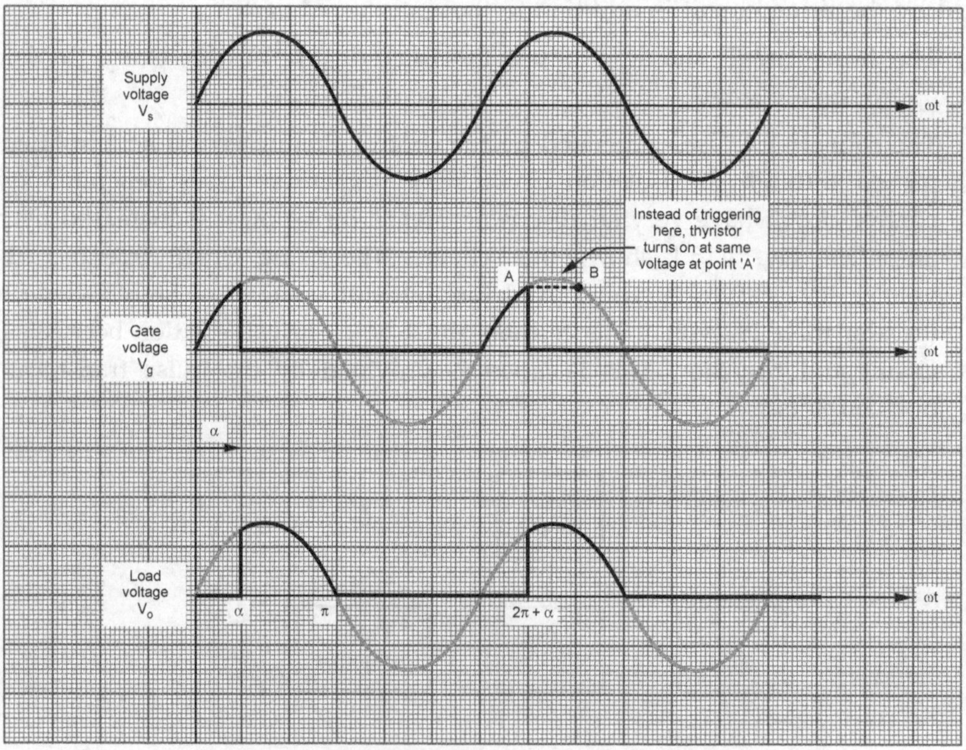

Fig. 1.15.3 Waveforms of R-firing circuit

The anode to cathode voltage and the gate current are in phase. Hence the triggering angle of T_1 cannot be delayed beyond 90°. In the above waveforms, observe that at points 'A' and 'B' gate voltage is same.

If it is desired to trigger thyristor at point 'B', similar voltage appears at point 'A'. Hence thyristor will turn-on at point 'A' only. Hence maximum triggering angle will be 90°. This is because the gate current and anode voltage are in phase.

1.15.3 RC Firing Circuit

Fig. 1.15.4 shows the circuit diagram of RC-firing circuit. In the negative half cycle, the capacitor charges through diode D_2 to negative supply voltage. The capacitor charges to $-V_m$ (i.e. negative peak) of the supply. This is shown in waveforms of Fig. 1.15.5. The capacitor then discharges (i.e. charges towards positive) through resistance R during the positive half cycle of the supply. The thyristor triggers when capacitor charges to value greater than $v_{g(\min)}$. Observe the capacitor voltage and load voltage waveforms in Fig. 1.15.5. The diode D_1 prevents the negative capacitor voltage appearing to gate of the thyristor. The triggering angle can be controlled from 0 to 180°. For zero output (i.e. maximum firing angle), the following relation holds :

$$RC \geq \frac{1.3}{2f} \qquad \ldots (1.15.1)$$

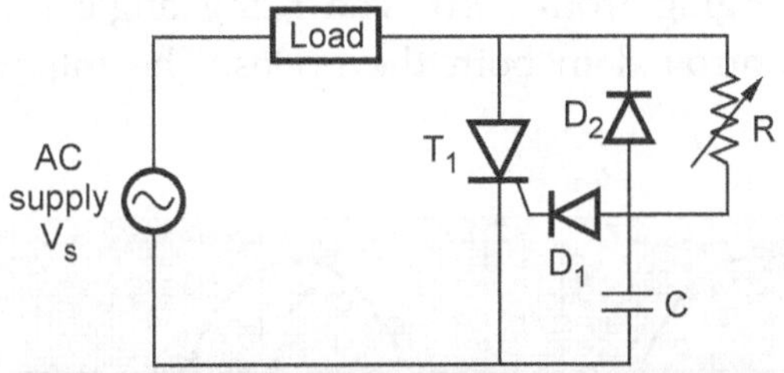

Fig. 1.15.4 RC half wave firing circuit

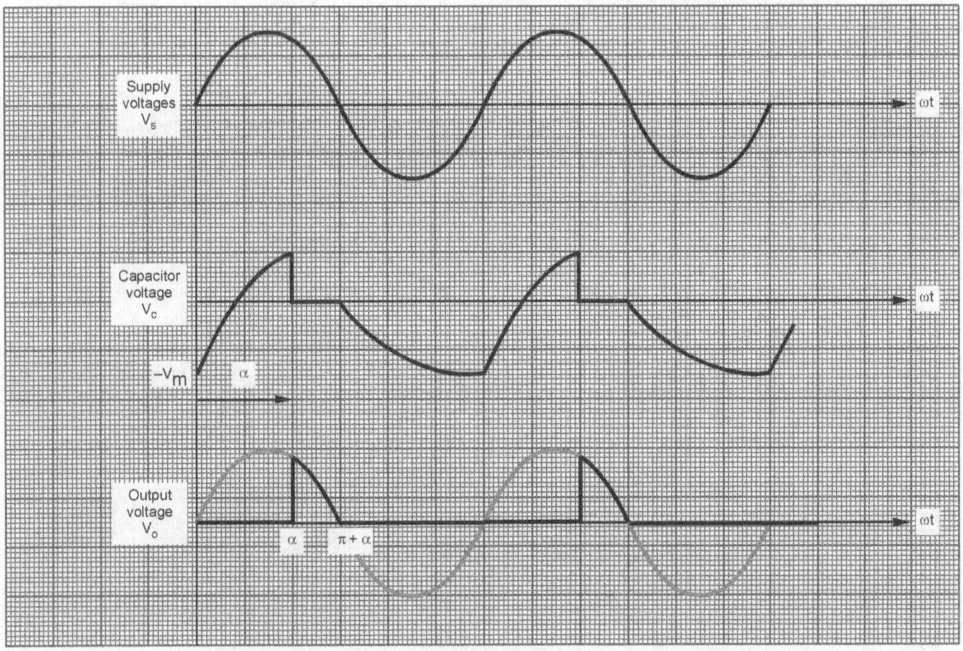

Fig. 1.15.5 Waveforms of half wave RC firing circuit

Here f is the supply frequency. Since triggering is controlled only in one half cycle of the supply, this circuit is also called *half wave RC firing circuit.*

1.15.4 Full Wave RC-Firing Circuit

Fig. 1.15.6 shows the full wave RC-firing circuit. The supply to the thyristor is given through the uncontrolled rectifier. Hence both the half cycles are positive half cycles to the thyristor. The capacitor starts charging in every half cycle at the beginning. Whenever the capacitor voltage reaches to the value greater than $v_{g(\min)}$, the thyristor turns-on. Fig. 1.15.7 shows the waveforms of this circuit. Once the thyristor turns-on, the capacitor voltage is clamped to zero, till next half cycle. The capacitor again starts charging from zero. The firing angle can be varied from 0 to 180°. The triggering is controlled in both the cycles. The following relation holds for maximum firing angle,

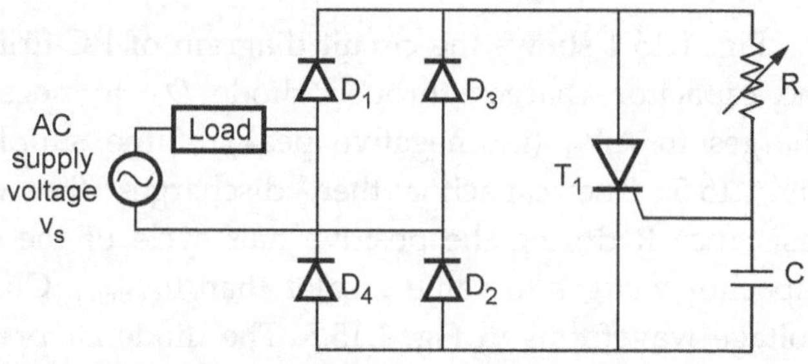

Fig. 1.15.6 Full wave RC firing circuit

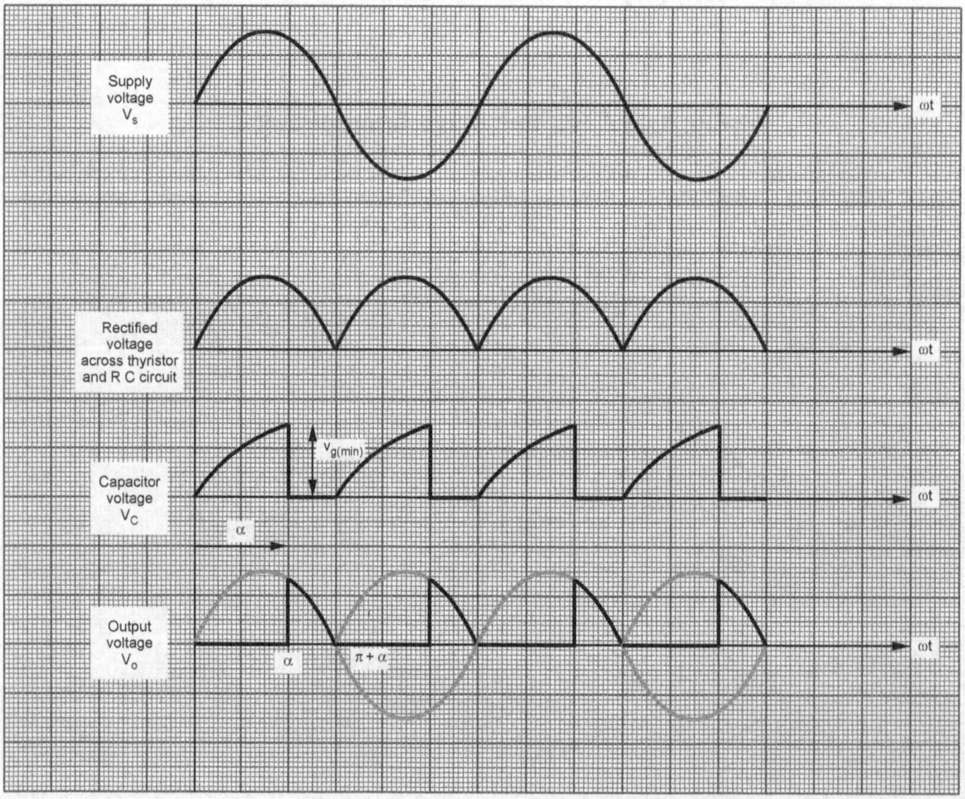

Fig. 1.15.7 Waveforms of full wave RC firing circuit

$$RC \geq \frac{0.157}{2\pi f} \quad \text{... (1.15.2)}$$

Here f is the frequency of the supply.

Example 1.15.1 : *Design a suitable RC triggering circuit for a thyristorised network operating on 220 V, 50 Hz supply. The specifications of the SCR are $V_{GT(min)}$ = 5 V, $I_{GT(max)}$ = 30 mA.*

Solution : Let us consider half wave RC firing circuit is used. This circuit is shown in Fig. 1.15.4. The given data is

Supply voltage, $V_s = 220$ V

$f = 50$ Hz

$V_{GT(min)} = 5$ V

$I_{GT(max)} = 30$ mA

The RC time constant of the half wave RC firing circuit is given by equation 1.15.1 as,

$$RC \geq \frac{1.3}{2f}$$

$$\geq \frac{1.3}{2 \times 50}$$

$$\geq 0.013$$

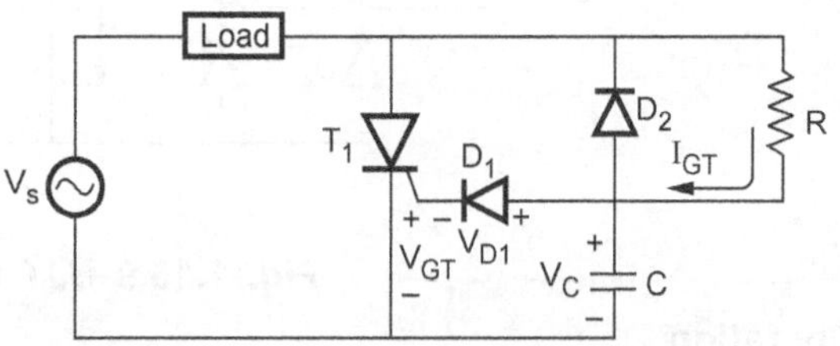

Fig. 1.15.8 RC half wave triggering circuit

The triggering circuit of Fig. 1.15.4 is reproduced below with various voltage drops.

The capacitor voltage is,

$$V_c = V_{GT} + V_{D1} \quad \text{... (1.15.3)}$$

The SCR will turn-on when,

$$V_s \geq I_{GT} R + V_c \quad \text{... (1.15.4)}$$

When above equation is satisfied, sufficient gate drive will be given to the SCR and it will turn-on. Putting expression for V_c from equation 1.15.3 in above equation,

$$V_s \geq I_{GT} R + V_{GT} + V_{D1}$$

$$\therefore \quad R \leq \frac{V_s - V_{GT} - V_{D1}}{I_{GT}} \quad \text{... (1.15.5)}$$

Here V_{GT} should be minimum and I_{GT} should be maximum to get maximum value of R. Putting values in above equation,

$$R \leq \frac{220 - 5 - 0.7}{30 \times 10^{-3}} \qquad \text{Here } V_{D1} = 0.7 \text{ V}$$

$\therefore \quad R \leq 7143.33\ \Omega$

$\therefore \quad C \geq \dfrac{0.013}{R}$

$\geq \dfrac{0.013}{7143.33}$

$\therefore \quad C \geq 1.82\ \mu F$

1.15.5 UJT Triggering Circuit

The Unijunction transistor (UJT) triggering circuit is used in most of the applications. Fig. 1.15.9 shows the circuit diagram of UJT triggering circuit.

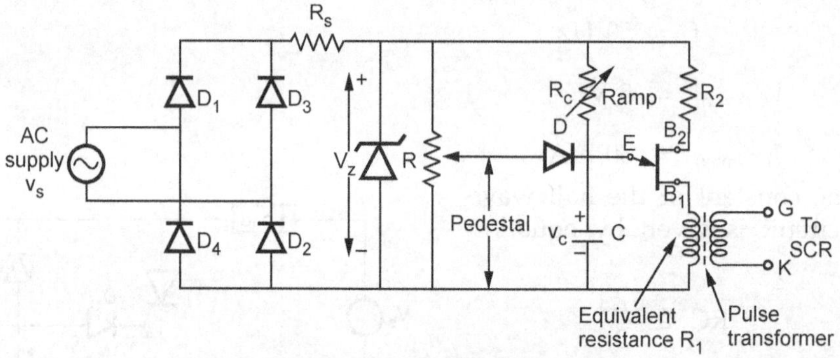

Fig. 1.15.9 UJT triggering circuit

Operation

- The supply voltage is rectified and given to the zener regulator. The voltage of zener diode is V_z. The zener diode clamps the rectified voltage to V_z as shown in the waveforms of Fig. 1.15.10. Hence voltage V_z is applied to the UJT circuit.
- The pedestal control indicates initial voltage level in the capacitor. It can be adjusted through resistance R. The ramp control indicates charging of capacitor from pedestal level. The waveforms of Fig. 1.15.10 shows these levels.

i) The capacitor charges through resistance R_c. When the capacitor voltage becomes equal to V_p, the peak voltage of the UJT, it turns-on. The capacitor discharges through emitter (E), base (B_1) and primary of pulse transformer. The UJT is turned-on when the capacitor discharges. Since current flows through the primary of pulse transformer, a pulse is generated. This pulse as shown in Fig. 1.15.10 is the gate triggering pulse.

ii) When the capacitor discharges to a voltage called valley voltage (V_v), the UJT turns-off and capacitor again starts charging from pedestal level. This mode of working of UJT is called *relaxation oscillator*.

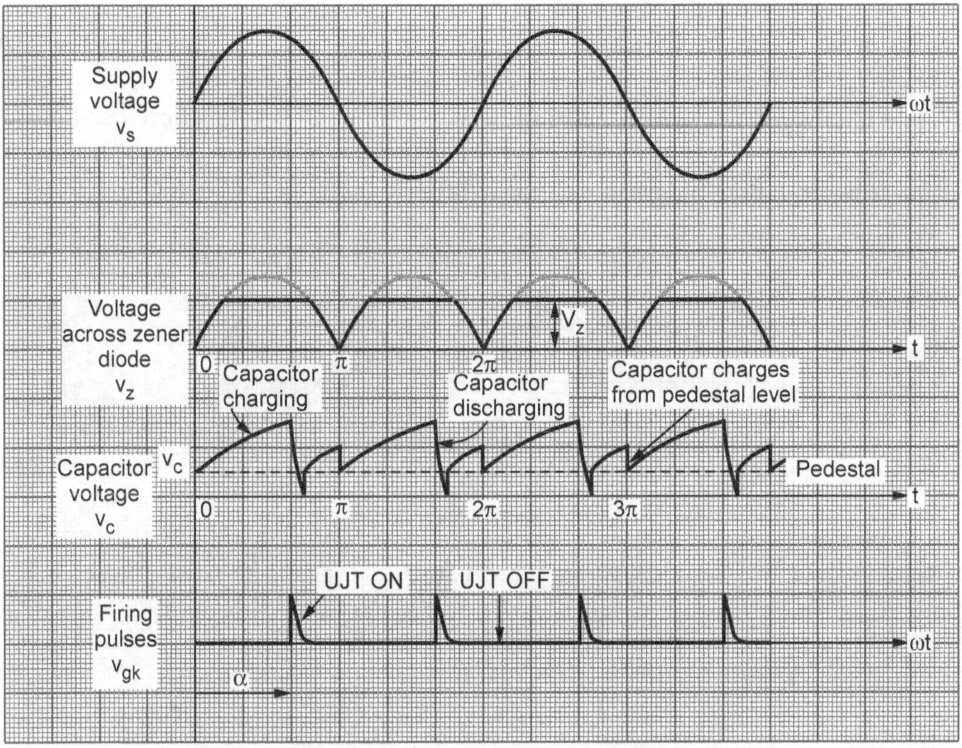

Fig. 1.15.10 Waveforms of UJT triggering circuit

iii) The delay angle 'α' is the angle when first triggering pulse is generated in the half cycle. The charging of the capacitor can be varied by resistance R_C. Hence delay angle can also be varied. The UJT trigger circuit has the firing angle range from 0 to 180°.

The zener voltage acts as a supply voltage for UJT relaxation oscillator. This voltage becomes zero at 0, π, 2π, 3 π, ...etc. The capacitor voltage also becomes zero at these instants. Thus synchronization with zero crossings is achieved. The UJT trigger circuit can be used to trigger SCRs in 1ϕ converters, 1ϕ AC regulators etc.

Mathematical analysis

The peak voltage at which UJT turns on is given as,

$$V_p = \eta V_{BB} + V_D \qquad \text{... (1.15.6)}$$

Here V_p is the peak voltage

V_{BB} is the supply voltage of UJT circuit

V_D is forward drop of UJT

η is intrinsic standoff ratio.

The intrinsic standoff ratio (η) depends upon the UJT. The period of oscillation of the UJT relaxation oscillator is given as

$$T = R_c \, C \, ln\left(\frac{1}{1-\eta}\right) \quad \text{... (1.15.7)}$$

Fig. 1.15.11 shows the waveforms of free running UJT relaxation oscillator. The capacitor voltage waveform and UJT output are shown in the above figure. From Fig. 1.15.10, it is clear that triggering angle will be,

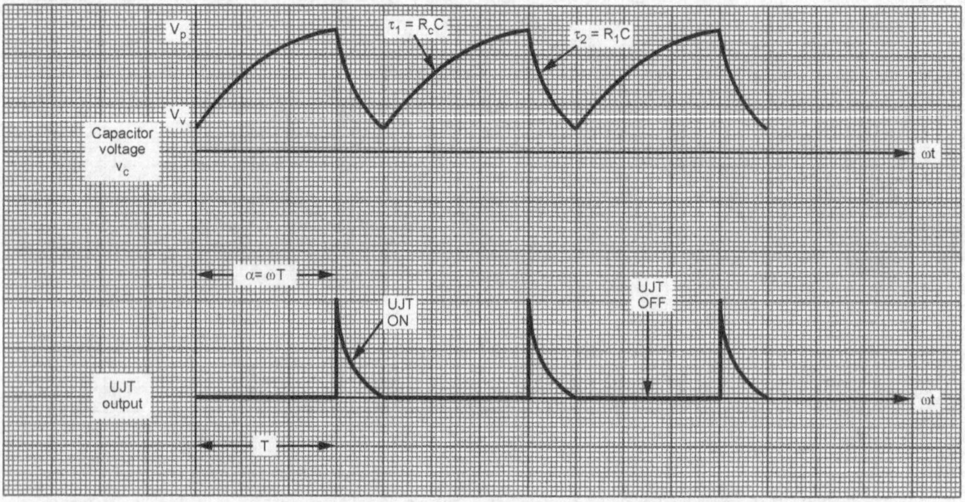

Fig. 1.15.11 Waveforms of free running UJT relaxation oscillator

$$\alpha = \omega T$$

Hence from equation 1.15.7 we can write,

$$\alpha = \omega \, R_c \, C \, ln\left(\frac{1}{1-\eta}\right) \quad \text{... (1.15.8)}$$

This equation gives firing angle of UJT triggering circuit. Here $\omega = 2\,\pi f$ and f is the frequency of UJT oscillator. The resistance R_2 should be selected as follows :

$$R_2 = \frac{0.7\,(R_{B2}+R_{B1})}{\eta \, V_{BB}} \quad \text{... (1.15.9)}$$

Here R_{B2} and R_{B1} are interbase resistance of the UJT. R_2 can also be calculated approximately as,

$$R_2 = \frac{10^4}{\eta \, V_{BB}} \quad \text{... (1.15.10)}$$

Note that this expression does not require R_{B1} and R_{B2}. Normally pulse transformer is connected at the base B_1 of UJT. Pulses are passed through pulse transformer. This provides isolation between SCR circuit and UJT triggering circuit. The resistance of pulse transformer primary can be denoted by R_1. This resistance controls width of the triggering pulse. From Fig. 1.15.11, this width is given as,

Width of triggering pulse, $\tau_2 = R_1 C$ ... (1.15.11)

More accurately this pulse width will be,

$$\tau_2 = (R_1 + R_{B1})C \quad \text{... (1.15.12)}$$

Here we have considered the interbase resistance R_{B1} also. If leakage current of UJT is given, then R_1 can be calculated using following equation,

$$V_{BB} = I_{leakage}(R_1 + R_2 + R_{B1} + R_{B2}) \quad \text{... (1.15.13)}$$

Here $I_{leakage}$ is the leakage current of UJT.

The maximum value of R_c is given as,

$$R_{c\,(max)} = \frac{V_{BB} - V_p}{I_p} \quad \text{... (1.15.14)}$$

and the minimum value of R_c is given as,

$$R_{c(min)} = \frac{V_{BB} - V_v}{I_v} \quad \text{... (1.15.15)}$$

Here V_p is peak voltage

I_p is peak current

V_v is valley voltage

I_v is valley current

1.15.6 Pedestal Circuit with Cosine Modified Ramp

Fig. 1.15.12 shows the pedestal circuit having cosine modified ramp control.

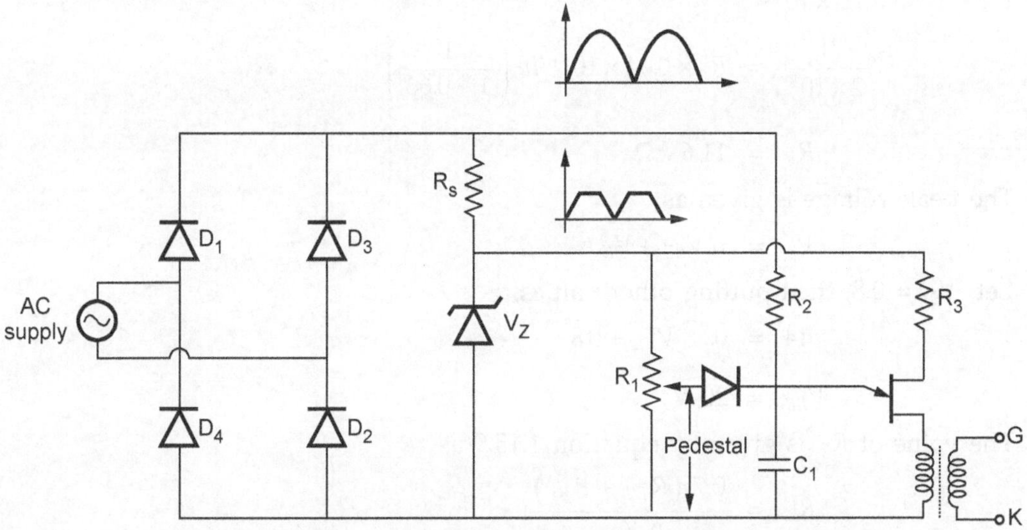

Fig. 1.15.12 Pedestal circuit with cosine modified ramp

- In the above circuit, the zener voltage is given to pedestal control and base of UJT.
- But the charging of capacitor C_1 takes place by rectified voltage.
- The pedestal voltage can be varied by resistance R_1.
- The charging of capacitor C_1 is cosine modified and its rate can be varied by resistance R_2.
- The firing angle can be controlled by controlling the pedestal voltage.

Advantages of cosine modified ramp control

(i) The control characteristics is linear.

(ii) Control gain is high.

Example 1.15.2 : *A UJT is used to trigger the thyristor whose minimum gate triggering voltage is 6.2 V. The UJT ratings are :*
$\eta = 0.66$, $I_p = 0.5\ mA$, $I_v = 3\ mA$,
$R_{B1} + R_{B2} = 5\ k\Omega$, *leakage current* = 3.2 mA
$V_p = 14\ V$ *and* $V_v = 1\ V$.
Oscillator frequency is 2 kHz and capacitor $C = 0.04\ \mu F$. *Design the complete circuit.*

Solution : From equation 1.15.7,

$$T = R_c C\ ln\left(\frac{1}{1-\eta}\right)$$

Here $T = \frac{1}{f} = \frac{1}{2\times10^3}$, since $f = 2$ kHz and putting other values,

$$\frac{1}{2\times10^3} = R_c \times 0.04\times10^{-6}\ ln\left(\frac{1}{1-0.66}\right)$$

$$R_c = 11.6\ k\Omega$$

The peak voltage is given as,

$$V_p = \eta\ V_{BB} + V_D$$

Let V_D = 0.8, then putting other values,

$$14 = 0.66\ V_{BB} + 0.8$$

$$V_{BB} = 20\ V$$

The value of R_2 is given by equation 1.15.9 as,

$$R_2 = \frac{0.7\left(R_{B2} + R_{B1}\right)}{\eta\ V_{BB}}$$

$$= \frac{0.7\left(5\times10^{3}\right)}{0.66\times 20}$$

$$\therefore \qquad R_2 = 265\ \Omega$$

Value of R_1 can be calculated by equation 1.15.13 as,

$$V_{BB} = I_{leakage}\left(R_1 + R_2 + R_{B1} + R_{B2}\right)$$

$$20 = 3.2\times10^{-3}\ (R_1 + 265 + 5000)$$

$$R_1 = 985\ \Omega$$

The value of $R_{c\,(\max)}$ is given by equation 1.15.14,

$$R_{c\,(\max)} = \frac{V_{BB} - V_p}{I_p}$$

$$= \frac{20 - 14}{0.5\times10^{-3}}$$

$$R_{c\,(\max)} = 12\ k\Omega$$

Similarly the value of $R_{c\,(\min)}$ is given by equation 1.15.15,

$$R_{c\,(\min)} = \frac{V_{BB} - V_v}{I_v} = \frac{20-1}{3\times10^{-3}}$$

$$R_{c\,(\min)} = 6.33\ k\Omega$$

Fig. 1.15.13 shows the completely designed circuit.

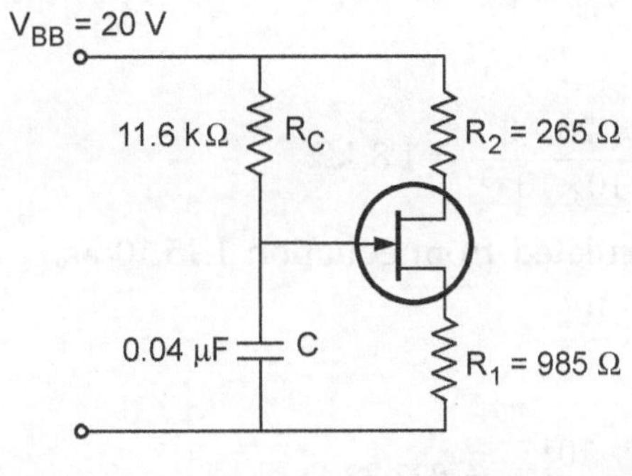

Fig. 1.15.13 UJT triggering circuit of example 1.15.2

Example 1.15.3 : *Design the UJT triggering circuit for SCR. Given -* V_{BB} *= 20 V,* $\eta = 0.6$*,* $I_p = 10\ \mu A$*,* $V_v = 2V$*,* I_v *= 10 mA. The frequency of oscillation is 100 Hz. The triggering pulse width should be 50 µs.*

Solution : The frequency $f = 100\ Hz$

$$\therefore \quad T = \frac{1}{f} = \frac{1}{100}$$

From equation 1.15.7,

$$T = R_c C\, ln\left(\frac{1}{1-\eta}\right)$$

Putting values in above equation,

$$\frac{1}{100} = R_c C\, ln\left(\frac{1}{1-0.6}\right)$$

$$\therefore \quad R_c C = 0.0109135$$

Let us select $C = 1\mu F$. Then R_c will be,

$$R_c = \frac{0.0109135}{1\times10^{-6}}$$

$$= 10.91\ k\Omega$$

The peak voltage is given as,

$$V_p = \eta V_{BB} + V_D$$

Let $V_D = 0.8$ and putting other values,

$$V_p = 0.6\times20+0.8 = 12.8\ V$$

The minimum value of R_c can be calculated from equation 1.15.15 as,

$$R_{c(\min)} = \frac{V_{BB}-V_v}{I_v}$$

$$= \frac{20-2}{10\times10^{-3}} = 1.8\ k\Omega$$

Value of R_2 can be calculated from equation 1.15.10 as,

$$R_2 = \frac{10^4}{\eta V_{BB}}$$

$$= \frac{10^4}{0.6\times20} = 833.33\ \Omega$$

Here the pulse width is given, i.e. 50 μs.

Hence, value of R_1 will be,

$$\tau_2 = R_1 C \quad \text{from equation 1.15.11}$$

The width τ_2 = 50 μsec and C = 1 μF, hence above equation becomes,

$$50\times10^{-6} = R_1\times1\times10^{-6}$$

$$\therefore \quad R_1 = 50\ \Omega$$

Thus we obtained the values of components in UJT triggering circuit as,

$$R_1 = 50\ \Omega, \quad R_2 = 833.33\ \Omega$$

$$R_c = 10.91\ \text{k}\Omega, \quad C = 1\ \mu F.$$

Example 1.15.4 : *A capacitor used in the UJT oscillator circuit is charged by a constant current source. The value of the capacitor is 0.5 μF and that of the constant current is 1 mA. The sawtooth voltage of oscillator is found to have a crest value of 8.5 volts and valley level of 2.5 V. Calculate the frequency of the oscillator.*

Solution : Given data is,

$$C = 0.5\ \mu F$$

$$i_c = 1\ \text{mA constant current charging}$$

$$V_p = 8.5\ \text{V}, \quad V_v = 2.5\ \text{V}$$

The waveform of capacitor is sawtooth. Hence discharge period can be neglected. The voltage across capacitor is given as,

$$V_p = \frac{1}{C}\int_0^T i_c\,dt + V_v$$

As given by above equation, the capacitor charges to V_p at the end of period T. The residual voltage on capacitor is V_v. Putting values in above equation,

$$8.5 = \frac{1}{0.5\times10^{-6}}\int_0^T 1\times10^{-3}\,dt + 2.5$$

$$\therefore \quad 6 = \frac{1\times10^{-3}}{0.5\times10^{-6}}\int_0^T dt$$

$$\therefore \quad \int_0^T dt = 3\times10^{-3}$$

$$\therefore \quad [t]_0^T = 3\times10^{-3}$$

$$\therefore \quad T = 3\times10^{-3}$$

$$\therefore \quad \text{Frequency, } f = \frac{1}{T}$$

$$= \frac{1}{3\times10^{-3}} = 333.33\ \text{Hz}$$

Example 1.15.5 : *For an SCR, the gate cathode characteristic is given by a straight line with a gradients of 16 volts/ampere passing through the origin. The maximum turn-on time is 4 μsec and the minimum gate current required to obtain this quick turn-on is 500 mA. If the gate source voltage is 15 V,*

a) calculate the resistance to be connected in series with the SCR gate

b) compute the gate power dissipation, given that pulse width is equal to turn-on time and average gate power dissipation is 0.3 W. Also compute the maximum triggering frequency that will be possible when pulse firing is used.

Solution : a) To calculate series resistance in gate

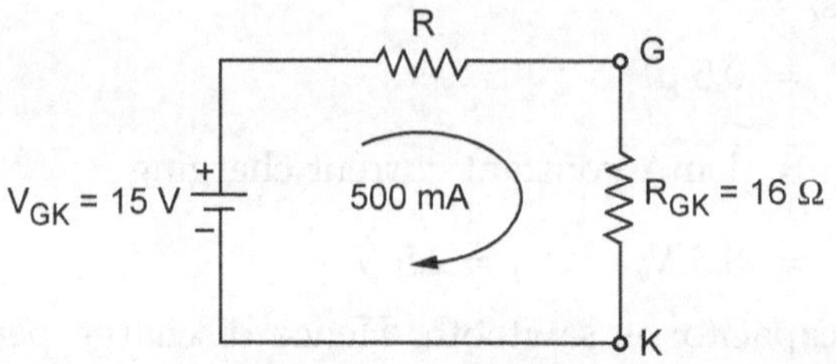

Fig. 1.15.14 Equivalent gate-cathode circuit

The gate-cathode characteristic has the gradient of 16 volts/Ampere. Hence gate cathode resistance is 16 Ω. Fig. 1.15.14 shows the equivalent gate-cathode circuit. In this circuit observe that gate-cathode is replaced by equivalent resistance of 16 Ω. The applied gate voltage is 15 V. A current of 500 mA passes through the circuit to trigger the SCR. By KVL to above circuit,

$$15 = (R + R_{GK}) \times 500 \times 10^{-3} = (R+16) \times 500 \times 10^{-3}$$

$$\therefore \quad R = 14\ \Omega$$

Thus a resistance of 14 Ω must be connected in series with the SCR gate.

b) To compute gate power dissipation and maximum pulse frequency

In Fig. 1.15.14, observe that the gate current of 500 mA passes through the gate of 16 Ω resistance. The firing pulse of width 4 μs is applied to the gate. Hence the gate power dissipation is,

$$P_g = \int_0^{t_{on}} i^2(t) R_{GK} dt = \int_0^{4 \mu s} (500 \times 10^{-3})^2 \times 16 dt$$

$$= 4 \int_0^{4 \mu s} dt = 16\ \mu W$$

The average gate power dissipation is 0.3 W. If we apply the multiple gate firing pulses, the gate power dissipation will increases. One pulse of 4 μs dissipates 16μW. Hence average 0.3W dissipation can be reached by,

$$f = \frac{0.3\times 1}{16\times 10^{-6}} = 18750$$

Thus the triggering frequency for pulse firing will be 18.75 kHz.

Example 1.15.6 : *A UJT is connected across a 20 V volts DC supply. The valley and peak point voltages are 1V and 15V. The period of UJT relaxation oscillator is 20 ms. Find the value of charging capacitor, if a charging resistor of 100 kΩ is used.*

Solution : The given data is,

$$V_{BB} = 220 \text{ V}$$

$$V_v = 1, \qquad V_p = 15 \text{ V}$$

$$T = 20\times 10^{-3}$$

$$R_C = 100 \text{ k}\Omega$$

The peak voltage of the UJT is given as,

$$V_p = \eta V_{BB} + V_D$$

Let $V_D = 0.8$ and putting values in above equation,

$$15 = \eta \times 20 + 0.8 \Rightarrow \eta = 0.71$$

From equation 1.15.7, $T = R_c C \; ln\left(\frac{1}{1-\eta}\right)$

Putting values in above equation,

$$20\times 10^{-3} = 10\times 10^{3} \times C \, ln\left(\frac{1}{1-0.71}\right)$$

$$\therefore \qquad C = 0.162 \; \mu F$$

Example 1.15.7 : *Find the values of charging circuit components if the line synchronised UJT circuit can be operated to get delay angles of 20° to 160°. Assume suitable data if required.*

Solution : The given data is,

$$20° \le \alpha \le 60° \text{ i.e. } 0.349 \le \alpha \le 2.793$$

Let the operating frequency be f = 50 Hz.

$$\therefore \qquad \omega = 2\pi f = 2\pi \times 50 = 100\,\pi$$

The firing angle is given by equation 1.15.8 as,

$$\alpha = \omega R_c C \, ln\left(\frac{1}{1-\eta}\right)$$

Let $\eta = 0.65$ and for $\alpha_1 = 0.349$,

$$0.349 = 100\pi \times R_{c1} C_1 \, ln\left(\frac{1}{1-0.65}\right)$$

$$\therefore \quad R_{c1} C_1 = 1.058 \times 10^{-3}$$

Similarly with $\alpha_2 = 2.793$,

$$2.793 = 100\pi \times R_{c2} C_2 \, ln\left(\frac{1}{1-0.65}\right)$$

$$\therefore \quad R_{c2} C_2 = 8.468 \times 10^{-3}$$

Let $C = C_1 = C_2 = 0.5\ \mu F$. Hence,

$$R_{c1} = \frac{1.058 \times 10^{-3}}{0.5 \times 10^{-6}} = 2.116\ k\Omega$$

$$R_{c2} = \frac{8.468 \times 10^{-3}}{0.5 \times 10^{-6}} = 16.936\ k\Omega$$

Thus the resistance R_c should be varied from 2.116 kΩ to 16.936 kΩ. Hence a variable resistance of 20 kΩ can be used.

Example 1.15.8 : *An UJT used in a relaxation oscillator circuit is having $\eta = 0.7$, $V_v = 1V$ and the supply voltage to the circuit is 15 V. Design the suitable values of R and C given that the frequency of oscillation is 1 kHz. Peak current is 1 mA and valley current is 8 mA.*

Solution : The given data is,

$$V_{BB} = 15\ V$$

$$\eta = 0.7$$

$$V_v = 1, \quad I_v = 8\ mA$$

$$f = 1000\ Hz$$

$$I_p = 1\ mA$$

The peak voltage is given as,

$$V_p = \eta V_{BB} + V_D$$

Let $V_D = 0.8$ and putting other values,

$$V_p = 0.7 \times 15 + 0.8 = 11.3 \text{ V}$$

The period of oscillation is given by equation 1.15.7 as,

$$T = R_c C \ln\left(\frac{1}{1-\eta}\right)$$

With $T = \frac{1}{f} = \frac{1}{1000}$ and putting $\eta = 0.7$,

$$\frac{1}{1000} = R_c C \ln\left(\frac{1}{1-0.7}\right)$$

$$\therefore \quad R_c C = 8.3058 \times 10^{-4}$$

$$\text{Let } C = 0.3 \ \mu F. \text{ Hence,}$$

$$R_c = \frac{8.3058 \times 10^{-4}}{0.3 \times 10^{-6}} = 2.768 \text{ k}\Omega$$

Now let us determine maximum and minimum values of R_c. From equation 1.15.14,

$$R_{c(max)} = \frac{V_{BB} - V_p}{I_p} = \frac{15 - 11.3}{1 \times 10^{-3}} = 3700 \ \Omega$$

And from equation 1.15.15,

$$R_{c(min)} = \frac{V_{BB} - V_v}{I_v}$$

$$= \frac{15 - 1}{8 \times 10^{-3}} = 1750 \ \Omega$$

Thus the calculated value of $R_c = 2.768$ kΩ lies in the angle of 1750 Ω to 3700 Ω.

Example 1.15.9 : *An UJT triggering circuit is connected across a 20 V zener. The valley and peak point voltages are 1 V and 15 V respectively The intrinsic stand-off ratio is 0.75. It operates at a frequency of 1200 Hz. Find the charging capacitor if R = 5.6 kΩ.*

Solution : The given data is,

$$V_{BB} = 20 \text{ V}$$

$$V_v = 1, \quad V_p = 15$$

$$\eta = 0.75$$

$$f = 1200 \text{ Hz} \qquad \text{Hence } T = \frac{1}{f} = \frac{1}{1200}$$

$$R_c = 5.6 \text{ k}\Omega$$

The period of oscillation of UJT relaxation oscillator is given by equation 1.15.7 as

$$T = R_C C \ln\left(\frac{1}{1-\eta}\right)$$

Putting values in above equation,

$$\frac{1}{1200} = 5.6\times 10^3 \times C \times \ln\left(\frac{1}{1-0.75}\right)$$

$$\therefore \quad C = 0.107\ \mu F$$

1.15.7 SCR/TRIAC Triggering using Diac

The diac is the two terminal and four layer device. It is mainly used for triggering triacs. Fig. 1.15.15 (a) shows the symbols of diac. The symbol shown in Fig. 1.15.15 (b) has arrows in both the directions. This means it conducts in either directions. The terminals are not named. It can be used in any direction. It is low power triggering device. There is no control terminal on the diac.

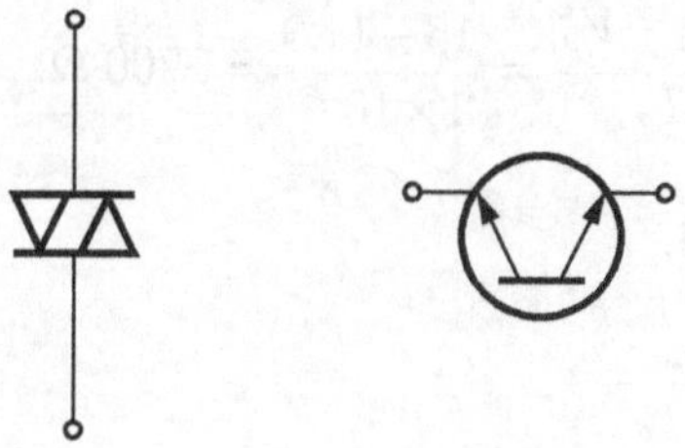

(a) (b)

Fig. 1.15.15 Two different symbols used for diac

1.15.7.1 V-I Characteristics of Diac

Fig. 1.15.16 shows the V-I characteristics of diac. The current through the diac is plotted with respect to voltage across it. The diac remains off till voltage is less than $\pm V_{BO}$. When voltage exceeds V_{BO}, diac turns 'on' and conducts heavily. Observe that the characteristic is exactly similar for positive as well as negative values of voltage and current. Once the diac turns on, the voltage across it drops to negligible value.

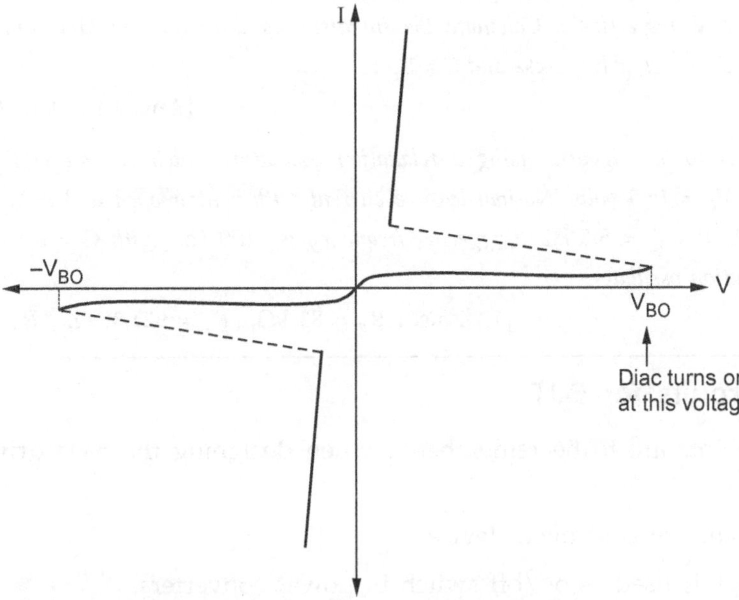

Fig. 1.15.16 V-I characteristic of diac

1.15.7.2 DIAC-TRIAC Phase Control

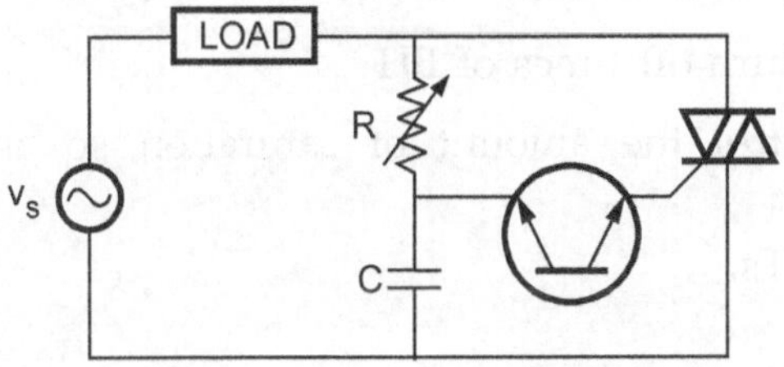

Fig. 1.15.17 Basic triac triggering using diac

Fig. 1.15.17 shows the basic triac triggering circuit using diac. In this diagram, when capacitor 'C' charges above the breakover voltage (V_{BO}) of diac, it starts conducting. Hence capacitor voltage is applied to triac gate and it turns-on. The charging of the capacitor can be controlled by the variable resistor 'R'. Which indirectly controls the firing delay of triac.

Review Questions

1. *Explain RC firing circuit for the SCR. Draw the necessary waveforms.*
2. *With the necessary circuit diagram and waveforms explain the operation of UJT triggering circuit.*
3. *How diac can be used to trigger triac ?*

Unsolved Examples

1. *A UJT trigger circuit is used to fire a PNPN device. It is supplied from a source across the SCR to be triggered through a 10 V zener. The valley and peak point voltages are found to be*

1 V and 7 V respectively. Calculate the intrinsic stand-off ratio of UJT and frequency of relaxation oscillator if R = 1 kΩ and C = 1 μ F.

[Ans. : η = 0.62, f = 1033 Hz]

2. *An SCR is to be triggered using a relaxation oscillator, which has an UJT with η = 0.7, I_p = 2 μ A, V_p = 16.5 volts. Normal leakage current with emitter open is 3 mA, V_v = 1 volts, I_v = 6 mA, $R_{B1\ B2}$ = 5.5 kΩ. Triggering frequency is 100 Hz. With C = 0.1 μF design the UJT relaxation oscillator.*

[Ans. : R_c = 83 kΩ, R_1 = 921.35 Ω, R_2 = 245.31 Ω]

1.16 Drive Circuits for BJT

Following points are to be remembered when designing the base drive circuit for transistor.

1. BJT is a current controlled device.
2. Power BJT is used as on/off switch in power converters.
3. Power BJT operates in saturation and cut-off when used as a switch.
4. Sufficient base current is required to drive BJT in saturation.
5. Amount of carrier injected in base region determine storage time of BJT.
6. Storage time determines turn-on and turn-off times of BJT.
7. There should be mechanism to control the amount of saturation so as to control storage time.

There are most important points about BJTs.

1.16.1 Base Drive Control During Turn-on

Fig. 1.16.1 shows the base drive circuit for turn-on of a transistor. Due to this circuit the 't_{on}' time is reduced.

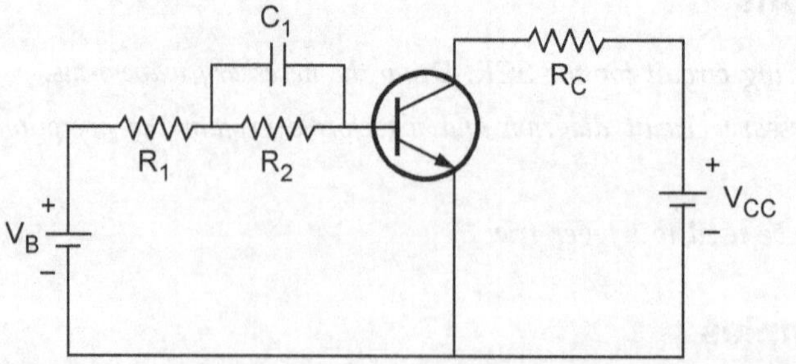

Fig. 1.16.1 Drive circuit to make base current high at beginning of t_{on}

As shown in this circuit, when base drive V_B is applied, the capacitor C_1 acts as a short. Hence R_2 is virtually by passes. Therefore an initial value of base current is only limited by R_1 and it is given as,

$$I_{B(peak)} = \frac{V_B - V_{BE}}{R_1}$$

This heavy base current drives transistor into saturation for quick turn-on. Once the transistor is turned on, there is no need of such large base current. This is taken care-off by R_2C_1 circuit. The capacitor C_1 starts charging and base current starts falling. This is shown in Fig. 1.16.2. Observe that there is peaking of base current at the beginning of turn-on. Then the current reduces to,

$$I_B = \frac{V_B - V_{BE}}{R_1 + R_2}$$

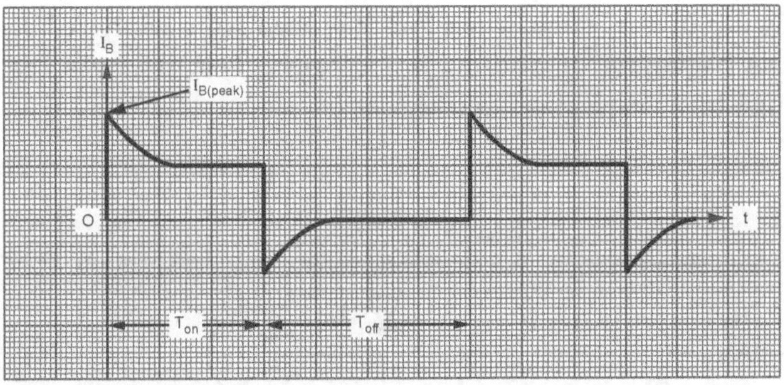

Fig. 1.16.2 Base current peaking at the beginning of turn-on

To turn-off the transistor, base voltage is made zero. Therefore capacitor voltage appears as negative voltage across base-emitter. Hence suddenly base current is reversed as shown in Fig. 1.16.2. This current slowly decays to zero after the stored charge in base region is removed. The capacitor C_1 then discharges through R_2. This discharge time constant is,

$$\tau_2 = R_2 C_1$$

The charging time constant of the capacitor is,

$$\tau_1 = \frac{R_1 R_2 C_1}{R_1 + R_2}$$

The T_{on} period of the transistor must be at least five times of τ_1. Similarly the T_{off} period must be five times of τ_2. The switching frequency of the BJT will then be,

$$f_s = \frac{1}{T_{on} + T_{off}}$$

We know that,

$$T_{on(min)} > 5\tau_1 \quad \text{and} \quad T_{off(min)} > 5\tau_2$$

This gives the maximum switching frequency of,

$$f_{s(max)} = \frac{1}{T_{on(\min)} + T_{off(\min)}}$$

$$= \frac{1}{5\tau_1 + 5\tau_2} = \frac{1}{5(\tau_1 + \tau_2)}$$

1.16.2 Base Drive Control During Turn-off

For quick turn-off, the stored charge in the base region must be removed fast. This can be done by applying negative base drive for turn-off. Due to $-V_B$, the base-emitter voltage becomes negative during turn-off. The voltage on capacitor C_1 also adds to this negative voltage. Therefore the negative peaking of base current (see Fig. 1.16.3) at the beginning of turn-off is increased. This removes the stored charge fast, and the transistor is turned-off quickly.

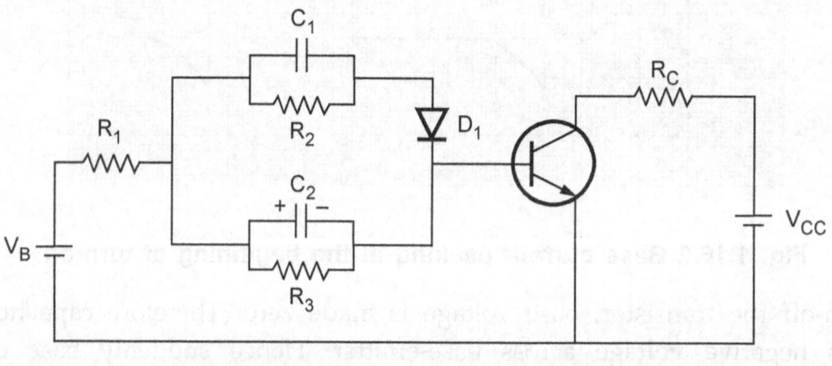

Fig. 1.16.3 Base drive circuit for positive and negative peaking of base current

It is possible to design the different turn-on and turn-off circuits for base current peaking. Fig. 1.16.3 shows one of such circuit. In this circuit, R_2C_1 is used for positive peaking of base current as discussed earlier. The diode D_1 acts as short when V_B is positive. Capacitor C_2 charges to the polarity as shown in above figure when V_B is positive. Normally R_3 is much higher than R_2, hence base current flows through R_2. The base voltage V_B is made negative to turn-off the transistor. Therefore the voltage of C_2 imposes additional negative input to the base. The negative base current flows through C_2, R_1 and V_B. Thus R_3C_2 becomes effective at the time of turn-off. No current flows through R_2C_1 during turn-off since diode D_1 is reverse biased. Once the base current is zero, then C_2 discharges through R_3. Similarly C_1 discharges through R_2. Thus it is possible to design separate peaking characteristics for turn-on and turn-off.

1.16.3 Proportional Base Control

The collector current changes as the load changes. If the load is reduced, then collector current reduces. Then the base current should also be reduced to avoid excess carries in the base region. Hence proportional base control can be used. Fig. 1.16.4 shows the circuit diagram of proportional base drive.

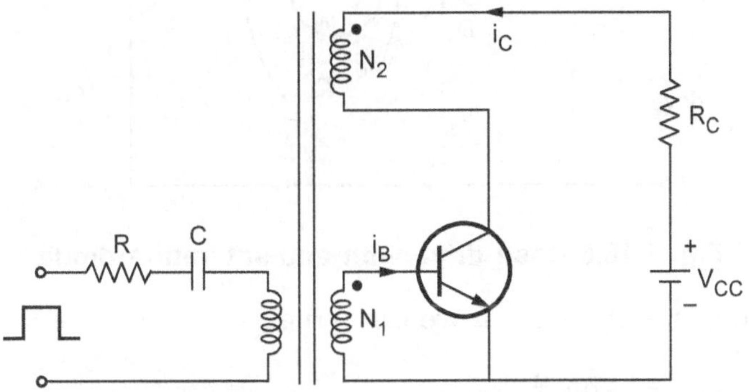

Fig. 1.16.4 Proportional base drive control

A short duration pulse is applied to turn-on the transistor. The transistor turns-on and collector current starts flowing. The collector current passes through the coil which is magnetically coupled to base coil. Hence collector current induces the current in the base coil also. This current acts as base drive to the BJT. The RC time constant determines the duration of the pulse to be applied externally to drive BJT into saturation. The turns ratio must be,

$$\frac{N_2}{N_1} = \frac{I_C}{I_B} = \frac{\beta I_B}{I_B} = \beta$$

The base current then varies according to variations in the collector current. The transistor can be turned off by applying negative pulse through RC circuit. This makes base current negative and BJT turns-off.

1.16.4 Anti-saturation Control

We know that excess base current increases storage time of the BJT. Hence the turn-off time increases. Such excess or heavy base drive is called hard saturation. Therefore transistor must be operated in soft saturation. This means base must be given the carriers which are sufficient to drive the transistor in just saturation (quasi saturation).

Fig. 1.16.5 shows the circuit diagram to achieve quasi saturation. In this circuit the base drive is applied at terminals a-b. We can write following equation for loop consisting of a − D_1 − B − E − b,

$$v_{a-b} = v_{D1} + v_{BE}$$

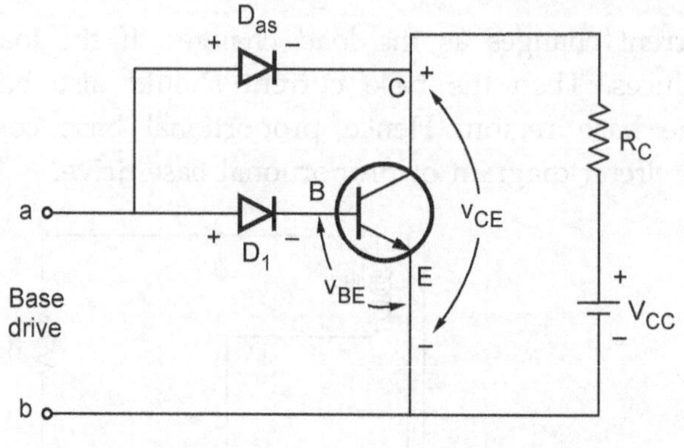

Fig. 1.16.5 Base drive with anti-saturation diode

Similarly for loop a – D_{as} – C – E we can write,

$$v_{a-b} = v_{Das} + v_{CE}$$

Hence equating the two equations,

$$v_{D1} + v_{BE} = v_{Das} + v_{CE}$$

Normally $v_{Das} = v_{D1}$. Hence above equation becomes,

$$v_{CE} = v_{BE}$$

This shows that the collector-emitter voltage will be equal to base-emitter voltage. When BJT turns on, the base-emitter voltage is nearly 0.7 and collector-emitter saturation voltage is 0.3. Because of anti-saturation diode (v_{Das}), the collector-emitter voltage is raised to v_{BE}, i.e. 0.7 V. Hence the BJT is no longer in saturation. It is just above saturation. This effect takes place because of anti-saturation diode v_{Das}. Since BJT is above saturation, there are no excess carriers in base and its storage time is reduced. This reduces 't_{off}' and hence switching time. The collector-emitter voltage can be further increased by putting additional diode in series of D_1.

Disadvantages of anti-saturation diode

The collector-emitter voltage is increased. This increases the on-state losses in BJT.

1.16.5 Typical Driver Circuit for Power BJT

Fig. 1.16.6 shows the typical driver circuit for power BJT that takes into consideration all the earlier requirements.

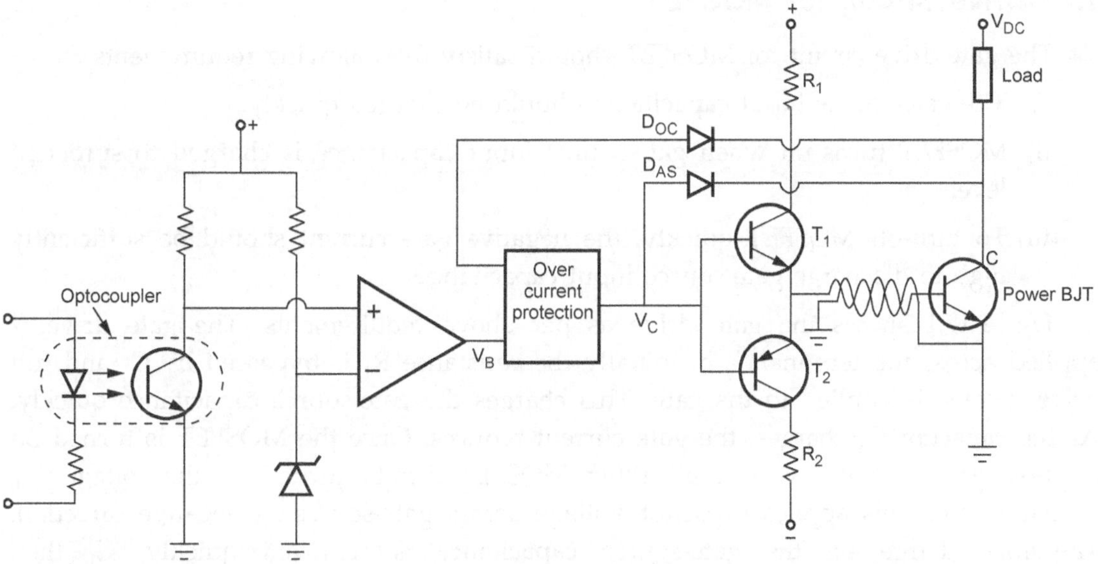

Fig. 1.16.6 Typical driver circuit for Power BJT

- The drive from control circuit is given to driver circuit through optocoupler.
- The signal is amplified, buffered by the comparator and then given to over current protection circuit. The collector voltage of main transistor is sensed through diode D_{OC}. If the collector current increases, then collector voltage will rise. This rise is sensed by over current protection circuit to disable the drive of BJT.
- The output from over current protection circuit is given to pair of pnp-npn transistors T_1 and T_2. These transistors provide the required base curent to power BJT to drive it in saturation.
- Note that the collector voltage of power BJT is monitored through diode D_{AS}. This provide anti-saturation control. It is called backer's clamp. It ensures that BJT always operates in quasi-saturation.
- Transistor T_1 provides large positive voltage and current to drive BJT in saturation of quasi-saturation. and transistor T_2 provides large negative voltage and current for faster turn-off of BJT.
- The base-emitter wires of power BJT are twisted to minimize stray inductance.

Review Questions

1. *Give different circuit configurations for base drive of BJT.*
2. *Draw the drive circuit for BJT. What is anti-saturation control ?*

1.17 Drive Circuit for MOSFET

The gate drive circuit for MOSFET should satisfy the following requirements :

(i) The gate-source input capacitance should be charged quickly.

(ii) MOSFET turns on when gate-source input capacitance is charged to sufficient level.

(iii) To turn-off MOSFET quickly, the negative gate current should be sufficiently high to discharge gate-source input capacitance.

Fig. 1.17.1 shows the gate drive as per above requirements. The gate drive is applied across the terminals a-b. Initially the resistance R_1 is bypassed by C_1 and full drive voltage is applied to the gate. This charges the gate-source capacitance quickly. As the capacitor C_1 charges, the gate current reduces. Once the MOSFET is turned on required gate current is very small. When MOSFET is to be turned off, the voltage v_{a-b} is made zero. This applies capacitor voltage across gate-source in negative direction. Therefore charge on the gate-source capacitance is removed quickly. C_1 then discharges through R_1. The resistance R_2 provides additional discharge path for gate-source capacitance.

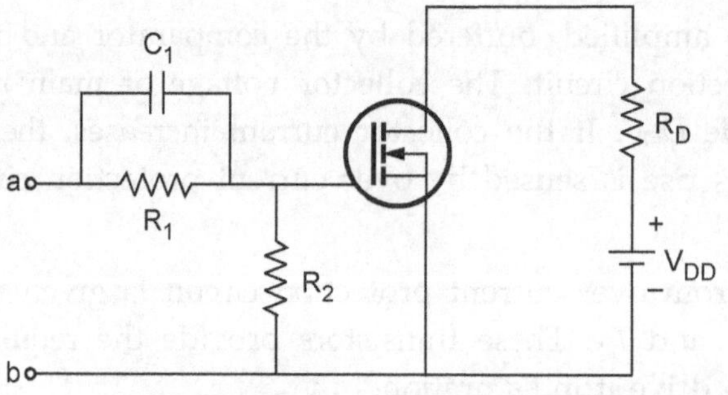

Fig. 1.17.1 Gate drive circuit

Review Questions

1. *Explain the typical gate drive circuit for MOSFET.*
2. *Draw the gate drive circuit for MOSFET.*

1.18 Driver Circuit for IGBT and MOSFET

Fig. 1.18.1 shows the driver circuit for IGBT which uses IR 2125 IC.

- Here IR 2125 is the high voltage, fast switching MOS gate driver with single floating gate driver channel. This IC can be used to drive N-channel power MOSFET or IGBT.

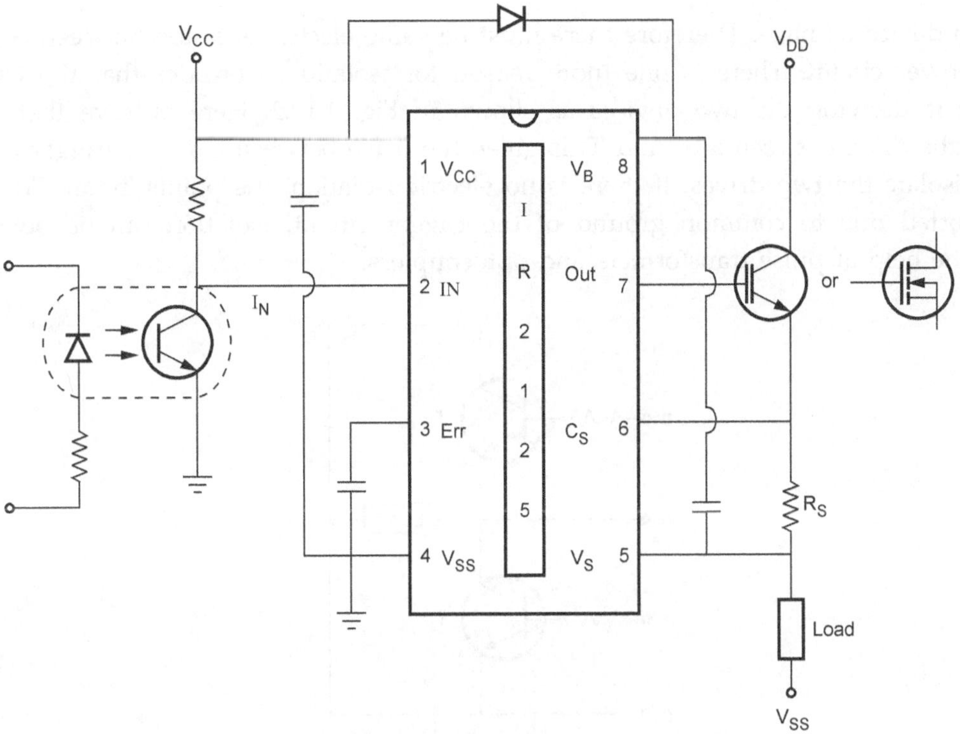

Fig. 1.18.1 Driver circuit for IGBT

- Over current flowing through the IGBT is detected through R_s and C_s terminal of the IC.
- The error pin of the IC indicates fault conditions.

1.19 Isolation of Gate and Base Drives

1.19.1 Necessity of Isolation

We know that driver circuits operate at very low power levels. Normally the signal levels are 3 to 12 volts. Sometimes digital circuits and microprocessors are also used in the triggering circuits. The gate and base drives are connected to power devices which operate at high power levels. Fig. 1.19.1 shows this situation. Observe that collector of BJT can have voltages of 200 V. But base is connected to trigger circuit that have voltages of 5 V. If BJT is damaged and collector-base gets shorted, then high voltage will get connected to trigger circuit. This will damage the trigger circuit also. This means trigger circuit is damage

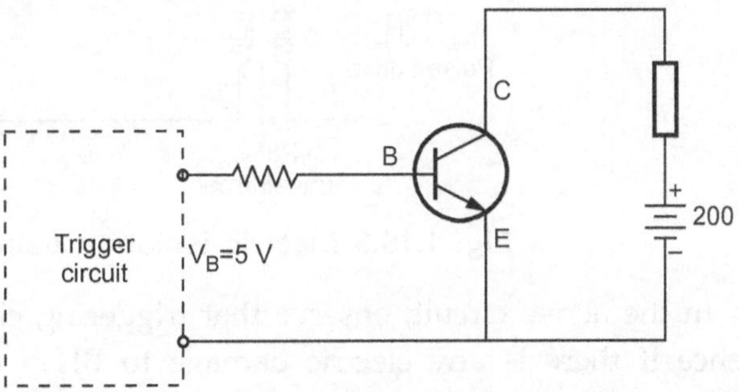

Fig. 1.19.1 Control / power levels

due to device damage. Therefore there must be some electric isolation between control and power circuit. There is one more reason for isolation. Consider that the trigger circuit is deriving the two devices as shown in Fig. 1.19.2. Here observe that T_1 is given the drive between a-b. And T_2 is given the drive between c-d. The trigger circuit must isolate the two drives. If there is no electric isolation, the points 'b' and 'd' may be shorted due to common ground of the trigger circuit. Isolation can be obtained with the help of pulse transformers and optocouplers.

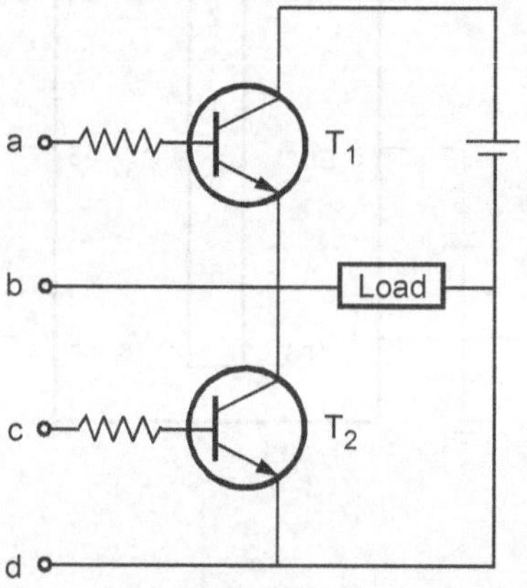

Fig. 1.19.2 Isolation of grounds

1.19.2 Isolation using Pulse Transformer

Pulse transformer has one primary and one or more secondary windings. It is normally used for pulsed mode of triggering Fig. 1.19.3 shows the isolation using pulse transformer.

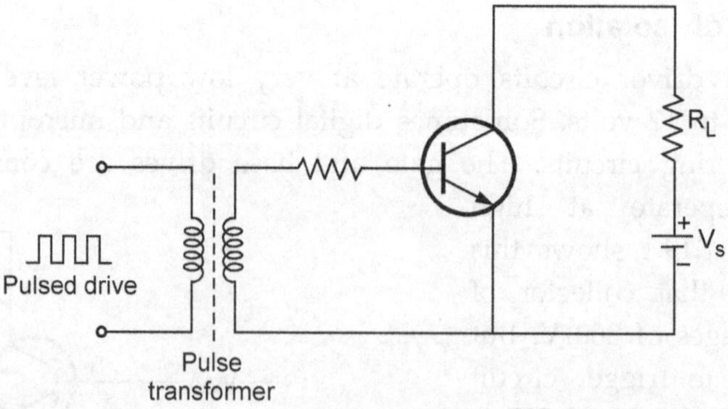

Fig. 1.19.3 Electric isolation using pulse transformer

In the above circuit, observe that triggering circuit is electrically isolated from BJT. Hence if there is any electric damage to BJT, there will be no effect on triggering circuit.

Advantages

i) Pulse transformer does not need external power for its operation.

ii) It is very simple to use.

Disadvantages

i) Pulse transformer saturates at low frequencies hence it can be used only for high frequencies.

ii) Due to magnetic coupling, the signal is distorted.

1.19.3 Isolation using Optocouplers

Optocoupler consists of a pair of infrared LED and phototransistor Fig. 1.19.4 shows the symbol of optocoupler. When the signal is applied to the infrared LED, it turns on. It's light falls on phototransistor. Therefore phototransistor also starts conducting. There is no electric connection between LED and phototransistor.

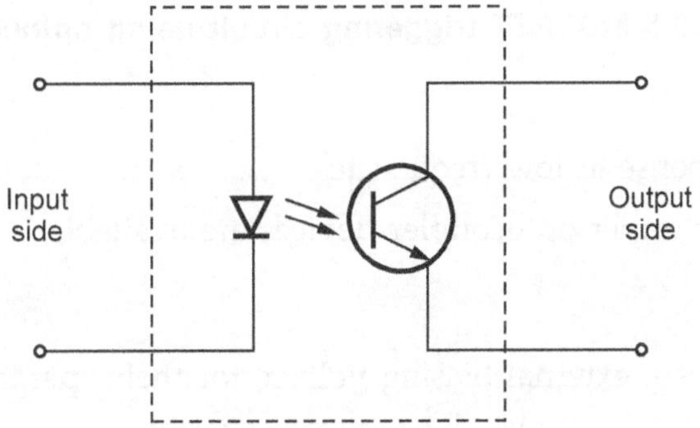

Fig. 1.19.4 Optocoupler

Fig. 1.19.5 shows the triggering circuit that uses optocoupler. In this circuit the triggering pulses are given to the input (LED) of optocoupler. When 'V_g' is positive, LED turns-on. It's light falls on phototransistor. Hence it turns on Therefore base of T_1 is connected to zero volts through phototransistor. Due to this, T_1 turns-on. Therefore the voltage V_{CC} is applied to gate of the MOSFET. Hence MOSFET turns on. When $V_g = 0$, the LED turns-off, therefore phototransistor also turn-off. Therefore base drive of T_1 goes to V_{CC} and it turn-off. When T_1 turns off, MOSFET gate voltage becomes zero. Therefore MOSFET turns-off. Thus gate drive circuit using optocoupler works.

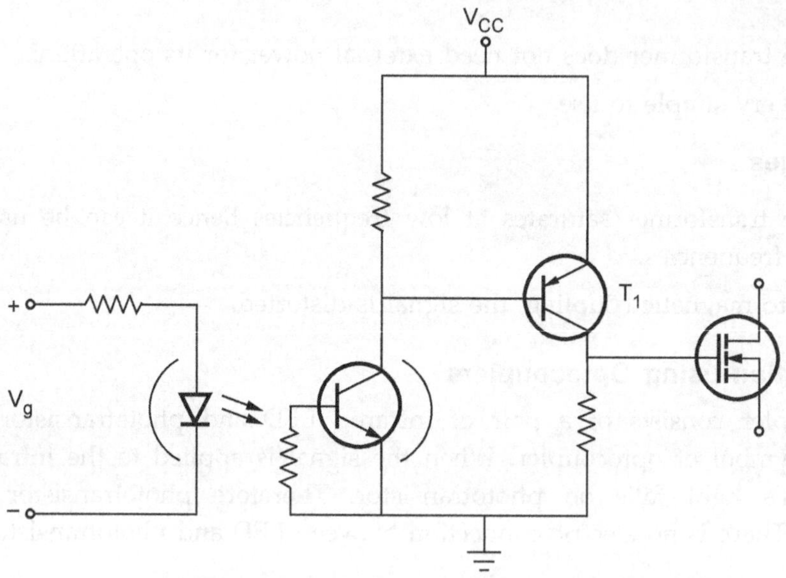

Fig. 1.19.5 MOSFET triggering circuit using optocoupler

Advantages

1) Very good response at low frequencies.
2) Compact and cheaper optocoupler devices are available.

Disadvantages

1) Optocoupler need, external biasing voltage for their operation.
2) High frequency response is poor.

Applications

Inverters, SMPS, Choppers, AC motor drives use optocouplers.

Review Question

1. *What is the necessity of isolation ? How it is implemented ?*

1.20 Protection Circuits

Due to unpredicted changes in the load and supply voltages, fault conditions are arised. Under these conditions, the voltage or current ratings of the power devices are exceeded. Hence there is a possibility of damage to the power device. Such damage can be avoided by using protection circuits. Following are the possible faults and required protection circuits :

i) High rate of change of voltage across the power device is called dv/dt. This normally turns-on the thyristor and damages BJTs. The value of dv/dt can be effectively reduced with the help of snubbers (RC circuit) and Metal oxide varistors (MOV).

ii) High rate of change of current in the device is called di/dt. The spread of current inside the device requires some time. If sufficient time is not given for current spread, then localized hot spot is created in side the device. The device is damaged due to this fault. The di/dt problem normally occurs in thyristors. This problem can be avoided by putting the current limiting inductor in series with the device. The inductor limits rapid changes in the current.

iii) Overcurrent normally occurs due to variations in the load. Overcurrent causes excessive heating of the device and it leads to damage. Fuses are normally used to limit overcurrents.

1.21 Protection Against dv/dt and Overvoltages

1.21.1 Snubber Circuits (Turn-off Snubber)

The transient overvoltages can switch on the thyristor. In some cases the thyristor can be damaged due to these transient voltages. These transient voltages are very common when the converter is having inductive loads. The thyristors can be protected against transient voltages by a RC network as shown in Fig. 1.21.1. This RC network is connected in parallel across the thyristor. It is called *snubber circuit*. The resistance has the value of few hundred ohms. Whenever there is a large spike or voltage transient across the thyristor, it is absorbed by the RC circuit. The RC circuit (snubber) acts as a lowpass filter for this voltage transient. The resistance has normally low value so that the transient is absorbed by the capacitor quickly. Thus the thyristor is protected against voltage transients. The RC snubber circuit is very commonly used for protection of thyristors against transient voltages (high frequency voltage spikes).

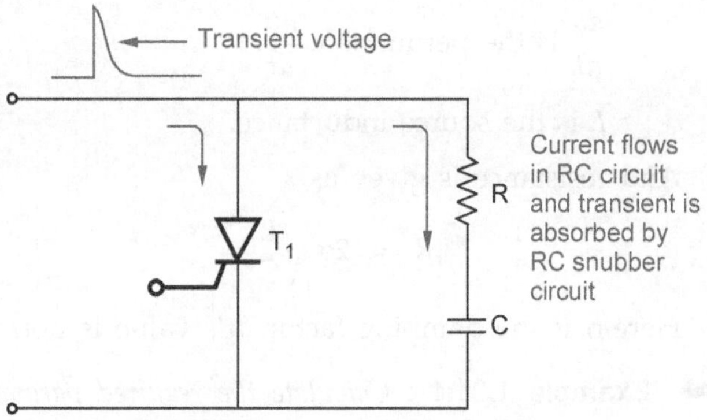

Fig. 1.21.1 A snubber (RC) network is used for transient voltage protection

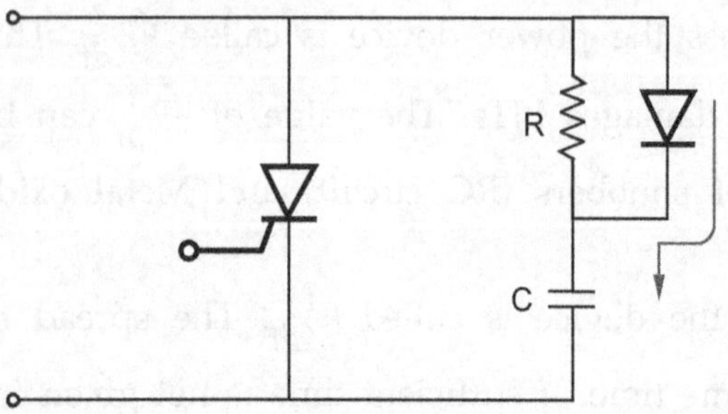

Fig. 1.21.2 Snubber is used for $\frac{dv}{dt}$ protection

$\frac{dv}{dt}$ also generates large voltage transients. These rapid voltage variations can also be suppressed by snubber circuit. The capacitor acts as a short for these $\frac{dv}{dt}$ variations. The snubber can be made more effective by connecting a diode across the resistance as shown in Fig. 1.21.2.

In case of voltage transient, the current flows through diode and capacitor. The capacitor acts as a short for the voltage transient. Thus it is suppressed. When thyristor turns-on, the capacitor discharges through resistance R. The R, C and diode snubber is more commonly used because it is very effective for $\frac{dv}{dt}$ and other voltage transients.

Design of snubber

The value of capacitor is given as,

$$C = \frac{1}{2L}\left(\frac{0.564\ V_m}{\frac{dv}{dt}}\right)^2 \qquad \text{... (1.21.1)}$$

Here V_m is the peak value of supply voltage

$\frac{dv}{dt}$ is the permissible $\frac{dv}{dt}$.

L is the source inductance.

And resistance is given as,

$$R = 2\sigma\sqrt{\frac{L}{C}} \qquad \text{... (1.21.2)}$$

Here σ is the damping factor. It's value is normally taken as 0.65.

Example 1.21.1 : *Calculate the required parameters for snubber circuit to provide $\frac{dv}{dt}$ protection to a SCR used in single phase bridge converter. The SCR has a maximum $\frac{dv}{dt}$ capability of 60 V / μ sec. The input line to line voltage has a peak value of 425 volts and the source inductance is 0.2 mH.*

Solution : Given :

$$\frac{dv}{dt} = 60 \ V / \mu\text{sec}$$

$$L = 0.2 \text{ mH}$$

$$V_m = 425 \text{ V}$$

From equation 1.21.1,

$$C = \frac{1}{2L}\left(\frac{0.564 \ V_m}{\frac{dv}{dt}}\right)^2 = \frac{1}{2 \times 0.2 \times 10^{-3}}\left(\frac{0.564 \times 425}{60} \times 10^{-6}\right)^2$$

$$= 0.04 \ \mu F$$

In the above equation observe that we have multiplied numerator by 10^{-6} inside the brackets. It comes from $60 \ V / \mu \ s$, i.e. $\frac{60}{1 \times 10^{-6}}$ to be substituted in equation for C.

Let the damping factor be $\sigma = 0.65$. From equation 1.21.2,

$$R = 2\,\sigma\sqrt{\frac{L}{C}} = 2 \times 0.65\sqrt{\frac{0.2 \times 10^{-3}}{0.04 \times 10^{-6}}}$$

$$R = 92 \ \Omega$$

1.21.2 Metal Oxide Varistors (MOVs)

High $dv/_{dt}$ and transient over voltages can also be suppressed with the help of Metal Oxide Varistors (MOV). MOVs are also called varistors or nonlinear voltage dependent resistor. MOVs consists of metal oxide particles, which are separated by an oxide film or insulation. When the applied voltage is less than specific value, then MOV offers high impedence. When the applied voltage is more than specific value, the oxide film becomes conductive and current starts flowing. Thus the voltage spike is suppressed by the MOV. Fig. 1.21.3 shows the characteristics and symbol of MOV.

(a) Symbol of MOV

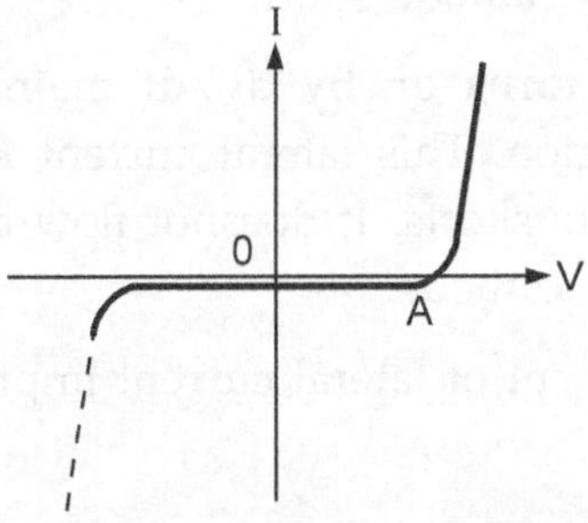

(b) V-I characteristics of MOV

Fig. 1.21.3

In the Fig. 1.21.3 observe that voltage increases above point 'A', the MOV starts conducting heavily.

The current and voltage of MOV are related as,

$$I = KV^{\alpha}$$

Here K is the device constant and 'α' lies in the range of 30 to 40.

Normally MOVs are connected across the supply lines over which the voltage transients are to be suppressed.

1.21.3 Improving *dv/dt* Rating with the Help of Cathode Short

The Fig. 1.21.4 shows the modification in the structure of SCR to improve dv/dt capability.

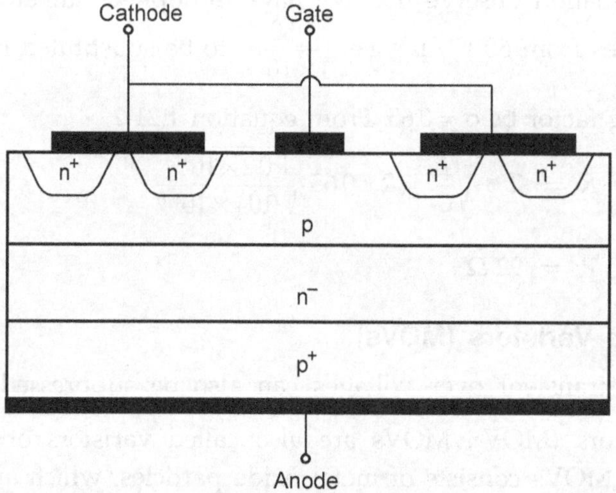

Fig. 1.21.4 Shorted emitter structure

- The cathode metallization is overlapped over the gate region, i.e. p2 region. It is called cathode short.
- The SCR turns on by dv/dt mainly due to lateral flow of current in the p-type region. This lateral current is intercepted in large amount mainly by the cathode shorts. It doesnot flow across the gate-cathode junction but flows directly to cathode.
- This intercept of lateral current improves dv/dt capability of the SCR.

Review Questions

1. *Explain the protection of power devices by snubber circuit.*
2. *Explain the use of Metal Oxide Varistors (MOV) for protection against overvoltages and voltage transients?*

Unsolved Example

1. *Calculate the required parameters for a snubber circuit to provide reliable* $\frac{dv}{dt}$ *protection to a SCR used in the single phase fully controlled bridge. The SCR has a maximum* $\frac{dv}{dt}$ *capability of 40V/μs. The input line to line voltage has a peak value of 325V and the source inductance is 0.1 mH.*

[Ans. : R = 41 Ω and C = 0.1 μF]

1.22 di/dt Protection with the Help of Inductor (Turn-on Snubber)

We know that at the time of turn-on, anode current increases rapidly. This rapid variation of anode current doesnot spread across the junction area of the thyristor. This creates the local hot-spots in the junction and increases the junction temperature. If the junction temperature exceeds permissible value, then the thyristor is damaged. The rapid variations of the thyristor current are also called $\frac{di}{dt}$. Every thyristor has maximum permissible value of $\frac{di}{dt}$.

The thyristor can be protected from excessive $\frac{di}{dt}$ by using an inductor in series as shown in Fig. 1.22.1. The inductance opposes for rapid current variations $\left(\frac{di}{dt}\right)$. Whenever there is rapid current variation, the inductor smooths it and protects the thyristor from damage.

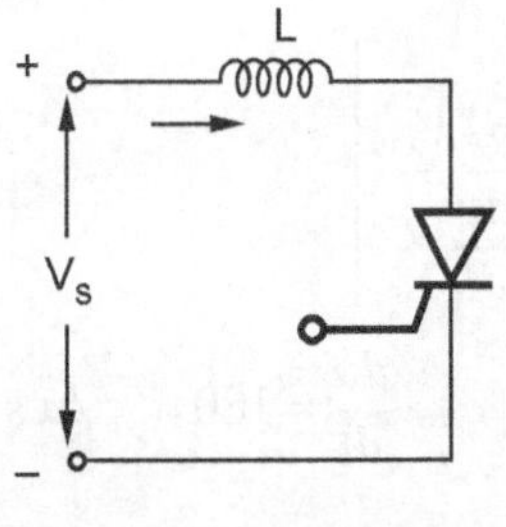

Fig. 1.22.1 An inductance in series with the thyristor provides protection against $\frac{di}{dt}$

The value of inductance can be calculated as,

$$L \geq \frac{V_s}{\frac{di}{dt}} \quad \text{... (1.22.1)}$$

Here $\frac{di}{dt}$ is the maximum value and L is the series inductance including stray inductance.

Example 1.22.1 : *Design the snubber circuit elements R_s and C_s connected across the SCR given that $\frac{dv}{dt}$(max) = 180 V/μ s and $\frac{di}{dt}$(max) = 45 A/μ s. An inductance L = 0.1 H and a resistance $R << R_s$ are in series with the SCR with a 300 V DC applied to the circuit.*

Solution : The value of $\frac{di}{dt}$ is given. Hence let us determine the required value of series inductance. From equation 1.22.1, it is given as,

$$L \geq \frac{V_s}{\frac{di}{dt}}$$

Here V_s = 300 V and $\frac{di}{dt}$ = 45 A/μ s. putting these values in above equation,

$$L \geq \frac{300}{45/10^{-6}}$$

$$\geq \frac{300}{45} \times 10^{-6}$$

$$\geq 6.667 \times 10^{-6} H$$

This inductance includes stray inductance also. There is an inductance of 0.1 H (given) in series with the SCR. Since this is more than 6.667×10^{-6} H, there is no need to connect extra inductance. Thus $\frac{di}{dt}$ protection is obtained through existing 0.1 H inductance. The capacitance of the snubber is given by equation 1.21.1. as,

$$C = \frac{1}{2L}\left[\frac{0.564\, V_m}{\frac{dv}{dt}}\right]^2$$

Here V_m = 300 V, $\frac{dv}{dt} = 180$ V/μ s and L = 0.1 H. Hence above equation becomes,

$$C = \frac{1}{2 \times 0.1}\left[\frac{0.564 \times 300}{180 / 10^{-6}}\right]^2$$

$$= \frac{1}{2 \times 0.1}\left[\frac{0.564 \times 300}{180} \times 10^{-6}\right]^2$$

$$= 4.418 \times 10^{-12} \text{ F or } 4.418 \text{ pF}$$

The resistance of the snubber is given by equation 1.21.2 as,

$$R = 2\sigma\sqrt{\frac{L}{C}}$$

Here $\sigma = 0.65$ (damping factor) and putting values of L and C calculated earlier,

$$R = 2 \times 0.65\sqrt{\frac{0.1}{4.418 \times 10^{-12}}}$$

$$= 195.58 \ k\Omega$$

It is mentioned in the example that series resistance is very very small than snubber resistance. Hence it can be neglected. Fig. 1.22.2 shows the snubber circuit.

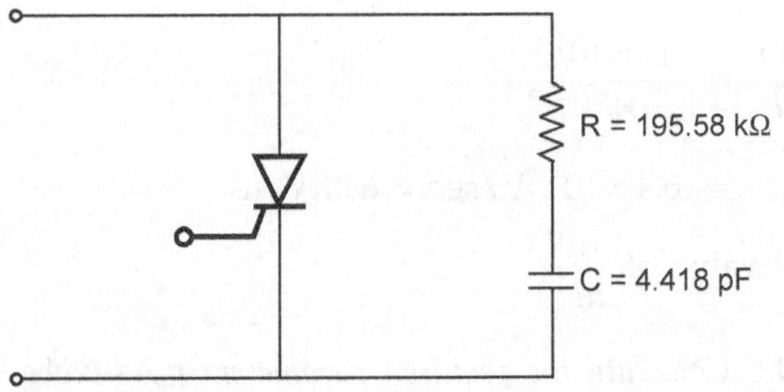

Fig. 1.22.2 Snubber circuit of example 1.21.2

Example 1.22.2 : *The capacitance of the reverse biased junction J_2 in a thyristor is 25 pF and can be assumed to be independent of the off-state voltage. The limiting value of the charging to turn-on the thyristor is 16 mA. Determine the critical value of $\frac{dv}{dt}$.*

Solution : The current through the junction J_2 capacitance is given as,

$$i = \frac{dQ}{dt}$$

$$= \frac{d}{dt}(C_{j2} V_{j2})$$

Here C_{j2} is capacitance of J_2 and

V_{j2} is voltage across J_2

$$\therefore \quad i = V_{j2}\frac{dC_{j2}}{dt} + C_{j2}\frac{dV_{j2}}{dt} \qquad \text{... (1.22.2)}$$

The capacitance is independent of the off-state voltage. Hence $\frac{dC_{j2}}{dt} = 0$. Hence above equation will be,

$$i = C_{j2}\frac{dV_{j2}}{dt}$$

Here V_{j2} is the voltage across J_2. This voltage is nearly equal to applied voltage V. Hence,

$$i = C_{j2}\frac{dv}{dt} \qquad \text{... (1.22.3)}$$

$$\therefore \quad \frac{dv}{dt} = \frac{i}{C_{j2}}$$

Here i = 16 mA is the limiting value of charging current and

C_{j2} = 25 pF. Therefore above equation becomes,

$$\frac{dv}{dt} = \frac{16\times10^{-3}}{25\times10^{-12}}$$

$$= 6.4\times10^{8}\ \text{V/sec} = 640\ \text{V}/\mu\text{s}$$

This is the critical value of $\frac{dv}{dt}$.

Example 1.22.3 : *Calculate the required parameters for snubber circuit to provide $\frac{dv}{dt}$ protection to a SCR used in a single phase bridge converter. The SCR has a maximum $\frac{dv}{dt}$ capability of 60 V/μ s. The input line to line voltage has a peak value of 425 volts and the source inductance is 0.2 mH.*

Solution : Given :

$$\frac{dv}{dt} = 60\ \text{V}/\mu\text{s}$$

$$V_m = 425\ \text{V}$$

$$L = 0.2\ \text{mH}$$

We have to calculate the values of R and C for the snubber. From equation 1.21.1, capacitor is given as,

$$C = \frac{1}{2L}\left[\frac{0.564V_m}{\frac{dv}{dt}}\right]^2$$

Putting the values in above equation,

$$C = \frac{1}{2\times 0.2\times 10^{-3}}\left[\frac{0.564\times 4.25}{60\times 10^{6}}\right]^2 = 0.04\ \mu F$$

And the value of resistance is given by equation 1.21.2 as,

$$R = 2\sigma\sqrt{\frac{L}{C}}$$

Value of σ is normally taken as 0.65. Putting for L and C,

$$R = 2\times 0.65\sqrt{\frac{0.2\times 10^{-3}}{0.04\times 10^{-6}}} = 92\ \Omega$$

Thus we obtained the values of snubber components as,

$$R = 92\ \Omega \quad \text{and} \quad C = 0.04\ \mu F$$

Example 1.22.4 : *The junction capacitance of the thyristor shown in Fig. 1.22.3 is 15 pF and is assumed to be independent of the off-state voltage. The value of charging current to turn-on the device is 5mA and the critical value of* $\frac{dv}{dt} = 200\ V/\mu s$. *Determine the value of* C_s *so that the thyristor will not be turned on due to* $\frac{dv}{dt}$.

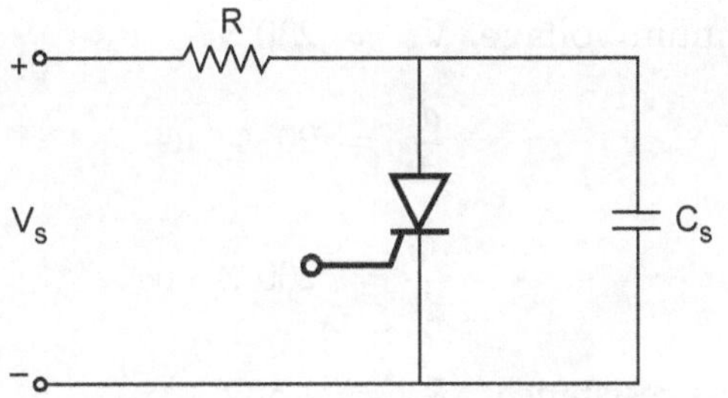

Fig. 1.22.3 Circuit of example 1.21.4

Solution : Here the junction capacitance of the thyristor and external capacitance (C_s) both will absorb the effect of $\frac{dv}{dt}$. These two capacitors will be in parallel. Hence charging current can be given as (using equation 1.22.3),

$$i = (C_s + C_j)\frac{dv}{dt}$$

The charging current should not exceed 5 mA. Hence above equation becomes,

$$5 \text{ mA} \leq (C_s + C_j)\frac{dv}{dt}$$

$$\text{or } C_s + C_j \geq \frac{5\,mA}{\frac{dv}{dt}}$$

$$\geq \frac{5\times10^{-3}}{200\times10^{6}}$$

$$\geq 25\ pF$$

Thus the total capacitance should be more than 25 pF. Out of this, $C_j = 15$ pF. i.e.,

$$C_s + 15\,pF \geq 25\ pF$$

$$\therefore \quad C_s \geq 10\,\text{pF}$$

Thus external capacitance of at least 10 pF is required to avoid false triggering due to $\frac{dv}{dt}$.

Example 1.22.5 : *Design the values of $\frac{di}{dt}$ inductor and RC snubber components for an SCR working in a 230 V system. Given $\frac{di}{dt}$ rating is 90 A/μs and $\frac{dv}{dt}$ rating is 200 V/ μs. Effective series resistance is 1.5 Ω. Take damping factor is as 0.6.*

Solution : The given data is,

$$\text{Maximum voltage, } V_s = 230 \text{ V}$$

$$\frac{di}{dt} = 90\ A/\mu s$$

$$\frac{dv}{dt} = 200\ A/\mu s$$

$$\text{Series resistance } R_s = 1.5\ \Omega$$

$$\text{Damping factor } \sigma = 0.6$$

The circuit diagram will look like the one shown below

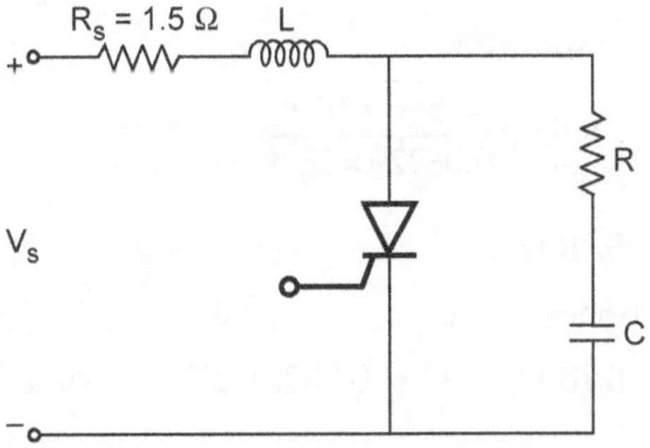

Fig. 1.22.4 Circuit of example 1.22.5

Let us calculate value of inductance for $\frac{di}{dt}$ protection. From equation 1.22.1 it is given as,

$$L \geq \frac{V_s}{\frac{di}{dt}}$$

$$\geq \frac{230}{90 \times 10^6}$$

$$= 2.556 \times 10^{-6} \text{ H}$$

$$\therefore \quad L = 2.556 \times 10^{-6} \text{ H} = 2.556 \ \mu\text{H}$$

Since there is no stray inductance given, this value of inductance must be connected for $\frac{di}{dt}$ protection.

Now the value of capacitance C can be calculated from equation 1.21.1 as,

$$C = \frac{1}{2L}\left[\frac{0.564 \, V_m}{\frac{dv}{dt}}\right]^2$$

Putting value in above equation,

$$C = \frac{1}{2 \times 2.556 \times 10^{-6}}\left[\frac{0.564 \times 230}{200 \times 10^6}\right]^2$$

$$= 0.08229 \ \mu F$$

In Fig. 1.22.4 observe that the resistance R_s and snubber resistance R are in series. These two resistors affect the charging rate of snubber capacitor. Hence, equation 1.22.2. must be written as,

$$(R_s + R) = 2\sigma\sqrt{\frac{L}{C}}$$

Putting values in above equation,

$$(1.5 + R) = 2\times 0.6\sqrt{\frac{2.556\times 10^{-6}}{0.08229\times 10^{-6}}}$$

$$\therefore \quad R = 5.18\ \Omega$$

Thus the snubber components are,

$$R = 5.18\ \Omega, \quad C = 0.08229\ \mu F \quad \text{and} \quad L = 2.556\ \mu H$$

Example 1.22.6 : *A SCR circuit operates from 300V DC supply, has series inductance of 4 μH. A resistance of 4 Ω and capacitance of 0.2 μF is connected across the SCR. Calculate the safe $\frac{dv}{dt}$ and $\frac{di}{dt}$ ratings of SCR.*

Solution : Here the given data is,

$$V_m = 300\ V$$

$$L = 4\ \mu H$$

$$R = 4\ \Omega$$

$$C = 0.2\ \mu F$$

The value of series inductance for $\frac{di}{dt}$ protection is given by equation 1.22.1 as,

$$L \geq \frac{V_s}{\frac{di}{dt}}$$

Here $V_s = V_m = 300$ and L = 4 μH. Hence above equation becomes,

$$4\ \mu H \geq \frac{300}{\frac{di}{dt}}$$

$$\therefore \quad \frac{di}{dt} = \frac{300}{4\times 10^{-6}} = 75\ A/\mu s$$

The value of snubber capacitor is given by equation 1.21.1 as,

$$C = \frac{1}{2L}\left[\frac{0.564\ V_m}{\frac{dv}{dt}}\right]^2$$

Putting values in above equation,

$$0.2\times10^{-6} = \frac{1}{2\times4\times10^{-6}}\left[\frac{0.564\times300}{\frac{dv}{dt}}\right]^2$$

$$\therefore \quad \frac{dv}{dt} = 133.76\ \text{V}/\mu\text{s}$$

Example 1.22.7 : *An SCR can be triggered with a* $\frac{dv}{dt}$ *of 220 V/μs. If the charging current flowing through the junction is 5 mA, calculate the equivalent capacitance of depletion layer.*

Solution : Equation 1.22.3 gives the charging current as,

$$i = C_{j2}\frac{dv}{dt}$$

Here i is the charging current and C_{j2} is the equivalent capacitance of depletion layer. Putting values in above equation,

$$5\times10^{-3} = C_{j2}\times220\times10^{6}$$

$$\therefore \quad C_{j2} = 22\ \text{pF}$$

Example 1.22.8 : *A SCR has a* $\frac{di}{dt}$ *= 120 A/μs and a* $\frac{dv}{dt}$ *of 300 V/μs. It operates on a 250 V DC source with a load resistance of 10 Ω. Find the suitable values for the components of the snubber circuit.*

Solution : The given data is,

$$\frac{di}{dt} = 120\ \text{A}/\mu\text{s}$$

$$\frac{dv}{dt} = 300\ \text{V}/\mu\text{s}$$

$$V_m = 250\ \text{V}$$

$$R_L = 10\ \Omega$$

The value of inductance for $\frac{di}{dt}$ protection is given by equation 1.22.1 as,

$$L \geq \frac{V_m}{\frac{di}{dt}}$$

$$\therefore \quad L \geq \frac{250}{120\times10^{6}}$$

$$\therefore \quad L \geq 2.08\ \mu H$$

The snubber circuit capacitor is given by equation 1.21.1 as,

$$C = \frac{1}{2L}\left[\frac{0.564\, V_m}{\frac{dv}{dt}}\right]^2$$

Putting values in above equation,

$$C = \frac{1}{2\times 2.08\times 10^{-6}}\left[\frac{0.564\times 250}{300\times 10^6}\right]^2$$

$$= 0.053\ \mu F$$

Here note that the load resistance is $R_L = 10\Omega$. This resistance appears in series with the snubber resistance. Hence the snubber resistance given by equation 1.21.2 can be given as,

$$R + R_L = 2\sigma\sqrt{\frac{L}{C}}$$

Let the damping factor $\sigma = 0.65$. Putting other values in above equation,

$$R + 10 = 2\times 0.65\sqrt{\frac{2.08\times 10^{-6}}{0.053\times 10^{-6}}}$$

$$\therefore \quad R = -1.856\ \Omega$$

The negative snubber resistance indicates that the load resistance is more than sufficient for snubber action. There is no need to connect additional snubber resistance.

Review Question

1. *How the devices are protected against* di/dt ?

1.23 Overcurrent Protection

1.23.1 Fuse

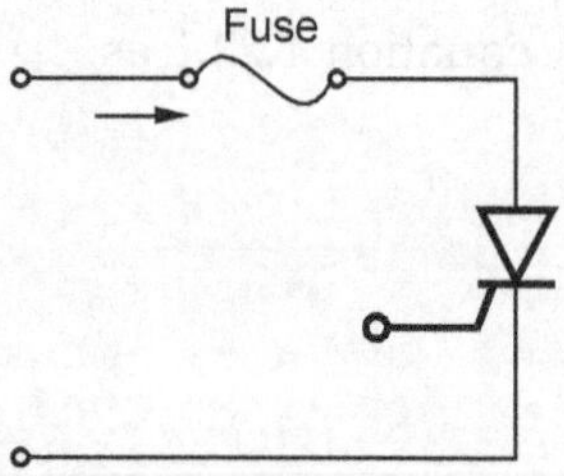

Fig. 1.23.1 Fast acting fuse is used to protect thyristor against overcurrent

The overcurrents flow in the thyristor circuits due to short circuits. The short circuits can take place because of short circuitted load, misalignment of firing pulses, failure of the thyristors due to overvoltages etc. The short circuit currents can be protected automatically because of load or supply transformers appear in the circuit. However the thyristors must be protected against overcurrents in the circuit.

Normally fast acting fuse are used for the protection of thyristor against overcurrents. These fuse melt at comparatively lower currents than current rating of the thyristor. Thus fuse melts and disconnects the circuit and the thyristor is protected. This is shown in Fig. 1.23.1. The fuse should be selected such that it should not melt or disconnect the circuit at normal load currents.

1.23.2 Semiconductor Fuses

Fig. 1.23.2 shows the internal structure of the semiconductor fuse. The fusing element is made of silver with one or more notches as shown in the Fig. 1.23.2. The cross-section of the fuse at the notch is less. When heavy current flows through the fuse, there is high current density at the notch. Because of high current density, temperature rises at the notch and arc is developed. This arc reduces the current flow. Therefore high voltage is formed across the notch. This further increases the temperature. Therefore the fuse element is vapourized and arc length is increased. Due to increased arc length, the current further reduces. At some stage the current is zero and the fuse becomes open. There is no arc and the complete voltage appears across the fuse. Note that the fuse has to withstand the voltage in open condition. The fusing element is embedded in quartz sand. This helps to conduct heat and serves as a quenching medium for the arc at the time of fusing.

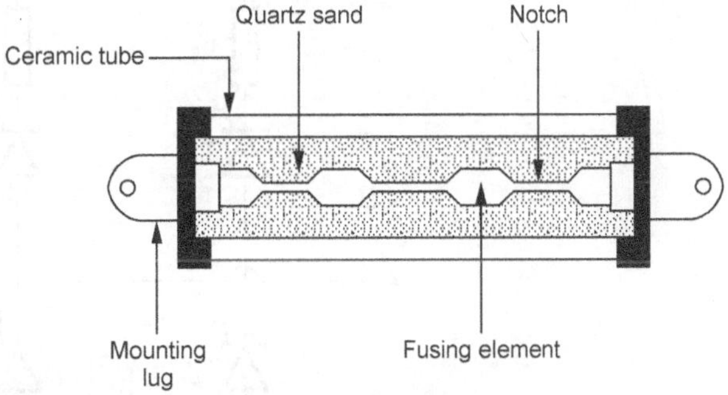

Fig. 1.23.2 Cross-section of semiconductor fuse

Fig. 1.23.3 shows the current versus time characteristic for fuse. At the peak let through value of the current, arcing starts at the notches in the fuse. This time is called melting time t_m. After t_m, the arcing starts at notches and current begins to reduce. The time 't_a' is the total arcing time. At the end of t_a, fuse

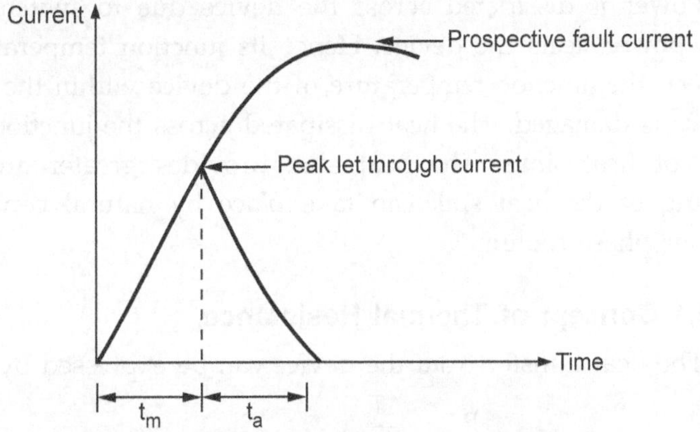

Fig. 1.23.3 Fuse current with respect to time

current is zero and it is completely open. The sum of t_m and t_a is called clearing time (t_c) of the fuse. At the peak let through current, the I^2t value of the fuse must be less than I^2t value of the device being protected. Therefore fuse protects the device before reaching to prospective fault current value.

Normally semiconductor fuses are used in series with each device being protected. Fig. 1.23.4 shows the placement of semiconductor fuses for single phase converter.

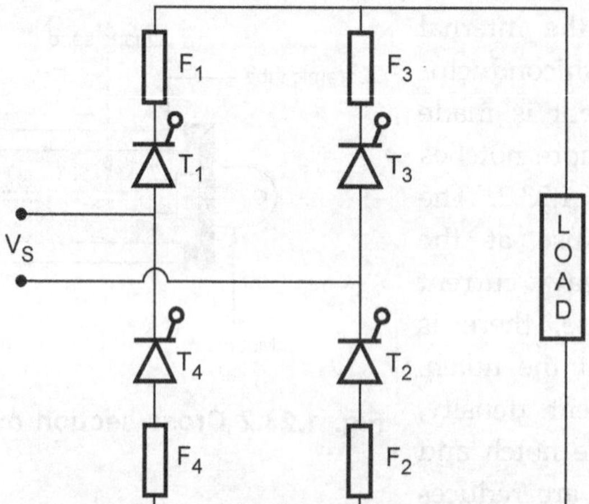

Fig. 1.23.4 Placement of semiconductor fuses for 1ϕ converter

Review Question

1. *Explain the working of semiconductor fuse. How it protects the device ?*

1.24 Cooling of Semiconductor Devices

Power is dissipated across the device due to switching and on-state conduction. This power heats the device. Hence its junction temperature increases. It is necessary to keep the junction temperature of the device within the specified limit, otherwise the device is damaged. The heat dissipated across the junction can be taken away with the help of heat sinks. The heat sink provides greater area for heat conduction. The cooling of the heat sink can take place by natural convection, forced air, liquid or vapour phase cooling.

1.24.1 Concept of Thermal Resistance

The heat transfer from the device can be expressed by following equation,

$$P = \frac{\Delta T}{R_\theta} \quad \text{... (1.24.1)}$$

Here P is the power disssipated in device

ΔT is the temperature difference between junction and ambient

R_θ is the thermal resistance to heat flow.

Here note that power/heat is dissipated at the junction of the device. This heat is carried to the case through the encapsulation. The resistance offered to the flow of heat is called thermal resistance.

1.24.2 Thermal Model of a Power Device

Now let us construct thermal model of the power device which is mounted on the heat sink. Fig. 1.24.1 shows such model.

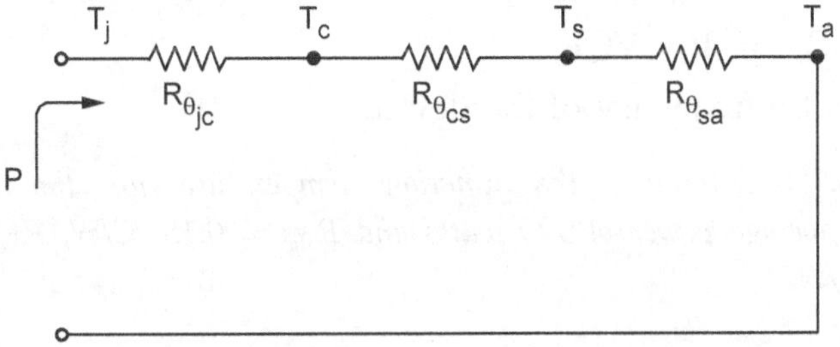

Fig. 1.24.1 Thermal model of a power device

In the above model, 'P' is the dissipated power at the junction. This power is conveyed in the form of heat to the ambient.

T_j is the junction temperature.

T_c is the case temperature.

T_s is the sink temperature.

T_a is the ambient temperature.

$R_{\theta jc}$ is the junction to case thermal resistance.

$R_{\theta cs}$ is the case to sink thermal resistance.

$R_{\theta sa}$ is the sink to ambient thermal resistance.

Here total R_θ will be,

$$R_\theta = R_{\theta jc} + R_{\theta cs} + R_{\theta sa}$$

and

$$\Delta T = T_j - T_a$$

Hence equation 1.24.1 can be written as,

$$P = \frac{T_j - T_a}{R_{\theta jc} + R_{\theta cs} + R_{\theta sa}} \quad \text{... (1.24.2)}$$

The thermal resistance normally depends upon the area of the device or sink. It is related to area by following equation,

$$R_\theta = \frac{1}{kA} \quad \text{... (1.24.3)}$$

Here 'A' is the area of the heat sink and 'k' is the heat transfer coefficient.

For high frequency switching, reverse recovery losses are also added to the conduction loss. The reverse recovery loss in watts/pulse is given as,

$$J = VQ$$

Here 'V' is the reverse voltage applied to the device after turn-off 'Q' is the reverse recovery charge.

Then the reverse recovery average power loss will be given as,

$$P_{rr} = Jf = VQF \quad \text{... (1.24.4)}$$

Here 'f' is the pulse frequency of the device.

Example 1.24.1 : *Determine the junction temperature for the thyristor which is dissipating the average power of 120 watts and $R_{\theta jc}$ = 0.15 °C/W, $R_{\theta cs}$ = 0.075 °C/W and $R_{\theta sa}$ = 0.45 °C/W.*

Solution :

Here P = 120 W

$$R_{\theta jc} = 0.15\ °C/W$$

$$R_{\theta cs} = 0.075\ °C/W$$

$$R_{\theta sa} = 0.45\ °C/W.$$

Assume ambient temperature as $T_a = 35\,°C$ consider equation 1.24.2,

$$P = \frac{T_j - T_a}{R_{\theta jc} + R_{\theta cs} + R_{\theta sa}}$$

$$\therefore \quad T_j = P(R_{\theta jc} + R_{\theta cs} + R_{\theta sa}) + T_a$$

Putting values in above equation,

$$T_j = 120(0.15 + 0.075 + 0.45) + 35$$

$$= 116\,°C$$

Thus the junction temperature of the device will be 116 °C.

Example 1.24.2 : *A power transistor develops a power loss of 3.42 watts and is mounted on a square heat sink. Transistor is linearly derated from 20°C to 200°C at 40°C/watt. Calculate the sink temperature.*

Solution :

Power dissipation, P = 3.42 W

$$T_1 = 200°C$$

$$T_2 = 20°C$$

$$R_\theta = 40°C/\text{watt}$$

$$\therefore \quad P = \frac{T_1 - T_{sink}}{R_\theta}$$

$$\therefore \quad 3.42 = \frac{200 - T_{sink}}{40}$$

$$T_{sink} = 63.2°C$$

Example 1.24.3 : *The maximum junction temperature of the thyristor is 150°C. Thermal resistance for the thyristor-sink combination is 0.015°C/W and 0.08°C/W. Determine the total average power loss in the thyristor-sink combination if the heat sink temperature is 60 °C. If the heat sink temperature is reduced to 50°C by forced air cooling, find the percentage increase in the device rating.*

Solution :

Here $R_{\theta js} = 0.015 + 0.08 = 0.095$ °C/W

$$P = \frac{T_j - T_s}{R_{\theta js}} = \frac{150 - 60}{0.095} = 947.36 \text{ W}$$

With forced air cooling $T_s = 50°C$. Hence power dissipation will be,

$$P = \frac{150 - 50}{0.095} = 1052.63 \text{ W}$$

Thyristor rating is proportional to square root of the average power loss.

Hence,

$$\% \text{ increase in thyristor rating} = \frac{\sqrt{1052.63} - \sqrt{947.36}}{\sqrt{947.36}}$$

$$= 5.4 \%$$

Thus thyristor rating increases approximatly by 54% due to forced air cooling.

➠ **Example 1.24.4 :** *A power device is used in a circuit and has the following data*
Total steady state thermal impedance of 0.3°C/W
Transient thermal impedance of 0.05°C/W for 100 ms
$T_{j(\max)} = 125°C$ *and* $T_a = 40°C$
What power loss the device can withstand following a steady state power loss of 200 W?

Solution : To calculate junction temperature for steady state operation

$$P = \frac{T_j - T_a}{R_{\theta ja}}$$

Here $R_{\theta ja} = 0.3°C/W$, $T_a = 40°C$ and P = 200 W, then above equation will be,

$$200 = \frac{T_j - 40}{0.3}$$

$$\therefore \quad T_j = 100°C$$

Thus in steady state the junction temperature will be 100°C.

To calculate power loss for transient operation

In steady state, the junction temperature is 100°C. The maximum junction temperature that the device can withstand is 125°C. The transient thermal impedance is 0.3°C/W. The the power dissipation will be,

$$P = \frac{T_{j(\max)} - T_j}{R_{\theta(transient)}} = \frac{125 - 100}{0.05} = 500 \text{ W}$$

Thus the transient power dissipation can be increased to 500 W for 100 ms.

➠ **Example 1.24.5 :** *Fig. 1.24.2 shows the switch waveform of the device.*
The duty cycle is 0.5 and switching frequency is 1 kHz. A power of 1000 W is dissipated when the device is ON. No power is dissipated when the device is off. The thermal impedance from junction to case is 0.035°C/W when the device is on, and it is 0.025°C/W when the device is off. Plot the difference temperature between junction and case.

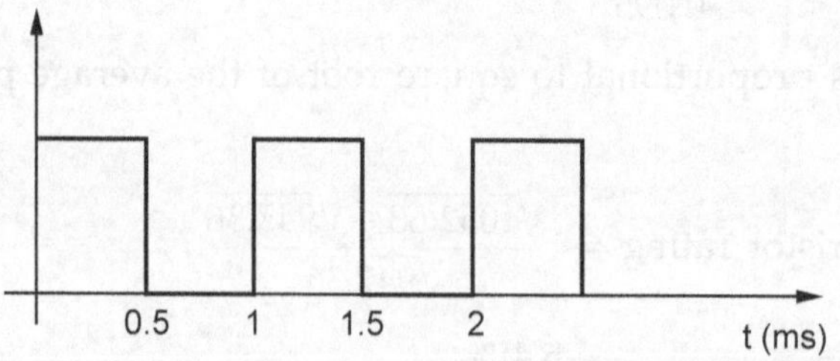

Fig. 1.24.2 Switching waveform of the device

Solution :

Here $$R_{\theta jc} = \begin{cases} 0.035\,°C / W \text{ when device is on} \\ 0.025\,°C / W \text{ when device is off} \end{cases}$$

$$P = \begin{cases} 1000\text{ W when device is on} \\ 0 \quad \text{when the device is off} \end{cases}$$

We know that, $$P = \frac{\Delta T}{R_\theta}$$

At 0.5 msec, $$1000 = \frac{\Delta T}{0.035}$$

$$\Delta T = 35°C$$

For $0.5 \le t \le 1$ msec there is no power dissipation from the device but available heat due to preceding 1000 W is reduced. Hence we can write,

At 1 msec, $$-1000 \times 0.025 + 35°C = \Delta T$$

$$\therefore \quad \Delta T = 10°C$$

Similarly at 1.5 msec,

$$1000 \times 0.035 + 10°C = \Delta T$$

$$\therefore \quad \Delta T = 45°C$$

At 2 msec, $$-1000 \times 0.025 + 45°C = \Delta T$$

$$\therefore \quad \Delta T = 20°C$$

Thus the difference temperature goes on increasing. Fig. 1.24.3 shows the plot of difference temperature.

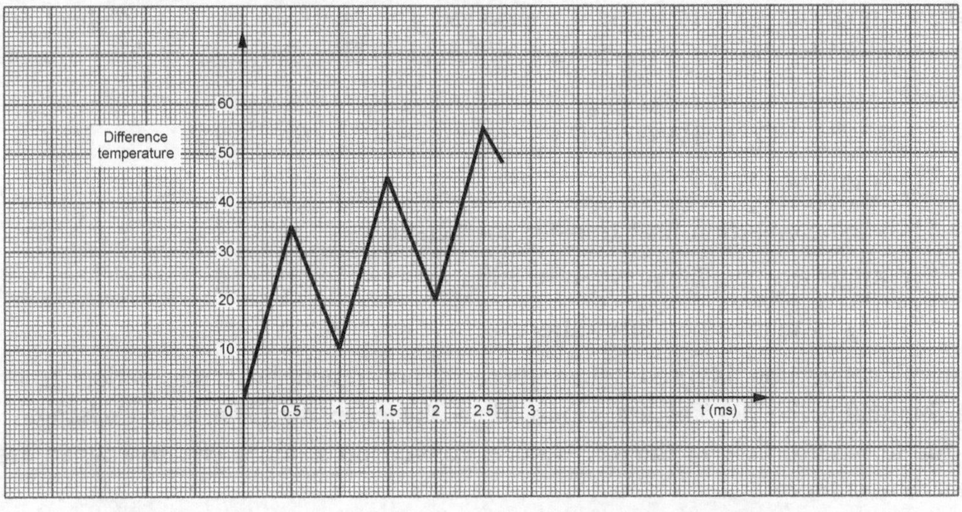

Fig. 1.24.3 Plot of junction to case difference temperature

Review Question

1. *What is the concept of thermal resistance ? Explain its model.*

Unsolved Example

1. *The power device has the steady state thermal resistance of 0.625 °C/W. The case temperature is 100 °C. The device carries a DC current of 25 A and voltage drop of 2 V. Determine its junction temperature.* **[Ans. : T_j = 131.25°C]**

1.25 Types of Cooling

We have seen that the heat is transmitted from device to the heat sink. Heat sink dissipates more heat due to its increased area. Normally heat sinks have fins to increase area. The heat can be dissipated into ambience from the surface of the heat sink by following ways :

i) Natural convection

ii) Forced air cooling

iii) Liquid cooling

iv) Vapour phase cooling

Let us now study these techniques in detail.

1.25.1 Natural Convection

The heat sink surfaces have some treatment to support heat transfer by radiation and natural convection. Copper fins are plated or painted. Aluminium fins are normally painted or anodized. Presently most of the heat sinks used are black anodized. The surface of heat sink under the device is free of paint or anodization. This gives minimum contact thermal resistance. The heat can move from heat sink to ambience naturally due to temperature difference. The equation of heat transfer is given as,

$$Q = hA(T_1 - T_2)$$

Here Q is the heat rate

h is the convection heat transfer coefficient

A is the area available for convection

$T_1 - T_2$ is the temperature difference.

Natural convection is used when there is not much heat dissipation from the device. Normally the devices are mounted in the open air. The heat exchange takes place due to natural flow of the air. Natural convection is more effective when area of the heat sink is more.

1.25.2 Forced Air Cooling

The circulation of air over the fins of heat sink can be increased with the help of fans or compressors. The hot air is taken away from the heat sink fins with the help of cooling fan. Therefore heat convection is fast and cooling of power device is more effective. The heat rate is similar to that of natural convection. But the heat transfer coefficient now depends upon,

$$h = 11.2\sqrt{\frac{v}{l}}\times 10^{-4}\, W / in^2\,°C$$

Here v is free steam linear cooling air velocity across fin surface.

l is length of the fin parallel to air flow.

Above equation shows that heating is more effective if air flow is more. Presently almost all the equipments right from computers to power supplies use forced air cooling. The exhaust fans are mounted on the equipments, then remove hot air. Forced air cooling is always better than natural convection.

1.25.3 Liquid Cooling

The power devices can also be cooled with the help of liquids. Normally oil or water is used as a coolent in liquid cooling. Liquid cooling can be of two types :

1) Oil-immersed natural cooling and (2) Forced liquid cooling.

1. **Oil-immersed natural cooling :** In the oil-immersed natural cooling the heat sink assembly is immersed in oil tank. The heat is transferred from the device to the heat sink fins and then to the oil. The heat is the transferred from the oil to tank surface and then to the atmosphere. The tanks also have fins on their surface. Hence the heat transfer from tank to atmosphere is also increased.
2. **Forced liquid cooling :** The liquid flows at some velocity around the heat sinks assembly. The liquid normally flows in a closed loop system. After passing over the heated device and heat sink, the liquid gets hot. Then it is passed through the narrow fins of the tank. These fins are open to the air. Therefore liquid gets cooled through the fins. Then again it is passed over the device. Forced liquid cooling is more effective. Water is not used as a liquid coolent, since it has corrosion problem. Other liquids like ethylene glycol-water 60% by weight, ethylene glycol 100% and mineral oil like mobiltherm 600 are normally used for forced liquid cooling.

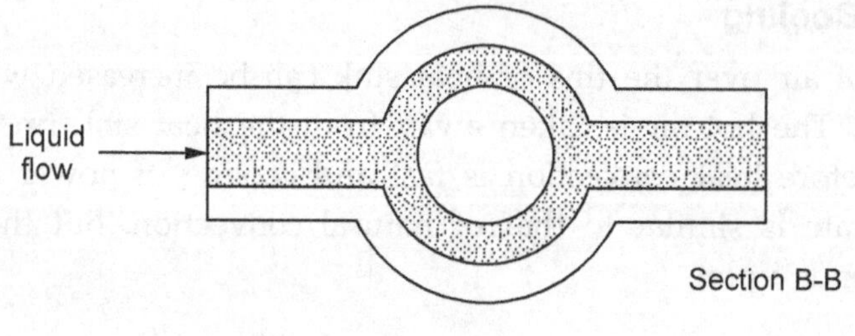

(a)

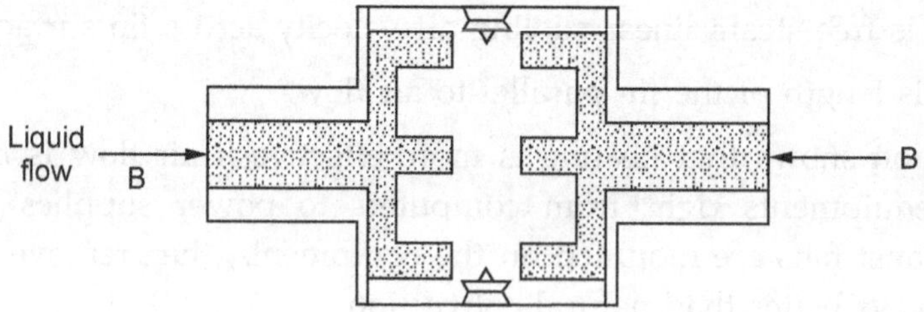

(b)

Fig. 1.25.1 Liquid cooling

Fig. 1.25.1 shows the forced liquid cooling heat exchanger.

Fig. 1.25.1 (a) shows the section of Fig. 1.25.1 (b) when it is cut along B-B. Fig. 1.25.1 (b) shows the vertical cut view of the device and the shaded area shows the passage for liquid. Observe that the liquid surrounds the complete device. It enters from one side and leaves the other side.

Relationship between liquid flow rate and thermal resistance

The thermal resistance of heat sink is inversely proportional to rate of liquid flow. Fig. 1.25.2 shows the relationship.

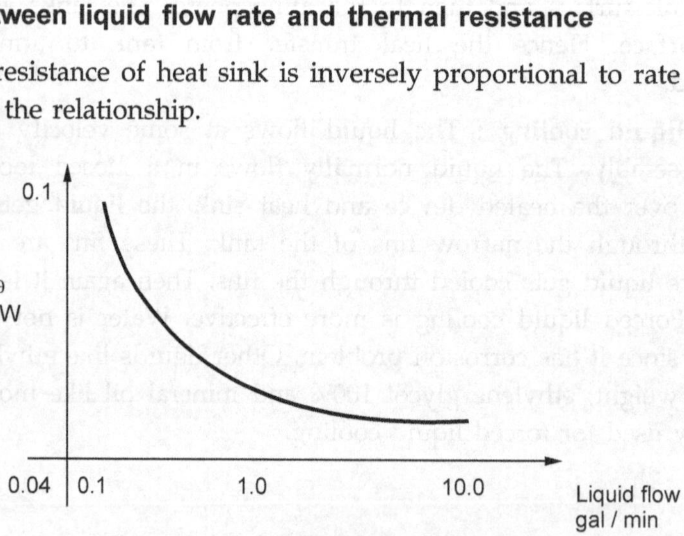

Fig. 1.25.2 Thermal resistance Vs liquid flow

The thermal resistance of the liquid cooled heat sink is much less compared to thermal resistance of copper heat sink.

1.25.4 Vapour Phase Cooling

Vapour phase cooling is normally used for high power devices. The heat produced by the device evaporates the liquid. This vapour then flows to the condenser. The condenser again converts the vapour to liquid and sents it to the device. Fig. 1.25.3 shows an arrangement for vapour-phase cooling.

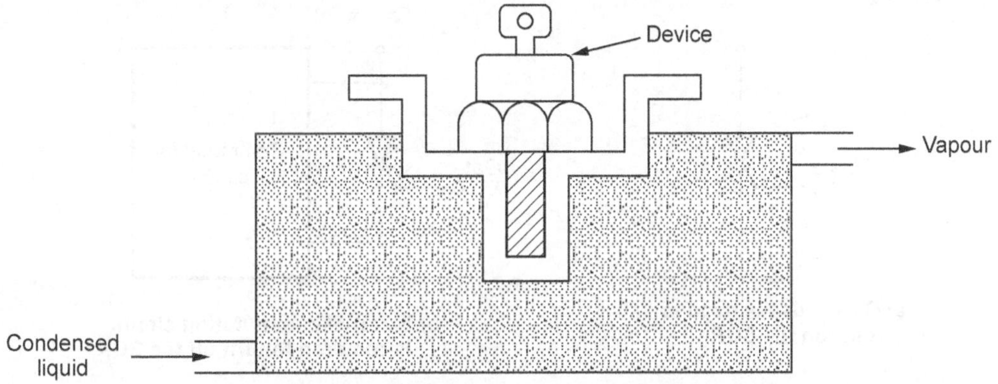

Fig. 1.25.3 Vapour phase cooling

The hermetically sealed copper tubing is used to transport liquid to and vapour from the device. Freon 113 is used as the coolant because it boils at 47°C. It is chemically inert and nontoxic. It has good dielectric constant.

Advantages of vapour phase cooling

i) Heat is exchanged through flow of vapour. Hence overall thermal resistance is reduced.

ii) Heat is uniformly spread or dispersed over large area condensor. Therefore thermal resistance is reduced.

iii) The thermal resistance can be made lower than 0.075 °C/W.

iv) The condenser and device can be placed at distance from each other. Hence equipment cabinet can be more compact.

Disadvantages

i) Additional condenser unit is required to condense the vapour to liquid.

ii) Overall cost of the cooling system is more.

Review Question

1. *Explain various cooling methods for power devices. Compare liquid cooling and vapour phase cooling.*

1.26 Commutation Techniques of Thyristor

1.26.1 Definition of Commutation

The SCR is used as a switch in many power electronic converters. It is turned on by applying a gate pulse. Once the SCR is latched, the gate has no control over the conduction of the SCR. By removing the gate pulse, SCR cannot be turned off. An external circuitry is required to turn-off the SCR. Such process of turning off the SCR is called *commutation*.

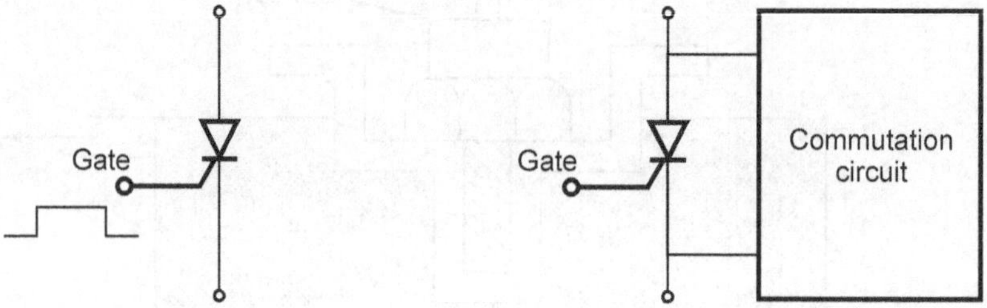

(a) Gate pulse is used to turn on the SCR

(b) External commutation circuit is required to turn off the SCR

Fig. 1.26.1 Commutation

The commutation circuits use LC components. They store energy and it is then used to turn-off the SCR. The commutation can be of two types :

(i) Natural commutation and

(ii) Forced commutation.

In this chapter we will study these commutation techniques in detail.

1.26.2 Natural Commutation (Class F : Line Commutation)

The natural commutation does not need any external components. It uses supply (mains) voltage for turning off the SCR. Hence it is also called as line commutation. Fig. 1.26.2 shows the circuit using natural commutation. It is basically half wave rectifier. The mains AC supply is applied to the input. The SCR is triggered in the positive half cycle at α. Since the SCR is forward biased, it starts conducting and load current i_o starts flowing. The waveforms of currents and voltages are shown in Fig. 1.28.3. Since the load is resistive,

$$i_o = \frac{v_o}{R}$$

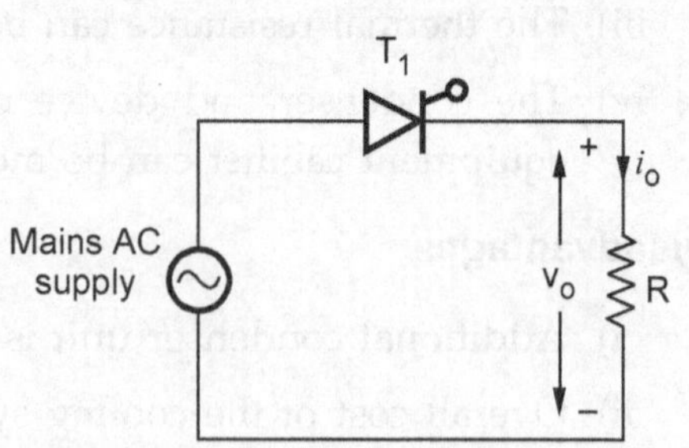

Fig. 1.26.2 A half wave rectifier uses natural commutation to turn

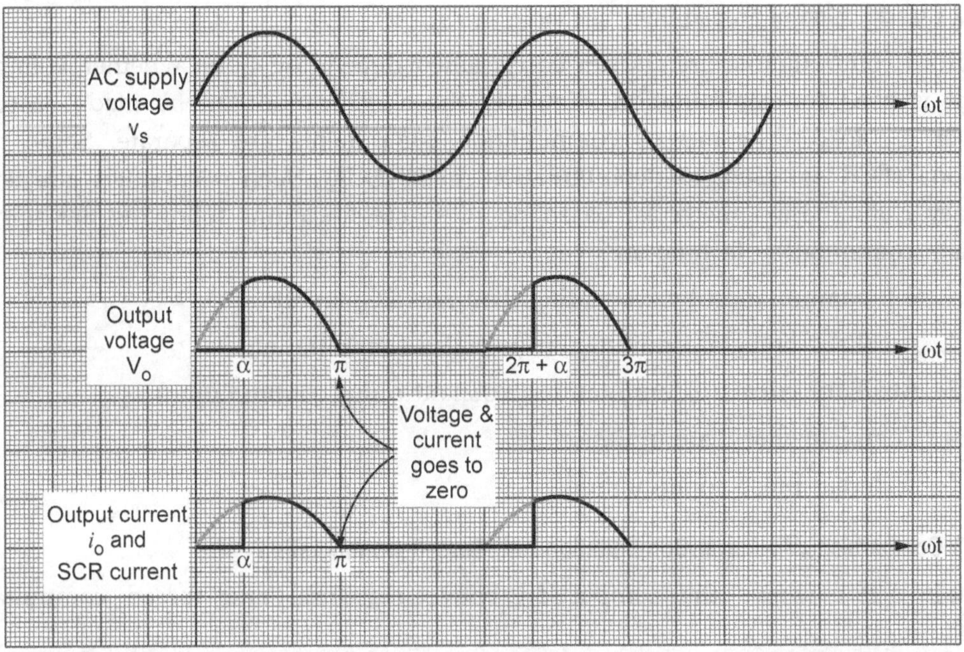

Fig. 1.26.3 Waveforms of half wave controlled rectifier to illustrate natural commutation

Hence the shape of the output current is same as output voltage. Observe that the output current is basically SCR current. At 'π' the supply voltage is zero. Hence current through SCR becomes zero. Therefore the SCR turns off. The supply voltage is then negative. This voltage appears across the SCRs and it does not conduct. Thus natural commutation takes place without any external components. Here note that natural commutation takes place only when the supply voltage is AC. Thus the controlled rectifiers use natural commutation.

1.26.3 Forced Commutation

1.26.3.1 Principle of Forced Commutation

Forced commutation is used when the supply is D.C. A commutation circuit is connected across the SCR as shown in Fig. 1.26.4.

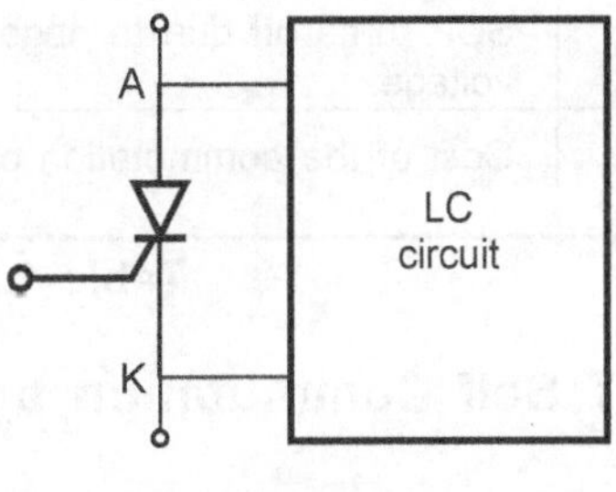

Fig. 1.26.4 Principle of forced commutation

The commutation circuit is normally LC circuit. The LC circuit stores energy when the SCR is ON. This energy is used to turn-off the SCR. The LC circuit imposes reverse bias across the SCR due to stored energy. Hence forward current of SCR is dropped below holding current and the SCR turns off.

There are different types of forced commutation circuits depending upon the way they are connected.

1.26.3.2 Classification of Forced Commutation

Forced commutation circuits can be classified depending upon whether voltage or current is used for commutation. Similarly the classification can be made based on whether the load resonates or commutation components are separate. Some times additional SCR is used for commutation main SCR. Such techniques are called auxiliary commutation methods. Based on these classifications following are some of the main commutation techniques :

1. Self commutation by resonating load and LC circuit
2. Auxiliary voltage commutation (impulse commutation)
3. Auxiliary current commutation (resonant pulse commutation)
4. Complementary commutation
5. External pulse commutation.

1.26.3.3 Comparison of Natural Commutation and Forced Commutation

Table 1.26.1 shows the comparison between natural and forced commutation techniques.

Sr. No.	Natural commutation	Forced commutation
1.	No external commutation components are required.	External commutation components are required.
2.	Requires AC voltage at the input.	Works on DC voltages at the input.
3.	Used in controlled rectifiers, AC voltage controllers etc.	Used in choppers, inverters etc.
4.	No power loss takes place during commutation.	Power loss takes place in commutating components.
5.	SCR turns off due to negative supply voltage.	SCR can be turned off due to voltage and current both.
6.	Cost of the commutation circuit is nil.	Cost of the commutation circuit is significant.

Table 1.26.1 Natural and forced commutation

1.27 Self Commutation by Resonating Load (Class A)

Fig. 1.27.1 shows the circuit diagram of self commutation. It uses inductance and capacitance for commutation. The SCR is turned off due to natural characteristics of LC circuit.

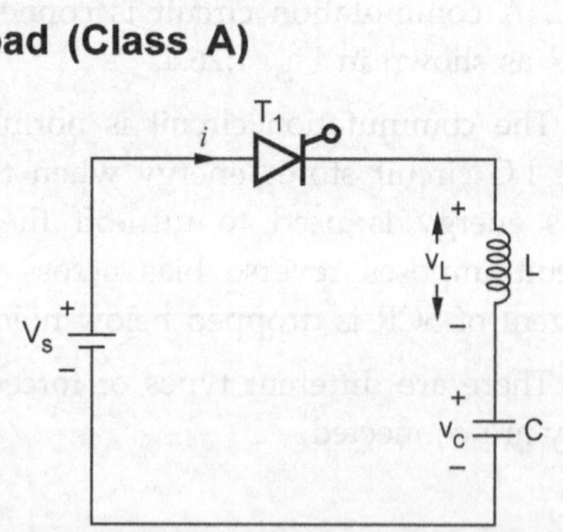

Fig. 1.27.1 Self commutation circuit

1.27.1 Operation

Let us consider that the SCR T_1 is triggered at instant t_0. The current starts flowing through T_1, L and C. Because of inductance, the current through T_1 increases slowly. Fig. 1.27.2 shows the waveforms of the above circuit. The capacitor then starts charging. At $t = t_1$, the capacitor voltage becomes equal to supply voltage V_s. As shown in Fig. 1.27.2, the current $i(t)$ is maximum and $v_c(t) = V_s$. But the current $i(t)$ is maintained in the same direction by inductance. The inductance voltage acts as forward bias for T_1. The current $i(t)$ keeps on decreasing after t_1. Since $i(t)$ flows through capacitor, it charges above V_s after t_1. At $t = t_2$, the current $i(t)$ becomes zero. Since the current through SCR goes to zero, it turns off at $t = t_2$. The capacitor voltage is greater than V_s. It acts as reverse bias on SCR_1. Hence it does not conduct. The voltage across the SCR is shown in Fig. 1.27.2.

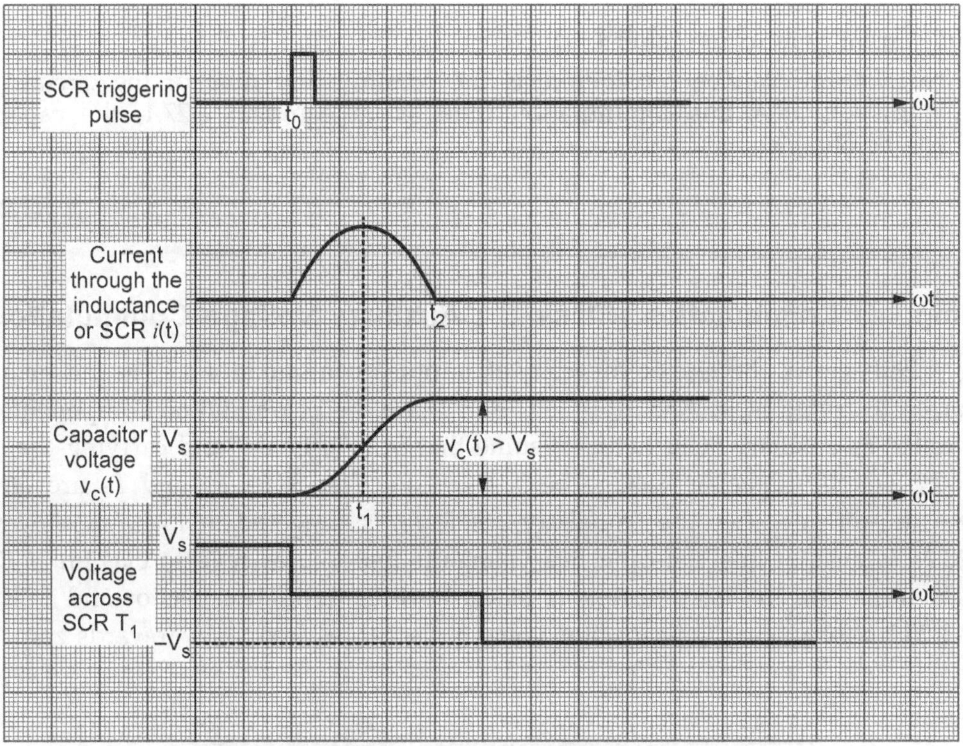

Fig. 1.27.2 Waveforms of self commutation

1.27.2 Mathematical Analysis

When the SCR is triggered, the KVL to the loop gives,

$$V_s = L\frac{di}{dt} + \frac{1}{C}\int i\, dt + v_c\,(t=0)$$

Let us assume that initial voltage on the capacitor is zero. i.e. $v_c\ (t=0)=0$. Then above equation becomes,

$$V_s = L\frac{di}{dt}+\frac{1}{C}\int i\,dt \qquad \text{... (1.27.1)}$$

The solution of above equation for current becomes,

$$i(t) = V_s\sqrt{\frac{C}{L}}\ sin\left(\frac{t}{\sqrt{LC}}\right) \qquad \text{... (1.27.2)}$$

Here note that when $t=\pi\sqrt{LC}$, above equation becomes,

$$i(t) = V_s\sqrt{\frac{C}{L}}\ sin\left(\frac{\pi\sqrt{LC}}{\sqrt{LC}}\right)$$

$$= 0$$

Thus current goes to zero after $t=\pi\sqrt{LC}$.

Hence $t=\pi\sqrt{LC}$ can be considered as the commutation time of the circuit. The SCR T_1 turns off after this commutation time. From equation 1.27.1, the capacitor voltage can be obtained as,

$$v_c(t) = V_s\left[1-cos\left(\frac{t}{\sqrt{LC}}\right)\right] \qquad \text{... (1.27.3)}$$

1.28 Impulse Commutation (Class D : Auxiliary Voltage Commutation)

Fig. 1.28.1 shows the circuit diagram of impulse commutation. T_1 is the main or load current carrying SCR. And T_2, T_3, C and L are the commutation components. T_2 is also called as auxiliary SCR. Let us assume that SCR T_1 is 'on' and carrying the load current I_o. Let the capacitor voltage be $v_c(t)=-V_s$. The SCRs T_2 and T_3 are off. To initiate turn-off of T_1, the auxiliary SCR T_2 is triggered at $t=t_0$ as shown in Fig. 1.28.2. Because of this, the capacitor voltage is imposed as reverse bias on SCR T_1. Hence T_1 turns off. The load current I_o starts flowing through capacitor and SCR T_2. In Fig. 1.28.2 observe that the capacitor voltage discharges from $-V_s$ towards zero.

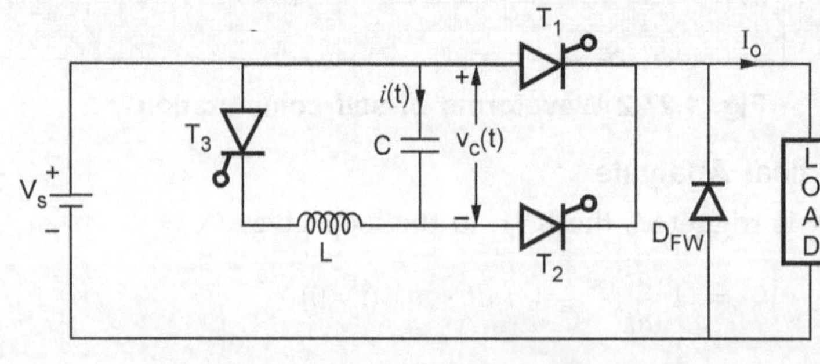

Fig. 1.28.1 Impulse commutation circuit

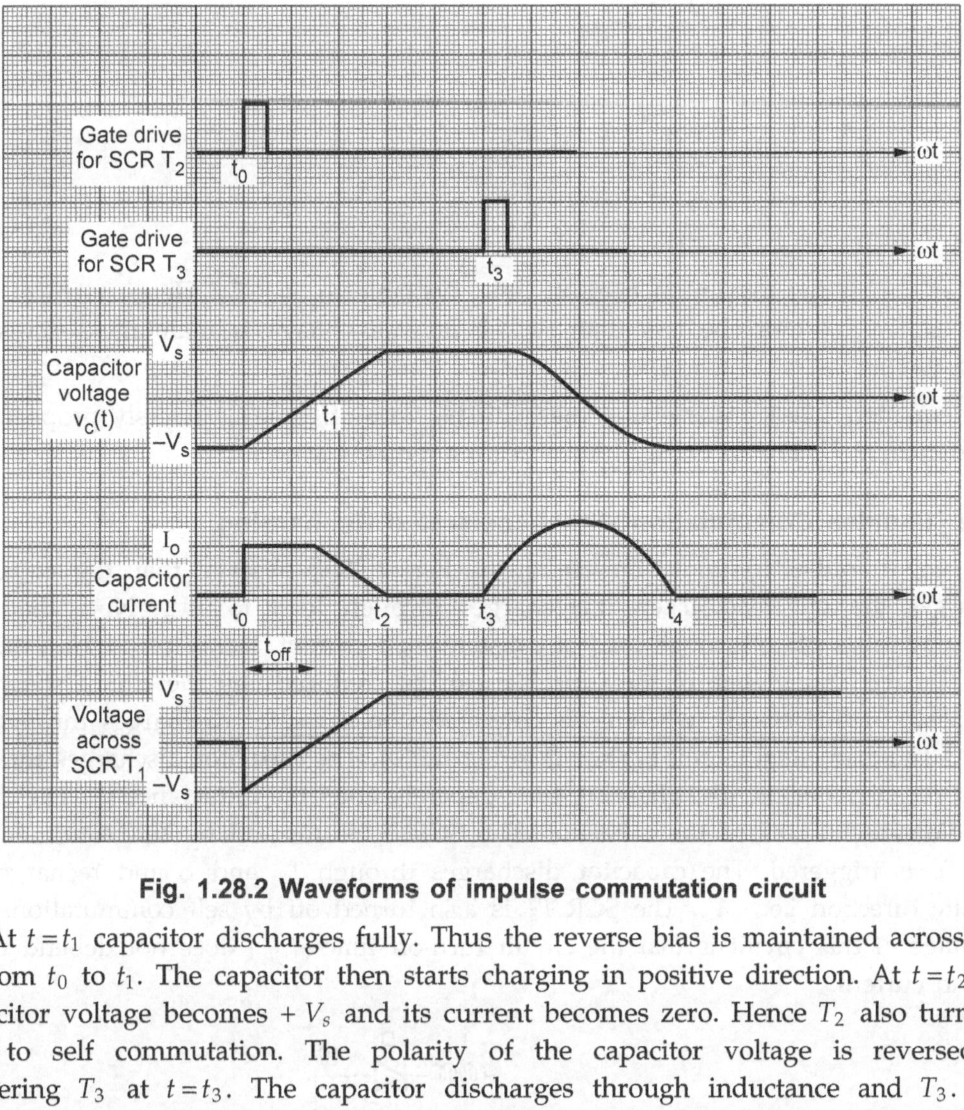

Fig. 1.28.2 Waveforms of impulse commutation circuit

At $t = t_1$ capacitor discharges fully. Thus the reverse bias is maintained across SCR T_1 from t_0 to t_1. The capacitor then starts charging in positive direction. At $t = t_2$, the capacitor voltage becomes + V_s and its current becomes zero. Hence T_2 also turns off due to self commutation. The polarity of the capacitor voltage is reversed by triggering T_3 at $t = t_3$. The capacitor discharges through inductance and T_3. This discharge current (capacitor current) is shown in Fig. 1.28.2. It becomes LC circuit. At $t = t_4$ capacitor voltage becomes – V_s and its current is zero. Hence T_3 turns off by self-commutation. In this commutation, the SCR T_1 turns off due to negative capacitor voltage. Hence it is also known as *voltage commutation*. Since auxiliary SCR $(T_2$ and $T_3)$ is used to turn-off main SCR, it is also known as *auxiliary voltage commutation*. The negative capacitor voltage remains across T_1 from t_0 to t_1. This period is called circuit turn-off time (t_{off}). This time must be greater than turn-off time 't_q' of the SCR T_1.

1.28.1 Mathematical Analysis

The discharge time of the capacitor depends upon the load current I_o. When the capacitor is discharging with a constant load current I_o, we can write,

$$V_s = \frac{1}{C}\int_0^{t_{off}} I_o \, dt$$

$$= \frac{I_o \, t_{off}}{C}$$

$$\therefore \quad t_{off} = \frac{V_s C}{I_o} \qquad \text{... (1.28.1)}$$

Thus 't_{off}' is the maximum turn-off time available to SCR T_1. Hence turn-off time of the SCR (T_q) must be less than t_{off}. A reverse voltage is maintained across T_1 during t_{off}.

In equation 1.28.1 observe that the available turn-off time is inversly proportional to load current I_o. For constant I_o, turnoff time remains constant.

1.28.2 Impulse Commutation with Accelerated Recharging

In the impulse commutation circuit of Fig. 1.28.1, observe that the capacitor discharges through load. Then it recharges through load to $+V_s$. Note that this recharging takes place through load at constant load current. Hence recharging time is long. The recharging time of the capacitor can be reduced by the circuit shown in Fig. 1.28.3. In this circuit, when T_2 is triggered, the capacitor discharges through D_1 and L_1. The equivalent circuit is shown in Fig. 1.28.4. The capacitor discharges through L_1 and recharges again. The charging time depends upon the inductance L_1. The diode D_1 avoids discharging of the capacitor. The SCR T_2 is turn-off by self commutation. Then T_3 is triggered. The capacitor discharges through T_3 and L and recharges in opposite direction i.e. $-V_s$. The SCR T_3 is also turned off by self commutation. The advantage of this circuit is that the circuit turn-off time (t_{off}) does not depend upon the load current.

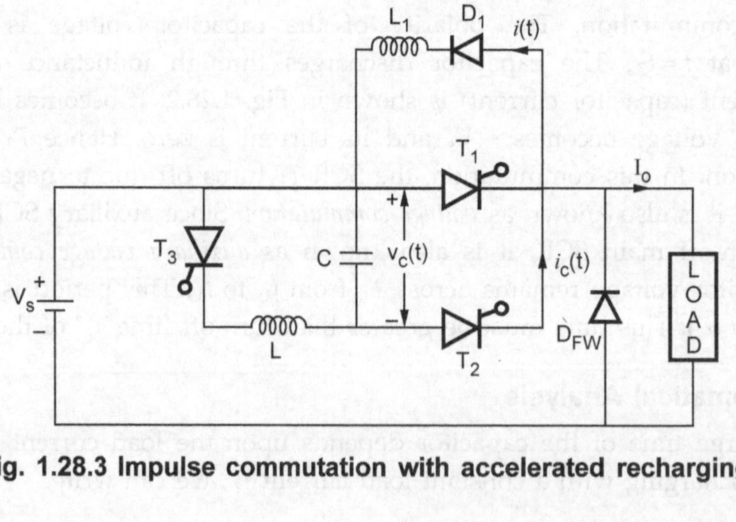

Fig. 1.28.3 Impulse commutation with accelerated recharging

Mathematical analysis

Fig. 1.28.3 shows the impulse commutated circuit with accelerated charging. From this figure we can write,

$$i_c(t) = i(t) + i_o \qquad \text{... (1.28.2)}$$

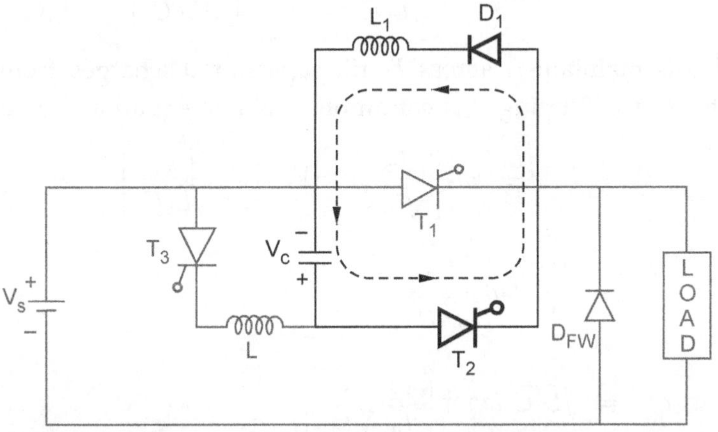

Fig. 1.28.4 Equivalent circuit showing discharging of the capacitor

Consider the loop formed by T_2, D_1, L_1 and C. If we neglect the drops in T_2 and D_1, then KVL to this loop gives,

$$v_c(t) = -L_1 \frac{di(t)}{dt} \qquad \text{... (1.28.3)}$$

From equation 1.28.2 we can write,

$$v_c(t) = -L_1 \frac{d\,i_c(t)}{dt} + \frac{d\,i_o}{dt}$$

Since i_o is constant, its derivative $\frac{d\,i_o}{dt}$ will be zero.

Hence above equation becomes,

$$v_c(t) = -L_1 \frac{d\,i_c(t)}{dt} \qquad \text{... (1.28.4)}$$

When T_2 is triggered, T_1 remains on till it is commutated. During this period T_1 and T_2 both conduct. Hence a loop is formed by T_1, T_2 and C. KVL to this loop gives,

$$v_c(t) = \frac{1}{C}\int i_c\,dt + v_c(t=0) \qquad \text{... (1.28.5)}$$

The initial capacitor current is $i_c(t) = I_o$ and initial capacitor voltage is $v_c(t=0) = -V_s$. Solving equation 1.28.4 and above equation with these initial conditions we get,

$$i_c(t) = V_s \sqrt{\frac{C}{L_1}} \sin\left(\frac{t}{\sqrt{L_1 C}}\right) + I_o \cos\left(\frac{t}{\sqrt{L_1 C}}\right) \quad \text{... (1.28.6)}$$

This expression gives capacitor current. And the voltage across capacitor will be

$$v_c(t) = I_o \sqrt{\frac{L_1}{C}} \sin\left(\frac{t}{\sqrt{L_1 C}}\right) - V_s \cos\left(\frac{t}{\sqrt{L_1 C}}\right) \quad \text{... (1.28.7)}$$

A reverse bias is maintained across T_1 till capacitor discharges from $-V_s$ to zero. Hence $v_c(t) = 0$ for $t = t_{off}$. Putting this condition in above equation we get,

$$0 = I_o \sqrt{\frac{L_1}{C}} \sin\left(\frac{t_{off}}{\sqrt{L_1 C}}\right) - V_s \cos\left(\frac{t_{off}}{\sqrt{L_1 C}}\right)$$

$$\therefore \quad \frac{V_s}{I_o} = \sqrt{\frac{L_1}{C}} \tan \frac{t_{off}}{\sqrt{L_1 C}}$$

$$\therefore \quad t_{off} = \sqrt{L_1 C} \tan^{-1}\left(\frac{V_s}{I_o}\sqrt{\frac{C}{L_1}}\right) \quad \text{... (1.28.8)}$$

1.28.3 Comparison between Self Commutation and Impulse Commutation

Sr.No.	Self Commutation	Impulse Commutation
1.	Commutation takes place due to resonating load.	Commutation takes place due to auxiliary voltage.
2.	Only one SCR is used.	Additional auxiliary SCR is used to turn-off main SCR.
3.	Load current flows through commutating components.	Load current flows through commutating components at the time of commutation.
4.	Commutation time depends on load.	Commutation time is independent of load.
5.	Less number of commutation components are required.	More number of commutation components are required.

Table 1.28.1 Self and impuse commutation

1.29 Resonant Pulse Commutation (Class D : Auxiliary Current Commutation)

Fig. 1.29.1 shows the circuit diagram of resonant pulse commutation. Here T_1 is the main SCR carrying load current. T_2, T_3, L and C are the commutating components. T_2 is the auxiliary SCR. It is used to turn-off T_1.

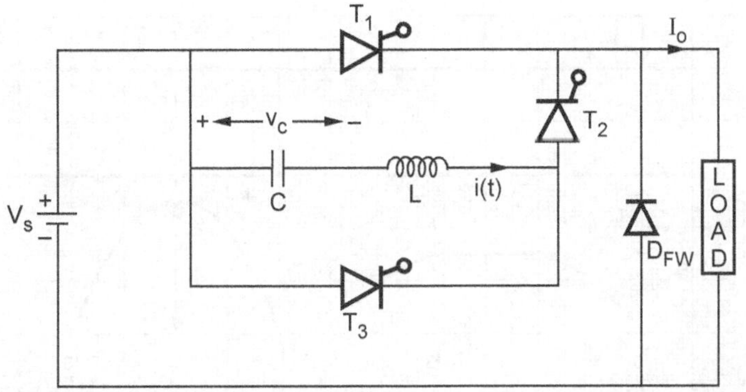

Fig. 1.29.1 Circuit diagram of resonant pulse commutation

1.29.1 Operation

Let us assume that capacitor is charged initially to $v_c(t) = -V_c$. This is shown in first (extreme left) equivalent circuit of Fig. 1.29.2. And SCR T_1 is carrying the load current. At $t = t_0$, SCR T_2 is triggered to initiate the turn-off of T_1. The capacitor discharges from $-V_c$ to $-V_1$. The capacitor discharge current $i(t)$ increases from zero as shown in waveforms of Fig. 1.29.2. At $t = t_1$ the capacitor discharge current reaches to I_0. During period $t_0 - t_1$, the SCRs T_1 and T_2 both carry the load current. This is shown in second equivalent circuit in Fig. 1.29.2. From t_1 to t_3, the load current is completely carried by SCR T_2 and T_1 carries no current. Hence T_1 is turned off. The available turnoff time is from t_1 to t_2. In this period the capacitor voltage is negative. At t_4, the capacitor current becomes zero and T_2 turns off by self commutation. Because of flow of current, capacitor overcharges to V_c. From t_3 to t_4, the energy (charge) stored in the inductance is transferred to capacitor. Hence it overcharges.

At t_5, SCR T_3 is triggered to reverse the capacitor voltage polarity. The capacitor voltage becomes $v_c(t) = -V_c$ and its current falls to zero. T_3 turns off by self commutation. In this circuit note that the resonant pulse of current (in LC circuit) is used to bypass forward current of T_1. Hence this commutation is called as current commutation. Since auxiliary SCR T_2 is used to carry the current of T_1, it is called as *auxiliary current commutation*. The circuit turn-off is $t_{off} = t_2$ to t_3.

1.29.2 Mathematical Analysis

A resonant circuit is formed by L, C, T_2 and T_1. The resonant current in such circuit is given as,

$$i(t) = V_c \sqrt{\frac{C}{L}} \sin\left(\frac{t}{\sqrt{LC}}\right) \quad \text{... (1.29.1)}$$

$$= I_p \sin\left(\frac{t}{\sqrt{LC}}\right) \quad \text{... (1.29.2)}$$

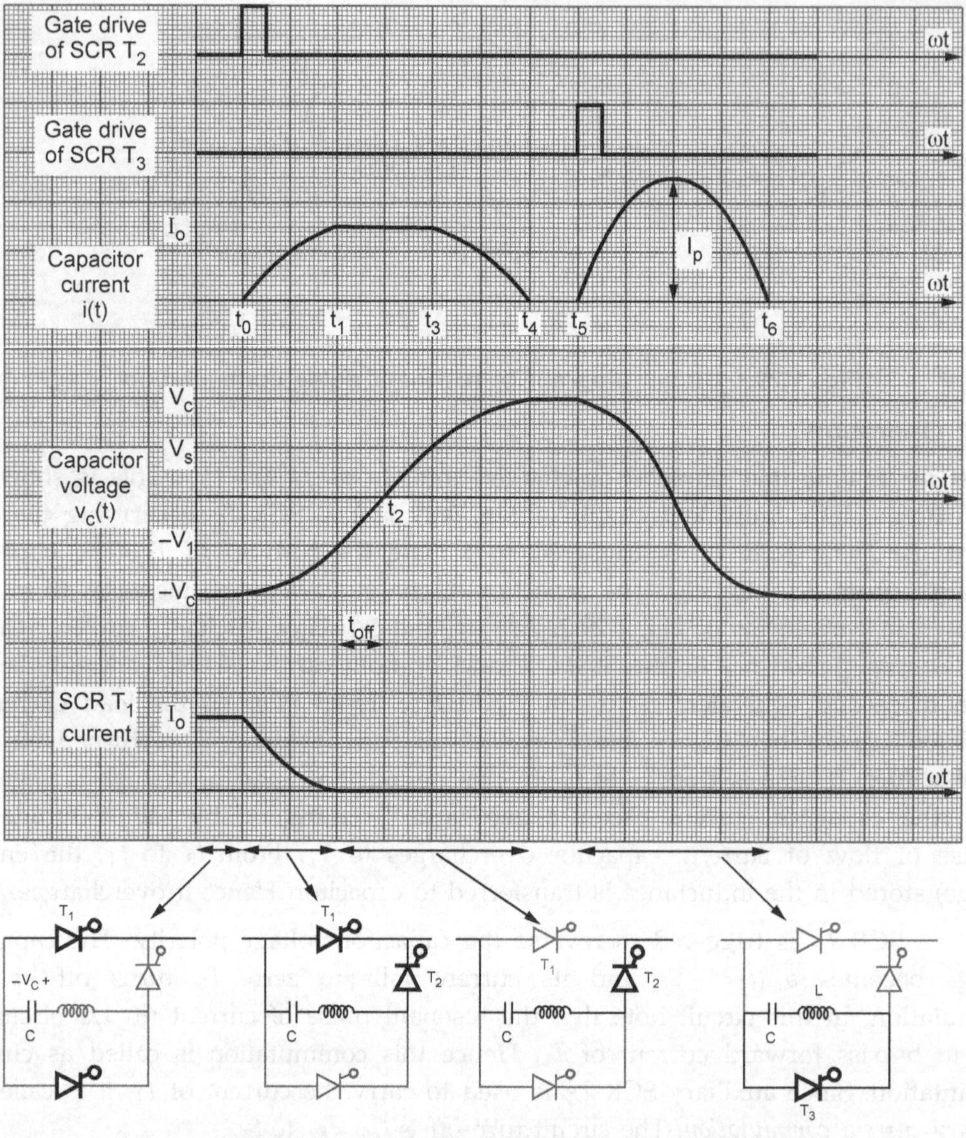

Fig. 1.29.2 Waveforms of resonant pulse commutation

Here I_p is the peak current of the resonant circuit. At $t = t_1$, the capacitor current $i(t) = I_o$, and forward current of T_1 falls to zero. With this condition we can write equation 1.29.1 as,

$$I_o = V_c \sqrt{\frac{C}{L}} \sin\left(\frac{t_1}{\sqrt{LC}}\right)$$

$$\therefore \quad t_1 = \sqrt{LC} \sin^{-1}\left(\frac{I_o}{V_c}\sqrt{\frac{L}{C}}\right) \qquad \text{... (1.29.3)}$$

The capacitor voltage for the LC circuit is given as,

$$v_c(t) = -V_c \cos \frac{t}{\sqrt{LC}} \qquad \text{... (1.29.4)}$$

At $t=t_1$ the forward current of T_1 becomes zero. From t_1 to t_2, the capacitor discharges from $-V_1$ to zero. This is the circuit turn-off time 't_{off}'. The capacitor is discharged at the constant load current I_o. Hence we can write,

$$V_1 = \frac{1}{C} \int_{t_{off}} I_o \, dt$$

$$= \frac{I_o \, t_{off}}{C}$$

or

$$t_{off} = \frac{C V_1}{I_o} \qquad \text{... (1.29.4(a))}$$

From equation 1.29.1 and equation 1.29.2, the peak current of LC circuit will be,

$$I_p = V_c \sqrt{\frac{C}{L}} \qquad \text{... (1.29.5)}$$

The ratio of peak resonance current to load current $\left(\text{i.e. } \frac{I_p}{I_o}\right)$ must be greater than 1. Otherwise T_1 does not turnoff. Hence,

$$\text{For successful commutation} \Rightarrow \frac{V_c}{I_o} \sqrt{\frac{C}{L}} > 1 \qquad \text{... (1.29.6)}$$

1.29.3 Resonant Pulse Commutation with Accelerating Diode

In resonant pulse commutation circuit of Fig. 1.29.1, the capacitor 'C' is charged/discharged with constant load current I_o. Hence the circuit turnoff time 't_{off}' is dependent on the load current (equation 1.29.4). This problem can be avoided by connecting an accelerating diode as shown in Fig. 1.29.3. Let us assume that the capacitor is charged to the voltage $v_c(t) = -V_c$ as shown in Fig. 1.29.3. Then SCR T_2 is triggered at t_0 to initiate the commutation of T_1. Refer Fig. 1.29.4.

The capacitor discharges through $T_2 - D - C - L$. As shown in the Fig. 1.29.4, the capacitor current $i(t) = I_o$ at t_1. From t_1 to t_2 $i(t) > I_o$. This is the resonant current of the LC circuit. Hence $t_1 - t_2$ is the available circuit turnoff 't_{off}' time. At t_2, $i(t) = I_o$ again. The capacitor then charges to + V_s at constant load current (I_o). From t_3 to t_4, the energy stored in the inductance is transferred to capacitor. Hence the capacitor is overcharged to + $V_c > V_s$. The SCR T_2 is turned off by self commutation.

The SCR T_3 is triggered at t_4 to reverse the voltage polarity on the capacitor. T_4 is turned off by self commutation. Due to accelerating diode, the discharge of the

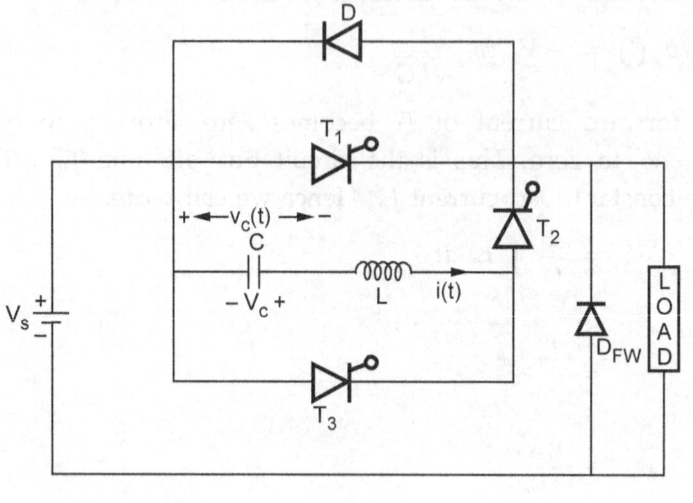

Fig. 1.29.3 Resonant pulse commutation with accelerating diode

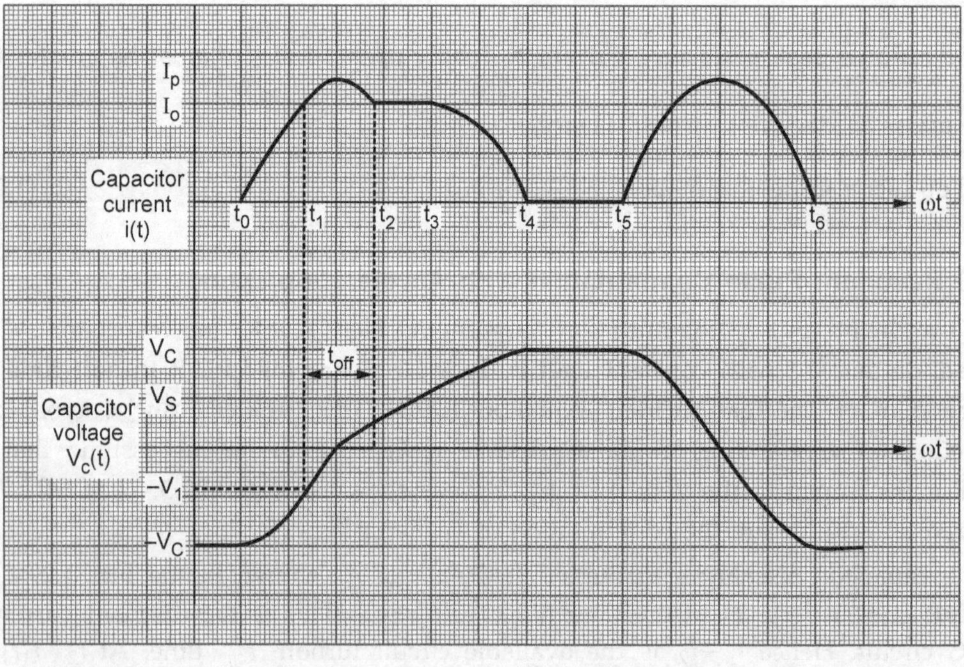

Fig. 1.29.4 Waveforms of resonant pulse commutation with accelerating diode

capacitor is fast. But the diode drop appears across T_1 as reverse bias. This reverse bias is too small and makes the SCR's recovery process slow.

1.30 Complementary Commutation (Class C)

Fig. 1.30.1 shows the circuit diagram of complementary commutation. There are two SCRs T_1 and T_2. These two SCRs are used to transfer the current between two loads R_1 and R_2. These SCRs turn-off each other. Hence it is called complementary commutation. Let us assume that the capacitor is charged to the voltage $v_c(t) = -V_s$ and SCR T_2 is conducting. The equivalent circuit for this operation is shown in Fig. 1.30.2 (circuit-I). Note that the current through capacitor is zero. SCR T_1 is triggered at $t = t_0$. Hence a negative voltage $-V_s$ is applied across T_1. The capacitor then discharges from $-V_s$ to zero. A reverse bias due to capacitor voltage is maintained from t_0 to t_1. This is the circuit turnoff time 't_{off}'. T_2 turns off and capacitor charges to $+V_s$ through SCR T_1. The SCR T_1 also carries load current of R_1. This is shown by equivalent circuit-II in Fig. 1.30.2. At t_2, the capacitor charges to $+V_s$ and its current becomes zero (refer to circuit-III in Fig. 1.30.2).

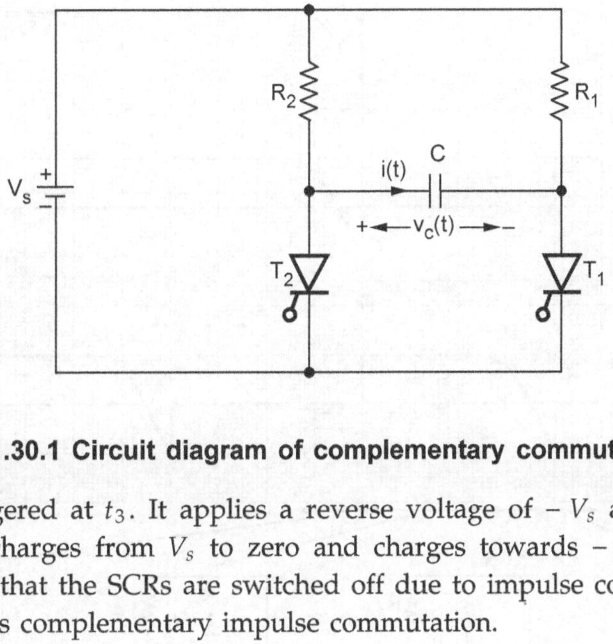

Fig. 1.30.1 Circuit diagram of complementary commutation

SCR T_2 is triggered at t_3. It applies a reverse voltage of $-V_s$ across SCR T_1. The capacitor then discharges from V_s to zero and charges towards $-V_s$. Thus the cycle repeats. Here note that the SCRs are switched off due to impulse commutation. Hence this is also called as complementary impulse commutation.

1.30.1 Mathematical Analysis

Consider the equivalent circuit-II of Fig. 1.30.2 when T_1 is triggered. It is given below in Fig. 1.30.3 for convenience.

Let us neglect the forward voltage drop of SCR T_1, and apply KVL to the loop shown in Fig. 1.30.3 ;

$$V_s = R\, i(t) + \frac{1}{C} \int i(t)\, dt + v_c(t = 0)$$

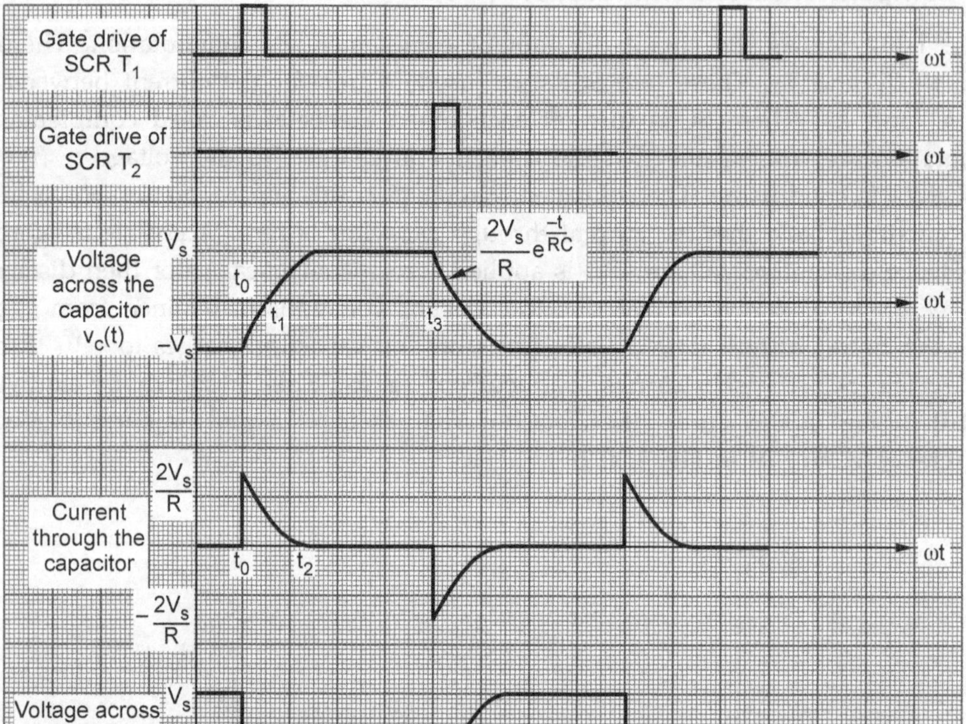

Fig. 1.30.2 Waveforms and operation of complementary commutation

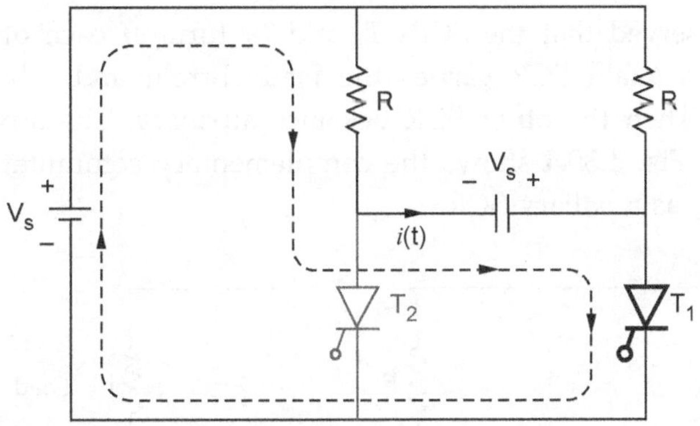

Fig. 1.30.3 Equivalent circuit, when T_1 is triggered

We know that $v_c(t=0) = -V_s$; then above equation becomes,

$$V_s = R\,i(t) + \frac{1}{C}\int i(t)\,dt - V_s$$

Solving the above equation for $i(t)$,

$$i(t) = \frac{2V_s}{R}\,e^{-\frac{t}{RC}} \qquad \text{... (1.30.1)}$$

From Fig. 1.30.3 we can write,

$$V_s = R\,i(t) + v_c(t)$$

Putting for $i(t)$ from equation 1.30.1

$$V_s = 2\,V_s\,e^{-\frac{t}{RC}} + v_c(t)$$

$$\therefore \quad v_c(t) = V_s\left(1 - 2\,e^{-\frac{t}{RC}}\right) \qquad \text{... (1.30.2)}$$

This is the equation for capacitor voltage. In the waveforms of Fig. 1.30.2 a negative capacitor voltage is applied across T_2 from t_0 to t_1. Hence circuit turn-off time (t_{off}) will be obtained from equation 1.30.2 when capacitor voltage becomes zero at t_1. Hence with $t = t_{off}$ and $v_c(t) = 0$ equation 1.30.2 will be,

$$0 = V_s\left(1 - 2\,e^{-\frac{t_{off}}{RC}}\right)$$

$$\therefore \quad t_{off} = 0.693\ \ RC \qquad \text{... (1.30.3)}$$

From the above equation, capacitor can be selected for given t_{off} time.

1.30.2 Modified Complementary Commutation Circuit

Just now we observed that the SCRs T_1 and T_2 turn-off each other. Some times it is required that only main SCR carries the load current and other SCR is used to turn-off main SCR. Then the other SCR becomes auxiliary. The auxiliary SCR should turn-off by its own. Fig. 1.30.4 shows the complementary commutation circuit with T_1 as main SCR and T_2 as auxiliary SCR.

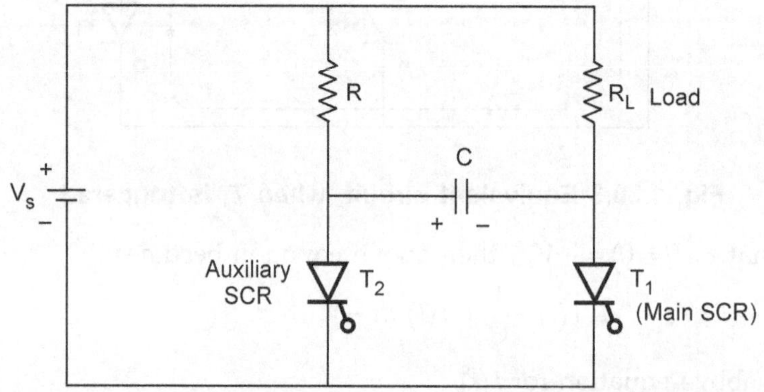

Fig. 1.30.4 Complementary commutation (Modified)

In this circuit, T_1 carries the load current through load R_L. Capacitor C is charged to the voltage polarity as shown. The auxiliary SCR T_2 is triggered to turn-off T_1. When T_2 starts conducting, the negative capacitor voltage appears across T_1 and it turns-off. Then capacitor starts charging through R_L. When capacitor charges fully, no current flows through load. Then current flows through R. The value of 'R' is adjusted such that current through it should be less than holding current of T_2. Hence T_2 turns off automatically. The value of R can be calculated as,

$$\frac{V_s}{R} < I_{H2}$$

or $$R > \frac{V_s}{I_{H2}} \qquad \text{... (1.30.4)}$$

The rest of the analysis is same as the circuit of Fig. 1.30.1.

Example 1.30.1 : *In Fig. 1.30.5, the source voltage $V_s = 100$ V and the current through R_1 and R_2 is 25 A. The turn-off time of both the SCRs is 40 μsec. Find the value of capacitor for successful commutation.*

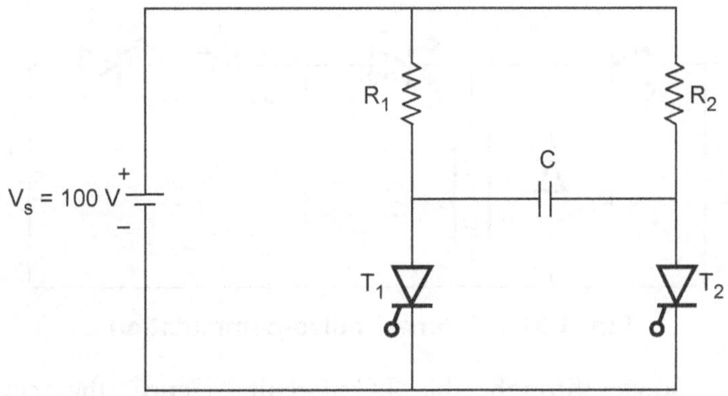

Fig. 1.30.5 Circuit diagram of example 1.30.1

Solution : This is complementary commutation. When T_1 is ON T_2 is OFF and vice versa. The values of R_1 and R_2 will be,

$$R_1 = R_2 = \frac{V_s}{I_o} = \frac{100\ V}{25\ A}$$

$$R = R_1 = R_2 = 4\ \Omega$$

For complementary commutation, the available turn-off time is given by equation 1.30.3 as,

$$t_{off} = 0.693\ \text{RC}$$

If t_q is the turn-off time of the SCRs, then $t_q \le t_{off}$ for successful commutation. Hence above equation can be written as,

$$t_q \le 0.693\ \text{RC}$$

$$\therefore \quad C \ge \frac{t_q}{0.693\ R}$$

The given data is $t_q = 40\ \mu sec$ and $R = 4\ \Omega$. Hence above equation becomes,

$$C \ge \frac{40 \times 10^{-6}}{0.693 \times 4}$$

$$\therefore \quad C \ge 14.43\ \mu F$$

Thus the capacitor must be more than 14.43 μF for successful commutation.

1.31 External Pulse Commutation (Class E)

Fig. 1.31.1 shows the circuit diagram of external pulse commutation. Here T_1 is the main SCR carrying load current. SCRs T_2, T_3 and L, C are commutation components. The supply voltage V_s supplies power to the load. The supply 'V' is the auxiliary supply used to turn-off the SCR. When the SCR T_3 is triggered, the capacitor charges from the auxiliary source. It is LC resonant circuit. A resonant current pulse of peak

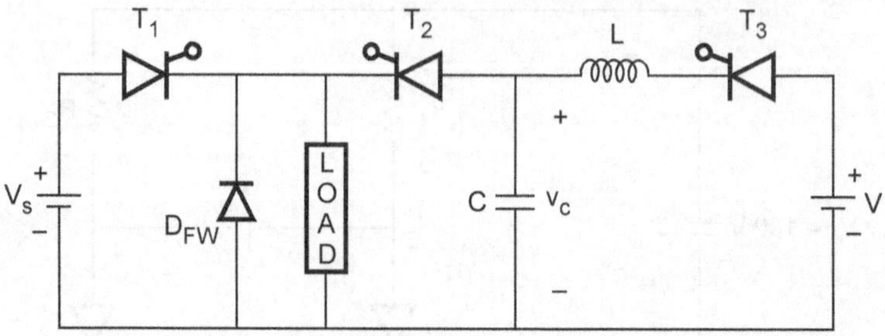

Fig. 1.31.1 External pulse commutation

amplitude $V\sqrt{\frac{C}{L}}$ flows through the LC circuit. Hence the capacitor charges approximately to voltage 2 V. The SCR T_3 turns off by self commutation. The waveforms are shown in Fig. 1.31.2. SCR T_2 is turned on at t_1 to initiate the commutation of T_1. The capacitor voltage is $v_c(t) = 2V$. This voltage is greater than supply voltage V_s. When T_2 is triggered, a reverse voltage of $(V_s - 2V)$ is applied across the SCR T_1. Hence T_1 turns off. The capacitor then discharges through the load at constant load current I_o. As shown in Fig. 1.31.2, a negative voltage is maintained across T_1 from t_1 to t_2, i.e. circuit turnoff time t_{off}. As capacitor voltage falls below V_s, the SCR T_1 is forward biased. Here the a pulse of current from capacitor is used to bypass the current of T_1. The capacitor supplies load current after $t = t_1$. Hence this is called external pulse commutation.

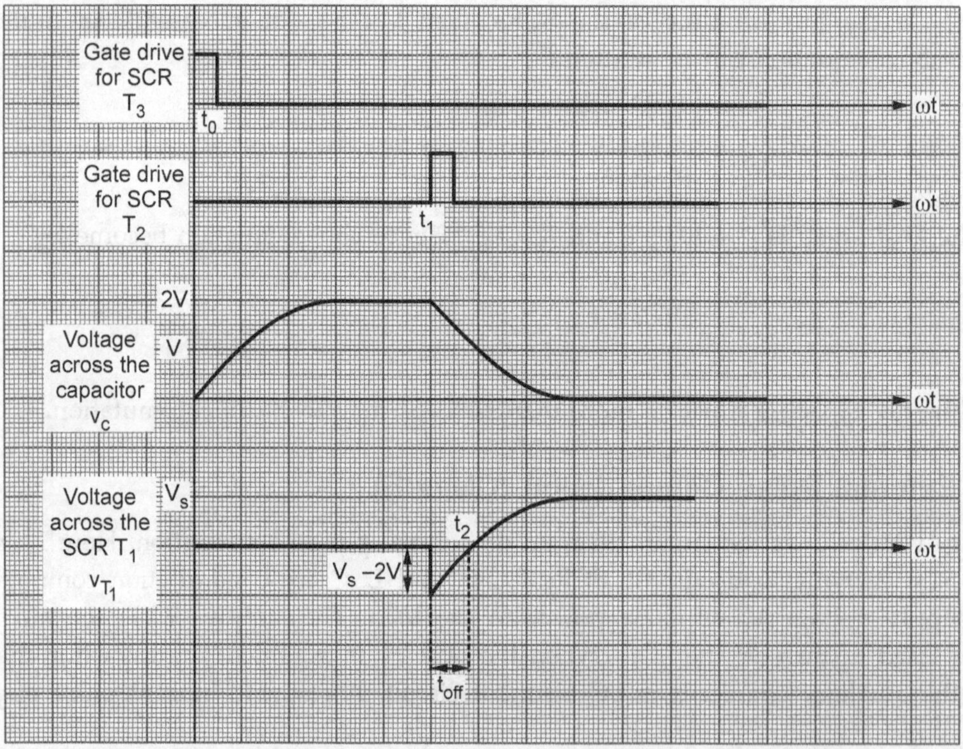

Fig. 1.31.2 Waveforms of external pulse commutation circuit

1.31.1 Mathematical Analysis

In the waveforms of Fig. 1.31.2, a reverse voltage of $(V_s - v_c(t)) = V_s - 2\,V$ is applied across the SCR T_1 at t_1. The capacitor voltage becomes zero at t_2. We know that $t_2 - t_1 = t_{off}$. The capacitor discharges at constant load current I_o.

Hence we can write,

$$V_s - 2V = \frac{1}{C}\int_{t_1}^{t_2} I_o \, dt$$

$$= \frac{I_o\,(t_2 - t_1)}{C} = \frac{I_o\, t_{off}}{C} \qquad \text{... (1.31.1)}$$

$$\therefore \qquad t_{off} = \frac{(V_s - 2V)\,C}{I_o} \qquad \text{... (1.31.2)}$$

Thus turnoff time depends upon the load current in this commutation. Some times the external current pulse can be applied through transformer to provide isolation.

❑❑❑

2 Single and Three-phase Controlled Converters

Objectives

- *Principle of controlled rectification.*
- *Single phase and 3 phase converters.*
- *Half wave and full wave converters.*
- *Bridge converters* → *semiconverter*, → *full bridge converter*
- *Resistive, inductive and motor (RLE) loads on converters.*
- *Continuous and discontinuous output current operation and its effects.*
- *Inverting operation (power flow from load to source) in case of full converters.*
- *Effects of feedback diode and freewheeling operation.*
- *Harmonic analysis of converters.*
- *Effect of source inductance.*

2.1 Introduction

2.1.1 Principle of AC/DC Conversion (Controlled Rectifier)

- Controlled rectifiers are basically AC to DC converters. The power transferred to the load is controlled by controlling triggering angle of the devices. Fig. 2.1.1 shows this operation.

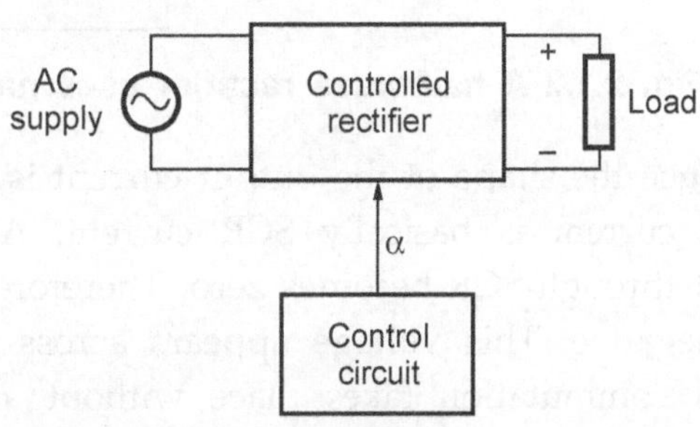

Fig. 2.1.1 Principle of operation of a controlled rectifier

- The triggering angle 'α' of the devices is controlled by the control circuit.
- The input to the controlled rectifier is normally AC mains. The output of the controlled rectifier is adjustable DC voltage. Hence the power transferred across the load is regulated.

Applications :

The controlled rectifiers are used in battery chargers, DC drives, DC power supplies etc. The controlled rectifiers can be single phase or three phase depending upon the load power requirement.

2.1.2 Concept of Line Commutation or Natural Commutation

Principle of line commutation

The natural commutation doesnot need any external components. It uses supply (mains) voltage for turning off the SCR. Hence it is also called as line commutation.

Explanation

Fig. 2.1.2 shows the circuit using natural commutation. It is basically half wave rectifier. The mains AC supply is applied to the input. The SCR is triggered in the positive half cycle at α. Since the SCR is forward biased, it starts conducting and load current i_o starts flowing. The waveforms of currents and voltages are shown in Fig. 2.1.3. Since the load is resistive,

$$i_o = \frac{v_o}{R}$$

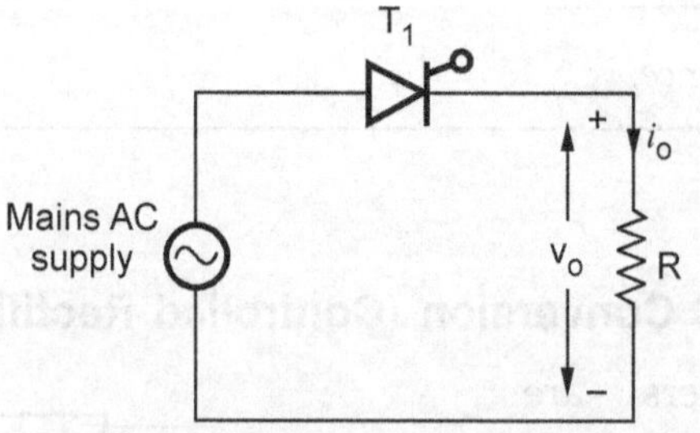

Fig. 2.1.2 A half wave rectifier uses natural commutation to turn off SCR

Hence the shape of the output current is same as output voltage. Observe that the output current is basically SCR current. At 'π' the supply voltage is zero. Hence current through SCR becomes zero. Therefore the SCR turns off. The supply voltage is then negative. This voltage appears across the SCRs and it doesnot conduct. Thus natural commutation takes place without any external components. Here note that natural commutation takes place only when the supply voltage is AC. Thus the controlled rectifiers use natural commutation.

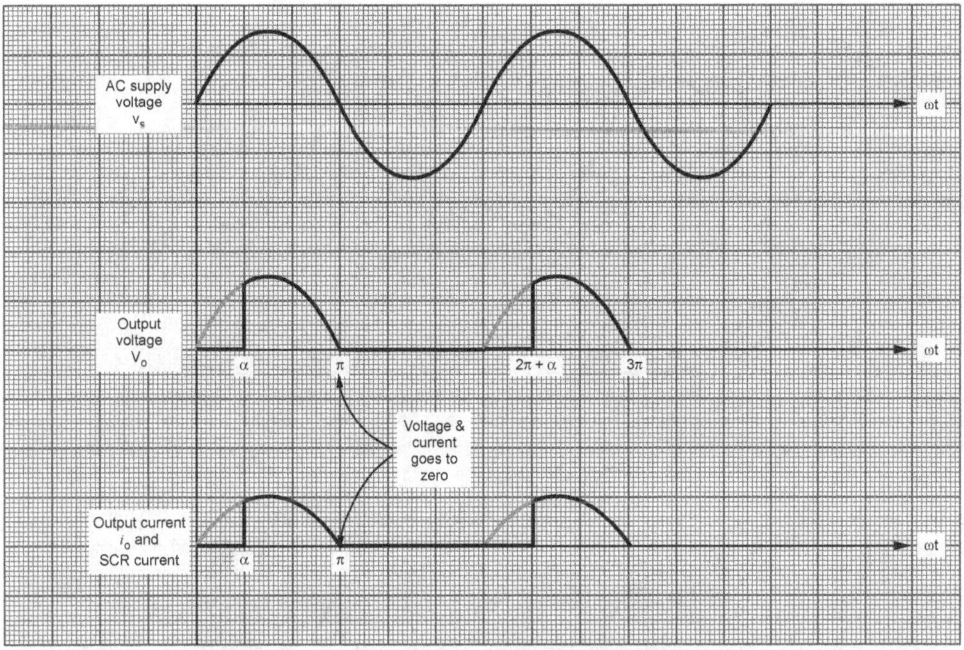

Fig. 2.1.3 Waveforms of half wave controlled rectifier to illustrate natural commutation

Review Questions

1. *Explain the principle of natural commutation. How it is used in controlled rectifiers ?*
2. *Explain the basic principle of phase controlled operation.*

2.2 Single Phase Half Wave Converter

2.2.1 Single Phase Half Wave Controlled Rectifier with Resistive Load

The principle of phase controlled operation can be explained with the help of half wave controlled rectifier shown in Fig. 2.2.1. The secondary of the transformer is connected to resistive load through thyristor or SCR T_1. The primary of the transformer is connected to the mains supply. In the positive cycle of the supply, T_1 is forward biased. T_1 is triggered at an angle α. This is also called as triggering or firing delay angle. T_1 conducts and secondary (i.e. supply) voltage is applied to the load. Current i_o starts flowing through the load. The output current and voltage waveforms are shown in Fig. 2.2.2.

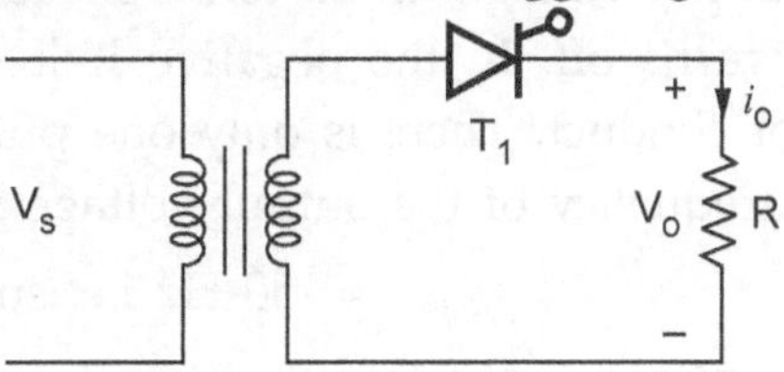

Fig. 2.2.1 Half wave controlled rectifier with R-load.

Since the load is resistive, output current is given as,

$$i_o = \frac{V_o}{R}$$

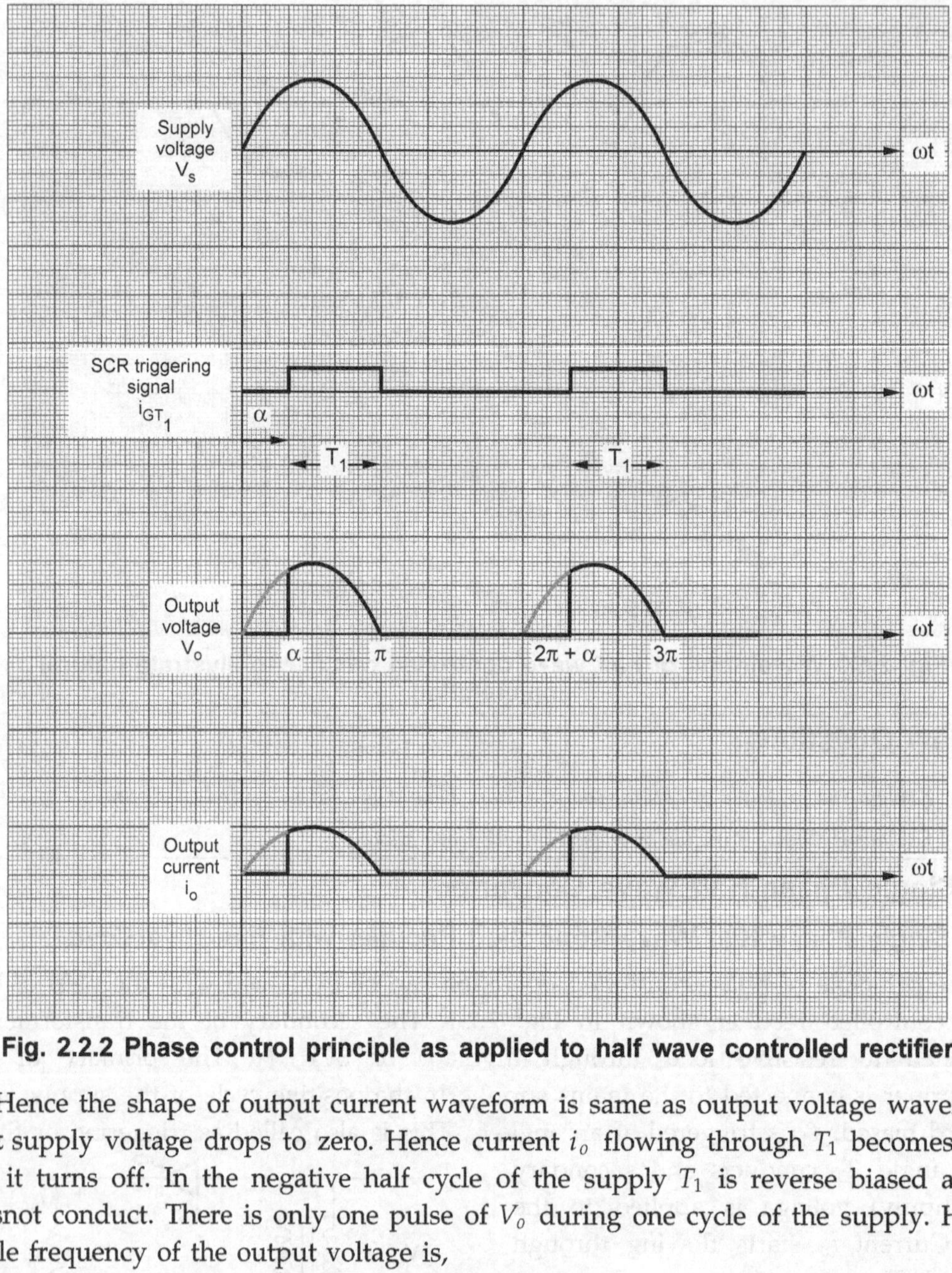

Fig. 2.2.2 Phase control principle as applied to half wave controlled rectifier

Hence the shape of output current waveform is same as output voltage waveform. At π supply voltage drops to zero. Hence current i_o flowing through T_1 becomes zero and it turns off. In the negative half cycle of the supply T_1 is reverse biased and it doesnot conduct. There is only one pulse of V_o during one cycle of the supply. Hence ripple frequency of the output voltage is,

$$f_{ripple} = 50 \text{ Hz i.e. supply frequency}$$

Mathematical analysis

The average value of output voltage is given as,

$$V_{o(av)} = \frac{1}{T}\int_0^T v_o\,(\omega t)\; d\,\omega t$$

The period of one pulse of $v_o(\omega t)$ can be considered as $T = 2\pi$. And $v_o(\omega t) = V_m \sin \omega t$ from α to π. For rest of the period $v_o(\omega t) = 0$. Hence above equation can be written as,

$$V_{o(av)} = \frac{1}{2\pi} \int_{\alpha}^{\pi} V_m \sin\omega t \, d\omega t$$

$$= \frac{V_m}{2\pi} [-\cos \omega t]_{\alpha}^{\pi}$$

$$\therefore \quad \boxed{V_{o(av)} = \frac{V_m}{2\pi} (1 + \cos \alpha)} \quad \text{... (2.2.1)}$$

The power transferred to the load will be,

$$P_{o(av)} = \frac{V_{o(av)}^2}{R}$$

Thus the output average voltage and power delivered by the controlled rectifier can be controlled by phase control (i.e. α). The phase control in converters means to control the delay (or triggering) angle α.

2.2.2 Half Wave Controlled Rectifier with RL Load

Now let us study the operation of single phase half wave controlled rectifier for inductive (RL) load. Normally motors are inductive load. L is the armature or field coil inductance and R is the resistance of these coils. Fig. 2.2.3 shows the circuit diagram of half wave controlled rectifier with RL load.

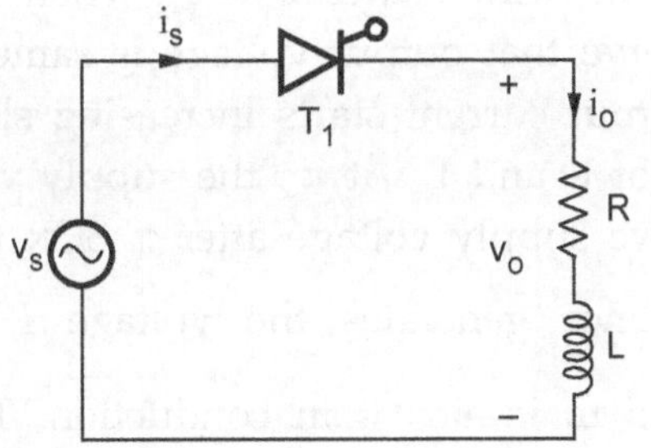

Fig. 2.2.3 Half wave controlled rectifier with RL load

The SCR will be forward biased in the positive half cycle of the supply. Hence SCR is applied with the firing pulses in the positive half cycle. The waveforms are shown in Fig. 2.2.4. Fig. 2.2.4(a) shows the supply voltage and Fig. 2.2.4(b) shows the firing pulses to the SCR.

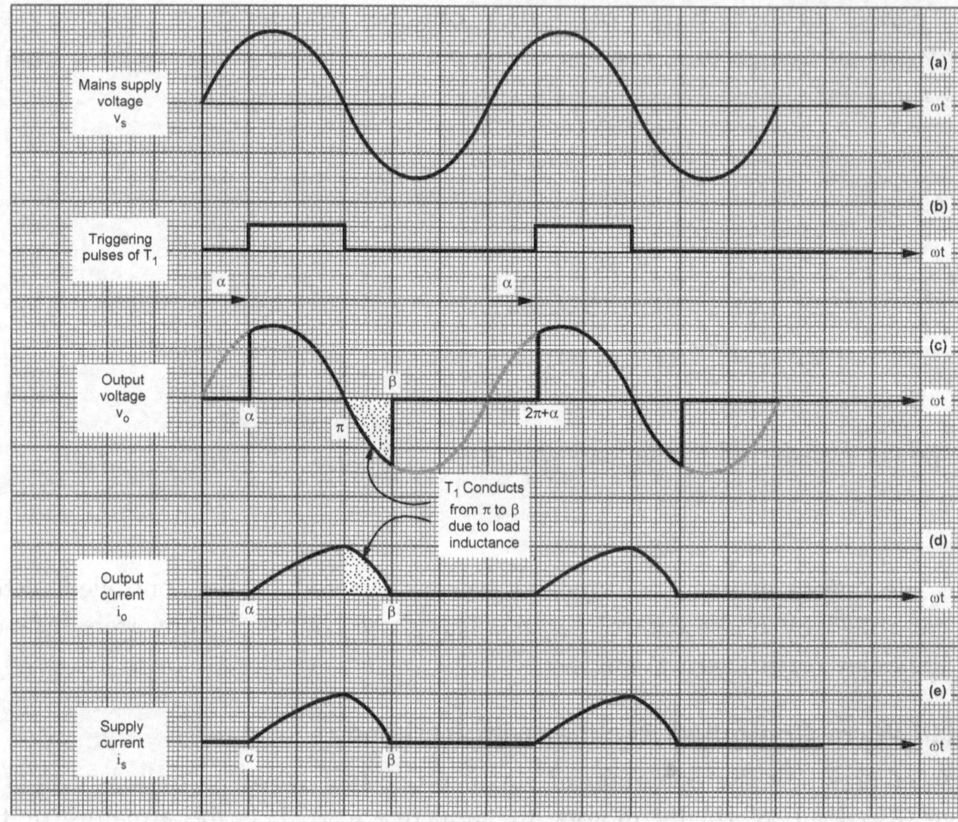

Fig. 2.2.4 Waveforms of half wave controlled rectifier for RL load

When the SCR is triggered, the supply voltage appears across load. We normally neglect small voltage drop in SCR. Hence $v_o = v_s$ when SCR is conducting. This is shown in Fig. 2.2.4(c). Observe that output voltage is same as supply voltage after α. Because of the RL load, output current starts increasing slowly from zero. The shape of i_o depends upon values of R and L. At π , the supply voltage becomes zero and i_o is maximum. Due to negative supply voltage after π, SCR tries to turn-off. But energy stored in the load inductance generates the voltage $L\frac{di_o}{dt}$. This induced voltage forward biases the SCR and maintains it in conduction. This is shown in Fig. 2.2.5. The basic property of inductance is that it opposes change in current. At π, the current i_o is maximum. As SCR tries to turn-off due to negative supply voltage, the output current i_o tries to go to zero. Such change in i_o is opposed by load inductance. Hence the energy stored in an inductance tries to maintain i_o. To maintain the flow of i_o, inductance generates the voltage $L\frac{di_o}{dt}$ with polarity as shown in Fig. 2.2.5. This voltage is higher than negative supply voltage. Hence T_1 is forward biased and it remains in conduction. The output current and supply current flow in the same loop.

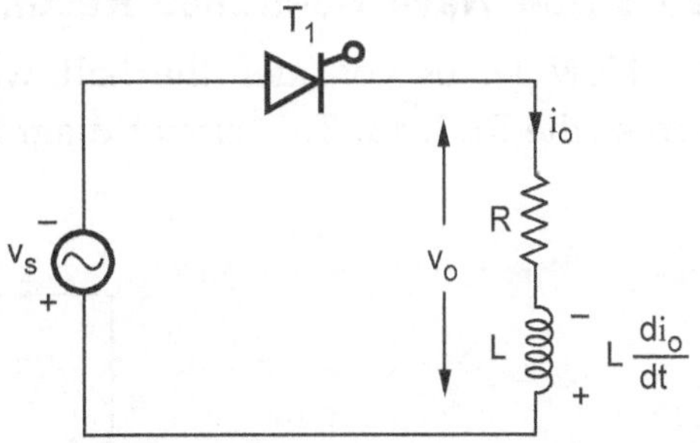

Fig. 2.2.5 SCR conducts due to inductance voltage after π

Hence $i_o = i_s$ all the time. The waveform of i_o is shown in Fig. 2.2.4(d) and i_s is shown in Fig. 2.2.4 (e). After π, i_o (i.e. i_s) flows against the supply. Hence energy is consumed in the supply. i_o flows due to load inductance energy. In other words, the inductance energy is partially fed to the mains and to the load it self. Therefore energy stored in inductance goes on reducing. Hence i_o also goes on reducing as shown in Fig. 2.2.4 (d). At β the energy stored in the inductance is finished. Hence i_o goes to zero. Therefore T_1 turns-off. In Fig. 2.2.4(c) observe that v_o is negative from π to β. Because T_1 conducts from π to β. Hence whenever T_1 conducts $v_o = v_s$.

The SCR is triggered again at $2\pi+\alpha$. Hence output voltage remains zero from β to $2\pi+\alpha$. Output current as well as supply current are also zero from β to $2\pi+\alpha$. At $2\pi+\alpha$, T_1 is triggered again and the cycle repeats. Here i_o goes to zero at β. Hence this is called *discontinuous conduction.*

Example 2.2.1 : *Derive an expression for average value of output voltage for 1 φ half wave controlled rectifier with RL load.*

Solution : For discontinuous conduction, the output voltage waveform is shown in Fig. 2.2.4(c). The output voltage waveform repeats at the period of T = 2π . The average value is given as,

$$V_{o(av)} = \frac{1}{T}\int_0^T v_o(\omega t)\, d\omega t \qquad \text{... (2.2.2)}$$

In Fig. 2.2.4 observe that,

$$v_o(\omega t) = \begin{cases} v_s = V_m \sin \omega t & \text{from } \alpha \text{ to } \beta \\ 0 & \text{from } 0 \text{ to } \alpha \text{ and } \beta \text{ to } 2\pi \end{cases}$$

Hence equation 2.2.2 can be written as,

$$V_{o(av)} = \frac{1}{2\pi}\int_\alpha^\beta V_m \sin \omega t\, d\omega t$$

$$= \frac{V_m}{2\pi}[-\cos \omega t]_\alpha^\beta$$

$$\therefore \quad \boxed{V_{o(av)} = \frac{V_m}{2\pi}(\cos \alpha - \cos \beta)} \qquad \text{... (2.2.3)}$$

This is an expression for average value of output voltage.

2.2.3 Half Wave Controlled Rectifier with Freewheeling Diode

Now let us consider the half wave controlled rectifier with freewheeling diode across the RL load. This circuit diagram is shown in Fig. 2.2.6.

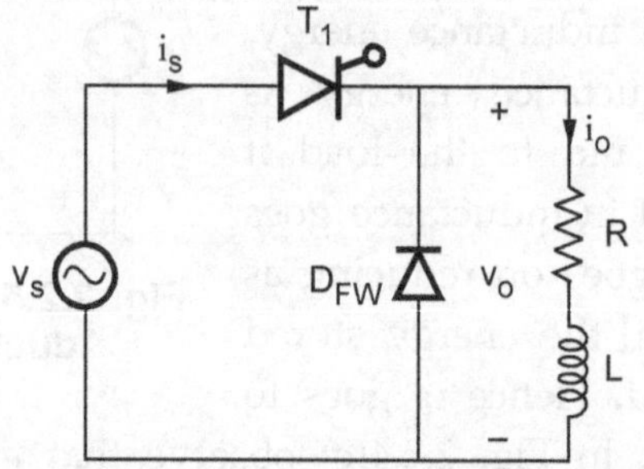

Fig. 2.2.6 Freewheeling diode in half wave controlled rectifier

The SCR is triggered at firing angle of α in positive half cycle of supply. Hence $v_o = v_s$. The waveform of v_o is shown in Fig. 2.2.7(c). Observe that from α to π, v_o is same as supply voltage v_s. The freewheeling diode (D_{FW}) is reverse biased, hence it does not conduct. The output current i_o increases from zero as shown in Fig. 2.2.7(d). This is shown in equivalent circuit-I in Fig. 2.2.7.

Please refer Fig. 2.2.7.

After π, the supply voltage becomes negative. Hence SCR tries to turn-off. Therefore i_o tries to go to zero. Observe that i_o is maximum at π. But the load inductance does not allow i_o to go to zero. The energy stored in inductance generates the voltage $L\frac{di_o}{dt}$ with polarity as shown in Fig. 2.2.8.

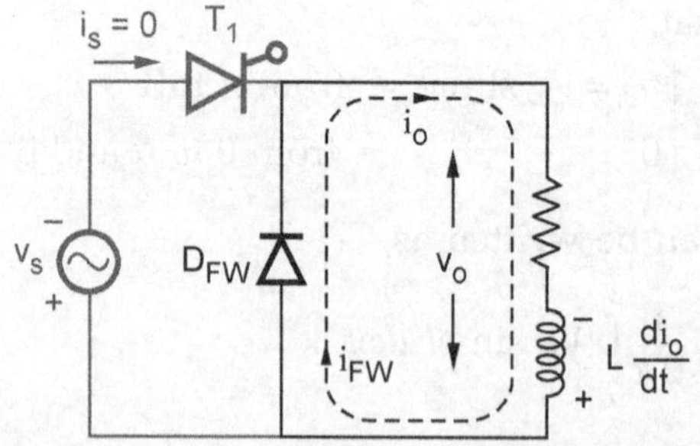

Fig. 2.2.8 Freewheeling action in half wave controlled rectifier

The induced inductance voltage forward biases freewheeling diode as well as SCR. But freewheeling diode (D_{FW}) is more forward biased. Hence it starts conducting. Therefore T_1 turns-off. The output current now flows through the freewheeling diode. In above figure observe that $i_o = i_{FW}$ when freewheeling diode conducts. Here i_{FW} is freewheeling current. Fig. 2.2.7(d) and (e) shown that $i_o = i_{FW}$ when freewheeling

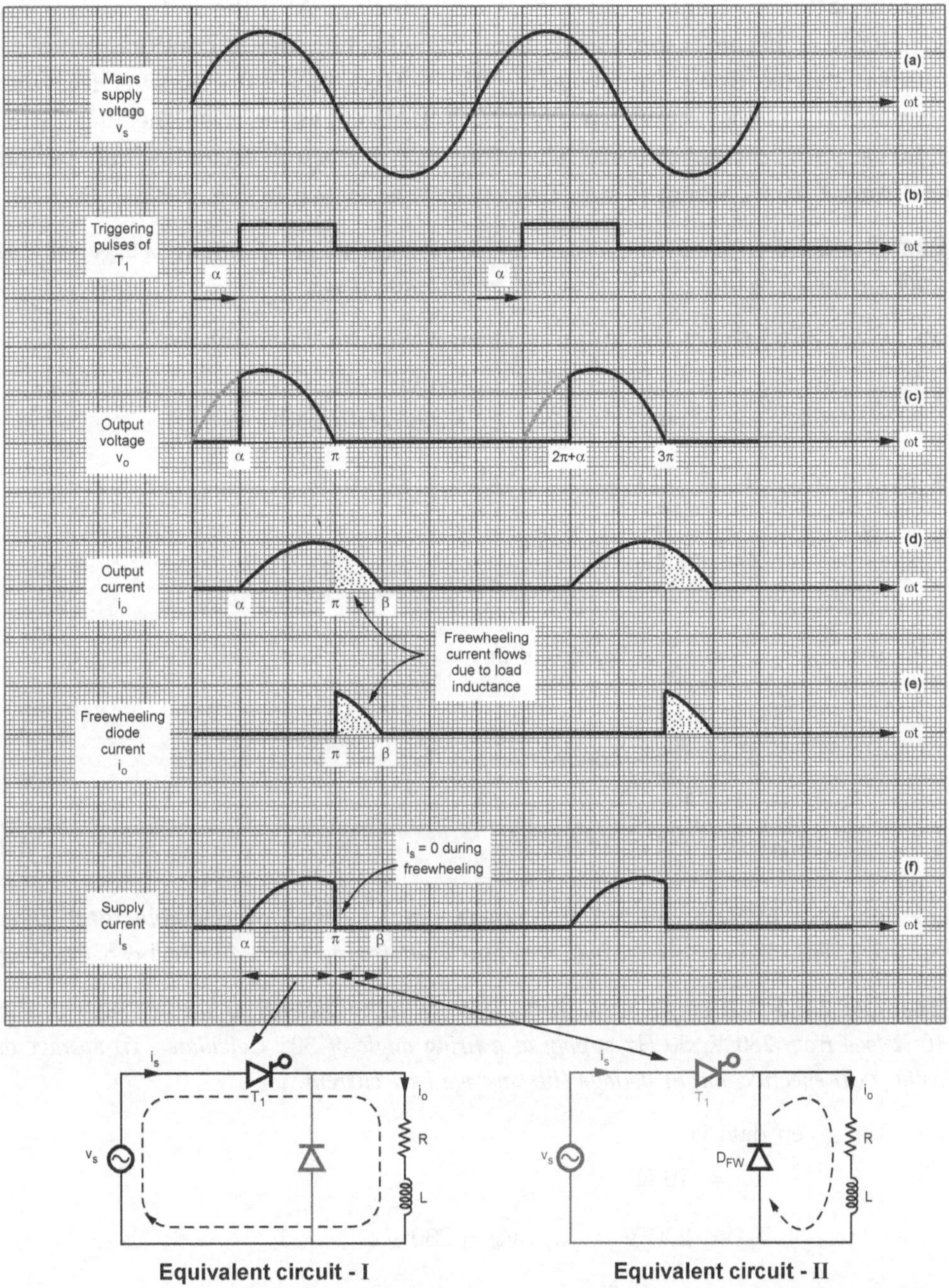

Fig. 2.2.7 Waveforms of half wave converter with freewheeling diode

diode conducts. The freewheeling current flows only due to energy stored in the load inductance. The output current flows in the load itself. Thus inductance energy is supplied back to the load itself. This process is called *freewheeling.* If load energy is fedback to the supply (mains), then it is called *feedback.* The energy of inductance goes on decreasing after π. Hence i_o also goes on reducing. At β the inductance energy is finished. Hence i_o becomes zero at β. Thus freewheeling diode conducts from π to β. The output is shorted due to freewheeling diode. Hence $v_o = 0$ whenever freewheeling diode conducts. This is shown in Fig. 2.2.7(c) also. During freewheeling T_1 is off. Hence no supply current flows. Therefore $i_s = 0$ during freewheeling period. T_1 conducts from α to π. Hence $i_o = i_s$ from α to π as shown in Fig. 2.2.7.

Example 2.2.2 : *Derive an expression for average value of output voltage for 1 ϕ half wave controlled rectifier for RL load and freewheeling diode.*

Solution : Fig. 2.2.7(c) shows the output voltage waveform. From this we can write,

$$v_o = \begin{cases} v_s = V_m \sin \omega t & \text{from } \alpha \text{ to } \pi \\ 0 & \text{from } 0 \text{ to } \alpha \text{ and } \pi \text{ to } 2\pi \end{cases}$$

The period of v_o is 2π. The average value is given as,

$$v_{o(av)} = \frac{1}{T}\int_0^T v_o(\omega t)\, d\omega t$$

$$= \frac{1}{2\pi}\int_\alpha^\pi V_m \sin \omega t\, d\omega t$$

$$= \frac{V_m}{2\pi}[-\cos \omega t]$$

$$\therefore \quad \boxed{v_{o(av)} = \frac{V_m}{2\pi}[1+\cos \alpha]} \quad \ldots(2.2.4)$$

Here note that average output voltage is same as that of resistive load given by equation 2.2.1. This is because output voltage waveforms are same in both the cases.

Example 2.2.3 : *A single phase half wave controlled rectifier is used to supply power to 10 Ω load from 230 V, 50 Hz supply at a firing angle of 30°. Calculate - (i) average output voltage (ii) effective output voltage (iii) average load current.*

Solution : The given data is,

$$R = 10\ \Omega$$

$$V_s = 230\text{ V} \quad \therefore \quad V_m = 230\sqrt{2}$$

$$\alpha = 30° = \frac{\pi}{6}$$

i) To obtain average output voltage $V_{o(av)}$

The load is resistive. For this load $V_{o(av)}$ is given by equation 2.2.1 as

$$V_{o(av)} = \frac{V_m}{2\pi}(1+\cos\alpha)$$

$$= \frac{230\sqrt{2}}{2\pi}\left(1+\cos\frac{\pi}{6}\right)$$

$$= 96.6 \text{ V}$$

ii) To obtain effective output voltage $V_{o(rms)}$

The rms value is given as,

$$V_{o(rms)} = \left[\frac{1}{T}\int_0^T v_o^2(\omega t)\, d\omega t\right]^{\frac{1}{2}}$$

From the output voltage waveform of Fig. 2.2.2 we can write,

$$V_{o(rms)} = \left[\frac{1}{2\pi}\int_\alpha^\pi v_m^2 \sin^2 \omega t\, d\omega t\right]^{\frac{1}{2}}$$

$$= \left[\frac{V_m^2}{2\pi}\int_\alpha^\pi \frac{1-\cos 2\omega t}{2}\, d\omega t\right]^{\frac{1}{2}}$$

$$= \left\{\frac{V_m^2}{4\pi}\left[\int_\alpha^\pi d\omega t - \int_\alpha^\pi \cos 2\omega t\, d\omega t\right]\right\}^{\frac{1}{2}}$$

$$= \left\{\frac{V_m^2}{4\pi}\left[[\omega t]_\alpha^\pi - \left[\frac{\sin 2\omega t}{2}\right]_\alpha^\pi\right]\right\}^{\frac{1}{2}}$$

$$= \frac{V_m}{2}\left[1-\frac{\alpha}{\pi}+\frac{\sin 2\alpha}{2\pi}\right]^{\frac{1}{2}} \quad \text{... (2.2.5)}$$

This is an expression for effective rms value of half wave controlled rectifier. Putting values in above equation,

$$V_{o\,(rms)} = \frac{230\sqrt{2}}{2}\left[1-\frac{\frac{\pi}{6}}{\pi}+\frac{\sin\left(2\frac{\pi}{6}\right)}{2\pi}\right]^{\frac{1}{2}}$$

$$= 160.27 \text{ V}$$

iii) To obtain average load current $I_{o(av)}$

The $I_{o(av)}$ can be calculated as,

$$I_{o(av)} = \frac{V_{o(av)}}{R}$$

$$= \frac{96.6}{10} = 9.66 \text{ A}$$

Example 2.2.4 : *A single phase half wave converter is operated from a 120 V, 50 Hz supply and the load resistance R = 10 Ω. If the average output voltage is 25% of the maximum possible average output voltage calculate -*

(i) Delay angle α

(ii) The rms and average output currents

(iii) The rms and average thyristor currents

(iv) The input power factor.

Solution : Given data

Supply voltage $V_s = 120$; hence $V_m = \sqrt{2} \times 120 = 169.7$ V Load resistance, R = 10 Ω

Average output voltage $V_{o(av)} =$ 25 % of $V_{o(av)}$ maximum

(i) To obtain delay angle α

The average output voltage of half wave converter is given by equation 2.2.1 as,

$$V_{o(av)} = \frac{V_m}{2\pi}(1+\cos\alpha)$$

$V_{o(av)}$ will be maximum at $\alpha = 0$. Hence above equation will be,

$$V_{o(av)max} = \frac{V_m}{2\pi}(1+\cos 0) = \frac{V_m}{\pi}$$

$$= \frac{169.7}{\pi} = 54 \text{ V}$$

It is given that the average output voltage is 25% of its maximum value. i.e.,

$$V_{o(av)} = 25 \text{ \% of } V_{o(av)max}$$

$$= 0.25 \times 54 = 13.5 \text{ V}$$

Consider again equation 2.2.1,

$$V_{o(av)} = \frac{V_m}{2\pi}(1+\cos\alpha)$$

Putting values in above equation,

$$13.5 = \frac{169.7}{2\pi}(1+\cos\alpha)$$

Solving above equation for α,

$$\alpha = 2.09 \text{ radians} = 120^\circ$$

(ii) To obtain rms and average output currents

Average value of output current is given as,

$$I_{o(av)} = \frac{V_{o(av)}}{R}$$

$$= \frac{13.5}{10} = 1.35 \text{ A}$$

The rms value of output voltage is given by equation 2.2.5 for half wave converter. i.e.,

$$V_{o(rms)} = \frac{V_m}{2}\left[1-\frac{\alpha}{\pi}+\frac{\sin 2\alpha}{2\pi}\right]^{\frac{1}{2}}$$

$$= \frac{169.7}{2}\left[1-\frac{2.09}{\pi}+\frac{\sin(2\times 2.09)}{2\pi}\right]^{\frac{1}{2}}$$

$$= 37.718 \text{ V}$$

Hence rms output current will be,

$$I_{o(rms)} = \frac{V_{o(rms)}}{R}$$

$$= \frac{37.718}{10} = 3.77 \text{ A}$$

(iii) RMS and average thyristor currents

The waveforms of half wave converter are given in Fig. 2.2.2. There is only one thyristor and output current flows through this thyristor. Hence thyristor current is same as output current. Therefore rms and average values of thyristor current will be same as these of output current. i.e.,

$$I_{T(av)} = I_{o(av)} = 1.35 \text{ A}$$

and $$I_{T(rms)} = I_{o(rms)} = 3.77 \text{ A}$$

(iv) To obtain input power factor

Since the load is resistive, the rms value of load current will be same as rms value of supply current. Note that the same current flows in supply and load. i.e.,

$$I_{s(rms)} = I_{o(rms)} = 3.77 \text{ A}$$

The total supply power will be,

$$\text{Total supply power} = V_{s(rms)} I_{s(rms)}$$

$$= 120 \times 3.77 = 452.4 \text{ VA}$$

The active load power will be,

$$\text{Active load power} = \frac{V_{o(av)}^2}{R}$$

$$= \frac{(13.5)^2}{10} = 18.225$$

$$\therefore \quad \text{Power factor} = \frac{\text{Active load power}}{\text{Total supply power}}$$

$$= \frac{18.225}{452.4} = 0.04 \text{ (lagging)}$$

2.2.4 Full Wave Mid-Point Converter

The full wave converter rectifies both, positive as well as negative half cycles of the supply. Fig. 2.2.9 shows the circuit diagram of the full wave mid-point converter.

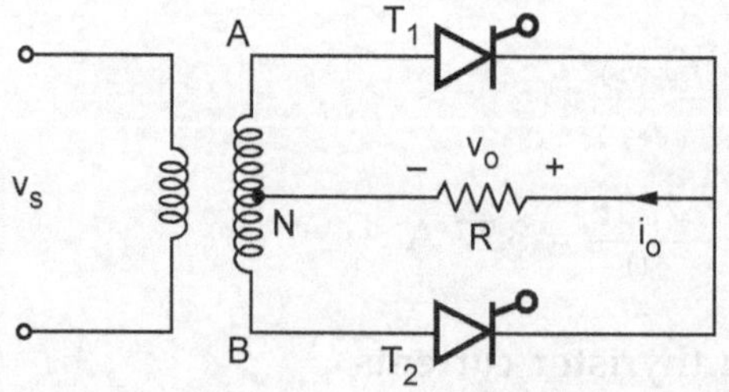

Fig. 2.2.9 Circuit diagram of full wave mid point converter

Operation : In the positive half cycle point A is positive and T_1 is forward biased. Hence it is triggered at α. The supply voltage appears across load as shown in the waveforms of Fig. 2.2.10. At 'π' the SCR T_1 turns-off and T_2 gets forward biased since point 'B' is more positive. At $\pi+\alpha$, SCR T_2 is triggered and it starts conducting. The supply voltage is applied across the load (inverted) an shown in above waveforms.

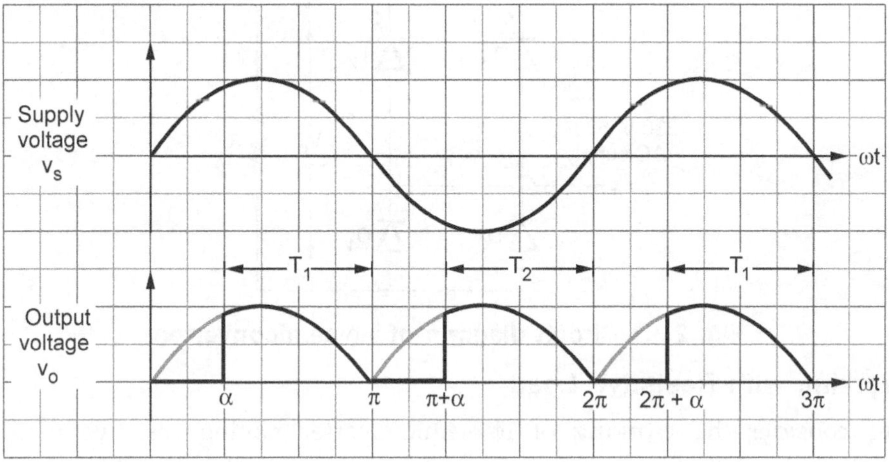

Fig. 2.2.10 Waveforms of full wave midpoint converter

Mathematical analysis

The ripple frequency is f_{ripple} = 100 Hz, since there are two pulses in one cycle of supply. Average output voltage can be obtained as,

$$V_{o(av)} = \frac{1}{T}\int_{o}^{T} v_o(\omega t)\, d\omega t$$

In Fig. 2.2.10 observe that, output voltage waveform repeats at $T = \pi$. Hence,

$$V_{o(av)} = \frac{1}{\pi}\int_{\alpha}^{\pi} V_m \sin\omega t\, d\omega t = \frac{V_m}{\pi}(1+\cos\alpha)$$

Unsolved Example

1. *Explain the operation of 1ϕ half wave convertor with the help of circuit diagram and waveforms.*

2.3 Single Phase Semiconverters (Half Bridge Converter)

The semi converter is also called as half bridge converter.

2.3.1 Circuit Diagram

Fig. 2.3.1 shows the circuit diagram of single phase semiconverter. Observe that the semiconverter has two SCRs T_1 and T_2. There are two diodes D_1 and D_2. The input is 230 V_1 50 Hz AC supply. The output V_o of the semiconverter is DC. The load 'R' is connected across the output.

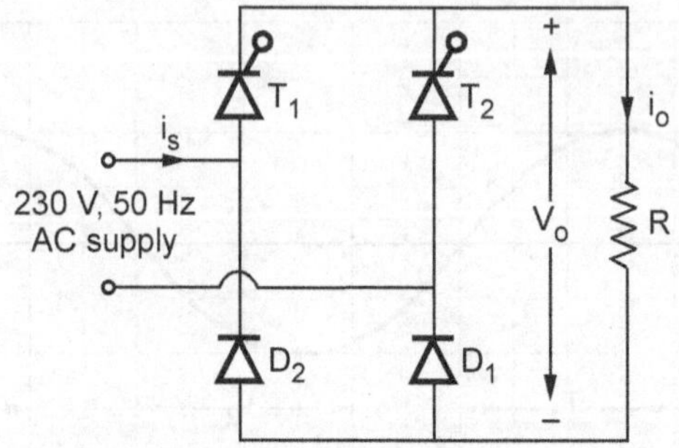

Fig. 2.3.1 Circuit diagram of 1 ϕ semiconverter

2.3.2 Working with Resistive Load

Let us consider the working of 1ϕ semiconverter having resistive load. In the positive half cycle of the supply, SCR T_1 and diode D_2 are forward biased. SCR T_1 is triggered at firing angle α. Current flows through the load. The equivalent circuit is shown below.

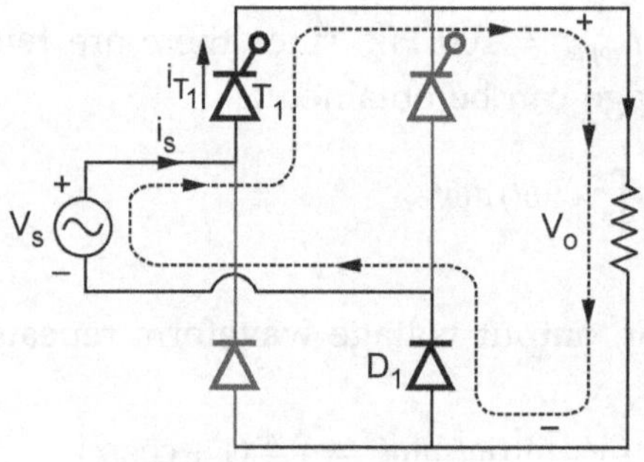

Fig. 2.3.2 Conduction of T_1 and D_1 in positive half cycle of the supply. Dotted line shows path of current flow

From the above figure, it is clear that, when $T_1 - D_1$ conducts,

$$V_o = V_s \quad \text{(i.e. supply voltage)} \qquad \text{... (2.3.1)}$$

and

$$i_o = \frac{V_o}{R} = \frac{V_s}{R} \qquad \text{... (2.3.2)}$$

Fig. 2.3.3 shows the waveforms of this circuit. The waveform of V_o is same as supply voltage V_s, when $T_1 - D_1$ conducts. Since the load is resistive, the output current waveform is same as voltage waveform. This is because,

$$i_o = \frac{V_o}{R}$$

Thus amplitude of V_o is only reduced by the factor 'R' to give i_o. But the shape of the current waveform doesnot change. In the Fig. 2.3.2 observe that i_{T1} is the SCR T_1 current, and i_s is the supply current. Basically i_o, i_{T1} and i_s is the same current. Hence,

$$i_o = i_s = i_{T1} \quad \text{(when } T_1 - D_1 \text{ conducts)}$$

These current are in the same direction and flow in the same loop. The waveforms of these currents are also shown in Fig. 2.3.3.

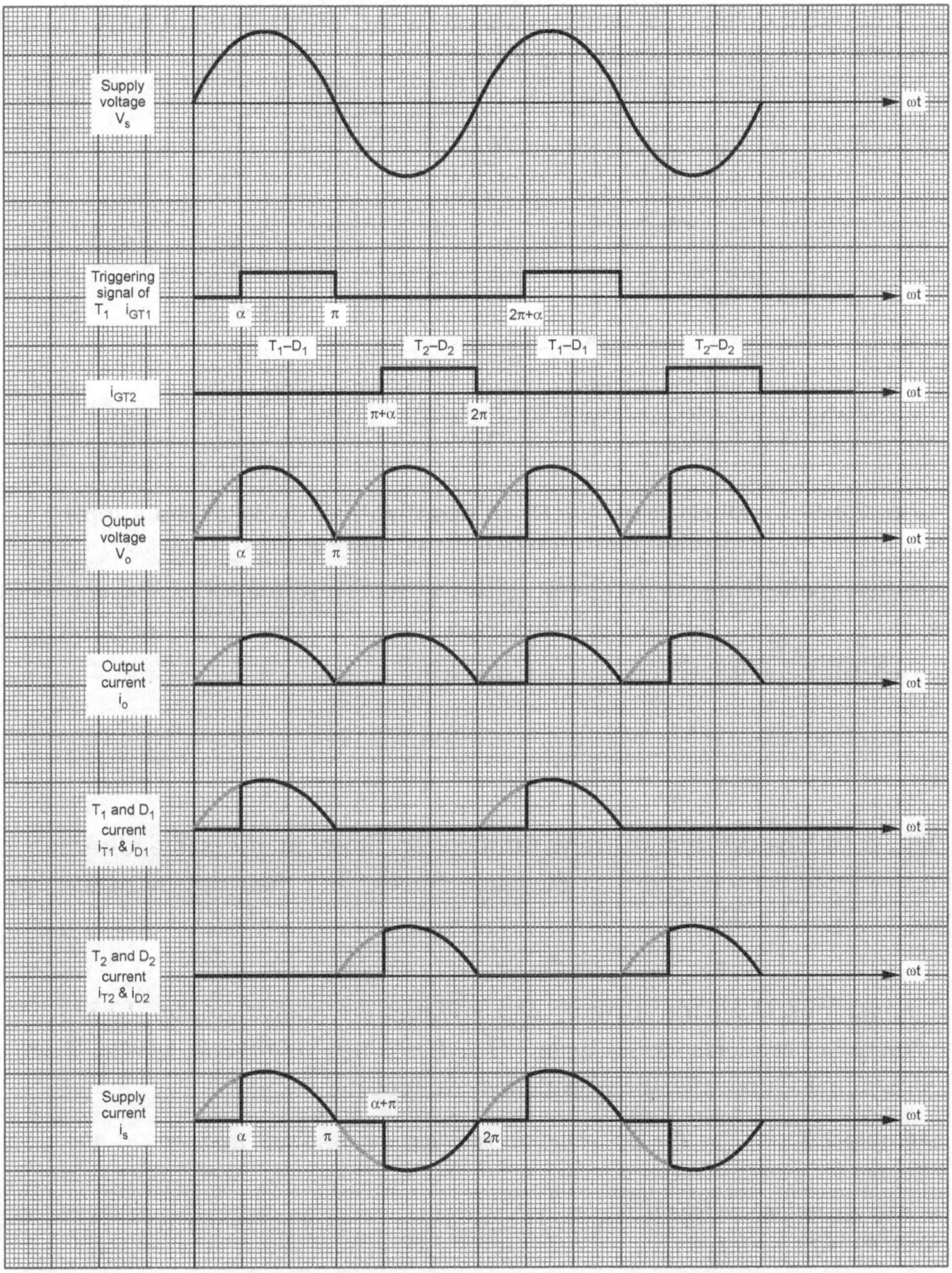

Fig. 2.3.3 Waveforms of semiconverter with R-load

SCR T_1 and diode D_1 conduct till π . At π supply voltage is zero. Hence current through SCR T_1 drops to zero. Hence T_1 turns off. After π, the supply voltage is negative and T_1 is reverse biased. Hence the output voltage V_o is also zero.

At $\pi+\alpha$, SCR T_2 is triggered. It starts conducting, since it is forward biased because of negative cycle of the supply. The current i_o flows through load, T_2 and D_2. Such equivalent circuit is shown in Fig. 2.3.4.

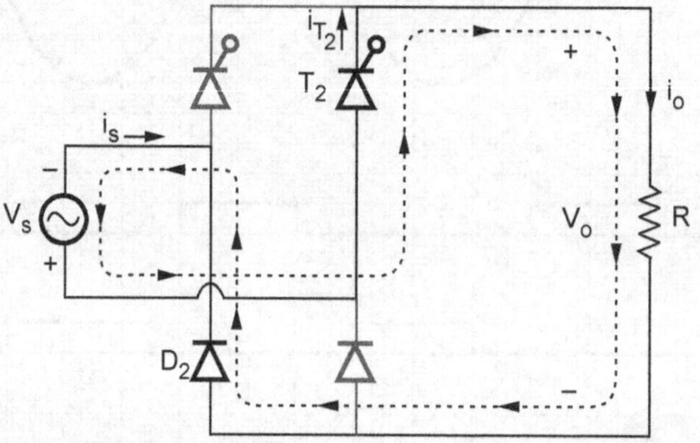

Fig. 2.3.4 Conduction of $T_2 - D_2$ in negative half cycle of the supply. Dotted line shows path of current flow

From the above equivalent circuit observe that positive of V_s is connected to positive of V_o. Hence V_o remains positive even if supply polarity (i.e. negative cycle) is reversed. Hence we can write,

$$V_o = -V_s \quad \text{... (2.3.3)}$$

and

$$i_o = \frac{V_o}{R} = -\frac{V_s}{R} \quad \text{... (2.3.4)}$$

In Fig. 2.3.4 observe that current through T_2 flows in the same direction as i_o. Hence $i_{T2} = i_o$. Similarly i_o and i_s is the same current, but their directions are opposite as shown in Fig. 2.3.4. Hence,

$$i_s = -i_o$$

The waveforms of all the currents and voltages are shown in Fig. 2.3.3. At 2π, the supply voltage is zero. Hence T_2 turns off. After 2π T_2 is reverse biased. Then T_1 is triggered again at $2\pi+\alpha$ and the complete cycle repeats.

Example 2.3.1 : *For the 1φ semiconverter having resistive load of 'R' determine the following :*

(i) Average output voltage $V_{o\,(av)}$

(ii) RMS output voltage $V_{o\,(rms)}$

Solution : (i) Average output voltage :

The average output voltage is given as,

$$V_{o\,(av)} = \frac{1}{T}\int_0^T V_o\,(\omega t)\,d\,\omega t$$

Observe the waveform of output voltage in Fig. 2.3.3. It has a period π. Hence above equation can be written as,

$$V_{o\,(av)} = \frac{1}{\pi}\int_\alpha^\pi V_m\ sin\ \omega t\ d\omega t$$

In the above equation $V_o\,(\omega t) = V_m\ sin\ \omega t$ from α to π. Solving the above integration we get,

$$\boxed{V_{o\,(av)} = \frac{V_m}{\pi}\,(1+cos\ \alpha)} \quad \text{... (2.3.5)}$$

(ii) RMS output voltage :

RMS output voltage is given as,

$$V_{o\,(rms)} = \left[\frac{1}{T}\int_0^T V_0^2\ (\omega t)\ d\omega t\right]^{1/2}$$

Putting the values in above equation,

$$V_{o\,(rms)} = \left[\frac{1}{\pi}\int_\alpha^\pi V_m^2\ sin^2\ \omega t\ d\omega t\right]^{\frac{1}{2}}$$

$$= \left\{\frac{V_m^2}{\pi}\int_\alpha^\pi \left[\frac{1-cos\ (2\omega t)}{2}\right] d\,\omega t\right\}^{\frac{1}{2}}$$

$$\therefore \quad \boxed{V_{o\,(rms)} = \left\{\frac{V_m^2}{2\,\pi}\left[\pi-\alpha+\frac{1}{2}\,sin 2\ \alpha\right]\right\}^{\frac{1}{2}}} \quad \text{... (2.3.6)}$$

This is the required derivation for rms value of output voltage.

2.3.3 Working with Inductive (R-L) Load

Normally the semiconverters are used to drive the DC motors. These motors are basically inductive (R-L) load. Hence it is necessary to consider the working of semiconverter with R-L load also. With the inductive load, the three modes are possible :

(i) Continuous load current

(ii) Discontinuous load current

(iii) Continuous and ripple free current for large inductive load.

2.3.3.1 Continuous Current Mode

In this mode, the current flows continuously in the load because of inductive effect. The waveforms of load current and load voltage are shown in Fig. 2.3.5. In these waveforms observe that SCR T_1 and diode D_1 conducts from α to π. Since the load is inductive current keeps on increasing (saturating) and it is maximum at π. At π, even though the supply voltage is zero, current doesnot go to zero. This is because

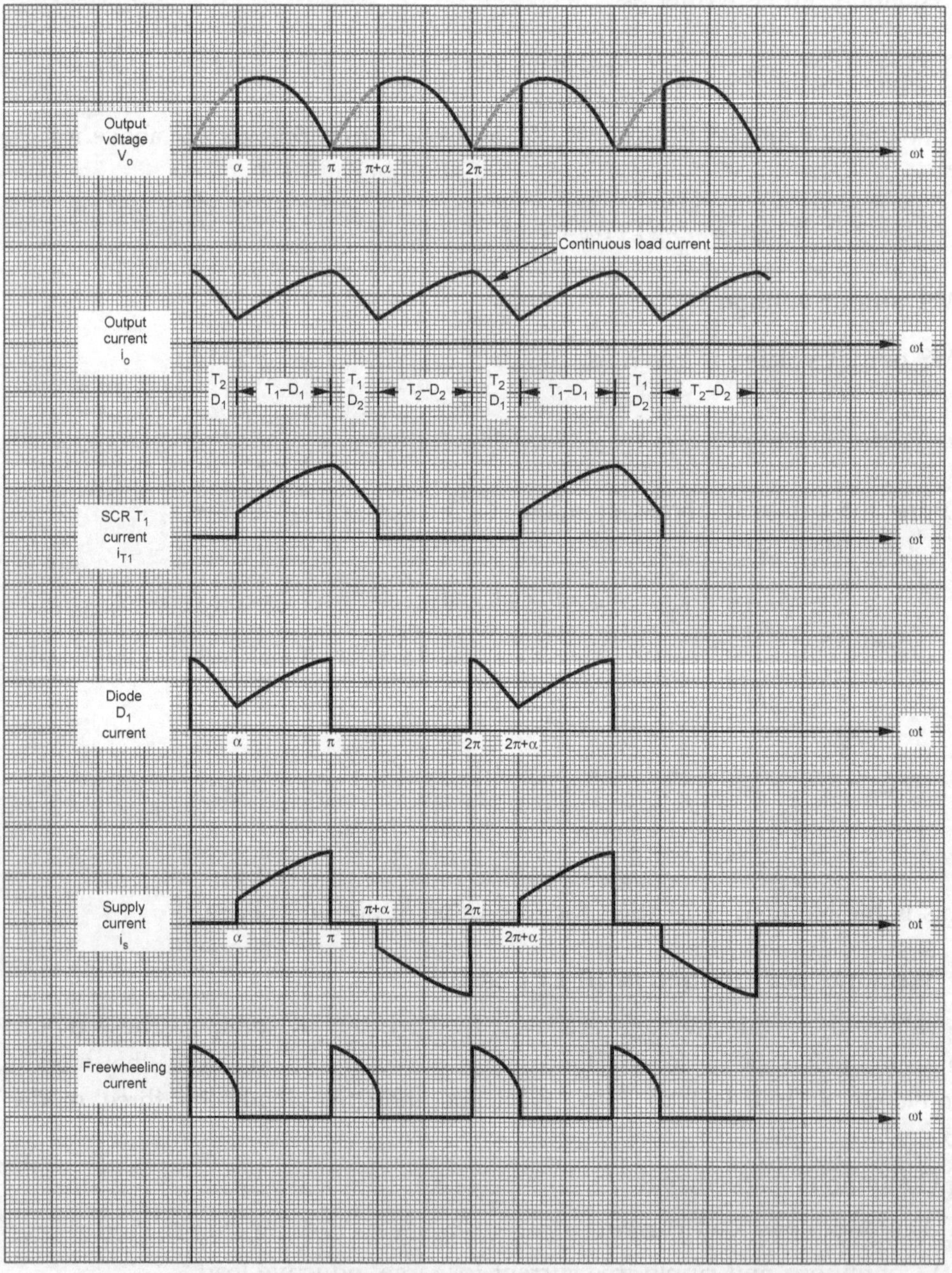

Fig. 2.3.5 Waveforms of 1ϕ semiconverter for continuous load current

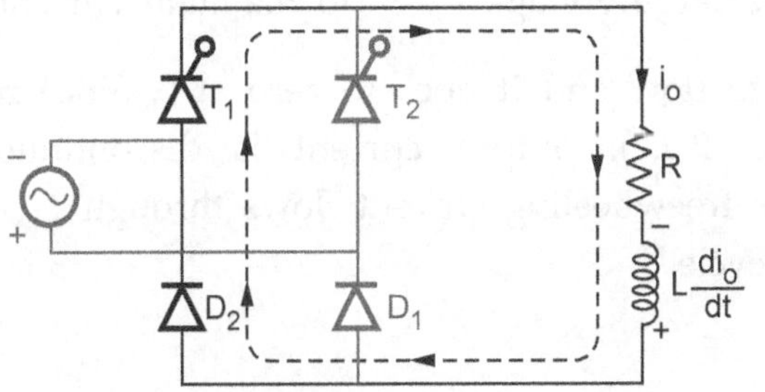

Fig. 2.3.6 Freewheeling action takes place through T_1- D_2

load inductance opposes this sudden change of current. The load inductance generates a large voltage so as to maintain load current. This current flows through T_1 and D_2. The equivalent circuit of this operation is shown in Fig. 2.3.6. The SCR T_1 conducts even after π, since it is forward biased due to voltage induced in the load inductance i.e. $L\frac{di}{dt}$. Diode D_2 is also forward biased due to this voltage. Hence current doesnot flow through supply i.e. i_s when freewheeling action takes place. Thus the energy stored in the load inductance is fedback to load itself in freewheeling action.

SCR T_2 is triggered at $\pi+\alpha$ and the output current starts increasing. Since the current i_o is continuous, it is called continuous current mode of semiconverter. The similar operation takes place when T_2 and D_2 conducts in negative half cycle of the supply. Fig. 2.3.5 shows supply current (i_s), freewheeling current and other waveforms for inductive load. Note that the output voltage waveform remains same. If there is freewheeling diode in semiconverter, then freewheeling current flows through this diode.

Average value of output voltage with inductive load

Compare the output voltage waveforms of Fig. 2.3.3 (resistive load) and Fig. 2.3.5 (inductive load). The voltage waveforms are same. Hence average and RMS values of output voltage are also same. i.e. for inductive load,

From equation 2.3.5 $$V_{o\,(av)} = \frac{V_m}{\pi}(1+\cos\alpha) \qquad \text{... (2.3.7)}$$

from equation 2.3.6 $$V_{o\,(rms)} = \left\{\frac{V_m^2}{2\pi}\left[\pi-\alpha+\frac{1}{2}\sin 2\alpha\right]\right\}^{\frac{1}{2}} \qquad \text{... (2.3.8)}$$

2.3.3.2 Discontinuous Current Mode

In this mode, the current through the load becomes zero for some duration. Hence it is called discontinuous current mode. Fig. 2.3.7 shows the waveforms of discontinuous current mode of semiconverter. (Refer Fig. 2.3.7).

As shown in above waveforms, T_1 - D_1 conducts from α to π and the load current i_o goes on increasing. At π supply voltage is zero. But because of inductance, i_0 does not go to zero. The load inductance induces a large voltage $L\frac{di_o}{dt}$ to maintain current in the same direction. Hence i_o continuous to flow and it goes to zero at β. Since next SCR T_2 is triggered at $\pi+\alpha$ (see Fig. 2.3.7), output current is discontinuous. Freewheeling takes place from π to β. The freewheeling current flows through T_1 and D_2. Similar operation repeats in next half cycle.

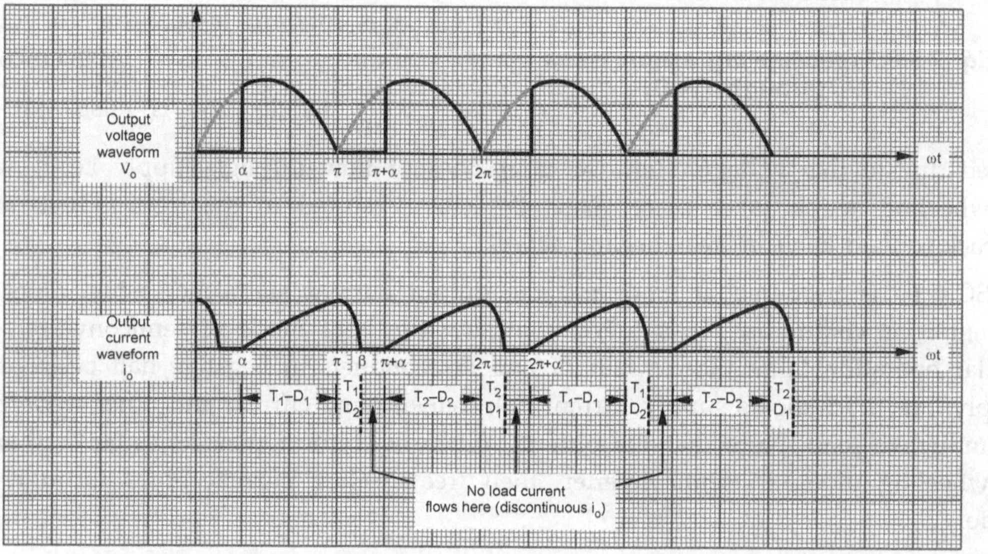

Fig. 2.3.7 Discontinuous mode of single phase semiconverter

Observe that the voltage waveform remains same in discontinuous mode also. Hence $V_{o(av)}$ and $V_{o(RMS)}$ are same as that of resistive load.

2.3.3.3 Continuous and Ripple Free Current for Large Inductive Load

As the load inductance increases, the ripple in i_o reduces. When the load inductance is very large, the ripple in i_o will be negligible. And i_o can be treated as continuous and ripple free. Fig. 2.3.8 shows the waveforms of 1φ semiconverter for large inductive load. The load current is continuous and ripple free. Observe that the output voltage waveform is same as resistive load. But the current waveforms are different.

The output current is constant DC of amplitude $I_{o(av)}$. The SCRs conduct for π radians. Hence SCR current is square wave. The supply current has the amplitudes of $\pm I_{o(av)}$. The supply current is zero whenever freewheeling action takes place.

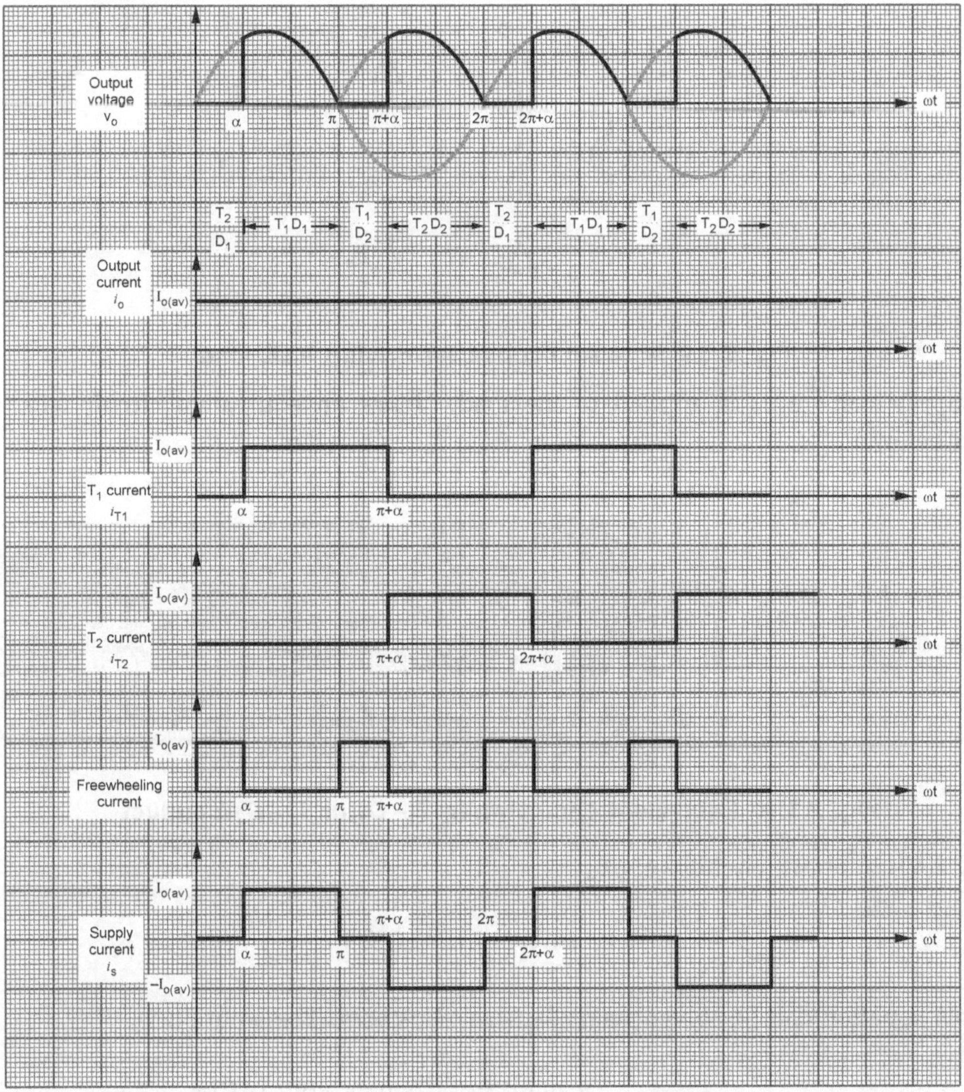

Fig. 2.3.8 Waveforms of 1ϕ semiconverter for highly inductive load

Example 2.3.2 : *Derive an expression for output current for RLE load driven by 1ϕ semiconverter. Assume continuous conduction.*

Solution : Fig. 2.3.9 shows the circuit diagram of 1 ϕ semiconverter for RLE load. See Fig. 2.3.9.

Normally, the RLE load is motor load. L is the inductance of the motor and R is the resistance of the inductance. E is an induced emf in the motor. The waveforms of this circuit will be similar to those shown in Fig. 2.3.5. From α to π, T_1- D_1 conducts

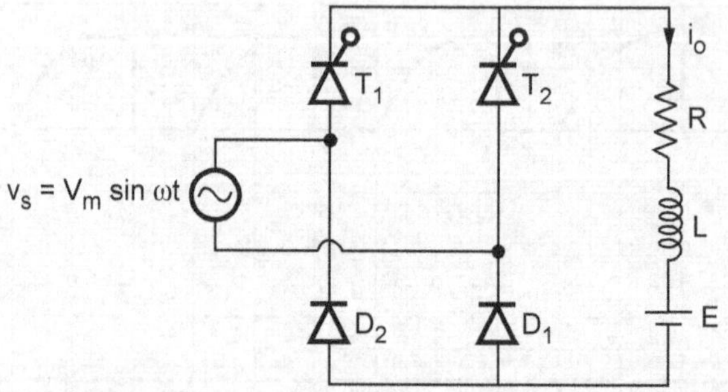

Fig. 2.3.9 A 1φ semiconverter driving RLE load

and supply voltage v_s is applied to the load. Hence an equivalent circuit will be as shown below :

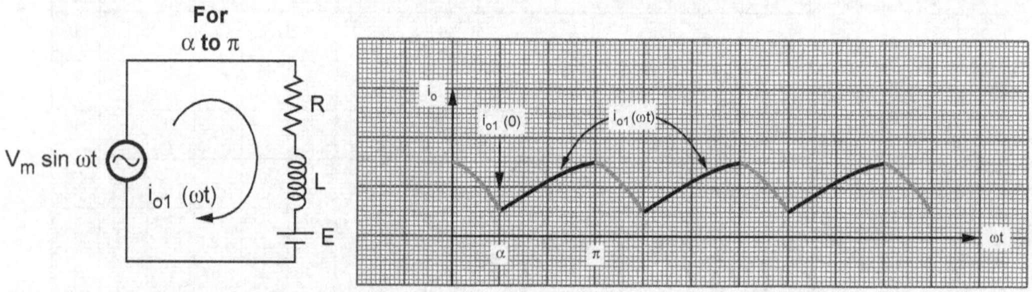

Fig. 2.3.10 Equivalent circuit when T_1- D_1 or T_2- D_2 conduct

By KVL in above circuit we get,

$$V_m \sin \omega t = R i_{o1}(\omega t) + L \frac{d i_{o1}(\omega t)}{dt} + E$$

This equation can be solved using laplace transform. The solution is,

$$i_{o1}(\omega t) = \frac{V_m}{Z} \sin(\omega t - \theta) + i_{o1}(0)\, e^{-t\frac{R}{L}} - \frac{E}{R} \quad \text{... (2.3.9)}$$

Here $\quad Z = \sqrt{R^2 (\omega L)^2}$

$$\theta = \tan^{-1}\left(\frac{\omega L}{R}\right)$$

$i_{o1}(0)$ is initial current at $\omega t = \alpha$.

From π to π+α freewheeling takes place. $T_1 D_2$ conduct in this duration. The equivalent circuit is shown in Fig. 2.3.11.

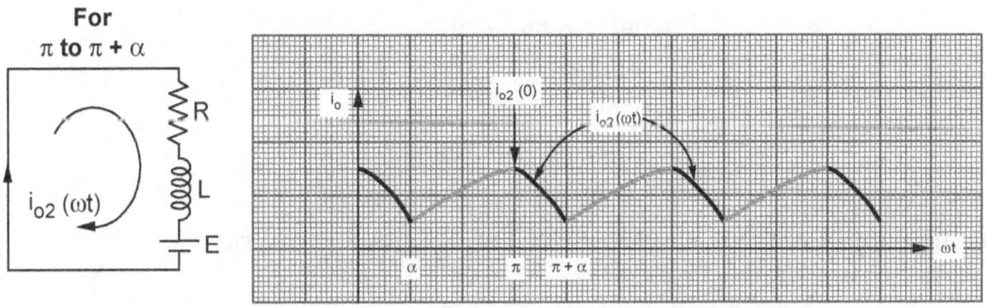

Fig. 2.3.11 Freewheeling action in T_1D_2 or T_2D_1

By KVL to this circuit we can write,

$$R i_{o2}(\omega t) + L\frac{d i_{o2}(\omega t)}{dt} + E = 0$$

This equation can be solved using laplace transform. The solution is,

$$i_{o2}(\omega t) = i_{o2}(0)e^{-t\frac{R}{L}} - \frac{E}{R}(1 - e^{-t\frac{R}{L}}) \qquad \text{... (2.3.10)}$$

Here $i_{o2}(0)$ is the initial current at $\omega t = \pi$. In the waveforms of Fig. 2.3.10 and Fig. 2.3.11 observe that,

$$\left.\begin{aligned} i_{o2}(0) &= i_{o1}(\omega t = \pi) \\ \text{and} \quad i_{o1}(0) &= i_{o2}(\omega t = \alpha) \end{aligned}\right\} \qquad \text{... (2.3.11)}$$

Putting the above two conditions in equation 2.3.9 and 2.3.10 we can get the initial values. Then two currents $i_{o1}(\omega t)$ and $i_{o2}(\omega t)$ are separately expressed for semiconverter.

Example 2.3.3 : *For a 1φ half bridge converter having highly inductive load, derive the following :*
i. Fourier series for supply current
ii. Fundamental component of supply current
iii. RMS value of supply current.

Solution : i. To determine fourier series

The general expression for Fourier series is given as,

$$i_s(\omega t) = I_{s(av)} + \sum_{n=1}^{\infty} c_n \sin(n\omega t + \phi_n)$$

where $c_n = \sqrt{a_n^2 + b_n^2}$

and $\phi_n = \tan^{-1}\left(\frac{a_n}{b_n}\right)$

Here $$a_n = \frac{2}{T}\int_0^T i_s(\omega t)\cos n\omega t\, d\omega t$$

$$= \frac{2}{2\pi}\int_0^{2\pi} i_s(\omega t)\cos n\omega t\, d\omega t$$

From the supply current waveform of Fig. 2.3.8 we can write,

$$a_n = \frac{2}{2\pi}\left[\int_\alpha^\pi I_{o(av)}\cos n\omega t\, d\omega t + \int_{\pi+\alpha}^{2\pi}(-I_{o(av)})\cos n\omega t\, d\omega t\right]$$

$$= \frac{I_{o(av)}}{\pi}\left[\int_\alpha^\pi \cos n\omega t\, d\omega t - \int_{\pi+\alpha}^{2\pi}\cos n\omega t\, d\omega t\right]$$

$$= -\frac{I_{o(av)}}{n\pi}\sin n\alpha(1-\cos n\pi)$$

$$\therefore \quad a_n = \begin{cases} \dfrac{-2I_{o(av)}}{n\pi}\sin n\alpha & \text{for } n=1,3,5,\ \ldots\ldots \\ 0 & \text{for } n=2,4,6,\ldots\ldots \end{cases} \quad \ldots (2.3.12)$$

The above equation shows that a_n is zero for even harmonics of supply current.

b_n is given as,

$$b_n = \frac{2}{T}\int_0^T i_s(\omega t)\sin n\omega t\, d\omega t$$

Putting values of $T = 2\pi$ and $i_s(\omega t)$ from supply current waveform of Fig. 2.3.8,

$$b_n = \frac{2}{2\pi}\left[\int_\alpha^\pi I_{o(av)}\sin n\omega t\, d\omega t + \int_{\pi+\alpha}^{2\pi}(-I_{o(av)})\sin n\omega t\, d\omega t\right]$$

$$= \frac{I_{o(av)}}{\pi}\left[\int_\alpha^\pi \sin n\omega t\, d\omega t - \int_{\pi+\alpha}^{2\pi}\sin n\omega t\, d\omega t\right]$$

$$= \frac{I_{o(av)}}{n\pi}(1+\cos n\alpha)(1-\cos n\pi)$$

$$\therefore \quad b_n = \begin{cases} \dfrac{2I_{o(av)}}{n\pi}(1+\cos n\alpha) & \text{for } n=1,3,5,\ \ldots\ldots \\ 0 & \text{for } n=2,4,6,\ldots\ldots \end{cases} \quad \ldots (2.3.13)$$

The above equation shows that b_n is zero for even harmonics of supply current.

Hence $$c_n = \sqrt{a_n^2 + b_n^2}$$

$$= \left\{ \left[-\frac{2I_{o(av)}}{n\pi} \sin n\alpha \right]^2 + \left[\frac{2I_{o(av)}}{n\pi} (1+\cos n\alpha) \right]^2 \right\}^{\frac{1}{2}}$$

$$= \frac{4\,I_o(av)}{n\pi} \cos\frac{n\alpha}{2} \qquad \text{for n = 1, 3, 5, ...} \qquad \text{... (2.3.14)}$$

This equation gives peak value of nth harmonic of supply current. And ϕ_n can be calculated as,

$$\phi_n = \tan^{-1}\frac{a_n}{b_n}$$

$$= \tan^{-1}\left\{ \frac{\frac{-2I_{o(av)}}{n\pi}\sin n\alpha}{\frac{2I_{o(av)}}{n\pi}(1+\cos n\alpha)} \right\}$$

$$= -\tan^{-1}\left\{ \tan\frac{n\alpha}{2} \right\}$$

$$= -\frac{n\alpha}{2} \qquad \text{... (2.3.15)}$$

Observe the supply current waveform of Fig. 2.3.8. It has symmetric positive and negative half cycles. Hence its average value is zero. This can also be varified mathematically as follows :

$$I_{s(av)} = \frac{1}{T}\int_0^T i_s(\omega t)\, d\omega t$$

Here T = 2π and putting values of $i_s(\omega t)$ from Fig. 2.3.8,

$$I_{s(av)} = \frac{1}{2\pi}\left[\int_\alpha^T I_{o(av)}\, d\omega t + \int_{\pi+\alpha}^{2\pi} -I_{o(av)}\, d\omega t \right]$$

$$= \frac{I_{o(av)}}{2\pi}\left[\int_\alpha^\pi d\omega t - \int_{\pi+\alpha}^{2\pi} d\omega t \right] = \frac{I_{o(av)}}{2\pi}\left\{ [\omega t]_\alpha^\pi - [\omega t]_{\pi+\alpha}^{2\pi} \right\}$$

$$= \frac{I_{o(av)}}{2\pi}\{\pi+\alpha-2\pi-(\pi+\alpha)\}$$

$$= 0$$

Thus the average value of symmetric waveform is zero.

Thus the fourier series can be written as,

$$i_s(\omega t) = \sum_{n=1,3,5,......}^{\infty} \frac{4 I_{o(av)}}{n\pi} \cos\frac{n\alpha}{2} \sin\left(n\omega t - \frac{n\alpha}{2}\right) \quad ... (2.3.16)$$

ii) Fundamental component of supply current

The fundamental component of supply current is given as,

$$I_{s1} = \frac{c_1}{\sqrt{2}}$$

From equation 2.3.14, $c_1 = \frac{4 I_{o(av)}}{\pi} \cos\frac{\alpha}{2}$ with n = 1.

Hence above equation becomes,

$$I_{s1} = \frac{4 I_{o(av)}}{\pi} \cos\frac{\alpha}{2} \times \frac{1}{\sqrt{2}}$$

$$= \frac{2\sqrt{2} I_{o(av)}}{\pi} \cos\frac{\alpha}{2} \quad ... (2.3.17)$$

iii) To obtain rms value of supply current

The rms value is given as,

$$I_{s(rms)} = \left[\frac{1}{T}\int_0^T i_s^2(\omega t)\, d\omega t\right]^{\frac{1}{2}}$$

With $T = 2\pi$ and putting for $i_s(\omega t)$ from supply current waveform of Fig. 2.3.8,

$$I_{s(rms)} = \left\{\frac{1}{2\pi}\left[\int_{\alpha}^{\pi} I_{o(av)}^2\, d\omega t + \int_{\pi+\alpha}^{2\pi} (-I_{o(av)})^2\, d\omega t\right]\right\}^{\frac{1}{2}}$$

$$= \left\{\frac{I_{o(av)}^2}{2\pi}\left[\int_{\alpha}^{\pi} d\omega t + \int_{\pi+\alpha}^{2\pi} d\omega t\right]\right\}^{\frac{1}{2}}$$

$$= I_{o(av)}\sqrt{\frac{\pi-\alpha}{\pi}} \quad ... (2.3.18)$$

The above equation shows that rms value of supply current depends on α.

Example 2.3.4 : *For a 1ϕ half controlled converter having highly inductive load, derive the following :*

(i) Displacement factor (DF) *(ii) Supply power factor (PF)*

(iii) Harmonic factor (HF) *(iv) Current distortion factor (CDF)*

Solution : i) Displacement factor

The displacement factor is given as,

$$DF = \cos\phi_1$$

From equation 2.3.15, $\phi_n = -\frac{n\alpha}{2}$; Hence $\phi_1 = -\frac{\alpha}{2}$.

$$\therefore \quad DF = \cos\left(-\frac{\alpha}{2}\right)$$

$$\therefore \quad \boxed{DF = \cos\frac{\alpha}{2}} \quad \text{... (2.3.19)}$$

ii) Supply power factor (PF)

The supply power factor is given as,

$$PF = \frac{I_{s1}}{I_{s(rms)}}\cos\phi_1$$

Putting the values of I_{s1} (equation 2.3.17), $I_{s(rms)}$ (equation 2.3.19) from previous example and ϕ_1 above we get,

$$PF = \frac{\frac{2\sqrt{2}\,I_{o(av)}}{\pi}\cos\frac{\alpha}{2}}{I_{o(av)}\sqrt{\frac{\pi-\alpha}{\pi}}}\cos\frac{\alpha}{2}$$

$$\therefore \quad \boxed{PF = \sqrt{\frac{8}{\pi(\pi-\alpha)}}\cos^2\frac{\alpha}{2}} \quad \text{... (2.3.20)}$$

iii) Harmonic factor

The harmonic factor (HF) is given as,

$$HF = \sqrt{\frac{I_{s(rms)}^2}{I_{s1}^2} - 1}$$

Putting values in above equation,

$$= \left\{\frac{I_{o(av)}^2\left(\frac{\pi-\alpha}{\pi}\right)}{\frac{8I_{o(av)}^2}{\pi^2}\cos^2\frac{\alpha}{2}} - 1\right\}^{\frac{1}{2}}$$

$$\therefore \quad \boxed{HF = \sqrt{\frac{\pi(\pi-\alpha)}{8\cos^2\frac{\alpha}{2}} - 1}} \qquad \text{... (2.3.21)}$$

This is an expression for harmonic factor of supply current.

iv) Current distortion factor (CDF)

The current distortion factor (CDF) is given as,

$$CDF = \frac{I_{s1}}{I_{s(rms)}}$$

$$= \frac{\frac{2\sqrt{2}\, I_{o(av)}}{\pi}\cos\frac{\alpha}{2}}{I_{o(av)}\sqrt{\frac{\pi-\alpha}{\pi}}}$$

$$= \frac{2\sqrt{2}\,\cos\frac{\alpha}{2}}{\sqrt{\pi(\pi-\alpha)}} \qquad \text{...(2.3.22)}$$

Example 2.3.5 : *For a 1 ϕ half controlled bridge having continuous and ripple free current, obtain, (i) active power and (ii) reactive power.*

Solution : (i) Active power

Active power is given as,

$$P_{active} = V_s\, I_{s1}\cos\phi_1$$

$$= V_s \cdot \frac{2\sqrt{2}\, I_{o(av)}}{\pi}\cos\frac{\alpha}{2}\cos\left(-\frac{\alpha}{2}\right) \qquad \text{since } \phi_1 = -\frac{\alpha}{2}$$

$$= \frac{\sqrt{2}\, V_s\, I_{o(av)}}{\pi}\, 2\cos^2\frac{\alpha}{2}$$

$$= \frac{V_m I_{o(av)}}{\pi}(1+\cos\alpha) \qquad \text{...(2.3.23)}$$

ii) Reactive power

Reactive power is given as,

$$P_{reactive} = V_s\, I_{s1}\sin\phi_1$$

$$= V_s \cdot \frac{2\sqrt{2}\, I_{o(av)}}{\pi}\cos\frac{\alpha}{2}\sin\left(-\frac{\alpha}{2}\right)$$

$$= -\frac{\sqrt{2}\,V_s\,I_{o(av)}}{\pi}\,2\sin\frac{\alpha}{2}\cos\frac{\alpha}{2}$$

$$= -\frac{V_m\,I_{o(av)}}{\pi}\sin\alpha \qquad \ldots(2.3.24)$$

The negative sign indicates that power is reactive.

Comments

i) Active power is consumed by the load.

ii) Reactive power is not consumed by the load. Hence its sign is negative.

iii) Reactive power fluctuates between load and source.

iv) Total power includes active as well as reactive power.

Example 2.3.6 : *Single phase semiconverter is operated from 120 V, 60 Hz supply. The load current with an average value of I_a is continuous with negligible ripple content. Turns ratio of transformer is unity. The delay angle $\alpha = \frac{\pi}{3}$. Calculate-*

(a) Harmonic factor of input current

(b) The displacement factor

(c) Input power factor

Solution : The given data is,

$$V_s = 120\text{ V}$$

$$I_{o(av)} = I_a$$

$$\alpha = \frac{\pi}{3}$$

(a) Harmonic factor is given by equation 2.3.21 as,

$$HF = \sqrt{\frac{\pi(\pi-\alpha)}{8\cos^2\frac{\alpha}{2}} - 1}$$

Putting the values in above equation,

$$HF = \left\{\frac{\pi\left(\pi-\frac{\pi}{3}\right)}{8\cos^2\left(\frac{\pi/3}{2}\right)} - 1\right\}^{\frac{1}{2}}$$

$$= 0.3108 \text{ or } 31.08\%$$

(b) The displacement factor is given by equation 2.3.19 as,

$$DF = \cos\frac{\alpha}{2}$$

$$= \cos\left(\frac{\pi/3}{2}\right)$$

$$= 0.866$$

(c) The input power factor is given by equation 2.3.20 as,

$$PF = \sqrt{\frac{8}{\pi(\pi-\alpha)}}\cos^2\frac{\alpha}{2}$$

$$= \sqrt{\frac{8}{\pi\left(\pi-\frac{\pi}{3}\right)}}\cos^2\left(\frac{\pi/3}{2}\right)$$

$$= 0.827 \text{ lagging}$$

2.3.4 Asymmetrical Half Bridge Converter

2.3.4.1 Operation with Resistive Load

Fig. 2.3.12 shows the circuit diagram of asymmetrical half bridge converter. Observe that both the SCRs appear on the same link.

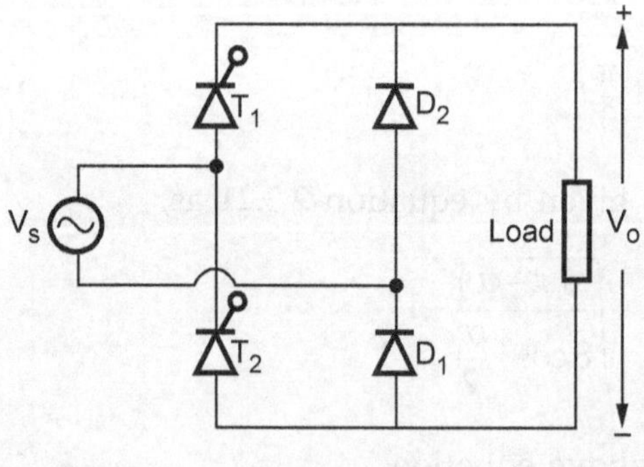

Fig. 2.3.12 Single phase controlled rectifier

When T_1 is triggered, current flows through T_1 and D_1. T_2 is triggered in the negative half cycle. Then current flows through T_2 and D_2.

Fig. 2.3.13 shows the waveforms of half bridge converter given in Fig. 2.3.12. These waveforms are shown for resistive load and $\alpha = \frac{\pi}{2}$. Observe that the output current

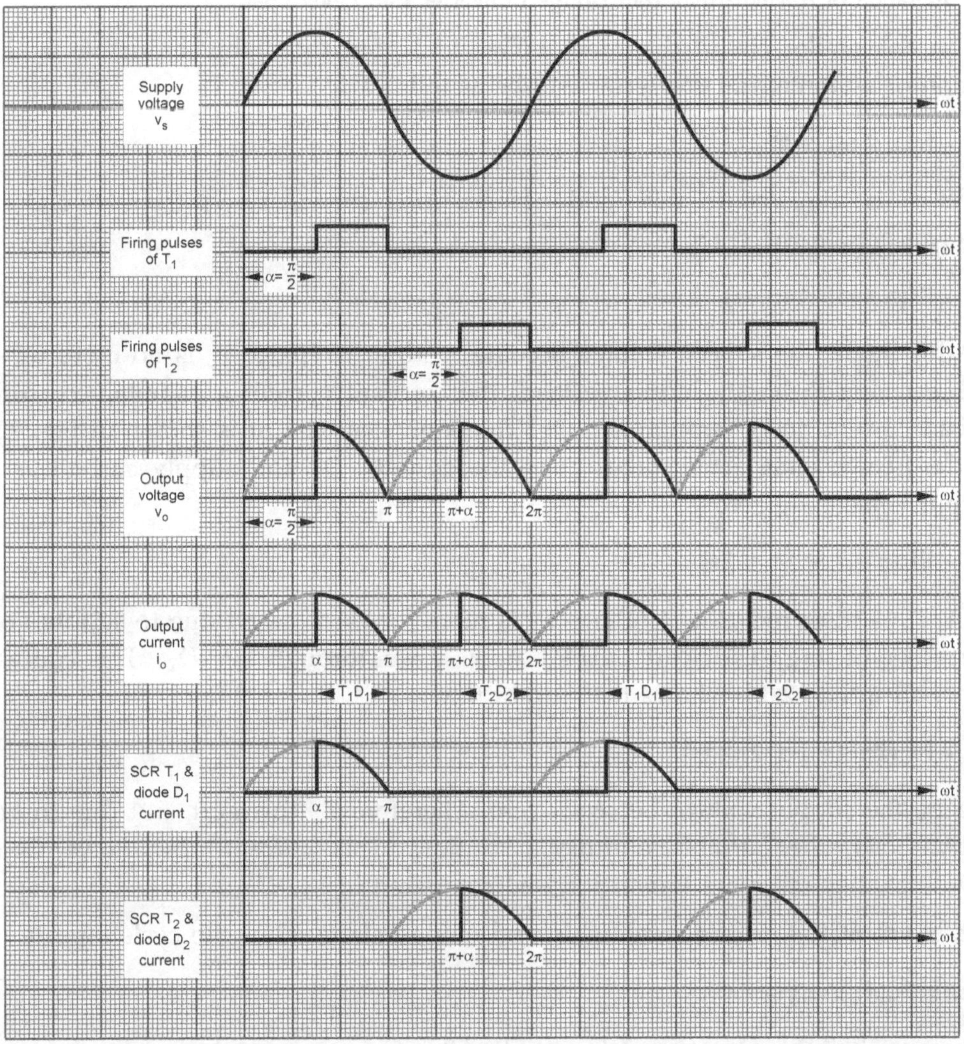

Fig. 2.3.13 Waveforms of half bridge converter of Fig. 2.3.12

waveform is similar to output voltage. Since T_1 and D_1 conduct simultaneously their current waveform is same. Similarly, the current waveform of T_2 and D_2 is same.

Since the above output voltage is same as that of single phase semiconverter, the rms and average values of output voltage will be,

$$V_{o(av)} = \frac{V_m}{\pi}(1+\cos\alpha)$$

$$V_{o(rms)} = \left\{\frac{V_m^2}{2\pi}\left[\pi-\alpha+\frac{1}{2}\sin 2\alpha\right]\right\}^{\frac{1}{2}}$$

2.3.4.2 Operation of Asymmetrical Half Bridge Converter with Level Load

With the similar circuit diagram of Fig. 2.3.12 but for highly inductive laod, the operation of asymmetrical converter will be as follows :

Mode - I ($\alpha \le \omega t \le \pi$)

SCR T_1 is triggered in this mode. Hence load current flows through $T_1 D_1$. The waveforms are shown in Fig. 2.3.14.

Mode - II ($\pi \le \omega t \le \pi + \alpha$)

In this mode, the supply voltage becomes zero at π. Hence T_1 is no more forward biased. But due to highly inductive load, the constant current is maintained to flow. This load current flows through D_1 - D_2. The equivalent circuit is shown in Fig. 2.3.14. Thus the freewheeling action takes place through D_1 - D_2 and supply current as well as output voltage are zero.

Mode - III ($\pi + \alpha \le \omega t \le 2\pi$)

SCR T_2 is triggered at $\pi + \alpha$. Since T_2 is more forward biased due to supply voltge, it starts conducting. The load current now flows through T_2 - D_2. The equivalent circuits - III in Fig. 2.3.14 shows the current flow.

Mode - IV ($2\pi \le \omega t \le 2\pi + \alpha$)

At 2π, the supply voltage becomes zero. Therefore T_2 turns-off. But due to heavy inductive load, the current continuous to flow. This current now flows through $D_1 - D_2$ since they are more forward biased compared to T_2 - D_2.

At $2\pi + \alpha$, SCR T_1 is triggered again and mode-I starts. Thus the cycle repeats.

Mathematical analysis

Observe that the waveform of output voltage is same as that of semiconvertor. Hence the rms and average values of its output voltage are,

$$V_{o(av)} = \frac{V_m}{\pi}(1 + \cos\alpha)$$

and
$$V_{o(rms)} = \left\{\frac{V_m^2}{2\pi}\left[\pi - \alpha + \frac{1}{2}\sin 2\alpha\right]\right\}^{\frac{1}{2}}$$

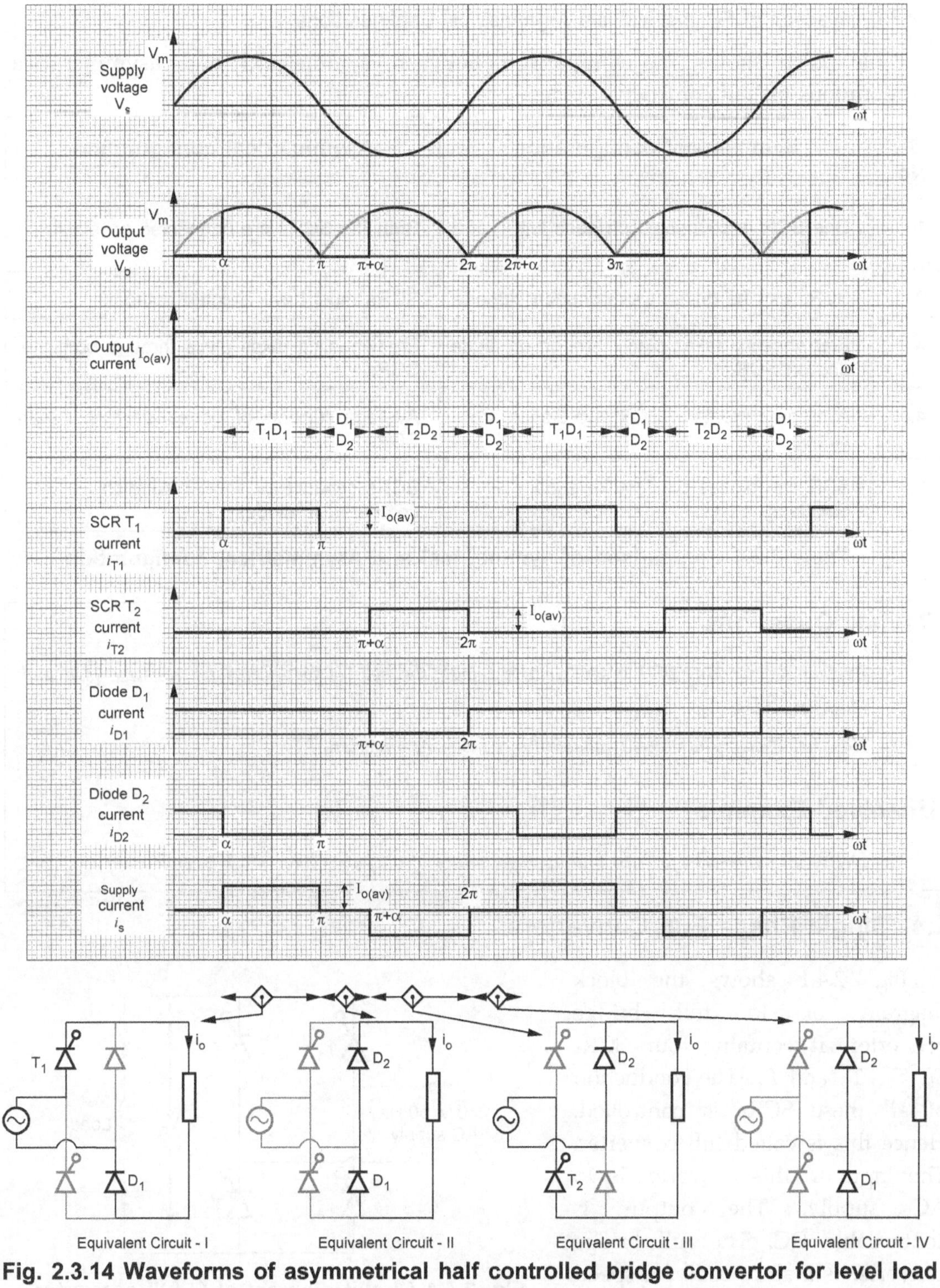

Fig. 2.3.14 Waveforms of asymmetrical half controlled bridge convertor for level load

2.3.4.3 Comparison of Symmetrical and Asymmetrical Configurations

Table 2.3.1 shows the comparison between symmetrical and asymmentrical configurations of half controlled bridge.

Sr. No.	Symmetrical configuration	Asymmetrical configuration
1.	One SCR is connected on each link.	Both the SCRs are connected on single link.
2.	SCRs can be driven with common cathode.	SCRs must have isolated cathods.
3.	Freewheeling takes place through on diode and on SCR.	Freewheeling takes place through both the SCRs.
4.	Average currents of SCR and diodes are same.	Average currents of diodes are higher than SCR.
5.	SCR and diodes conduct for equal durations.	SCRs conduct for shorter duration compared to diodes.

Table 2.3.1 Comparison of symmetrical and asymmetrical configuration

Review Questions

1. *Explain the working of 1 ϕ semiconverter with the help of waveforms for resistive load and inductive loads.*
2. *How freewheeling is present inherently in the semiconverters ?*

Unsolved Example

1. *Derive an expression for rms and average values of output voltage for 1 ϕ semiconvertor.*

2.4 Single Phase Full Converters

Fig. 2.4.1 shows the block diagram of 1ϕ full bridge converter. It contains four SCRs T_1, T_2, T_3 and T_4. The conduction of all these SCRs is controlled. Hence this is called full converter. The input to this converter is 1ϕ AC supply. The output is controllable DC. The full bridge converter is mainly used for speed control of DC motors.

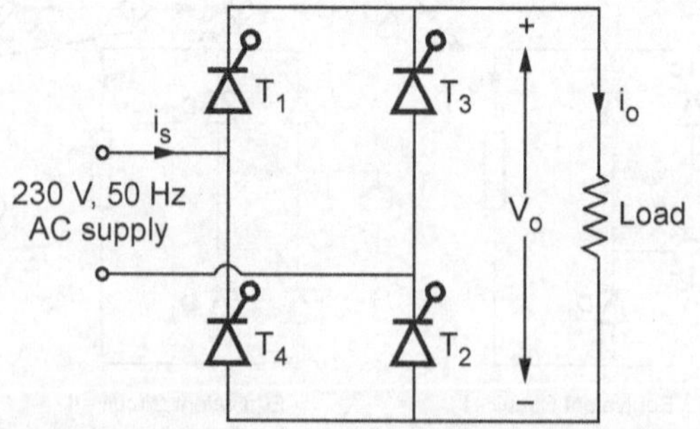

Fig. 2.4.1 Circuit diagram of 1ϕ full converter

2.4.1 Working with Resistive Load

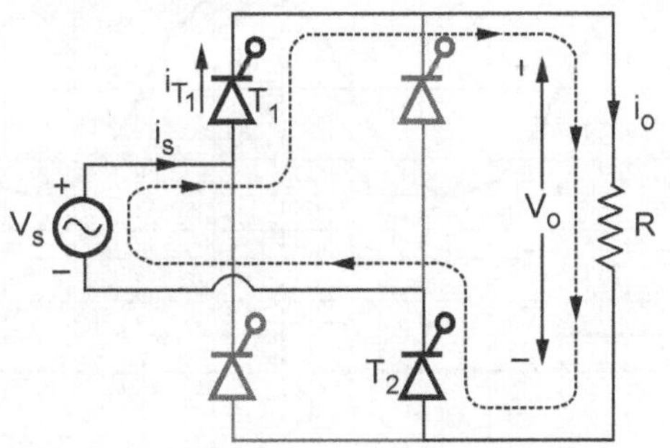

Fig. 2.4.2 Conduction of T_1 and T_2 in positive half cycle of the supply. Dotted line shows path of current flow

Let us consider the working of 1ϕ bridge (Full) converter with resistive load. In the positive half cycle of the supply SCRs T_1 and T_2 are triggered at firing angle α. Hence current starts flowing through the load. The equivalent circuit for this operation is shown in Fig. 2.4.2.

It is clear from Fig. 2.4.2 that, when T_1 and T_2 conducts,

$$V_o = V_s \quad \text{(i.e. supply voltage)} \qquad \text{... (2.4.1)}$$

and,

$$i_o = \frac{V_o}{R} = \frac{V_s}{R} \qquad \text{... (2.4.2)}$$

Fig. 2.4.3 shows the waveforms of this circuit. Observe that load voltage is same as supply voltage from α to π. Since the load is resistive, waveforms of V_o and i_o are same. The supply current i_s and i_o are in the same direction hence $i_s = i_o$. T_1 and T_2 turn off when supply voltage becomes zero at π. In the negative half cycle T_3 and T_4 are triggered at $\pi + \alpha$.

Fig. 2.4.4 shows the equivalent circuit when T_3 and T_4 conduct.

In the adjacent figure observe that supply current i_s and load current i_o flow through the same loop. But directions of i_s and i_o are opposite hence

$$i_s = -i_o$$

The supply current waveform is also shown in Fig. 2.4.3. T_3 and T_4 turn off when supply voltage becomes zero at 2π. At $2\pi + \alpha$, T_1 and T_2 are triggered again and the cycle repeats.

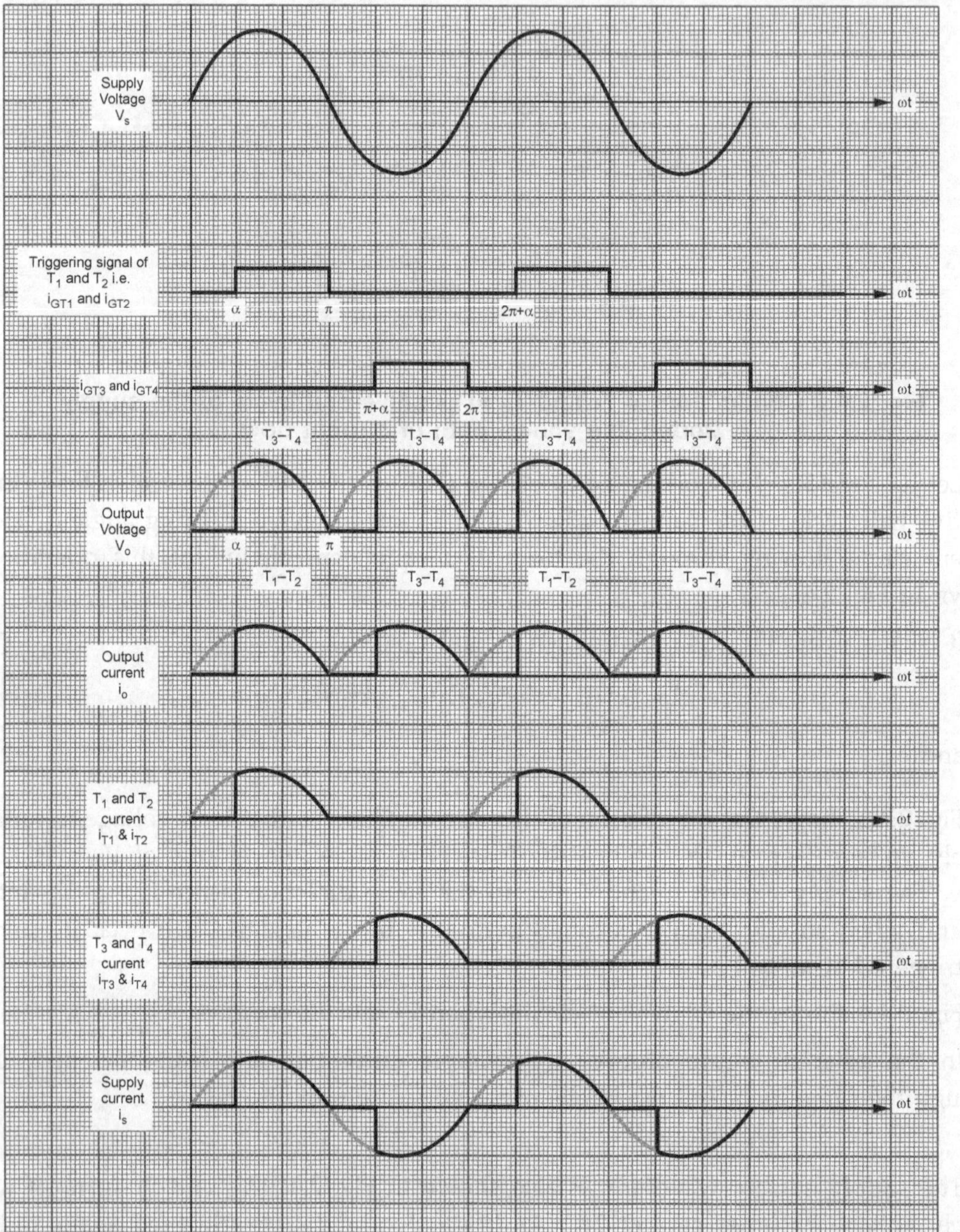

Fig. 2.4.3 Waveforms of full bridge converter for resistive load

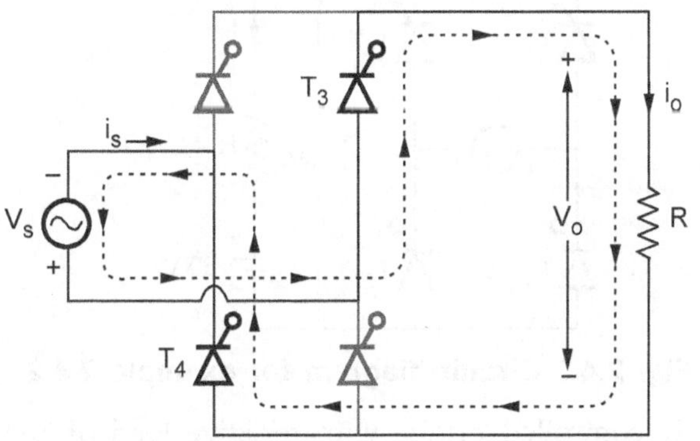

Fig. 2.4.4 Conduction of $T_3 - T_4$ in negative half cycle of the supply. Dotted line shows current path

Example 2.4.1 : *For the 1 ϕ fully controlled bridge converter having load of 'R' determine the following :*
i) Average output voltage $V_{o(av)}$ *ii) RMS output voltage* $V_{o(rms)}$
If supply voltage is 230 V, 50 Hz and firing angle is 60°, determine average output voltage.

Solution : Compare the waveforms of 1ϕ semiconverter given in Fig. 2.3.3 and that of 1ϕ full converter given in Fig. 2.4.3. The waveforms are exactly same. Thus the operation of semiconverter and full converter is exactly same for resistive load. Hence their average and RMS output voltages are also same. Hence from equation 2.3.5, the average output voltage of full converter is,

$$\boxed{V_{o(av)} = \frac{V_m}{\pi}(1+\cos\alpha)} \quad \text{... (2.4.3)}$$

The supply voltage is 230 V. Hence,

$$V_s = 230 \text{ V} \qquad \therefore \quad V_m = 230\sqrt{2}$$

and $\alpha = 60°$

Putting values in equation 2.4.3,

$$V_{o(av)} = \frac{230\sqrt{2}}{\pi}(1+\cos 60°) = 155.3 \text{ V}$$

From equation 2.3.6, RMS output voltage of full converter is,

$$\boxed{V_{o(rms)} = \frac{V_m^2}{2\pi}\left[\pi-\alpha+\frac{1}{2}\sin 2\alpha\right]^{\frac{1}{2}}} \quad \text{... (2.4.4)}$$

Example 2.4.2 : *For the circuit shown in Fig. 2.4.5, find the current through 100 Ω load, if the SCRs are triggered at 30° delay. Supply voltage is 200 V, 50 Hz.*

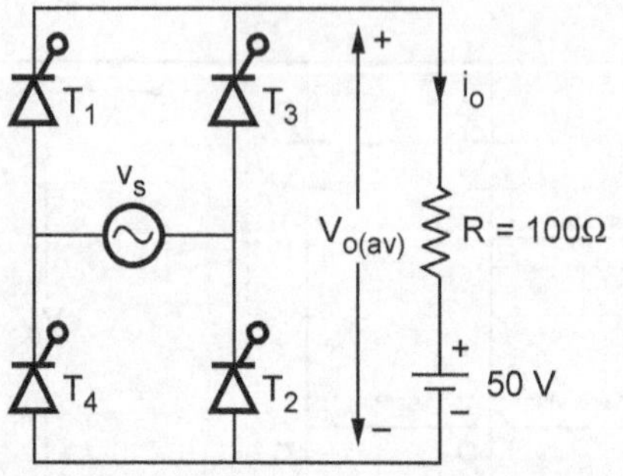

Fig. 2.4.5 Circuit diagram for example 2.4.2

Solution : This is a fully controlled bridge with resistive load of 100 Ω in series with the battery of 50 V. Hence output voltage of the converter appears across resistance of 100 Ω and battery of 50 V. Hence let us first calculate average value of output voltage. The given data is,

$$\alpha = 30°$$

$$V_s = 220 \text{ V} \qquad \therefore \qquad V_m = 220\sqrt{2}$$

The average output voltage for resistive load is given by equation 2.4.3 as,

$$V_{o(av)} = \frac{V_m}{\pi}(1+\cos\alpha)$$

$$= \frac{230\sqrt{2}}{\pi}(1+\cos 30°)$$

$$= 184.8 \text{ V}$$

This voltage is applied to the load. Fig. 2.4.6 shows the equivalent circuit.

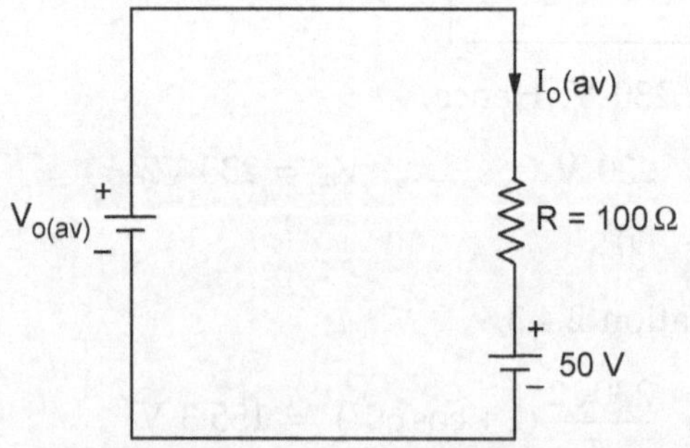

Fig. 2.4.6 Equivalent load circuit

By applying KVL to above circuit,

$$V_{o(av)} = I_{o(av)}R + 50$$

$$\therefore \quad 184.8 = I_{o(av)} \times 100 + 50$$

$$I_{o(av)} = 1.348 \text{ A}$$

Thus the current through the load is 1.348 A.

2.4.2 Working with Inductive Load

The inductive load means resistance and inductance in the load. Such loads are DC motors. Because of the inductive (R-L) load, the load current shape is changed. Hence operation of the full bridge converter can be discussed into three modes :

(i) Continuous load current

(ii) Continuous and ripple free current for large inductive load

(iii) Discontinuous load current

2.4.2.1 Continuous Load Current

In the continuous load current, the load or output current i_0 flows continuously. The waveforms are shown in Fig. 2.4.7.

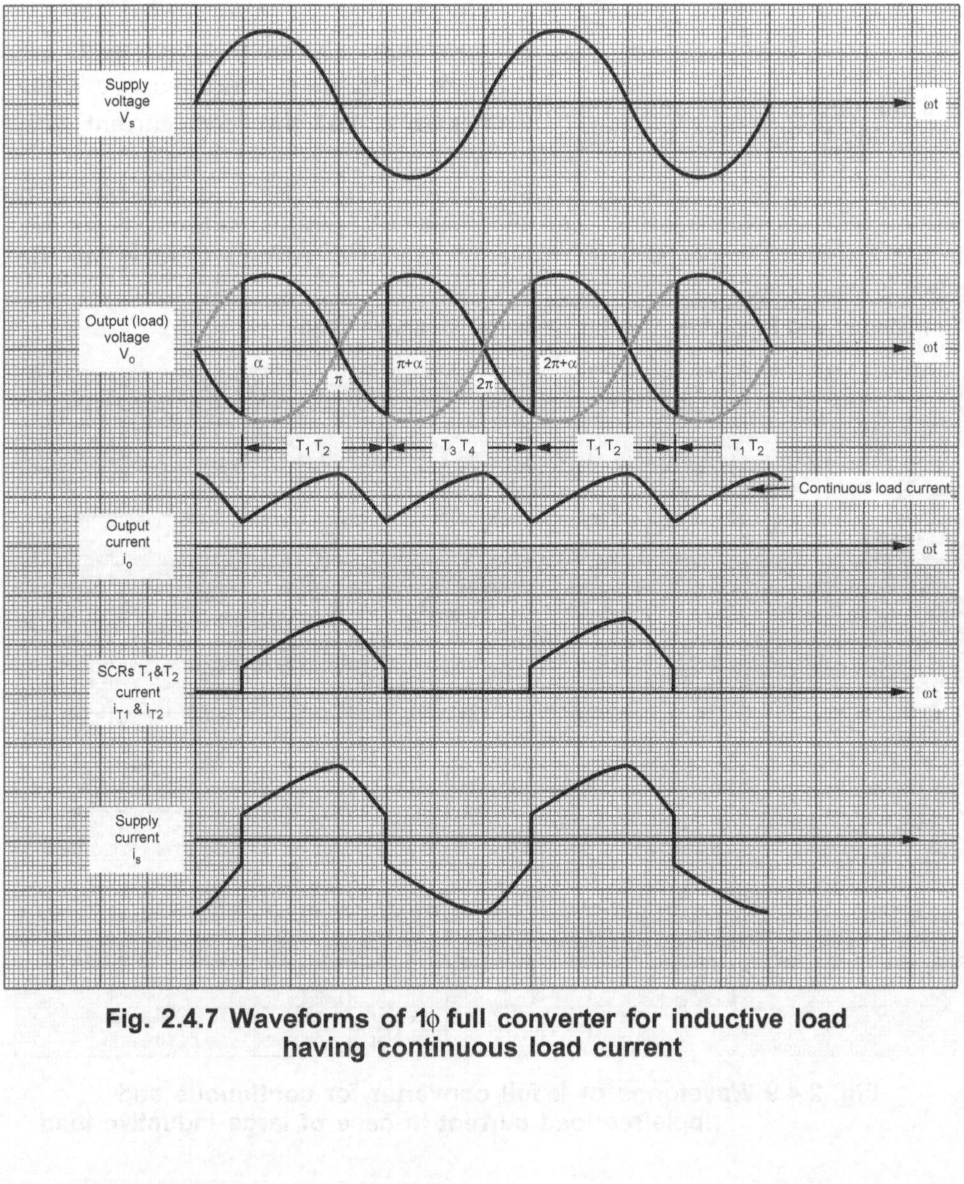

Fig. 2.4.7 Waveforms of 1ϕ full converter for inductive load having continuous load current

As shown in the waveforms of Fig. 2.4.7, T_1 and T_2 conduct from α to π. The nature of the load current depends upon values of R and L in the inductive load. Because of the inductance, i_o keeps on increasing and becomes maximum at π. At π, the supply voltage reverses but SCRs T_1 and T_2 does not turn off. This is because, the load inductance does not allow the current i_o to go to zero instantly. The load inductance generates a large voltage $L\frac{di_o}{dt}$.

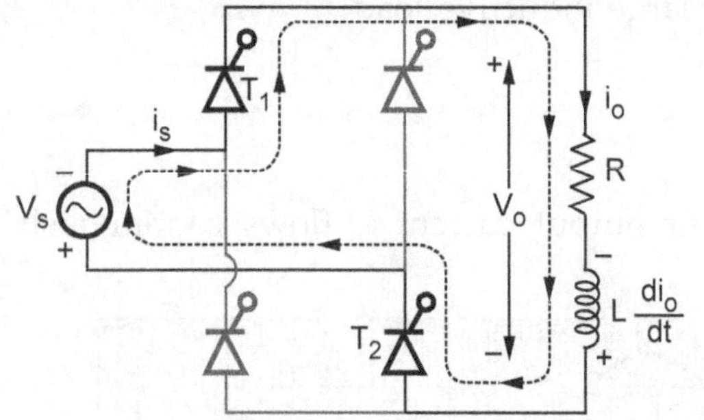

Fig. 2.4.8 Conduction of T_1 and T_2 from π to $\pi+\alpha$ due to inductance voltage

This voltage forward biases T_1 and T_2 as shown in Fig. 2.4.8. In this figure observe that the load current flows against the supply voltage. The energy stored in the load inductance is supplied partially to the mains supply and to the load itself. Hence this is also called as *feedback* operation. The output voltage is negative from π to $\pi+\alpha$ since supply voltage is negative. But the load current keeps on reducing.

At $\pi+\alpha$, SCRs T_3 and T_4 are triggered. The load current starts increasing. The load current remains continuous in the load. The similar operation repeats. The ripple in the load current reduces as the load inductance is increased.

2.4.2.2 Continuous and Ripplefree Current for Large Inductive Load

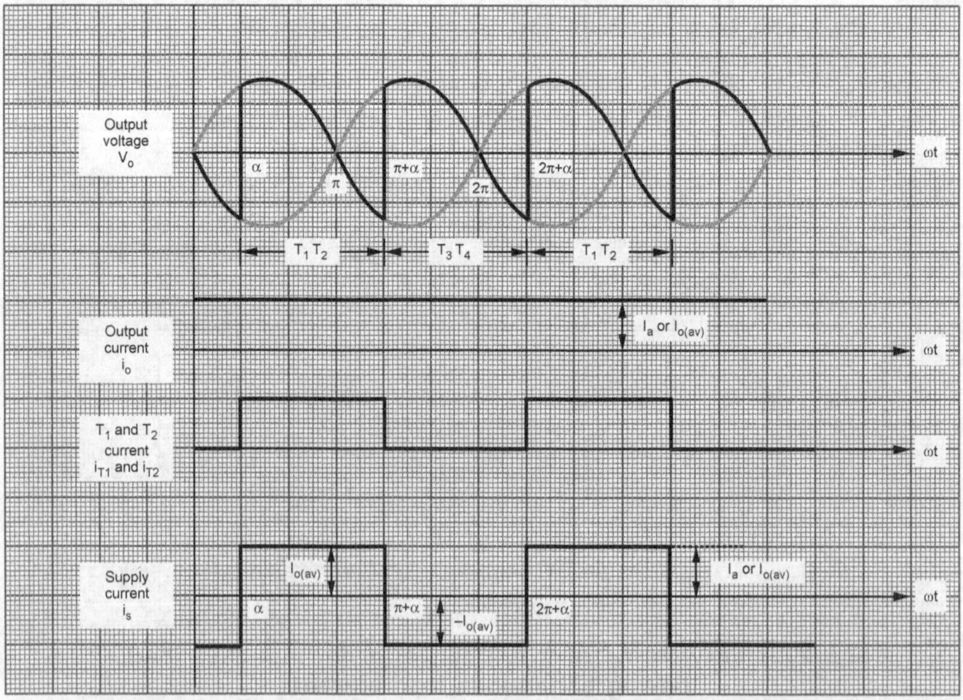

Fig. 2.4.9 Waveforms of 1φ full converter for continuous and ripplefree load current in case of large inductive load

Now let us consider the case when there is large inductance in the load. Because of the large inductance, the ripple in the load current is very small and it can be neglected. Hence load current will be totally DC as shown in Fig. 2.4.9.

In the waveforms shown in Fig. 2.4.9, there is no effect on output voltage waveform for large inductive load. The supply current waveform (i_s) is square wave for large inductive load.

Example 2.4.3 : *For the 1φ full converter having inductive load and continuous load current, obtain the following :*
(i) Average output voltage $V_{o\,(av)}$
(ii) RMS output voltage $V_{o\,(rms)}$

Solution : (i) Average output voltage for inductive load

The average output voltage is given as,

$$V_{o\,(av)} = \frac{1}{T}\int_0^T v_o\,(\omega t)\,d\omega t$$

Observe the waveforms of 1φ full converter for inductive load given in Fig. 2.4.7 and Fig. 2.4.9. The output voltage waveform has a period from α to π+α ; i.e. π. And $v_o\,(\omega t) = V_m \sin \omega t$ during this period. Hence above equation becomes,

$$V_{o\,(av)} = \frac{1}{\pi}\int_{\alpha}^{\pi+\alpha} V_m \sin \omega t\; d\omega t$$

$$= \frac{V_m}{\pi}\left[-\cos \omega t\right]_{\alpha}^{\pi+\alpha}$$

$$\therefore \quad \boxed{V_{o\,(av)} = \frac{2\,V_m}{\pi}\cos \alpha} \qquad \text{... (2.4.5)}$$

This is the expression for average load voltage of 1φ full converter for inductive load.

Plot of $V_{o(av)}$ versus firing angle (α)

Following table lists the values of $V_{o(av)}$ with firing angle (α)

α	$V_{o(av)} = \frac{2V_m}{\pi}\cos\alpha$
0	$\frac{2V_m}{\pi} = 0.637\ V_m$
30°	$0.55\ V_m$
60°	$0.318\ V_m$

90°	0
120°	$-$ 0.318 V_m
150°	$-$ 0.55 V_m
180°	$-$ 0.637 V_m

Table 2.4.1 $V_{o(av)}$ with respect to α

Observe that $V_{o\ (av)}$ is positive for $\alpha < 90°$. Hence it is called rectification. For $\alpha > 0$, $V_{o\ (av)}$ is negative. Hence it is called inverting mode of operation. In inverting mode, output energy is fedback to the source.

(ii) RMS value of output voltage for inductive load

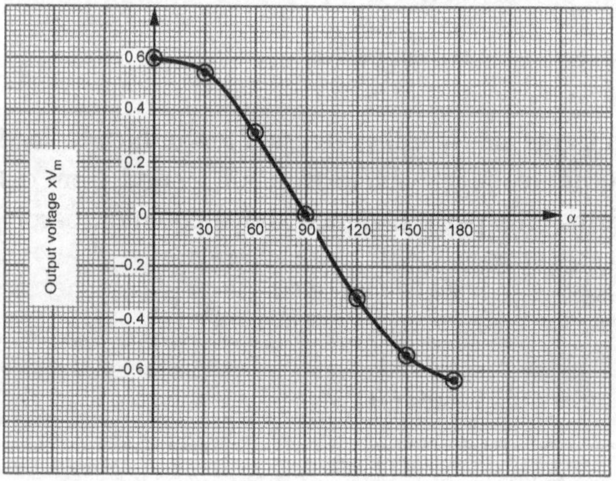

Fig. 2.4.10 Variation of $V_{o\ (av)}$ with respect to α

The rms value is given as,

$$V_{o\ (rms)} = \left[\frac{1}{T}\int_0^T v_o^2\ (\omega t)\ d\,\omega t\right]^{\frac{1}{2}}$$

$$= \left[\frac{1}{\pi}\int_\alpha^{\pi+\alpha} V_m^2\ sin^2\ \omega t\ d\,\omega t\right]^{\frac{1}{2}}$$

$$= \left[\frac{V_m^2}{\pi}\int_\alpha^{\pi+\alpha} \frac{1-cos\ 2\omega\ t}{2}\ d\,\omega t\right]^{\frac{1}{2}}$$

$$= \left[\frac{V_m^2}{2}\right]^{\frac{1}{2}}$$

$$\therefore \quad \boxed{V_{o\,(rms)} = \frac{V_m}{\sqrt{2}} = V_s} \quad \text{... (2.4.6)}$$

Thus the rms value of load voltage is same as rms value of the AC supply voltage.

Example 2.4.4 : *Draw the circuit arrangement of a single phase full converter feeding a general load comprising of R, L and E. Sketch the AC supply voltage o/p voltage and the load current waveforms. Assuming continuous load current operation, derive an expression for DC output voltage.*

A single phase full converter feeding an RLE load is fed by 230 V, 50 Hz mains.

If R = 0.5 Ω, L = 8 mH and E = 50 volts, assuming that conduction is continuous and firing angle is 40°, find average value of load current.

Solution : Circuit diagram and waveforms

Fig. 2.4.11 shows the circuit diagram of full converter supplying RLE load.

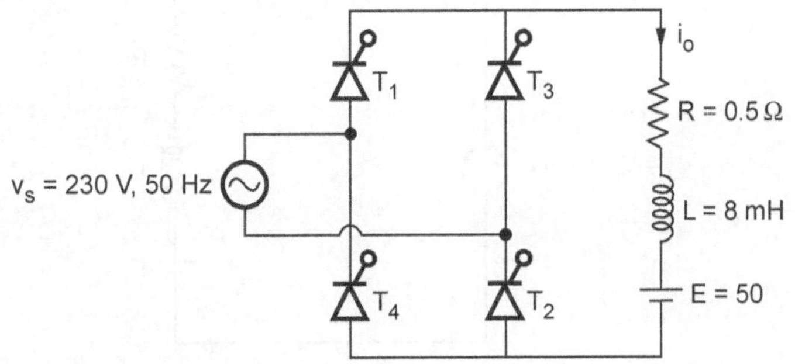

Fig. 2.4.11 1ϕ full converter feeding RLE load

The RLE load is normally motor load. 'R' is the resistance and 'L' is an inductance of armature winding of the motor. 'E' is the induced emf of the motor. When the load current is continuous, then waveforms of this circuit will be similar to that of RL load. Hence with small ripple in output current, the waveforms of this circuit will be similar to those shown in Fig. 2.4.7. Note that 'E' is not reflected in the waveforms as long as output current (i_o) is continuous.

If output current (i_o) is constant and ripple free, then the waveforms will be similar to those shown in Fig. 2.4.9.

RMS and average output voltage

The output voltage waveform remains same with RL load and RLE load when i_o is continuous. Therefore the rms and average values of output voltage will be same as those derived in previous example for RL load. i.e.,

$$V_{o(av)} = \frac{2V_m}{\pi}\cos\alpha$$

$$V_{o(rms)} = \frac{V_m}{\sqrt{2}} = V_s$$

Second part : To obtain average load current

The ripple in the load current (i_o) depends upon values of R, L and E. If load inductance is small, then i_o can become discontinuous. In Fig. 2.4.7, observer that i_o repeats at the intervals of π . The waveform of i_o remains same whenever T_1-T_2 or T_3T_4 conducts. Hence in any interval (i.e. $\alpha \le \omega t \le \pi+\alpha$ or $\pi+\alpha \le \omega t \le 2\pi+\alpha$) the equivalent circuit will be as shown below.

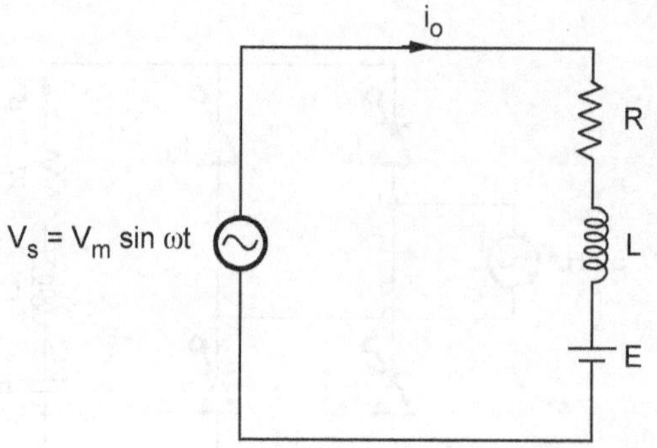

Fig. 2.4.12 Equivalent circuit when T_1-T_2 or T_3-T_4 conduct

By applying KVL to above circuit,

$$V_m \sin\omega t = Ri_o + L\frac{di_o}{dt} + E$$

This equation can be solved using laplace transform. The solution of above equation is,

$$i_o(\omega t) = \frac{V_m}{Z}\sin(\omega t-\theta) - \frac{E}{R} + \left\{i_o(0) + \frac{E}{R} - \frac{V_m}{Z}\sin(\alpha-\theta)\right\} e^{\frac{R}{\omega L}(\alpha-\omega t)} \quad \text{... (2.4.7)}$$

Here $Z = \sqrt{R^2 + (\omega L)^2}$

and $\theta = \tan^{-1}\left(\frac{\omega L}{R}\right)$

$i_o(0)$ is the initial value of the output current. This value is same at $\alpha, \pi+\alpha, 2\pi+\alpha, \ldots\ldots$ and so on.

Hence $\quad i_o(\omega t=\alpha) = i_o(0)$

or $\quad i_o(\omega t=\pi+\alpha) = i_o(0)$

Putting this value in equation 2.4.7 and solving for $i_o(0)$ we get,

$$i_o(0) = -\frac{V_m}{Z}\sin(\alpha-\theta)\left[\frac{1+e^{-\frac{\pi R}{\omega L}}}{1-e^{-\frac{\pi R}{\omega L}}}\right]-\frac{E}{R} \qquad \ldots\ldots (2.4.8)$$

The given data is,

$V_s = 230 \quad \therefore \quad V_m = \sqrt{2}\times 230 = 325.27$ V

f = 50 Hz $\therefore \omega = 2\pi f = 2\pi\times 50 = 314.159$

R = 0.5 Ω

L = 8 mH

E = 50 V

$\alpha = 40°$

$$\therefore \quad \omega L = 314.154\times 8\times 10^{-3} = 2.513$$

$$\frac{\pi R}{\omega L} = \frac{\pi\times 0.5}{2.513} = 0.625$$

$$\therefore \quad e^{-\frac{\pi R}{\omega L}} = e^{-0.625} = 0.5352$$

$$\theta = \tan^{-1}\left(\frac{\omega L}{R}\right) = \tan^{-1}\left(\frac{2.513}{0.5}\right) = 1.3744$$

$$\alpha = 40° = 0.698 \text{ radians}$$

$$\text{and } Z = \sqrt{R^2+(\omega L)^2}$$

$$= \sqrt{(0.5)^2+(2.513)^2} = 2.5622$$

Putting values in equation 2.4.8 we get $i_o(0)$ as

$$i_o(0) = -\frac{325.27}{2.5622}\sin(0.698-1.3744)\left[\frac{1+0.5352}{1-0.5352}\right]-\frac{50}{0.5}$$

$$= 162.48 \text{ A}$$

This is the minimum value of output current. If this value becomes negative, then it indicates discontinuous operation.

Putting values in equation 2.4.7 we get equation for $i_o(\omega t)$. i.e.,

$$i_o(\omega t) = \frac{325.27}{2.5622}\sin(\omega t - 1.3744) - \frac{50}{0.5}$$

$$+\left\{162.48 + \frac{50}{0.5} - \frac{325.27}{2.5622}\sin(0.698 - 1.3744)\right\} e^{\frac{0.5}{2.513}(0.698 - \omega t)}$$

$$= 126.95 \sin(\omega t - 1.3744) - 100 + 392.89\, e^{-0.1989\,\omega t} \qquad \text{... (2.4.9)}$$

This is the equation for output current from α to $\pi+\alpha$. This waveform has period of π and it repeats at $\pi+\alpha$. Hence average value of i_o will be given as,

$$I_{o(av)} = \frac{1}{\pi}\int_{\alpha}^{\pi+\alpha} i_o(\omega t)\, d\omega t$$

$$= \frac{1}{\pi}\int_{0.698}^{\pi+0.698} [126.95\sin(\omega t - 1.3744) - 100 + 392.89 e^{-0.1989\,\omega t}]\, d\omega t$$

$$= \frac{126.95}{\pi}\int_{0.698}^{3.839} \sin(\omega t - 1.3744)\, d\omega t - \frac{100}{\pi}\int_{0.698}^{3.839} d\omega t + \frac{392.89}{\pi}\int_{0.698}^{3.839} e^{-0.1989\omega t}\, d\omega t$$

$$= 217.28 \text{ A}$$

This is the average value of output current.

Example 2.4.5 : *If a freewheeling diode is added across the highly inductive load in 1φ full converter, derive an expression for average load voltage.*

Solution : We know that freewheeling action does not take place in 1φ full converter inherently. In the positive half cycle, T_1 and T_2 conduct from α to π as usual. But from π to $\pi+\alpha$ freewheeling diode starts conducting. This is shown in Fig. 2.4.13. The freewheeling diode is more forward biased compared to T_1 and T_2. Hence freewheeling diode conducts. The freewheeling diode is connected across the output V_o. Hence $V_o = 0$ during freewheeling. The energy stored in the load inductance is circulated back in the load itself. Fig. 2.4.14 shows the waveforms of this operation. The output voltage becomes zero in the freewheeling periods. Compare the load voltage waveform of Fig. 2.4.13 with that of 1φ full converter with resistive load (Fig. 2.4.3). They are same. Hence the average load voltage can be obtained from equation 2.4.3. i.e.,

$$V_{o(av)} = \frac{V_m}{\pi}(1 + \cos\alpha) \qquad \text{... (2.4.10)}$$

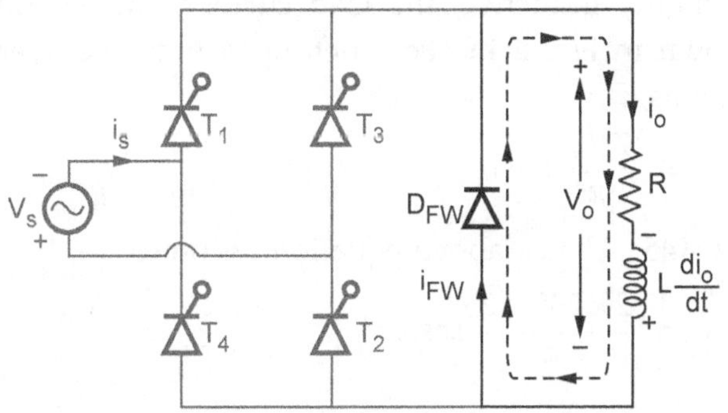

Fig. 2.4.13 Freewheeling diode conducts from π to $\pi + \alpha$ due to inductive load

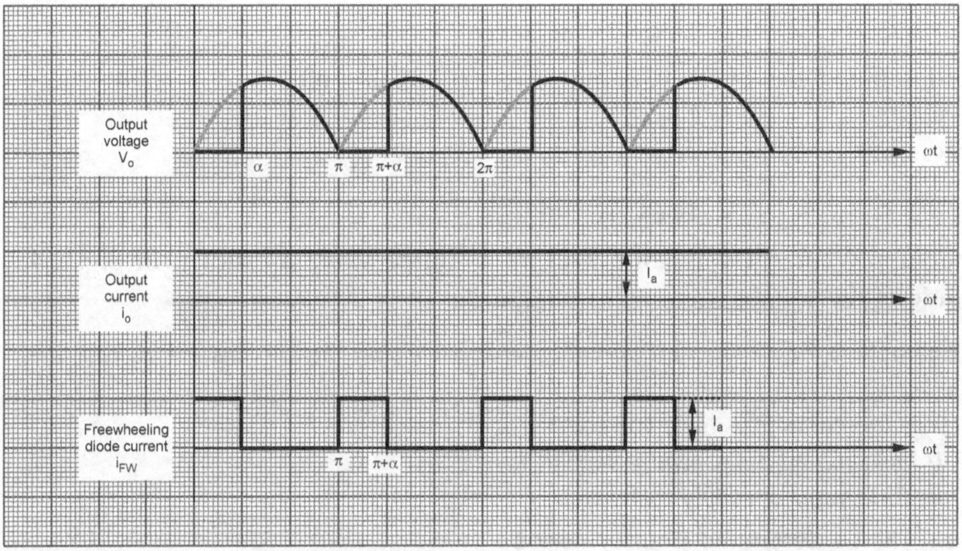

Fig. 2.4.14 Waveforms of 1φ full converter for highly inductive load and freewheeling diode across the load

Example 2.4.6 : *A single phase fully controlled bridge rectifier is fed from 230 V - 50 Hz supply. The load is highly inductive. Find the average load voltage and current if the load resistance is 10 Ω and firing angle is 45°. Draw the supply current waveform.*

Solution : The rms value of the supply voltage is,

$$V_{s\,(rms)} = 230 \text{ V}$$

Hence peak value of supply voltage is,

$$V_m = V_{s\,(rms)}\sqrt{2}$$

$$= 230\sqrt{2}$$

Since the load is highly inductive, the load current can be considered continuous and ripple free as shown in Fig. 2.4.9. For such operation, the average load voltage is given by equation 2.4.5 as,

$$V_{o\,(av)} = \frac{2V_m}{\pi} \cos\alpha$$

The firing angle $\alpha = 45°$. Hence above equation becomes

$$V_{o\,(av)} = \frac{2 \times 230\sqrt{2}}{\pi} \cos 45°$$

$$= 146.42 \text{ volts}$$

The average load current $I_{o(av)}$ or I_a is given as,

$$I_{o\,(av)} = \frac{V_{o\,(av)}}{R}$$

Putting the values of $R = 10\ \Omega$ and $V_{o\,(av)} = 146.42$ volts,

$$I_{o\,(av)} = \frac{146.42}{10}$$

$$= 14.64 \text{ A}$$

The supply current waveform will be a square wave as shown in Fig. 2.4.9. The amplitude of the square wave will be $I_{o\,(av)}$ i.e. 14.64 A.

Example 2.4.7 : *For a 1φ full converter having highly inductive load derive the following :*

(i) Fourier series for supply current

(ii) Fundamental component of supply current

(iii) RMS value of supply current

Solution : (i) To determine Fourier series

The general expression for Fourier series is given as,

$$i_s(\omega t) = I_{s\,(av)} + \sum_{n=1}^{\infty} c_n \sin(n\omega t + \phi_n)$$

Where $$c_n = \sqrt{a_n^2 + b_n^2}$$

and $$\phi_n = \tan^{-1}\left(\frac{a_n}{b_n}\right)$$

Here, $$a_n = \frac{2}{T}\int_0^T i_s(\omega t) \cos n\omega t\, d\omega t$$

$$= \frac{2}{2\pi}\int_0^{2\pi} i_s(\omega t) \cos n\omega t\, d\omega t$$

From the supply current waveform of Fig. 2.4.9 we can write,

$$a_n = \frac{2}{2\pi}\left[\int_{\alpha}^{\pi+\alpha} I_{o(av)} \cos n\omega t \, d\omega t + \int_{\pi+\alpha}^{2\pi+\alpha} \left(-I_{o(av)}\right) \cos n\omega t \, d\omega t\right]$$

$$= \frac{I_{o(av)}}{\pi}\left[\int_{\alpha}^{\pi+\alpha} \cos n\omega t \, d\omega t - \int_{\pi+\alpha}^{2\pi+\alpha} \cos n\omega t \, d\omega t\right]$$

$$= \frac{2 I_{o(av)}}{n\pi} \sin n\alpha \, [\cos n\pi - 1]$$

$$= \begin{cases} \dfrac{-4 I_{o(av)}}{n\pi} \sin n\alpha & \text{for } n = 1,3,5,\ldots\ldots. \\ 0 & \text{for } n = 0,2,4,\ldots\ldots. \end{cases} \quad \ldots (2.4.11)$$

Similarly, $$b_n = \frac{2}{T}\int_{0}^{T} i_s(\omega t) \sin n\omega t \, d\omega t$$

$$= \frac{2}{2\pi}\int_{0}^{2\pi} i_s(\omega t) \sin n\omega t \, d\omega t$$

From supply current of Fig. 2.4.9,

$$b_n = \frac{2}{2\pi}\left[\int_{\alpha}^{\pi+\alpha} I_{o(av)} \sin n\omega t \, d\omega t + \int_{\pi+\alpha}^{2\pi+\alpha} \left(-I_{o(av)} \sin n\omega t \, d\omega t\right)\right]$$

$$= \frac{I_{o(av)}}{\pi}\left[\int_{\alpha}^{\pi+\alpha} \sin n\omega t \, d\omega t - \int_{\pi+\alpha}^{2\pi+\alpha} \sin n\omega t \, d\omega t\right]$$

$$= \frac{2 I_{o(av)}}{n\pi} \cos n\alpha \, [1 - \cos n\pi]$$

$$= \begin{cases} \dfrac{4 I_{o(av)}}{n\pi} \cos n\alpha & \text{for } n = 1,3,5,\ldots\ldots \\ 0 & \text{for } n = 0,2,4,6,\ldots\ldots \end{cases} \quad \ldots (2.4.12)$$

Hence $$c_n = \sqrt{a_n^2 + b_n^2}$$

$$= \left\{\left(\frac{4 I_{o(av)}}{n\pi}\right)^2 \left[\sin^2 n\alpha + \cos^2 n\alpha\right]\right\}^{\frac{1}{2}}$$

$$= \frac{4\, I_{o(av)}}{n\,\pi} \quad \text{for} \quad n = 1, 3, 5, \ldots\ldots \qquad \ldots (2.4.13)$$

And $\phi_n = tan^{-1} \dfrac{a_n}{b_n} = -n\,\alpha$ from equation 2.4.11 and equation 2.4.12.

Thus $\phi_n = -n\,\alpha$ ... (2.4.14)

The average value of supply current is zero. i.e. $I_{s(av)} = 0$. This is clear from Fig. 2.4.9.

Therefore Fourier series is,

$$i_s\,(\omega t) = \sum_{n=1,3,5,\ldots}^{\infty} \frac{4\, I_{o(av)}}{n\pi}\, sin\,(n\omega t - n\,\alpha) \qquad \ldots (2.4.15)$$

(ii) Fundamental component of the supply current

The fundamental component of the supply current is given as,

$$I_{s1} = \frac{C_1}{\sqrt{2}}$$

From equation 2.4.13, $C_1 = \dfrac{4\, I_{o(av)}}{\pi}$ with $n = 1$. Hence above equation will be,

$$\boxed{I_{s1} = \frac{4\, I_{o\,(av)}}{\pi} \times \frac{1}{\sqrt{2}} = \frac{2\sqrt{2}\; I_{o\,(av)}}{\pi}} \qquad \ldots (2.4.16)$$

(iii) To obtain rms value of supply current

The rms value is given as,

$$I_{s\,(rms)} = \left[\frac{1}{T}\int_0^T I_s^2\,(\omega t)\; d\,\omega t\right]^{1/2}$$

From supply current waveform of Fig. 2.4.9,

$$I_{s(rms)} = \left\{\frac{1}{2\pi}\left[\int_{\alpha}^{\pi+\alpha} I_{o(av)}^2\, d\,\omega t + \int_{\pi+\alpha}^{2\pi+\alpha} \left(-I_{o(av)}\right)^2 d\,\omega t\right]\right\}^{\frac{1}{2}}$$

$$\therefore \quad \boxed{I_{s\,(rms)} = I_{o\,(av)}} \qquad \ldots (2.4.17)$$

Example 2.4.8 : *For a 1ϕ full converter having highly inductive load, derive the following :*

(i) Displacement factor (DF)

(ii) Supply power factor (PF)

(iii) Harmonic factor (HF)

iv) Current distortion factor (CDF)

Solution : (i) Displacement factor

The displacement factor (DF) is given as,

$$DF = \cos \phi_1 \qquad \text{... (2.4.18)}$$

From equation 2.4.14 $\phi_n = -n\,\alpha$; Hence $\phi_1 = -\alpha$.

$$\therefore \quad DF = \cos(-\alpha)$$

$$\therefore \quad \boxed{DF = \cos \alpha} \qquad \text{... (2.4.19)}$$

(ii) Supply power factor (PF)

The supply power factor is given as,

$$PF = \frac{I_{s1}}{I_{s(rms)}} \cos \phi_1$$

From result of previous example and equation 2.4.18,

$$PF = \frac{\dfrac{2\sqrt{2}\, I_{o(av)}}{\pi}}{I_{o(av)}} \cos \alpha$$

$$\therefore \quad \boxed{PF = \frac{2\sqrt{2}}{\pi} \cos \alpha} \qquad \text{... (2.4.20)}$$

(iii) Harmonic factor

The harmonic factor (HF) is given as,

$$HF = \sqrt{\left(\frac{I_{s(rms)}}{I_{s1}}\right)^2 - 1} = \left[\frac{I^2_{o(av)}}{\left(\dfrac{2\sqrt{2}\, I_{o(av)}}{\pi}\right)^2} - 1\right]^{\frac{1}{2}}$$

$$\therefore \quad \boxed{HF = 0.4834 \quad \text{or} \quad 48.34\%} \qquad \text{... (2.4.21)}$$

Thus the harmonic factor of supply current is fixed to 0.4834, irrespective of triggering angle.

(iv) Current distortion factor (CDF)

The current distortion factor (CDF) is given as,

$$CDF = \frac{I_{s1}}{I_{s(rms)}}$$

$$= \frac{\frac{2\sqrt{2}\, I_{o(av)}}{\pi}}{I_{o(av)}}$$

$$= \frac{2\sqrt{2}}{\pi} = 0.9 \qquad \text{...(2.4.22)}$$

Example 2.4.9 : *For a 1 ϕ fully controlled bridge having continuous and ripple free current obtain, (i) active power and (ii) reactive power.*

Solution : (i) Active power

Active power is given as,

$$P_{active} = V_s\, I_{s1} \cos\phi_1$$

$$= V_s \cdot \frac{2\sqrt{2}\, I_{o(av)}}{\pi} \cos(-\alpha), \qquad \text{since } \phi_1 = -\alpha$$

$$= 2 \cdot \frac{\sqrt{2}\, V_s\, I_{o(av)}}{\pi} \cos(\alpha)$$

$$= \frac{2 V_m\, I_{o(av)}}{\pi} \cos\alpha \qquad \text{...(2.4.23)}$$

(ii) Reactive power

Reactive power is given as,

$$P_{reactive} = V_s\, I_{s1} \sin\phi_1$$

$$= V_s \cdot \frac{2\sqrt{2}\, I_{o(av)}}{\pi} \sin(-\alpha)$$

$$= -2 \cdot \frac{\sqrt{2}\, V_s\, I_{o(av)}}{\pi} \sin\alpha$$

$$= -\frac{2 V_m\, I_{o(av)}}{\pi} \sin\alpha \qquad \text{...(2.4.24)}$$

The negative sign indicates that the power is reactive.

Comment

Compare the reactive powers of full converter and half converter. They are as follows :

$$P_{reactive}\,(\text{HCB}) = -\frac{V_m\, I_{o(av)}}{\pi} \sin\alpha$$

$$P_{reactive}\,(\text{FCB}) = -\frac{2 V_m\, I_{o(av)}}{\pi} \sin\alpha$$

From above two equations we have,

$$P_{reactive}\text{(FCB)} = 2 \times P_{reactive}\text{ (HCB)}$$

- Thus half controlled bridge draws 50% reactive power compared to that of full controlled bridge.

Example 2.4.10 : *A single phase full converter is operated from a 120 V, 60 Hz supply. The load current with an average value of I_a is continuous, with negligible ripple current. If the turns ratio of the transformer is unity, if the delay angle is*

$\alpha = \frac{\pi}{3}$. *Calculate the*

(i) HF of input current

(ii) DF

(iii) PF

Solution : Given data

Supply voltage, $V_s = 120$

Delay angle, $\alpha = \frac{\pi}{3}$

(i) Harmonic factor (HF)

For continuous load current, the harmonic factor is fixed for full converter. And it is given by equation 2.4.21 as,

$$HF = 0.4834 \text{ or } 48.34 \%$$

(ii) Displacement factor (DF)

For 1 ϕ full converter, DF is given by equation 2.4.19 as,

$$DF = \cos\alpha = \cos\left(\frac{\pi}{3}\right) = \frac{1}{2}$$

(iii) Power factor (PF)

For 1 ϕ full converter, PF is given by equation 2.4.20 as,

$$PF = \frac{2\sqrt{2}}{\pi}\cos\alpha$$

$$= \frac{2\sqrt{2}}{\pi}\cos\left(\frac{\pi}{3}\right)$$

$$= 0.45$$

This is lagging PF, since current lags the voltage.

2.4.3 Inversion in 1ϕ Full Converter

The waveforms of 1ϕ full converter for inductive load are given in Fig. 2.4.9. Observe that the output voltage v_o goes negative for some duration. These intervals are 0 to α, π to π+α, ... and so on. The output current i_o remains positive always. Thus output instantaneous power becomes negative in such intervals. In other words, load power flows to source when v_o goes negative. The average output voltage is given by equation 2.4.5. i.e.,

$$V_{o(av)} = \frac{2V_m}{\pi} \cos\alpha$$

The variation of $V_{o(av)}$ with respect to α is shown in Fig. 2.4.15. In this figure observe that the $V_{o(av)}$ is positive from 0 to $\frac{\pi}{2}$. For $\alpha = 90°$ or $\frac{\pi}{2}$, the $V_{o(av)}$ is zero.

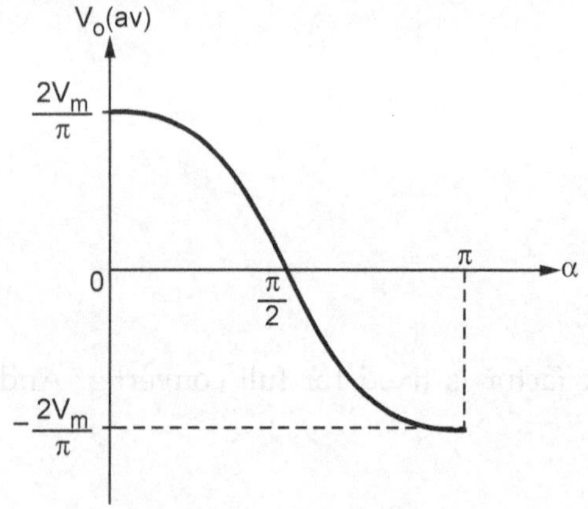

Fig. 2.4.15 Variation of output average voltage with respect to firing angle

Fig. 2.4.15 shows the waveform of v_o for $\alpha = 90°$ or $\frac{\pi}{2}$. The SCRs conduct, current flows in the load but $V_{o(av)} = 0$. This means power fluctuates between load and the source. No power is consumed by the load. The load inductance stores power from source when v_o is positive (i.e. rectification). And this stored power is fed back to the source when v_o is negative (i.e. inversion). When the firing angle is increased above 90°, the average output voltage becomes negative as shown in Fig. 2.4.15. Hence net power is fed from output (load) to the source. But where does this power comes from ? Because load inductance cannot supply more power than it stores. At $\alpha = 90°$, stored power and power supplied to the source are equal. For $\alpha > 90°$, the stored power is less and more power needs to be supplied to the source. Hence an external DC source is to be connected in the load load as shown in Fig. 2.4.17. This DC source maintains the forward bias on the SCRs. Hence they keep on conducting even though $\alpha > 90°$. Such output voltage waveform is shown in Fig. 2.4.16. The average output voltage $V_{o(av)}$ is negative. Hence power flows from load side to the source. The DC supply provides this power. Thus DC power at the output is converted to AC power at the source or input. This is called as *inverting* operation of converter. It is also called as inversion. Note that for $\alpha < 90°$, converter operates in rectification mode. For $\alpha > 90°$, converter operates in inverting and at $\alpha = 90°$, output voltage is zero. Table 2.4.2 gives these details.

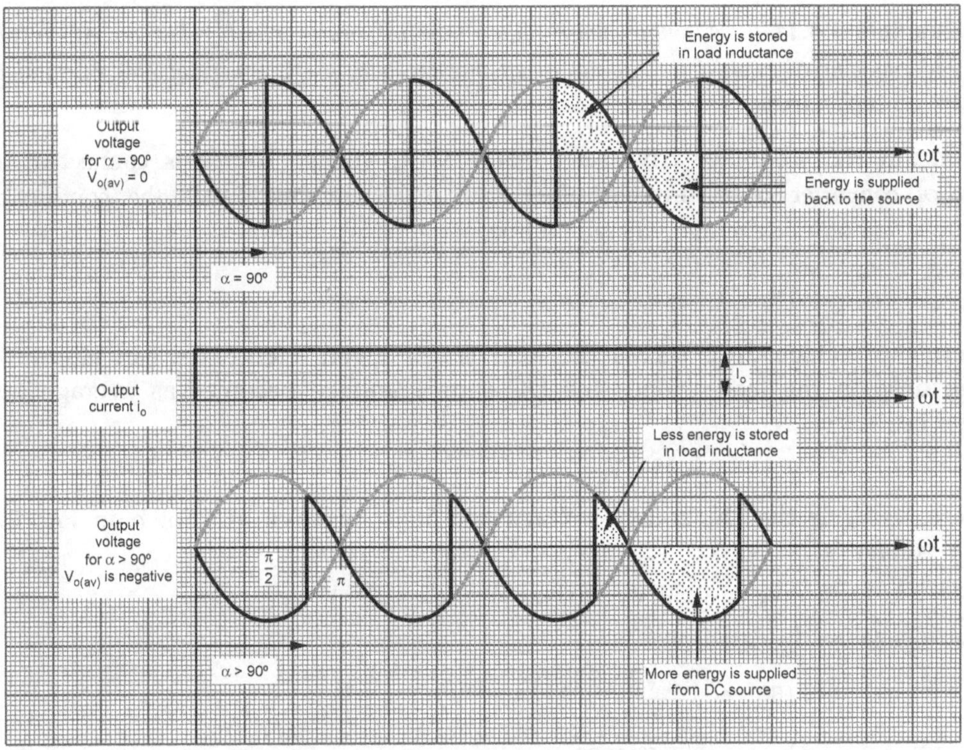

Fig. 2.4.16 Inversion in 1ϕ full converters

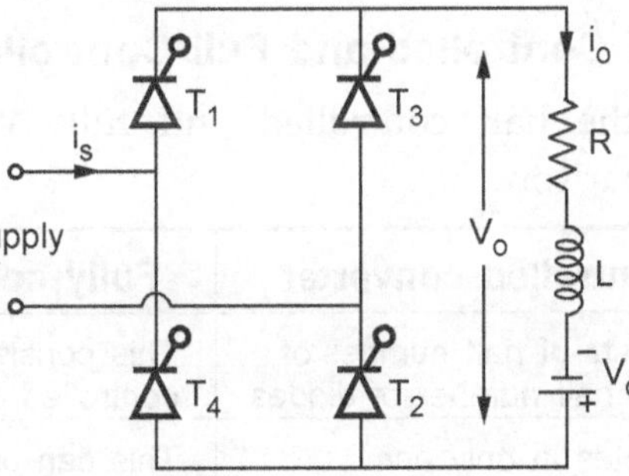

Fig. 2.4.17 Inverting operation in 1ϕ full converter

Triggering angle	Mode of operation	Power flow	Output voltage
$\alpha < 90°$	Rectification	Source to load	$V_{o(av)}$ positive
$\alpha = 90°$	None	None	$V_{o(av)} = 0$
$\alpha > 90°$	Inversion	Load to source	$V_{o(av)}$ negative

Table 2.4.2 Operating modes of 1ϕ full converter

Example 2.4.11 : *A single phase fully controlled bridge rectifier has an AC voltage of 230 rms applied to it. If it is to act as an inverter with a DC source of 150 V, estimate the trigger angle delay.*

Solution : In Fig. 2.4.16 we have seen that the DC voltage source is connected at the output of converter. The converter then acts as an inverter. Since the conduction is continuous the output average voltage will be,

$$V_{o(av)} = \frac{2V_m}{\pi}\cos\alpha \quad \text{... (2.4.25)}$$

Here the voltage drop across the load is not given. Hence we can assume zero voltage drop in the load. Hence the DC source voltage becomes as average output voltage. i.e.,

$$V_{o(av)} = -150 \text{ V}$$

Here 150 V is the DC source voltage and it is negative as shown in Fig. 2.4.16. Supply voltage is,

$$V_s = 230 \text{ V} \quad \text{Hence } V_m = \sqrt{2} \times 230 \ V$$

Putting values in equation 2.4.21,

$$-150 = \frac{2 \times \sqrt{2} \times 230}{\pi}\cos\alpha$$

$$\therefore \quad \cos\alpha = -0.724383$$

$$\therefore \quad \alpha = 136.41°$$

2.4.4 Comparison of Half Controlled and Full Controlled Rectifiers

Now let us compare the half controlled and fully controlled bridge rectifiers. Table 2.4.3 shows this comparison.

Sr. No.	Half controlled converter	Fully controlled converter
1.	This consists of half number of SCRs and half number of diodes.	This consists of all the SCRs as controlled devices.
2.	This operates in only one quadrant.	This can operate in two quadrants.
3.	Output voltage is always positive.	Output voltage can be negative in case of inductive loads.
4.	Inherent freewheeling action is present.	External freewheeling diode is to be connected for freewheeling.
5.	Power factor is better.	Power factor is poor than half converter.
6.	Inversion is not possible.	Inversion is possible.
7.	Used for battery chargers, lighting and heater control.	Used for DC motor drives.

Table 2.4.3 Comparison of half and fully controlled bridges

Review Question

1. *Explain the inversion in 1ϕ full converter.*

Unsolved Examples

1. *A 1ϕ full converter is operated from 230 V/ 50 Hz mains and is delivering power to the resistance R_L = 10 Ω in series wtih a large smoothing inductor. Find out the follwing if α = 45°.*

   i) $V_{o(av)}$ *ii)* $V_{o(rms)}$ *iii) FF iv) RF*

   [Ans. : $V_{o(av)}$ = 146.75, $V_{o(rms)}$ = 230, FF = 1.57, RF = 1.21]

2. *With the help of neat circuit diagram and waveforms explain the operation of 1 ϕ full converter.*

3. *Derive an expression for average and rms output voltage for 1 ϕ full bridge convertor.*

2.5 Three Phase Half Wave Converters

In the preceding sections we discussed 1ϕ semi and full converters. The output power from 1ϕ converters is small. When high power is required, then 3ϕ converters are used. In this section we will discuss the basic of operation of 3ϕ half wave converters. Fig. 2.5.1 shows the circuit diagram of 3ϕ half wave converter.

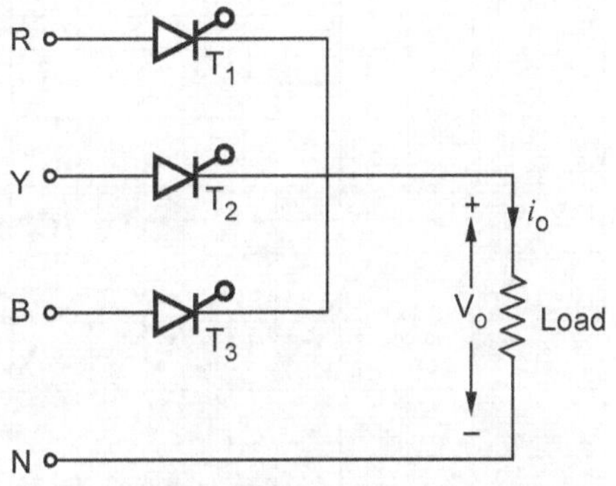

Fig. 2.5.1 3ϕ half wave converter

2.5.1 Operation with Resistive Load

Let us consider the operation of 3ϕ half wave converter having resistive load. The SCRs T_1, T_2 and T_3 are connected to supply R, Y and B respectively. One end of the load is connected to the neutral N. Fig. 2.5.2 shows the phasor diagram of the 3ϕ supply. Observe that all the three phases are separated from each other by 120°. Fig. 2.5.3 shows the waveforms of R, Y and B. In the figure, observe that Y is lagging with respect to R by $\frac{2\pi}{3}$ radians

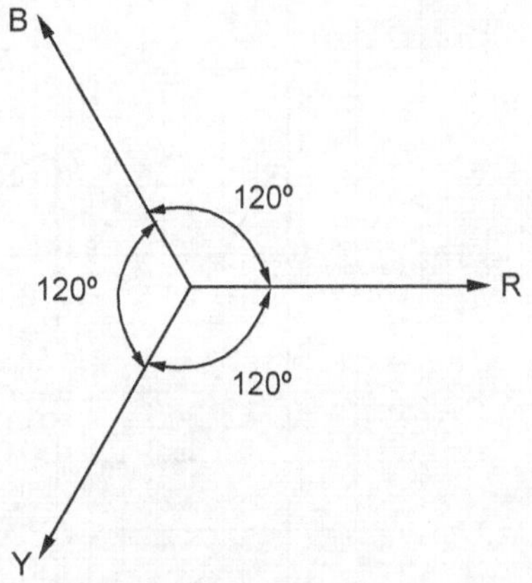

Fig. 2.5.2 Phasor diagram of 3ϕ supply

(i.e. 120°). Similarly B is lagging with respect to Y by $\frac{2\pi}{3}$ radians (i.e. 120°). In the waveforms of Fig. 2.5.3, SCRs are triggered at firing angle of $\alpha = 30°$. For each phase $\alpha = 0$ is actually $\frac{\pi}{6}$ i.e. 30° from zero crossing of that phase. As shown in the output voltage waveform, SCR T_1 conducts from $\left(\frac{\pi}{6} + \alpha\right)$ to π.

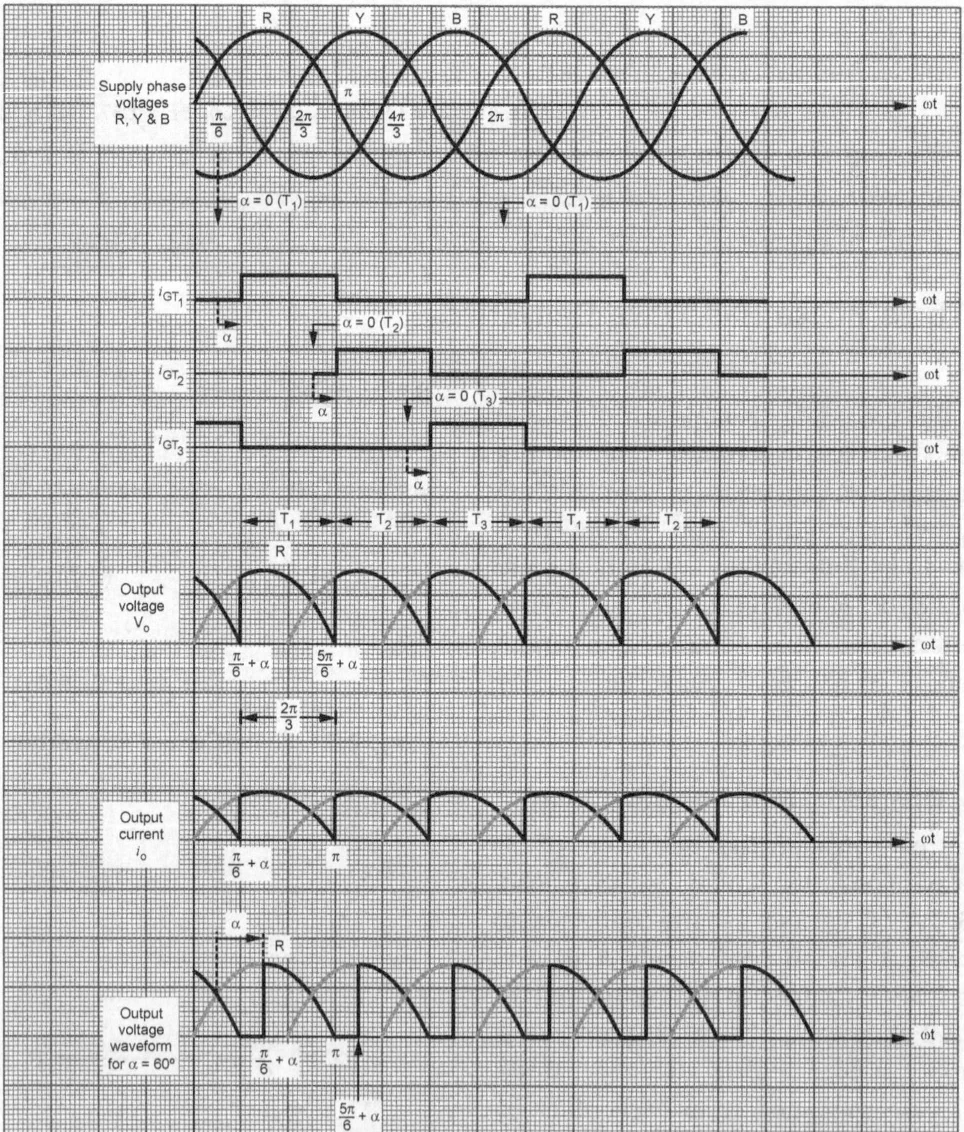

Fig. 2.5.3 Waveforms of 3φ half wave converter for $\alpha = 30°$ for resistive load

Fig. 2.5.4 shows an equivalent circuit when SCR T_1 conducts. T_2 and T_3 are reverse biased because of T_1 conduction. The Y phase has zero crossing at $\frac{2\pi}{3}$. For T_2, $\alpha = 0$ is at $\frac{2\pi}{3}+\frac{\pi}{6}=\frac{5\pi}{6}$. And T_2 is triggered at $\frac{5\pi}{6}+\alpha$. In the waveforms of Fig. 2.5.3, observe that each SCR conducts for the maximum duration of $\frac{2\pi}{3}$ (i.e. 120°). The output current is given as,

$$i_o = \frac{V_o}{R} \quad \text{for resistive load}$$

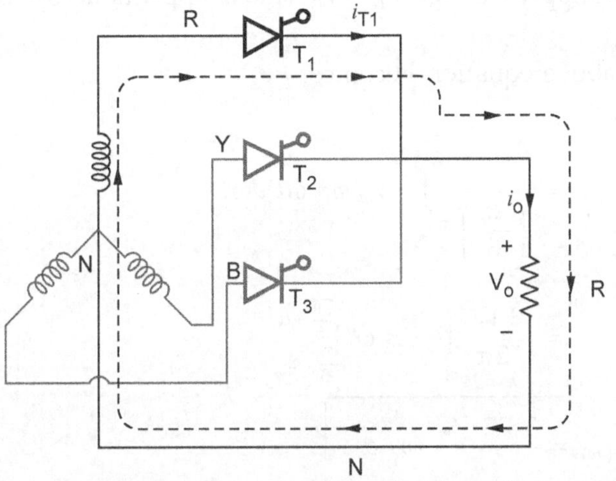

Fig. 2.5.4 Conduction of T_1 in 3ϕ half wave converter. Dotted line shows current path

Hence the shape of the output current is same as output voltage.

The period of each pulse in the output voltage waveform is $\frac{2\pi}{3}$ (i.e. 120°). The supply has the frequency of 50 Hz. There are three pulses of output voltage during one cycle of supply voltage. Hence ripple frequency of output voltage is three times of supply frequency. i.e.,

$$f_{ripple} = 3\times 50 = 150 \text{ Hz}$$

Sometimes this converter is also called as 3-pulse converter.

Example 2.5.1 : *Derive an equation for average output voltage of 3ϕ half wave converter.*

OR

With the help of circuit diagram and relevant waveform, describe the operation of 3ϕ half wave converter and hence device expressions for (i) V_{dc} (ii) V_{dm} and (iii) V_{rms} (Assume load being resistive.)

Solution : (i) To obtain average value of output voltage, $V_{o(av)}$ or V_{dc}

Let the voltage of the R-phase be represented as,

$$V_R = V_m \sin \omega t$$

Here V_m is the peak value of R-phase voltage.

In Fig. 2.5.3, observe that T_1 conducts from $\left(\frac{\pi}{6}+\alpha\right)$ to $\left(\frac{5\pi}{6}+\alpha\right)$. It is the one period of the load ripple. Thus $T = \frac{2\pi}{3}$. Average value of load voltage is given as,

$$V_{o\,(av)} = \frac{1}{T}\int_0^T v_o\,(\omega t)\; d\omega t \qquad \text{... (2.5.1)}$$

Here $T = \frac{2\pi}{3}$ and supply voltage $V_R = V_m \sin \omega t$ appears across output from $\left(\frac{\pi}{6}+\alpha\right)$ to $\left(\frac{5\pi}{6}+\alpha\right)$. Hence above equation becomes,

$$V_{o\,(av)} = \frac{1}{\left(\frac{2\pi}{3}\right)} \int_{\frac{\pi}{6}+\alpha}^{\frac{5\pi}{6}+\alpha} V_m \sin \omega t\; d\omega t$$

$$= \frac{3\,V_m}{2\pi}\left[-\cos \omega t\right]_{\frac{\pi}{6}+\alpha}^{\frac{5\pi}{6}+\alpha}$$

$$\boxed{V_{o\,(av)} = \frac{3\sqrt{3}\,V_m}{2\pi}\cos \alpha} \qquad \text{... (2.5.2)}$$

Note that the above equation is valid only for $\alpha \le 30°$.

Now let us consider the case when $\alpha = 60°$ ($\alpha > 30°$). The last waveform in Fig. 2.5.3 is drawn for $\alpha = 60°$.

SCR T_1 conducts from $\left(\frac{\pi}{6}+\alpha\right)$ to π The output voltage remains zero from π to $\left(\frac{5\pi}{6}+\alpha\right)$. At $\left(\frac{5\pi}{6}+\alpha\right)$ SCR T_2 is triggered. At 'π' T_1 turns off since supply voltage V_R becomes negative. Putting the values in equation 2.5.1 we get,

$$V_{o\,(av)} = \frac{1}{\frac{2\pi}{3}} \int_{\frac{\pi}{6}+\alpha}^{\pi} V_m \sin \omega t\, d\omega t$$

$$= \frac{3\,V_m}{2\pi}\left[-\cos \omega t\right]_{\frac{\pi}{6}+\alpha}^{\pi}$$

$$= \frac{3\,V_m}{2\pi}\left[1+\cos\left(\frac{\pi}{6}+\alpha\right)\right] \qquad \text{... (2.5.3)}$$

Note that this equation is valid only for $\alpha > 30°$, i.e. discontinuous operation for resistive load.

(ii) To obtain peak value of output voltage (V_{dm})

Peak value of output will be obtained when $\alpha = 0$. i.e.,

$$V_{dm} = V_{o(av)}\big|_{\alpha=0}$$

$$= \frac{3\sqrt{3}V_m}{2\pi}\cos 0$$

$$= \frac{3\sqrt{3}V_m}{2\pi}$$

(iii) To obtain rms value of output voltage $V_{o(rms)}$

RMS value is given as,

$$V_{o(rms)} = \left[\frac{1}{T}\int_0^T v_0^2(\omega t)\, d\omega t\right]^{\frac{1}{2}} = \left[\frac{1}{\frac{2\pi}{3}}\int_{\frac{\pi}{6}+\alpha}^{\frac{5\pi}{6}+\alpha} V_m^2 \sin^2 \omega t\, d\omega t\right]^{\frac{1}{2}}$$

$$= \left[\frac{3V_m^2}{2\pi}\int_{\frac{\pi}{6}+\alpha}^{\frac{5\pi}{6}+\alpha} \frac{1-\cos 2\omega t}{2}\, d\omega t\right]^{\frac{1}{2}}$$

$$= V_m\sqrt{1+\frac{3\sqrt{3}}{8\pi}\cos 2\alpha}$$

This is the required expression for rms value of ouput. Above equation is valid for continuous operation, i.e., $\alpha \le 30°$.

Example 2.5.2 : *The 3ϕ half wave converter is operated from a 3 ϕ wye connected 220 V, 60 Hz supply and load resistance R = 10 Ω. If the average output voltage is 25% of the maximum possible average output voltage, calculate.*

(a) Delay angle

(b) rms and average output currents.

Solution : (a) To obtain delay angle (α)

The output average voltage for $\alpha \le 30$ is given by equation 2.5.2 as,

$$V_{o(av)} = \frac{3\sqrt{3}\,V_m}{2\pi}\cos\alpha$$

Hence maximum value of average output voltage will be,

$$V_{o(av)\,max} = \frac{3\sqrt{3}\,V_m}{2\pi}$$

Here V_m is the peak value of phase voltage.

The line voltage is given as 220 V. Hence,

$$V_{ph} = \frac{V_{line}}{\sqrt{3}} = \frac{220}{\sqrt{3}}$$

And peak value of phase voltage will be,

$$V_m = \sqrt{2}\,V_{ph} = \sqrt{2} \times \frac{220}{\sqrt{3}} = 179.63 \text{ V}$$

Hence maximum average output voltage will be,

$$V_{o(av)\,max} = \frac{3\sqrt{3}\,V_m}{2\pi}$$

$$= \frac{3\sqrt{3} \times 179.63}{2\pi} = 148.55 \text{ V}$$

Average output required is 25% of the maximum.

$$\therefore \quad V_{o(av)} = 0.25 \times V_{o(av)\,max}$$

$$= 0.25 \times 148.55 = 37.138 \text{ V}$$

Now let us calculate the output voltage for $\alpha = 30$ using equation 2.5.2 i.e.,

$$V_{o(av)} = \frac{3\sqrt{3} \times 179.63}{2\pi} \cos{(30^\circ)}$$

$$= 128.65 \text{ V}$$

Here observe that 25% output is less than output for $\alpha = 30^\circ$. Hence delay angle must be greater than 30°. For resistive load, output current is discontinuous for $\alpha \geq 30$. Hence we have to use equation 2.5.3 for discontinuous operation. i.e.,

$$V_{o(av)} = \frac{3V_m}{2\pi}\left[1 + \cos\left(\frac{\pi}{6} + \alpha\right)\right]$$

Putting values in above equation,

$$37.138 = \frac{3 \times 179.63}{2\pi}\left[1 + \cos\left(\frac{\pi}{6} + \alpha\right)\right]$$

$$\therefore \quad \alpha = 94.54^\circ \text{ or } 1.65 \text{ radians.}$$

Thus the delay angle is more than 30°, is required to get 25% of maximum output.

(b) rms and average output currents

Average output current can be calculated as,

$$I_{o(av)} = \frac{V_{o(av)}}{R}$$

$$= \frac{37.138}{10} = 3.7138 \text{ A}$$

The rms output voltage for $\alpha > 30^\circ$ can be calculated as,

$$V_{o(rms)} = \left[\frac{1}{\frac{2\pi}{3}} \int_{\frac{\pi}{6}+\alpha}^{\pi} V_m^2 \sin^2 \omega t \, d\omega t\right]^{\frac{1}{2}} d\omega t = \left[\frac{3V_m^2}{2\pi} \int_{\frac{\pi}{6}+\alpha}^{\pi} \sin^2 \omega t \, d\omega t\right]^{\frac{1}{2}} d\omega t$$

$$= \frac{3V_m^2}{2\pi} \int_{\frac{\pi}{6}+\alpha}^{\pi} \frac{1-\cos 2\omega t}{2} d\omega t$$

$$= \left\{\frac{3V_m^2}{4\pi}\left[\frac{5\pi}{6} - \alpha + \frac{1}{2}\sin\left(\frac{\pi}{3} + 2\pi\right)\right]\right\}^{\frac{1}{2}} \quad \text{...(2.5.4)}$$

Putting values in above equation,

$$= \left\{\frac{3 \times 179.63^2}{4\pi}\left[\frac{5\pi}{6} - 1.65 + \frac{1}{2}\sin\left(\frac{\pi}{3} + 2 \times 1.65\right)\right]\right\}^{\frac{1}{2}}$$

$$= 62.12 \text{ V}$$

$\therefore$ rms output current will be,

$$I_{o(rms)} = \frac{V_{o(rms)}}{R} = \frac{62.12}{10} = 6.212 \text{ A.}$$

Example 2.5.3 : *A 3ϕ half wave converter is operated from 3ϕ star connected 208 V, 60 Hz supply and the load resistance is $R = 10\,\Omega$. If it is required to obtain an average output voltage of 50% of the maximum possible output voltage, calculate -*

(i) Delay angle

(ii) rms and average o/p currents

(iii) average and rms thyristor currents

(iv) rectification efficiency

(v) TUF

(vi) Input power factor

Solution : (i) To obtain delay angle

The converter is operated from star connected supply. Hence line voltage is given. i.e.

$$V_{line} = 208 \text{ V}$$

$$\therefore \quad V_{ph} = \frac{208}{\sqrt{3}} \text{ V.}$$

$\therefore$ Peak value of phase voltage is,

$$V_m = \sqrt{2}\, V_{ph}$$

$$= \sqrt{2} \times \frac{208}{\sqrt{3}} = 169.83 \text{ V}$$

Maximum average output can be calculated from equation 2.5.2 as,

$$V_{o(av)\max} = \frac{3\sqrt{3}\, V_m}{2\pi}$$

$$= \frac{3\sqrt{3} \times 169.83}{2\pi} = 140.45$$

Hence 50% of maximum average voltage will be,

$$V_{o(av)} = 0.5 \times 140.45 = 70.225$$

Now let us calculate output for $\alpha = 30$. Hence from equation 2.5.2,

$$V_{o(av)} = \frac{3\sqrt{3}\, V_m}{2\pi} \cos\alpha = \frac{3\sqrt{3} \times 169.83}{2\pi} \cos 30^\circ$$

$$= 121.63 \text{ V}$$

Since this voltage is more than required value of 70.225, the delay angle must be more than 30°. Hence we have to use equation 2.5.3, which is applicable for $\alpha > 30^\circ$. i.e.,

$$V_{o(av)} = \frac{3V_m}{2\pi}\left[1 + \cos\left(\frac{\pi}{6} + \alpha\right)\right]$$

Putting the values in this equation,

$$70.225 = \frac{3 \times 169.83}{2\pi}\left[1 + \cos\left(\frac{\pi}{6} + \alpha\right)\right]$$

$$\therefore \quad \alpha = 67.69^\circ \text{ or } 1.18156 \text{ radians}$$

(ii) rms and average o/p currents

Average output current will be,

$$I_{o(av)} = \frac{V_{o(av)}}{R} = \frac{70.225}{10\ \Omega} = 7.0225 \text{ A.}$$

The rms value of this converter for $\alpha > 30°$ is given by equation 2.5.4 as,

$$V_{o(rms)} = \left\{\frac{3V_m^2}{4\pi}\left[\frac{5\pi}{6} - \alpha + \frac{1}{2}\sin\left(\frac{\pi}{3} + 2\alpha\right)\right]\right\}^{\frac{1}{2}}$$

$$= \left\{\frac{3\times(169.83)^2}{4\pi}\left[\frac{5\pi}{6} - 1.18156 + \frac{1}{2}\sin\left(\frac{\pi}{3} + 2\times 1.18156\right)\right]\right\}^{\frac{1}{2}}$$

$$= 94.7447 \text{ V}$$

$\therefore$ rms output current will be,

$$I_{o(rms)} = \frac{V_{o(rms)}}{R} = \frac{94.7447}{10} = 9.47 \text{ A}$$

(iii) average and rms thyristor currents

There are three *SCRs*. They conduct equally. Hence average SCR current will be,

$$I_{T(av)} = \frac{I_{o(av)}}{3} = \frac{7.0225}{3} = 2.34 \text{ A.}$$

And the rms SCR current will be,

$$I_{T(rms)} = \frac{I_{o(rms)}}{\sqrt{3}} = \frac{9.47}{\sqrt{3}} = 5.467 \text{ A.}$$

(iv) rectification efficiency

Rectification efficiency is given as,

$$\eta = \frac{\text{DC power in load}}{\text{rms power in load}} = \frac{V_{o(av)} I_{o(av)}}{V_{o(rms)} I_{o(rms)}}$$

$$= \frac{70.225 \times 7.0225}{94.7447 \times 9.47} = 0.5496 \text{ or } 54.96\ \%$$

(v) TUF

The rms input line current is same as the thyristor rms current, i.e. 5.467 A. Therefore VA of one phase will be,

VA of one phase $= V_{ph} \times I_{ph}$

$$= \frac{208}{\sqrt{3}} \times 5.467 = 656.525$$

$$\therefore \quad \text{VA of } 3\phi = 3 \times 656 = 1969.577$$

$$\text{Load DC power} = V_{o(av)} I_{o(av)} = 70.225 \times 7.045 = 493.155$$

$$\therefore \quad \text{TUF} = \frac{Load\ DC\ Power}{Transformer\ VA} = \frac{493.155}{1969.577}$$

$$= 0.25 \text{ or } 25\ \%.$$

(vi) Input power factor

The output power is,

$$P_o = I^2_{o(rms)} R$$

$$= (9.47)^2 \times 10 = 996.8 \text{ W}$$

$$\therefore \quad \text{Input power factor} = \frac{output\ power}{supply\ VA}$$

$$= \frac{996.8}{1969.577} = 0.455 \text{ lagging.}$$

2.5.2 Operation with Inductive Load

Fig. 2.5.6 (See on next page) shows the waveforms of 3ϕ half wave converter for highly inductive load. Hence the load current i_o is ripplefree and continuous. SCRT_1 is triggered at $\alpha = 60°$. It starts conducting and supply voltage V_R is applied across the load. The equivalent circuit for this operation is shown in Fig. 2.5.5.

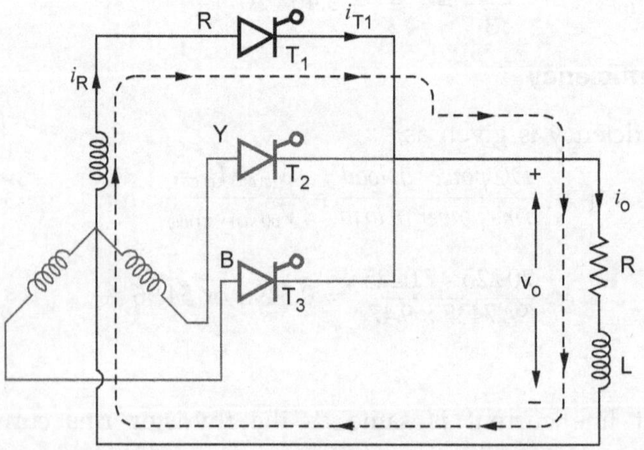

Fig. 2.5.5 3ϕ half wave converter with inductive load

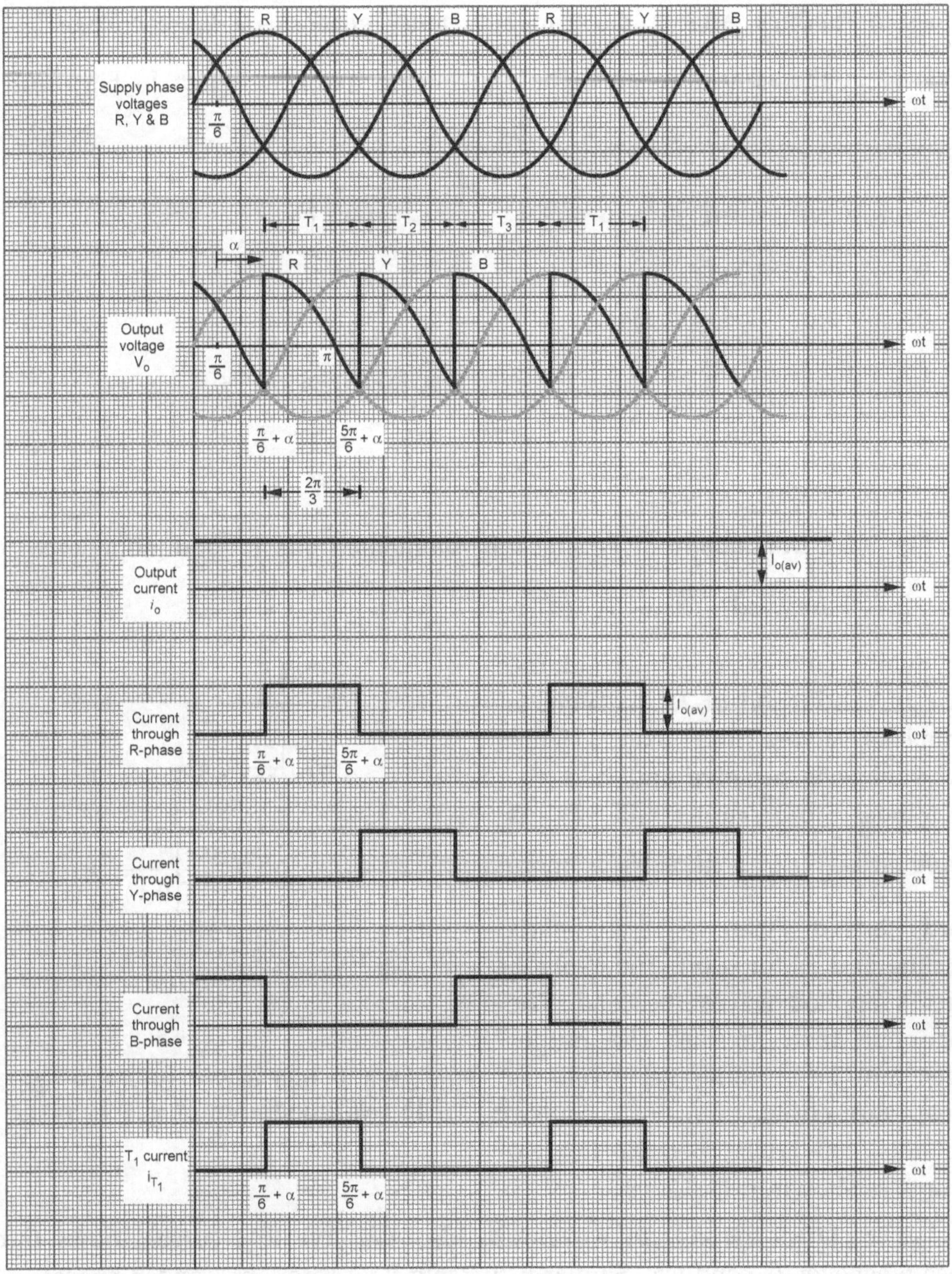

Fig. 2.5.6 Waveforms of 3ϕ half wave converter for highly inductive load

At π, the supply voltage V_R becomes zero. And it becomes negative after π. But T_1 continues to conduct. After π, the load inductance generates a large voltage $L \frac{di_o}{dt}$ to maintain i_o in the same direction. The equivalent circuit is shown in Fig. 2.5.7. The

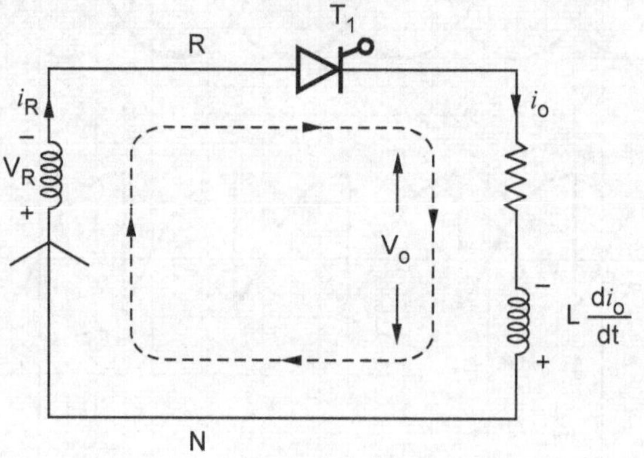

Fig. 2.5.7 Conduction of T_1 after π is maintained due to load inductance

voltage $L \frac{di_o}{dt}$ forward biases T_1 (it is more than V_R). Hence T_1 conducts. Because of conduction of T_1, the voltage V_R appears across the load. Hence output voltage is negative and it is V_R. T_1 continues to conduct till T_2 is triggered at $\left(\frac{5\pi}{6}+\alpha\right)$. Since T_2 is more forward biased than T_1, it starts conducting and T_1 turns off. The Y-phase voltage is applied across the load. The similar cycle then repeats. The output voltage can be positive as well as negative. But current remains positive. Hence this converter operates in two quadrants as shown in Fig. 2.5.8.

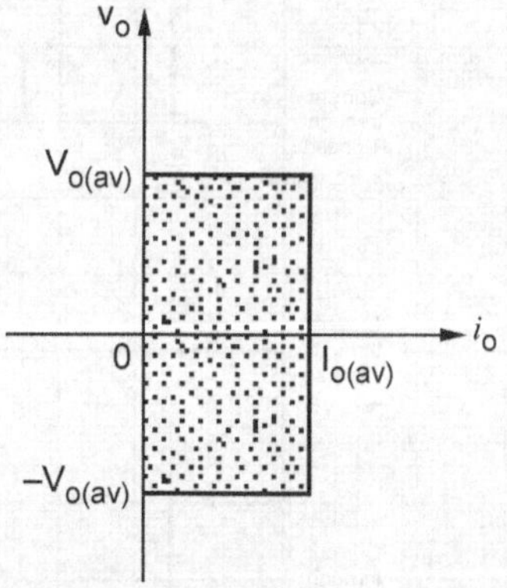

Fig. 2.5.8 3ϕ half wave converter operates in two quadrants of $v_o - i_o$ plane

Fig. 2.5.6 shows the current waveforms also. Observe that current flows through R-phase whenever T_1 conducts. Similarly currents flow through Y-and B-phases.

Example 2.5.4 : *Derive the expression for average output voltage of 3φ half wave converter for highly inductive load.*

Solution : With highly inductive load, the output current can be considered continuous and ripplefree. Observe the output voltage waveform of Fig. 2.5.6. It has the ripple period of $T = \frac{2\pi}{3}$. The supply voltage $V_R = V_m \sin \omega t$ appears across output from $\left(\frac{\pi}{6}+\alpha\right)$ to $\left(\frac{5\pi}{6}+\alpha\right)$. Hence we can write,

$$V_{o(av)} = \frac{1}{T}\int_0^T v_o(\omega t)\, d\omega t$$

$$= \frac{1}{2\pi/3}\int_{\frac{\pi}{6}+\alpha}^{\frac{5\pi}{6}+\alpha} V_m \sin \omega t\, d\omega t$$

$$= \frac{3\,V_m}{2\pi}\left[-\cos \omega t\right]_{\frac{\pi}{6}+\alpha}^{\frac{5\pi}{6}+\alpha}$$

$$\therefore \quad \boxed{V_{o\,(av)} = \frac{3\sqrt{3}\,V_m}{2\pi}\cos\alpha} \qquad \text{... (2.5.5)}$$

Observe that this equation is same as equation 2.5.2 The above equation is valid for continuous conduction in general.

Example 2.5.5 : *A three phase half wave controlled rectifier circuit is operated from 3φ star connected, supply transformer with a line to line ac supply voltage of 440 volts rms, 50 Hz. If the thyristors are triggered at a delay angle of $\alpha = 30°$, load resistance $R_L = 20\ \Omega$, sketch the waveform of*

i) 3 φ, phase supply voltage

ii) Load voltage

iii) Thyristor current through T_1 connected to line A

Derive the expression for the average output voltage for continuous, ripple free load current and calculate :

i) V_{dc} ii) $V_{dc\,(max)}$ iii) I_{dc} iv) $I_{\tau\,(av)}$

Solution : Waveforms for continuous load current

Fig. 2.5.6 shows the waveforms of 3φ half wave controlled rectifier for $\alpha = 60°$. Similar waveforms can be drawn for $\alpha = 30°$. The current flowing through line 'R' flows through thyristor T_1 also. Hence T_1 current will be same as phase R current.

i) To obtain V_{dc} or $V_{o(av)}$

The average output voltage for continuous ripple free output current is given by equation 2.5.5 as,

$$V_{o(av)} = \frac{3\sqrt{3}V_m}{2\pi} \cos\alpha$$

Here V_m is peak value of supply phase voltage. We have

$$V_{line\ (rms)} = 440 \text{ V}$$

$$\therefore \quad V_{ph\ (rms)} = \frac{V_{line}}{\sqrt{3}} = \frac{440}{\sqrt{3}} = 254 \text{ V}$$

$$\therefore \quad V_m = \sqrt{2}\, V_{ph(rms)} = \sqrt{2} \times 254 = 359.26 \text{ V}$$

$$\therefore \quad V_{o(av)} = \frac{3\sqrt{3} \times 359.26}{2\pi} = \cos 30^\circ$$

$$= 257.3 \text{ V}$$

ii) To obtain $V_{dc(max)}$

The maximum value of average output will be obtained at $\alpha = 0$. i.e.,

$$V_{dc(max)} = \frac{3\sqrt{3}\, V_m}{2\pi}$$

$$= \frac{3\sqrt{3} \times 359.26}{2\pi}$$

$$= 297 \text{ V}$$

iii) To obtain I_{dc} or $I_{o(av)}$

Averge output current will be,

$$I_{o(av)} = \frac{V_{o(av)}}{R}$$

$$= \frac{257.3}{20}$$

$$= 12.86 \text{ A}$$

iv) To obtain $I_{T\ (av)}$

Fig 2.5.6 shows the current waveform of thyristor T_1. The average value can be calculated as,

$$I_{T(av)} = \frac{1}{T}\int_0^T i_{T1(\omega t)}\, d\omega t$$

$$= \frac{1}{\pi} \int_{\frac{\pi}{6}+\alpha}^{5\frac{\pi}{6}+\alpha} I_{o(av)} d\omega t \quad = \frac{I_{o(av)}}{\pi} [\omega t]_{\frac{\pi}{6}+\alpha}^{\frac{5\pi}{6}+\pi}$$

$$= \frac{I_{o(av)}}{\pi} \left[\frac{5\pi}{6} + \alpha - \frac{\pi}{6} - \alpha\right]$$

$$= \frac{2}{3} I_{o(av)}$$

Putting the value of $I_{o(av)}$ = 12.86 in above equation,

$$I_{T(av)} = \frac{2}{3} \times 12.86 = 8.57 \text{ A}$$

Review Question

1. *Explain the operation of 3 φ half wave converter. Compare it with bridge converters.*

2.6 Three Phase Semiconverters

2.6.1 Operation with Resistive Load

We have discussed 1φ semiconverter earlier. The 3φ semiconverter delivers more power. It uses three SCRs T_1, T_3 and T_5 and three diodes D_4, D_6 and D_2. Fig. 2.6.1 shows the circuit diagram of 3φ semiconverter. Fig. 2.6.3 (a) shows the waveforms of supply phase voltages R, Y and B. Note that these are phase voltages. These are the voltages with respect to neutral N. In Fig. 2.6.1 (3φ semiconverter), when any SCR and diode conducts, line voltage is applied to the load. Hence it is necessary to draw the line voltage waveforms.

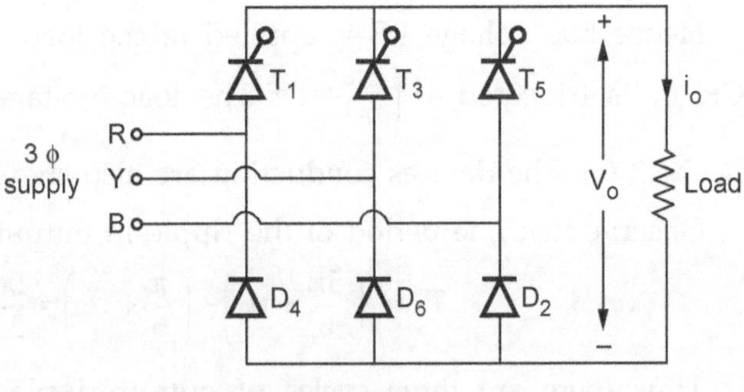

Fig. 2.6.1 3φ semiconverter or half bridge converter

Fig. 2.6.2 shows the phasor diagram of supply phase and line voltages. In this diagram observe that line voltage RB lags phase R by 30°. This is clear from waveforms of Fig. 2.6.3 (b) also. The phase shift between two line voltages is 60°.

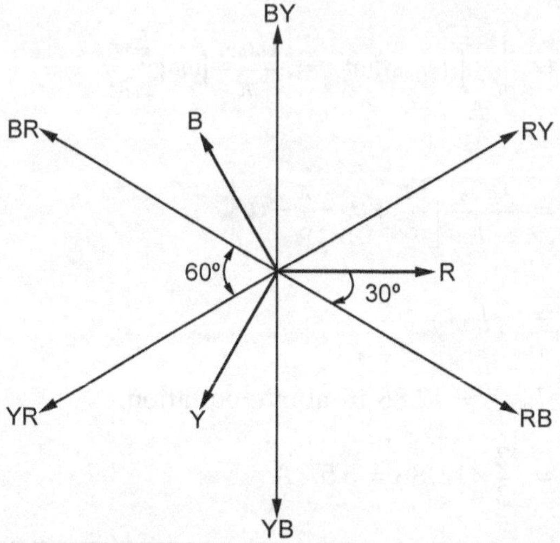

Fig. 2.6.2 Phasor diagram showing the relationship between phase and line voltages of 3φ supply

When α < 60°

T_1 is triggered at $\alpha = 30°$ (see Fig. 2.6.3 (c)). SCR T_1 and diode D_6 conducts. Hence line voltage RY is applied to the load from $\left(\frac{\pi}{6}+\alpha\right)$ to $\frac{\pi}{2}$. At $\frac{\pi}{2}$, diode D_2 is more forward biased and hence it starts conducting.

Hence line voltage RB is applied to the load. T_1 D_2 keeps on conducting till next SCR T_3 is triggered at $\left(\frac{5\pi}{6}+\alpha\right)$. The load voltage waveform for $\alpha = 30°$ is shown in Fig. 2.6.3 (c). The devices conducting are also shown in respective intervals.

Observe that one period of the ripple in output voltage waveform is,

$$T = \left(\frac{5\pi}{6}+\alpha\right)-\left(\frac{\pi}{6}+\alpha\right)=\frac{2\pi}{3} \quad \text{... (2.6.1)}$$

Thus there are three cycles of output ripple in one cycle of the supply. Hence ripple frequency is three times of the supply frequency. i.e.,

$$f_{ripple} = 3\times 50 = 150 \text{ Hz}$$

In the Fig. 2.6.3 (c), observe that each SCR conducts for the maximum duration of $120°\left(\text{i.e. } \frac{2\pi}{3}\right)$.

The current waveform for resistive load will be similar to voltage waveform since,

$$i_o = \frac{V_o}{R}$$

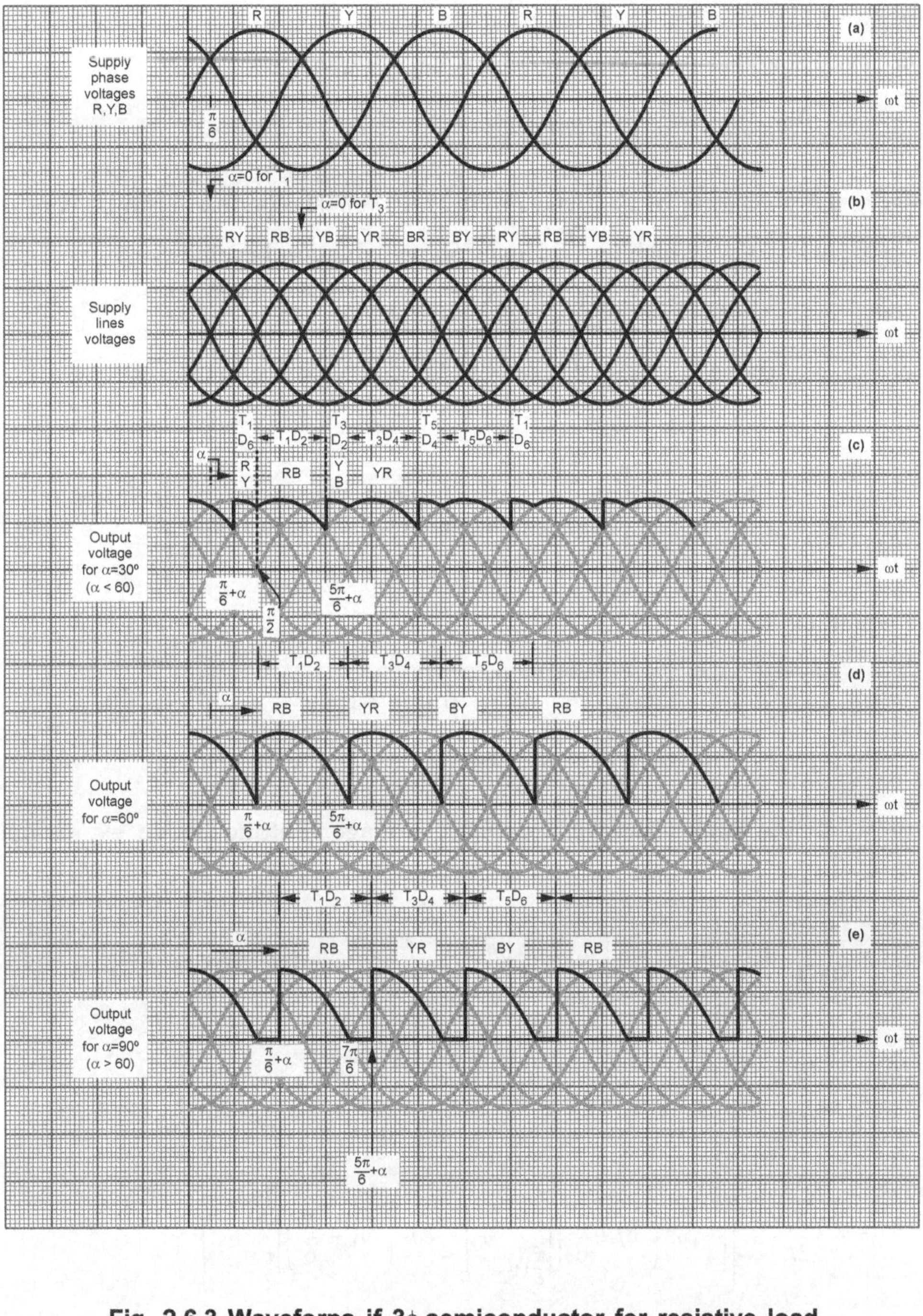

Fig. 2.6.3 Waveforns if 3ϕ semiconductor for resistive load

Example 2.6.1 : *Derive an expression for the average output voltage of 3φ semiconverter having resistive load for α ≤ 60°.*

Solution : We know that the average output voltage is given as,

$$V_{o(av)} = \frac{1}{T}\int_0^T v_o(\omega t)\, d\omega t$$

Observe the waveform of V_o given in Fig. 2.6.3 (c). The period T can be considered from $\left(\frac{\pi}{6}+\alpha\right)$ to $\left(\frac{5\pi}{6}+\alpha\right)$ which is $\frac{2\pi}{3}$. Hence above equation becomes,

$$V_{o(av)} = \frac{1}{\left(\frac{2\pi}{3}\right)}\int_{\frac{\pi}{6}+\alpha}^{\frac{5\pi}{6}+\alpha} v_o(\omega t)\, d\omega t$$

$$= \frac{3}{2\pi}\left\{\int_{\frac{\pi}{6}+\alpha}^{\frac{\pi}{2}} V_{RY}(\omega t)\, d\omega t + \int_{\frac{\pi}{2}}^{\frac{5\pi}{6}+\alpha} V_{RB}(\omega t)\, d\omega t\right\} \quad \text{... (2.6.2)}$$

The equations for V_{RY} and V_{RB} can be written from Fig. 2.6.3 (b) as follows,

$$\left.\begin{aligned} V_{RY}(\omega t) &= \sqrt{3}\, V_m \sin\left(\omega t+\frac{\pi}{6}\right) \\ V_{RB}(\omega t) &= \sqrt{3}\, V_m \sin\left(\omega t-\frac{\pi}{6}\right) \end{aligned}\right\} \quad \text{... (2.6.3)}$$

Here V_m is the peak value of the phase voltage. Putting above expressions in equation 2.6.2,

$$V_{o(av)} = \frac{3}{2\pi}\left\{\int_{\frac{\pi}{6}+\alpha}^{\frac{\pi}{2}} \sqrt{3}\, V_m \sin\left(\omega t+\frac{\pi}{6}\right) d\omega t + \int_{\frac{\pi}{2}}^{\frac{5\pi}{6}+\alpha} \sqrt{3}\, V_m \sin\left(\omega t-\frac{\pi}{6}\right) d\omega t\right\}$$

$$= \frac{3\sqrt{3}\, V_m}{2\pi}\left\{\int_{\frac{\pi}{6}+\alpha}^{\frac{\pi}{2}} \sin\left(\omega t+\frac{\pi}{6}\right) d\omega t + \int_{\frac{\pi}{2}}^{\frac{5\pi}{6}+\alpha} \sin\left(\omega t-\frac{\pi}{6}\right) d\omega t\right\}$$

$$= \frac{3\sqrt{3}\, V_m}{2\pi}\left\{\left[-\cos\left(\omega t+\frac{\pi}{6}\right)\right]_{\frac{\pi}{6}+\alpha}^{\frac{\pi}{2}} + \left[-\cos\left(\omega t-\frac{\pi}{6}\right)\right]_{\frac{\pi}{2}}^{\frac{5\pi}{6}+\alpha}\right\}$$

$$= \frac{3\sqrt{3}\, V_m}{2\pi}\left\{-\cos\left(\frac{\pi}{2}+\frac{\pi}{6}\right) + \cos\left(\frac{\pi}{6}+\alpha+\frac{\pi}{6}\right) - \cos\left(\frac{5\pi}{6}+\alpha-\frac{\pi}{6}\right) + \cos\left(\frac{\pi}{2}-\frac{\pi}{6}\right)\right\}$$

$$= \frac{3\sqrt{3}\,V_m}{2\pi}(1+\cos\,\alpha) \qquad \text{... (2.6.4)}$$

This is the expression for average output voltage for $\alpha \le 60$.

When $\alpha = 60$

Fig. 2.6.3 (d) shows the output voltage waveform for $\alpha = 60°$. Observe that the voltage waveform is just continuous. When T_1 is triggered at $\alpha = 60°$, the line voltage RB is applied across the load. T_1 and D_2 conducts from $\left(\frac{\pi}{6}+\alpha\right)$ to $\left(\frac{5\pi}{6}+\alpha\right)$. Similarly next SCR T_3 is triggered at $\left(\frac{5\pi}{6}+\alpha\right)$ and T_3 D_4 conduct. The current waveform will be similar to voltage waveform for resistive load. The average output voltage is given by equation 2.6.4, since voltage waveform is continuous.

When $\alpha > 60$

Fig. 2.6.3 (e) shows the output voltage waveform for $\alpha = 90°$. SCR T_1 is triggered at $\left(\frac{\pi}{6}+\alpha\right)$. T_1 and D_2 conducts and line voltage RB is applied across the load. In the waveform observe that, output voltage becomes zero at $\frac{7\pi}{6}$. In Fig. 2.6.3 (b) observe that line voltage RB becomes zero at $\frac{7\pi}{6}$. Hence SCR T_1 is turned off. Since T_3 is not triggered, the output voltage becomes zero. At $\left(\frac{5\pi}{6}+\alpha\right)$ T_3 is triggered and T_3 D_4 conducts. Line voltage YR is applied across the load. Thus for $\alpha > 60$, the output voltage is discontinuous. Since the load is resistive, the current is also discontinuous. The current waveform will be similar to voltage waveform.

Example 2.6.2 : *Derive an expression for average output voltage of 3ϕ semiconverter having resistive load for $\alpha > 60°$.*

Solution : Fig. 2.6.3 (e) shows the waveform of output voltage for $\alpha = 90°$ (i.e. $\alpha > 60°$). Observe that the period of ripple cycle is

$$T = \left(\frac{5\pi}{6}+\alpha\right)-\left(\frac{\pi}{6}+\alpha\right) = \frac{2\pi}{3}$$

When $T_1 - D_2$ conducts voltage V_{RB} is applied across the load from $\left(\frac{\pi}{6}+\alpha\right)$ to $\frac{7\pi}{6}$.

Hence

$$v_o = V_{RB} \text{ from } \frac{\pi}{6}+\alpha \text{ to } \frac{7\pi}{6}$$

From Fig. 2.6.3, we can write an equation for V_{RB} as,

$$V_{RB} = \sqrt{3}\, V_m \sin\left(\omega t - \frac{\pi}{6}\right)$$

Output average voltage is given as,

$$V_{o\,(av)} = \frac{1}{T}\int_0^T v_o\,(\omega t)\; d\,\omega t$$

We know that $v_O\,(\omega t) = V_{RB}$, hence above equation can be written as,

$$V_{o\,(av)} = \frac{1}{2\pi/3}\int_{\frac{\pi}{6}+\alpha}^{\frac{7\pi}{6}} \sqrt{3}\, V_m \sin\left(\omega t - \frac{\pi}{6}\right) d\omega t$$

$$= \frac{3\sqrt{3}\, V_m}{2\pi}\left[-\cos\left(\omega t - \frac{\pi}{6}\right)\right]_{\frac{\pi}{6}+\alpha}^{7\pi/6}$$

$$= \frac{3\sqrt{3}\, V_m}{2\pi}\left[-\cos\left(\frac{7\pi}{6} - \frac{\pi}{6}\right) + \cos\left(\frac{\pi}{6} + \alpha - \frac{\pi}{6}\right)\right]$$

$$\therefore \quad \boxed{V_{o\,(av)} = \frac{3\sqrt{3}\, V_m}{2\pi}\,(1+\cos\,\alpha)} \qquad \text{... (2.6.5)}$$

Observe that the above equation is same as equation 2.6.4 for $\alpha \le 60$. Thus above equation can be used for continuous as well as discontinuous mode of operation of 3ϕ semiconverter.

Example 2.6.3 : *With the circuit diagram and waveforms explain operation of a 3ϕ half controlled full wave bridge rectifier. The SCRs are triggered at a delay angle of 40°. The input voltage is 440 V, 3ϕ, 50 Hz. Find average DC voltage available at the bridge terminals. Derive any formula used.*

Solution : This is 3ϕ semiconverter. Since the load is not mentioned, we will assume resistive load. The operation of this converter is explained with waveforms in section 2.6.1. The given data is,

$$\alpha = 40°$$

$$V_{s(line)} = 440 \text{ V}$$

$$\therefore \quad V_{s(ph)} = \frac{440}{\sqrt{3}}$$

$$\therefore \quad V_m = \sqrt{2}\, V_{s(ph)} = \sqrt{2}\cdot\frac{440}{\sqrt{3}} = \sqrt{\frac{2}{3}}\; 440 \text{ V}$$

We have derived an expression for average value of output voltage earlier. For $\alpha \leq 60$, $V_{o(av)}$ is given by equation 2.6.5 as,

$$V_{o(av)} = \frac{3\sqrt{3}\, V_m}{2\pi}(1 + \cos\alpha)$$

Here V_m is peak value of phase voltage. Putting values in above equation,

$$V_{o(av)} = \frac{3\sqrt{3}\cdot\sqrt{\frac{2}{3}}\cdot 440}{2\pi}(1+\cos 40°)$$

$$= 524.7 \text{ volts.}$$

2.6.2 Operation with Inductive Load

We have discussed the operation of 3ϕ semiconverter with resistive load. Now let us consider the operation with highly inductive load. Hence the load current i_o can be considered continuous and ripplefree. Fig. 2.6.4 shows the circuit diagram of 3ϕ semiconverter driving highly inductive load.

Consider that SCRs are triggered at firing angle of 90°. Fig. 2.6.5 shows the waveforms of this converter. Fig. 2.6.5 (a) shows the supply phase voltages R, Y and

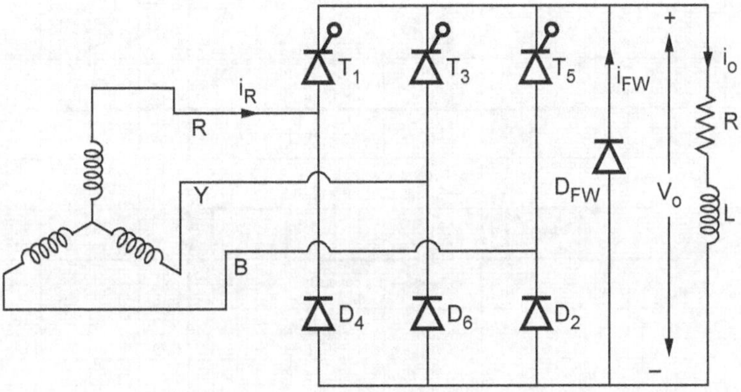

Fig. 2.6.4 3ϕ semiconverter driving inductive load

B. Fig. 2.6.5 (b) shows the supply line voltages. These waveforms are drawn according to the phasor diagram of 3ϕ supply shown in Fig. 2.6.2. Fig. 2.6.5 (c) shows the gate drives of SCRs T_1, T_2 and T_3 for $\alpha = 90°$. Fig. 2.6.5 (d) shows the output voltage waveform v_o. Observe that T_1 is triggered at $\left(\frac{\pi}{6}+\alpha\right)$. Line voltage V_{RB} is maximum at the time of triggering. Hence $T_1 - D_2$ conducts and line voltage V_{RB} is applied across

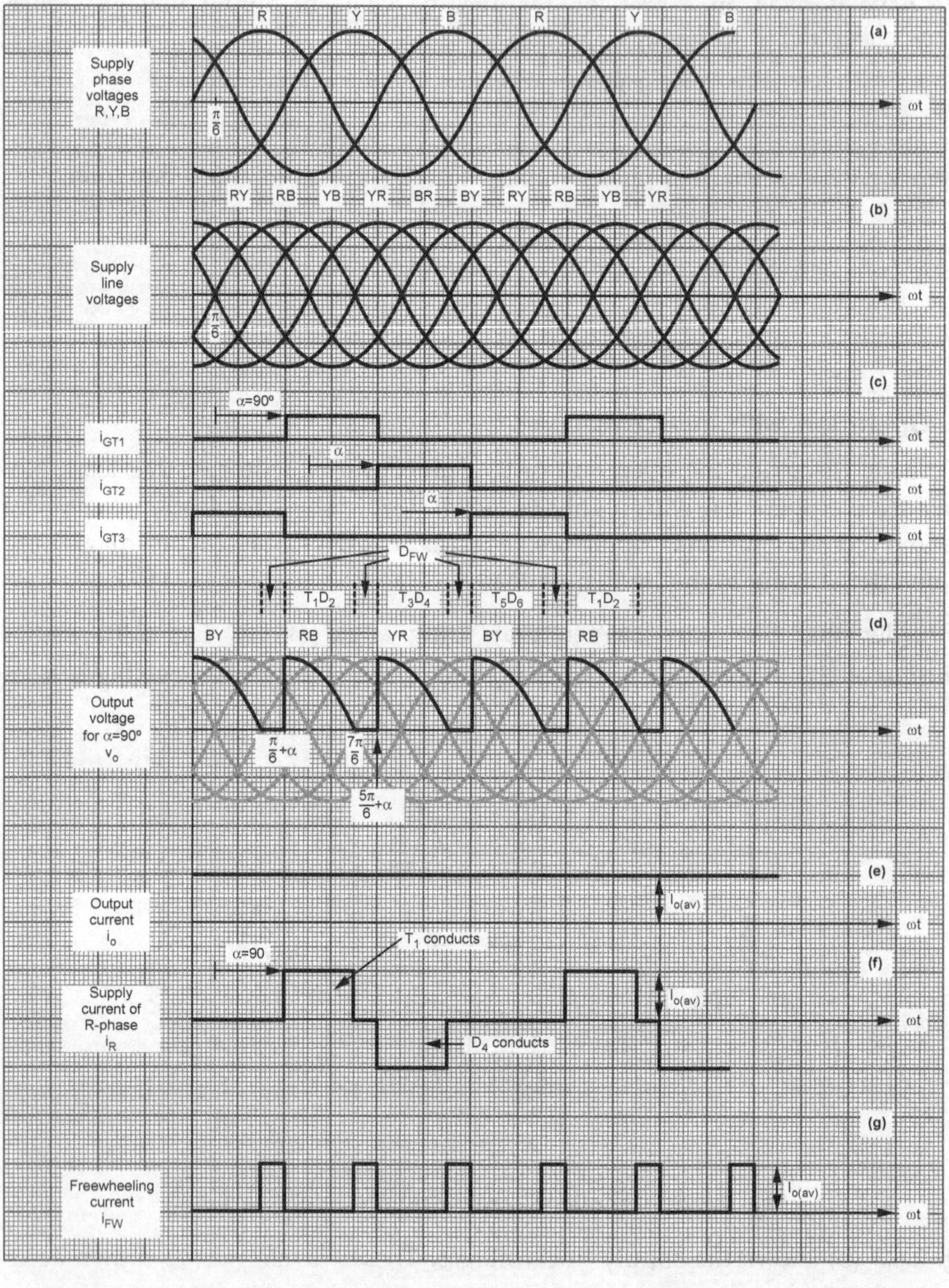

Fig. 2.6.5 Waveforms of 3ϕ semiconverter for highly inductive load

the load. At $\frac{7\pi}{6}$, the line voltage $V_{RB} = 0$, but the output current i_o [Fig. 2.6.5 (e)] is not zero. The output current waveform is continuous and ripplefree. The load inductance generates a very large voltage $L\frac{di_o}{dt}$ to maintain i_o continuous. This situation is shown in Fig. 2.6.6.

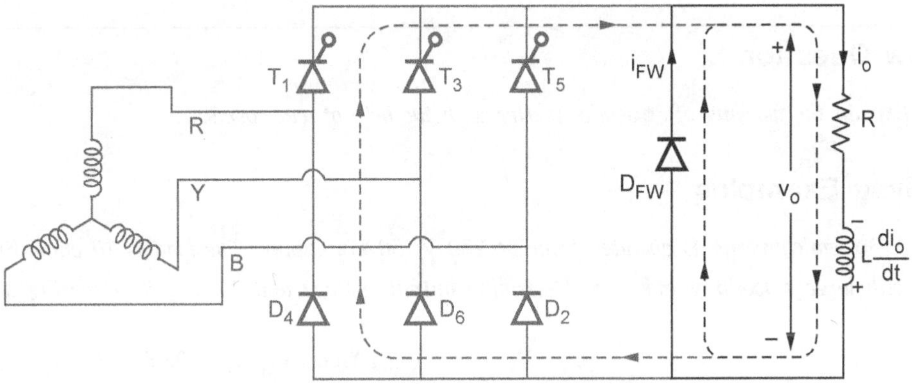

Fig. 2.6.6 Freewheeling action in 3φ semiconverter. Dotted lines shows freewheeling current paths

The freewheeling diode D_{FW} is forward biased by the load inductance voltage $L\frac{di_o}{dt}$. The freewheeling current i_{FW} is basically i_o. Thus the energy stored in the load inductance is fed back to the load itself. Fig. 2.6.5 (g) shows the waveform of freewheeling current. SCR T_1 turns off at $\frac{7\pi}{6}$ as soon as freewheeling diode starts conducting. This is because freewheeling diode is more forward biased compared to T_1 and D_1 after $\frac{7\pi}{6}$. If extra freewheeling diode D_{FW} is not connected, then freewheeling current flows through T_1 and D_4. This is shown by 'thin dotted line' in Fig. 2.6.6. Thus freewheeling action is inherent in 3φ semiconverter.

If we neglect the drop of freewheeling diode, then output voltage during freewheeling period is zero. As shown in Fig. 2.6.5 (f), no supply current flows during freewheeling period. Compare the output voltage waveform of Fig. 2.6.3 (e) (resistive load) and Fig. 2.6.5 (d) (inductive load). Both the waveforms are same.

Example 2.6.4 : *Derive an expression for average output voltage of 3φ semiconverter having highly inductive load.*

Solution : We have seen that the output voltage waveforms of 3φ semiconverter are same for resistive as well as inductive loads. Hence their average values are also same. Hence from equation 2.6.5 we have,

$$V_{o\,(av)} = \frac{3\sqrt{3}\,V_m}{2\pi}(1+\cos\alpha) \qquad \text{... (2.6.6)}$$

This is the expression for average output voltage for resistive as well as inductive loads and $0 \le \alpha \le \pi$.

Review Question

1. *Explain the working of 3ϕ semiconverter with the help of waveforms.*

Unsolved Examples

1. *A 3ϕ semiconverter is operated from 3ϕ 230 V, 50 Hz supply. The load is 10 ohms in series with large smoothing inductor. Determine output voltage and current if triggering angle is 60°.*

[Ans. : $V_{o(av)} = 403.5\,V$, $I_{o(av)} = 40.35\,A$]

2. *Derive an expression for average value of output voltage for 3ϕ semiconverter.*

2.7 Three Phase Full Converters

Three phase half converters operate only in first quadrant of $v_o - i_o$. The output voltage v_o is always positive for resistive as well as inductive loads. The output current i_o is also always positive. Hence 3ϕ semiconverter operates in first quadrant only. Three phase full converters can operate in two quadrants. The output voltage of 3ϕ full converter can be positive as well as negative. It uses six SCRs as shown in Fig. 2.7.1.

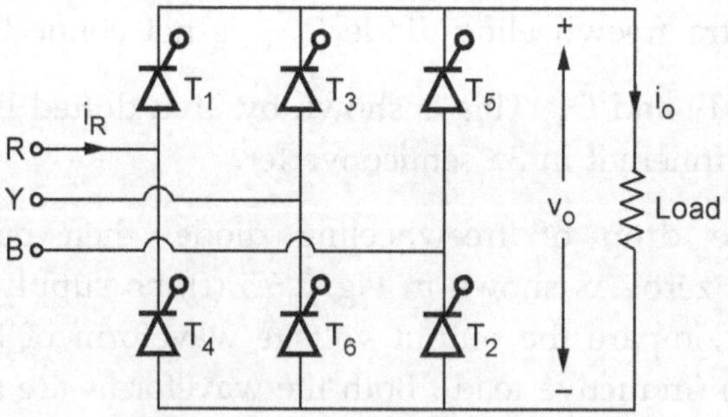

Fig. 2.7.1 3ϕ full converter

2.7.1 Operation with Resistive Load

Let us consider the operation of 3ϕ full converter having resistive load. Fig. 2.7.2 shows the waveforms of 3ϕ full converter having resistive load. Fig. 2.7.2 (a) shows the supply phase voltages R, Y and B. Fig. 2.7.2 (b) shows the supply line voltages. These supply voltage waveforms are drawn according to the phasor diagram shown in

Fig. 2.6.2. Fig. 2.7.2 (c) shows the gate drives for $\alpha = 30°$. For six SCRs, there are six gate drives.

In Fig. 2.7.2, observe that in interval-I, gate drives are given to SCRs T_6 and T_1. Hence line voltage V_{RY} is applied across the load. The equivalent circuit for this interval is shown in Fig. 2.7.3.

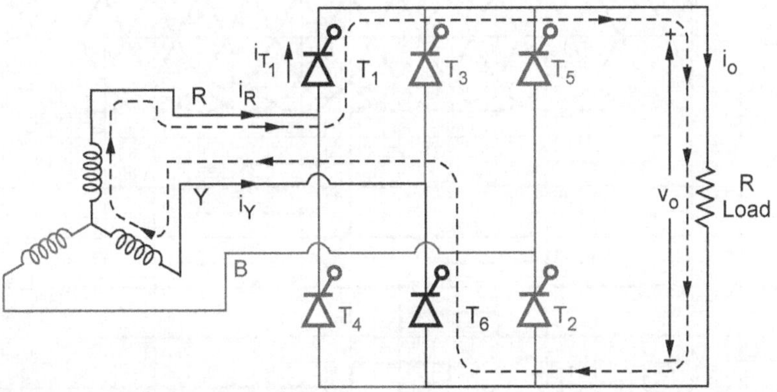

Fig. 2.7.3 Operation of 3ϕ full converter in interval-I ($T_6 - T_1$)

In the above figure observe that SCRs T_6 and T_1 (normally written as 6 – 1) conduct. Hence,

$$v_o = v_{RY}$$

Observe that output current i_o and R-phase current i_R flows in the same direction. Hence,

$$i_R = i_o$$

Similarly observe that Y-phase current i_y and output current i_o are in opposite directions. Hence,

$$i_Y = -i_o$$

The SCR pair $T_6 - T_1$ conduct from $\left(\frac{\pi}{6} + \alpha\right)$ to $\left(\frac{\pi}{2} + \alpha\right)$. Line voltage V_{RY} is applied during this period. At $\left(\frac{\pi}{2} + \alpha\right)$ SCR T_2 is triggered (Fig. 2.7.2 (c)). Here note that T_6 turns off, since T_2 is triggered. Hence $T_1 - T_2$ starts conducting and it is marked as interral II. In this interval supply line voltage V_{RB} is applied across the load. At $\left(\frac{5\pi}{6} + \alpha\right)$, T_3 is triggered. Hence T_1 turns off and $T_2 - T_3$ starts conducting. Therefore line voltage V_{YB} is applied across the load. It is marked as interval III.

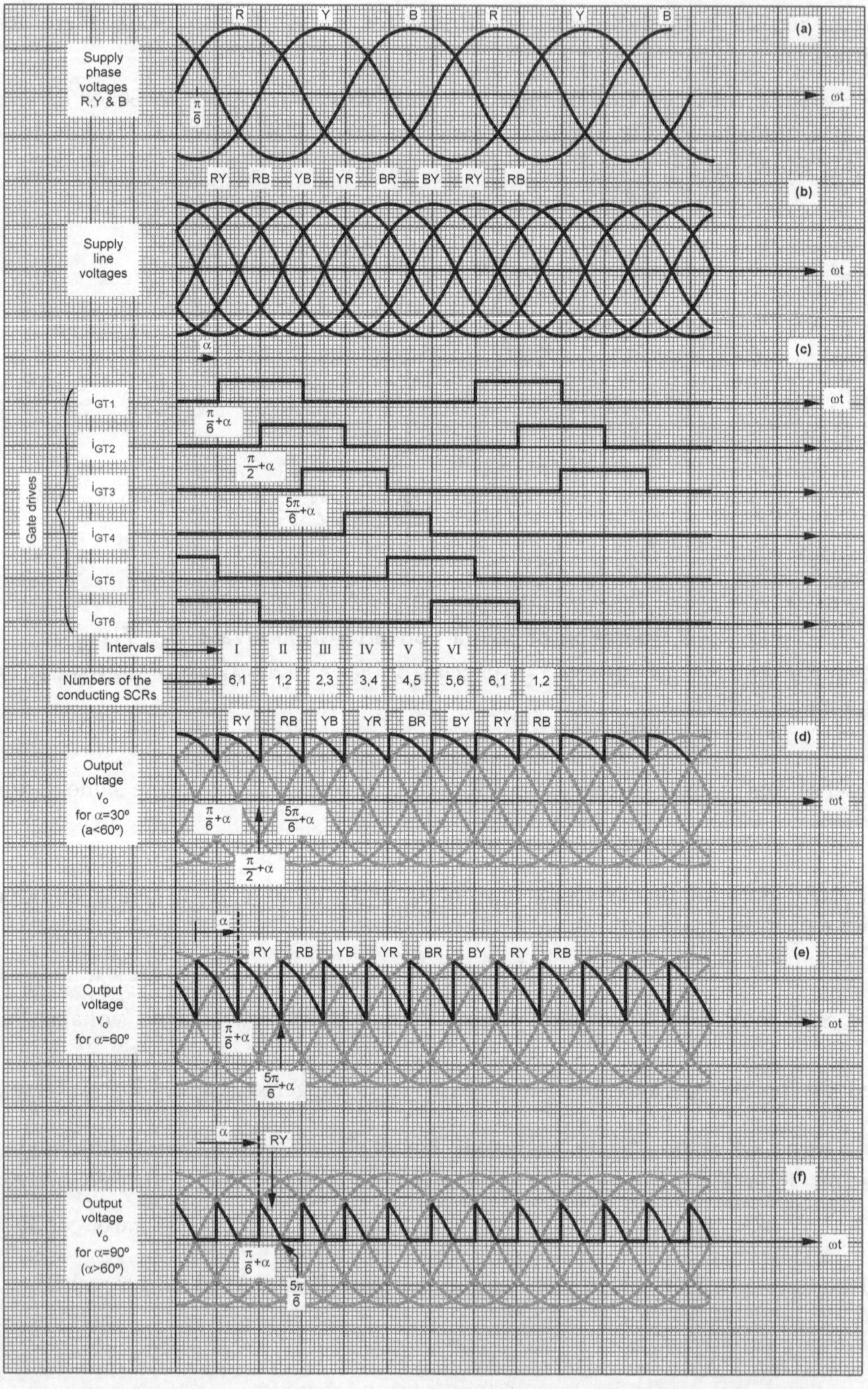

Fig. 2.7.2 Waveforms of 3φ full converter having resistive load

The following points are important about 3ϕ full converter operation.

(i) Only two SCRs conduct in any interval.

(ii) Each SCR conduct for 120°.

(iii) Each SCR pair conduct for one interval of 60°.

(iv) SCRs are triggered in following sequence :

$........T_1 - T_2 - T_3 - T_4 - T_5 - T_6 - T_1 - T_2 ....$

The triggering delay between individual SCRs is 60°.

(v) The output voltage waveforms is continuous upto $\alpha = 60°$, and it is discontinuous for $\alpha > 60°$. Since the load is resistive, output current waveform is similar to output voltage waveform.

Example 2.7.1 : *For the 3ϕ full converter having resistive load find the following :*
(i) ripple frequency f_{ripple}
(ii) output average voltage $V_{o\,(av)}$

Solution : (i) To determine ripple frequency :

In the load voltage waveforms of Fig. 2.7.2 observe that line voltage V_{RY} is applied across the load from $\left(\frac{\pi}{6}+\alpha\right)$ to $\left(\frac{\pi}{2}+\alpha\right)$ i.e. $\frac{\pi}{3}$. Similarly other line voltages are also applied across the load for the period of $\frac{\pi}{3}$. The load voltage has the ripple period of $\frac{\pi}{3}$. In one cycle of supply, six such ripple cycles are present in output waveform. Hence ripple frequency must be six times of the supply frequency. i.e.,

$$f_{ripple} = 6 \times 50 = 300 \text{ Hz}$$

Thus ripple frequency of 3ϕ full converter is higher than 3ϕ semiconverter. This is true for all the firing angles.

(ii) To determine output average voltage

We know that average value is given as,

$$V_{o(av)} = \frac{1}{T}\int_0^T v_o(\omega t)\, d\omega t \qquad ... (2.7.1)$$

Case - I : $\alpha \le 60°$

Consider the case when $\alpha \le 60°$ i.e. for continuous output voltage waveform. Fig. 2.7.2 (d) shows the output voltage waveform for $\alpha = 30°$. Consider the period $\left(\frac{\pi}{6}+\alpha\right)$ to $\left(\frac{\pi}{2}+\alpha\right)$ when voltage V_{RY} is applied across the load. This period is,

$$T = \left(\frac{\pi}{2}+\alpha\right)-\left(\frac{\pi}{6}+\alpha\right) = \frac{\pi}{3}$$

From Fig. 2.7.2 (b) we can write an equation for line voltage V_{RY} as,

$$V_{RY} = \sqrt{3}\, V_m \sin\left(\omega t+\frac{\pi}{6}\right)$$

V_m is peak value of the phase voltage.

Putting the above values in equation 2.7.1,

$$V_{o(av)} = \frac{1}{\pi/3}\int_{\frac{\pi}{6}+\alpha}^{\frac{\pi}{2}+\alpha} \sqrt{3}\, V_m \sin\left(\omega t+\frac{\pi}{6}\right) d(\omega t)$$

$$= \frac{3\sqrt{3}\, V_m}{\pi}\left[-\cos\left(\omega t+\frac{\pi}{6}\right)\right]_{\frac{\pi}{6}+\alpha}^{\frac{\pi}{2}+\alpha}$$

$$\therefore \quad \boxed{V_{o\,(av)} = \frac{3\sqrt{3}\, V_m}{\pi}\cos\alpha} \qquad \text{... (2.7.2)}$$

Case II : $\alpha > 60°$

Now consider the case when $\alpha > 60°$. For this firing angle, the output voltage waveform is discontinuous. Fig. 2.7.2 (f) shows the output voltage waveform for $\alpha = 90°$. Putting the integration limits and V_{RY} in equation 2.7.1 for this case.

$$V_{o(av)} = \frac{1}{\pi/3}\int_{\frac{\pi}{6}+\alpha}^{\frac{5\pi}{6}} \sqrt{3}\, V_m \sin\left(\omega t+\frac{\pi}{6}\right) d\,\omega t$$

$$= \frac{3\sqrt{3}\, V_m}{\pi}\left[-\cos\left(\omega t+\frac{\pi}{6}\right)\right]_{\frac{\pi}{6}+\alpha}^{\frac{5\pi}{6}}$$

$$\therefore \quad \boxed{V_{o\,(av)} = \frac{3\sqrt{3}\, V_m}{\pi}\left[1+\cos\left(\frac{\pi}{3}+\alpha\right)\right]} \qquad \text{... (2.7.3)}$$

This is the equation for average output voltage for $\alpha > 60°$.

Example 2.7.2 : *For a 3ϕ fully controlled SCR bridge converter operating from 400 V, 3 phase AC supply, calculate the average DC output voltage for a firing angle of 45°.*

Solution : The given data is,

$$V_{line} = 400 \text{ V (rms)}$$

$$\alpha = 45°$$

Hence rms value of phase voltage is,

$$V_{ph} = \frac{400}{\sqrt{3}}$$

Hence peak value of phase voltage is,

$$V_m = \sqrt{2}\,V_{ph} = \sqrt{2}\cdot\frac{400}{\sqrt{3}}\,V$$

Here firing angle is 45°. Hence the conduction will be continuous for resistive as well as inductive load. Therefore the average DC output is given by equation 2.7.2 i.e.,

$$V_{o(av)} = \frac{3\sqrt{3}\,V_m}{\pi}\cos\alpha$$

Putting values in above equation ,

$$V_{o(av)} = \frac{3\sqrt{3}\times\sqrt{2}\cdot\frac{400}{\sqrt{3}}}{\pi}\cos 45°$$

$$= 382 \text{ volts.}$$

2.7.2 Operation with Highly Inductive Load

Let us consider the operation of 3ϕ full converter with highly inductive load. The output current will be continuous and ripplefree. In the waveforms of Fig. 2.7.2, observe that voltage waveform is continuous till $\alpha = 60°$. But with inductive load, voltage waveform is continuous for any value of α. Fig. 2.7.4 shows the waveforms of 3ϕ full converter for highly inductive load. Fig. 2.7.4 (c) shows the output voltage waveform for $\alpha = 60°$. Observe that this waveform is same as that of resistive load shown in Fig. 2.7.2 (e). Fig. 2.7.4 (d) shows the continuous and ripplefree output current. Fig. 2.7.4 (e) shows supply phase current waveforms i_R , i_Y and i_B. Observe that the R-phase current is positive whenever T_1 conducts and it is negative whenever T_4 conducts. All the three current waveforms are of the same nature (quasi square wave) having 120° phase shift with respect to each other.

Fig. 2.7.4 (f) shows the output voltage waveform for $\alpha = 90°$. The waveform goes negative for same period, because of inductive load. The load inductance generates a large voltage to maintain the load current in the same direction. Hence SCRs continue to conduct and load voltage becomes negative occasionaly. Note that there is no freewheeling in full converter.

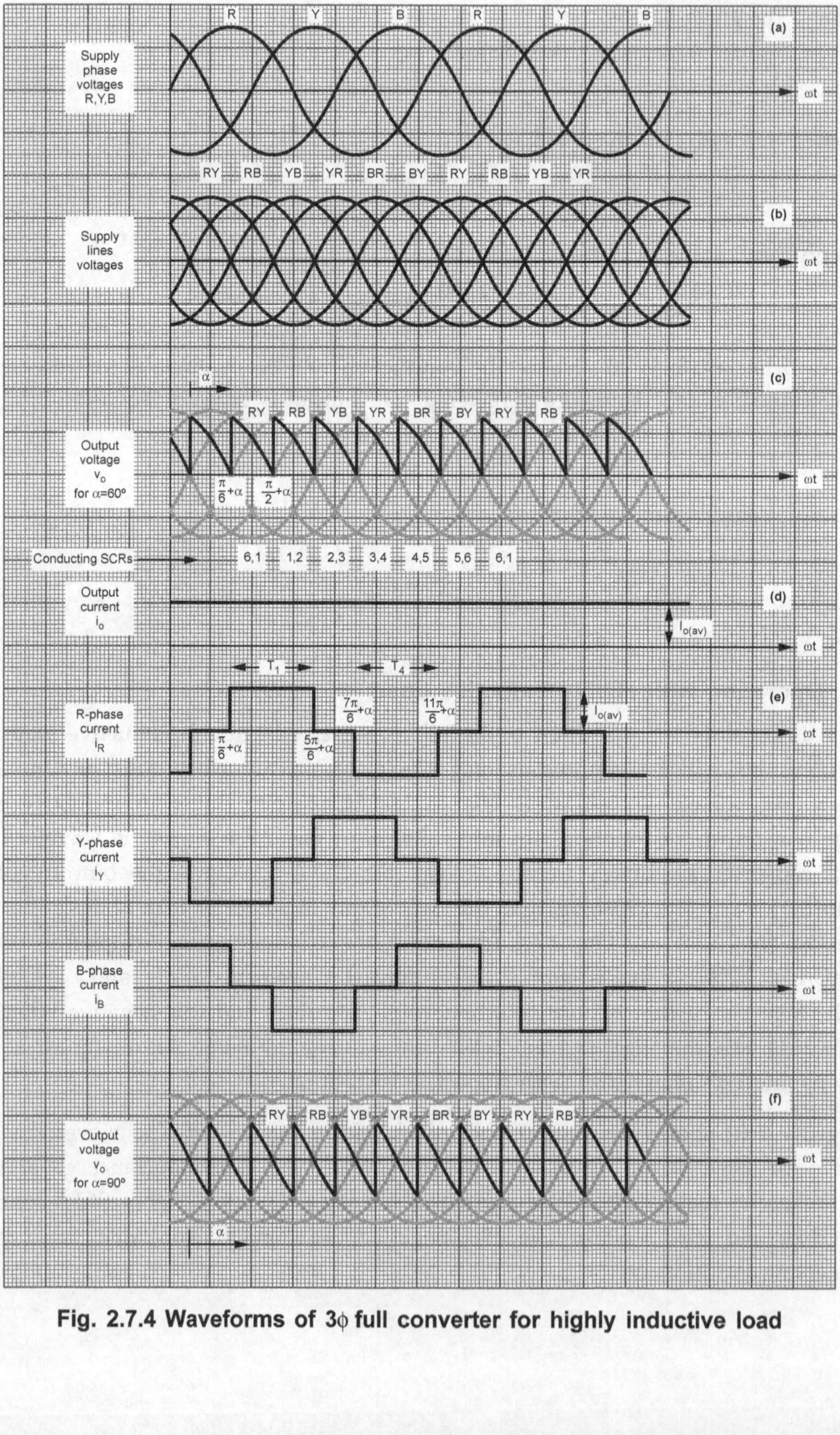

Fig. 2.7.4 Waveforms of 3ϕ full converter for highly inductive load

Example 2.7.3 : *Derive an expression for average output voltage of 3φ full converter having highly inductive load.*

Solution : In the Fig. 2.7.3 observe that output voltage waveform is continuous for complete range of α. Hence single expression can be derived. In Fig. 2.7.3 (c), observe that one ripple period of output voltage can be,

$$T = \left(\frac{\pi}{2}+\alpha\right)-\left(\frac{\pi}{6}+\alpha\right)=\frac{\pi}{3}$$

During this period line voltage V_{RY} is applied across the load. From Fig. 2.7.3 (b), V_{RY} will be,

$$V_{RY} = \sqrt{3}\, V_m \sin\left(\omega t+\frac{\pi}{6}\right)$$

Here V_m is the peak value of supply phase voltage.

The average output voltage is given as,

$$V_{o(av)} = \frac{1}{T}\int_0^T v_o(\omega t)\, d\omega t$$

$$= \frac{1}{\pi/3}\int_{\frac{\pi}{6}+\alpha}^{\frac{\pi}{2}+\alpha} v_{RY}(\omega t)\, d\omega t$$

$$= \frac{1}{\pi/3}\int_{\frac{\pi}{6}+\alpha}^{\frac{\pi}{2}+\alpha} \sqrt{3}\, V_m \sin\left(\omega t+\frac{\pi}{6}\right) d\omega t$$

$$= \frac{3\sqrt{3}\, V_m}{\pi}\left[-\cos\left(\omega t+\frac{\pi}{6}\right)\right]_{\frac{\pi}{6}+\alpha}^{\frac{\pi}{2}+\alpha}$$

$$\therefore \quad \boxed{V_{o(av)} = \frac{3\sqrt{3}\, V_m}{\pi}\cos\alpha} \quad \text{... (2.7.4)}$$

This equation holds for complete range of α.

Example 2.7.4 : *A 3 φ full converter operated from 3 φ – Y connected 208 V, 60 Hz, supply with R_L = 10 Ω. It is required to obtain 50 % of the maximum possible output voltage. Calculate -*

(i) Delay angle α

(ii) rms and average currents

(iii) rms and average thyristor ratings

(iv) η of rectification

(v) PF

Solution : Given data

$$V_{line} = 208 \text{ V. Hence } V_{ph} = \frac{V_{line}}{\sqrt{3}} = \frac{208}{\sqrt{3}} = 120 \text{ V}$$

$$\therefore \quad V_m = \sqrt{2}\,V_{ph} = \sqrt{2}\times 120 = 169.7 \text{ V}$$

load resistance, $R = 10\ \Omega$

output voltage $V_{o(av)} = 50\ \%$ of $V_{o(av)max}$

(i) To obtain delay angle α

Average output voltage of 3 φ full converter is given by equation 2.7.2 as,

$$V_{o(av)} = \frac{3\sqrt{3}V_m}{\pi}\cos\alpha$$

Maximum value of output voltage will be obtained when $\alpha = 0$. i.e.,

$$V_{o(av)max} = \frac{3\sqrt{3}V_m}{\pi}$$

$$= \frac{3\sqrt{3}\times 169.7}{\pi} = 280.68 \text{ V}$$

Since output voltage is 50 % of its maximum valve,

$$V_{o(av)} = 0.5\ V_{o(av)max}$$

$$= 0.5 \times 280.68 = 140.34$$

Consider the formula for output voltage,

$$V_{o(av)} = \frac{3\sqrt{3}V_m}{\pi}\cos\alpha$$

Putting values in this equation,

$$140.34 = \frac{3\sqrt{3}\times 169.7}{\pi}\cos\alpha$$

$$\therefore \quad \alpha = 60^\circ$$

(ii) Average and rms output currents

Average output current is given as,

$$I_{o(av)} = \frac{V_{o(av)}}{R} = \frac{140.34}{10} = 14.034 \text{ A}$$

To obtain rms current, we have to obtain rms output voltage, for $\alpha \le 60$, the output voltage waveform is continuous as shown in Fig. 2.7.2 (d). Consider the period from $\left(\frac{\pi}{6}+\alpha\right)$ to $\left(\frac{\pi}{2}+\alpha\right)$ when voltage V_{RY} is applied arrows the load. This period is,

$$T = \left(\frac{\pi}{2}+\alpha\right)-\left(\frac{\pi}{6}+\alpha\right)=\frac{\pi}{3}$$

From Fig. 2.7.2 (b) we can write an equation for line voltage V_{RY} as,

$$V_{RY} = \sqrt{3}V_m \sin\left(\omega t+\frac{\pi}{6}\right)$$

Here V_m is the peak value of phase voltage. RMS value is given as,

$$V_{o(rms)} = \left[\frac{1}{T}\int_0^T V_0^2(\omega t)\, d\omega t\right]^{\frac{1}{2}}$$

$$= \left[\frac{1}{\pi/3}\int_{\frac{\pi}{6}+\alpha}^{\frac{\pi}{2}+\alpha}\left[\sqrt{3}V_m \sin\left(\omega t+\frac{\pi}{6}\right)\right]^2 d\omega t\right]^{\frac{1}{2}}$$

$$= \left[\frac{3}{\pi}\cdot 3V_m^2\int_{\frac{\pi}{6}+\alpha}^{\frac{\pi}{2}+\alpha}\sin^2\left(\omega t+\frac{\pi}{6}\right)d\omega t\right]^{\frac{1}{2}}$$

$$= \left[\frac{9V_m^2}{\pi}\int_{\frac{\pi}{6}+\alpha}^{\frac{\pi}{2}+\alpha}\frac{1-\cos 2\left(\omega t+\frac{\pi}{6}\right)}{2}d\omega t\right]^{\frac{1}{2}}$$

$$\therefore \quad \boxed{V_{o(rms)} = 3V_m\sqrt{\frac{\frac{\pi}{3}+\frac{\sqrt{3}}{2}\cos 2\alpha}{2\pi}}} \qquad \text{... (2.7.5)}$$

Putting values in above equation,

$$V_{o(rms)} = 3\times 169.7\sqrt{\frac{\frac{\pi}{3}+\frac{\sqrt{3}}{2}\cos(2\times 60^{\circ})}{2\pi}}$$

$$= 159.17\ \text{V}$$

Therefore rms output current will be,

$$I_{o(rms)} = \frac{V_{o(rms)}}{R} = \frac{159.17}{10} = 15.917 \text{ A}$$

(iii) rms and average thyristor ratings

In the waveforms of Fig. 2.7.2, observe that two thyristors conduct at a time. The load current is carried equally by three thyristors in one cycle. These three thyristors are T_1, T_3, T_5 and T_4, T_6, T_2. Hence the average output current is shared by these three thyristors. Hence average current of single thyristor becomes,

$$I_{T(av)} = \frac{I_{o(av)}}{3} \quad \text{... (2.7.6)}$$

$$= \frac{14.034}{3} = 4.678 \text{ A}$$

This is the average current carried by each thyristor.

The rms current is also shared by three thyristors. Hence we can write,

$$I^2_{o(rms)} = I^2_{T(rms)} + I^2_{T(rms)} + I^2_{T(rms)}$$

Above equation shows the relationship between output rms current and rms current of three thyristors. RMS current of each thyristor is same. Hence,

$$I^2_{o(rms)} = 3I^2_{T(rms)}$$

$$\therefore \quad I_{T(rms)} = \frac{I_{o(rms)}}{\sqrt{3}} \quad \text{... (2.7.7)}$$

$$= \frac{15.917}{\sqrt{3}} = 9.189 \text{ A}$$

(iv) Rectification efficiency (η)

Rectification efficiency is given as,

$$\eta = \frac{\text{Average or dc load power}}{\text{rms load power}}$$

$$= \frac{V_{o(av)}\, I_{o(av)}}{V_{o(rms)}\, I_{o(rms)}}$$

$$= \frac{140.34 \times 14.034}{159.17 \times 15.917} = 0.777 \text{ or } 77.7 \text{ \%}$$

(v) To obtain power factor

The active load power is the power consumed in the load. It can be calculated as,

Active load power = $I^2_{o(rms)} \times R$

$$= (15.917)^2 \times 10$$
$$= 2533.5 \text{ W}$$

At any time instant two thyristors conduct. Hence the supply current can be given in terms of thyristor currents as,

$$I^2_{s(rms)} = 2I^2_{T(rms)}$$
$$= 2 \cdot \frac{I^2_{o(rms)}}{3}$$

$$\therefore \quad I_{s(rms)} = I_{o(rms)} \sqrt{\frac{2}{3}} \quad \text{... (2.7.8)}$$

$$= 15.917 \sqrt{\frac{2}{3}}$$

$$= 13 \text{ A}$$

For 3 ϕ supply, the total supply volt-ampere will be,

$$\text{supply VA} = 3 \, V_s \, I_s$$
$$= 3 \times 120 \times 13 = 4680 \text{ VA}$$

The power factor is given as,

$$\text{PF} = \frac{\textit{Active load power}}{\textit{Total supply power (VA)}}$$

$$= \frac{25335}{4680} = 0.5413 \text{ (lagging)}$$

2.7.3 Comparison between 3ϕ and 1ϕ Converters

Sr.No.	Parameter	1ϕ converter	3ϕ converters
1	Ripple content in output	More	Less
2	Output power	Less upto 5 kW	More than 5 kW
3	Supply current waveform	Square wave for 1ϕ full converter	Quasi square wave for 3ϕ full converter
4	Ripple frequency	100 Hz	150 Hz and 300 Hz
5	Control and complexity	Less complex and easy control	Complex control and implementation
6	Maximum supply power factor	0.9	0.955
7	Supply and load derating	Higher	Less

Table 2.7.1 Comparison of 3ϕ and 1ϕ converters

It shows that it is preferable to use 3ϕ converters for better power efficiency. Hence for higher load power requirement 3ϕ converters are always preferred. But for simple and low power applications 1ϕ converters are used because of their simplicity of implementation.

Review Question

1. *Explain the working of 3ϕ full converter for highly inductive load with the help of waveforms.*

Unsolved Example

1. *A 3ϕ full converter operates from 3ϕ, 415 V, 50 Hz supply. The load is highly inductive. Determine the triggering angle of the converter to get average output voltage of 300 volts. The load resistance is $R = 10\ \Omega$. Determine the load current and power.*

[Ans. : $\alpha = 57.63°$, $I_{o(av)} = 30 A$, $P_{o(av)} = 9 kW$]

2.8 Effect of Source Inductance

Normally the converters are fed from supply transformer. The inductance of the supply transformer becomes input to converter. This inductance is called source inductance. It affects the performance of converter.

2.8.1 Effect of Source Inductance in 1ϕ Full Converter

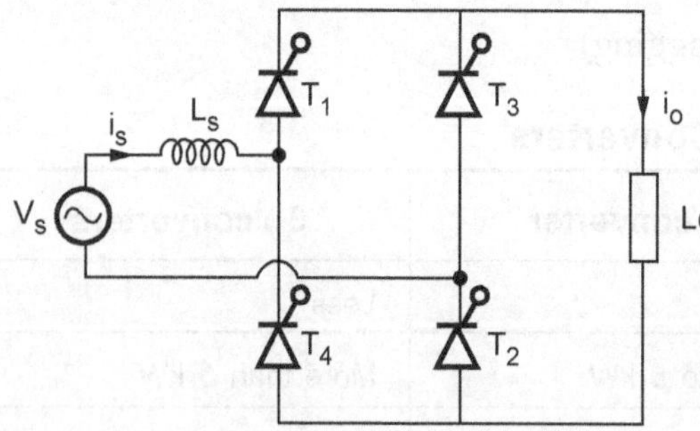

Fig. 2.8.1 1ϕ FCB with source inductance L_s

Fig. 2.8.1 shows the circuit diagram of 1ϕ full converter with source inductance. Note that the source inductance 'L_s' is not physically connected. It is equivalent inductance of the source.

Operation of the Converter

Here we will assume that the load is highly inductive. Hence the output current is continuous and ripple free.

Mode-I ($\alpha + \mu$ to $\pi + \alpha$)

Fig. 2.8.2 shows the waveforms of the circuit. Equivalent circuit -1 shows the current path for the duration ($\alpha + \mu$ to $\pi + \alpha$). SCRs T_1 and T_2 are conducting. The complete load current is carried by T_1 and T_2. The supply current flows through source inductance. It stores the energy. This mode ends at ($\pi + \alpha$).

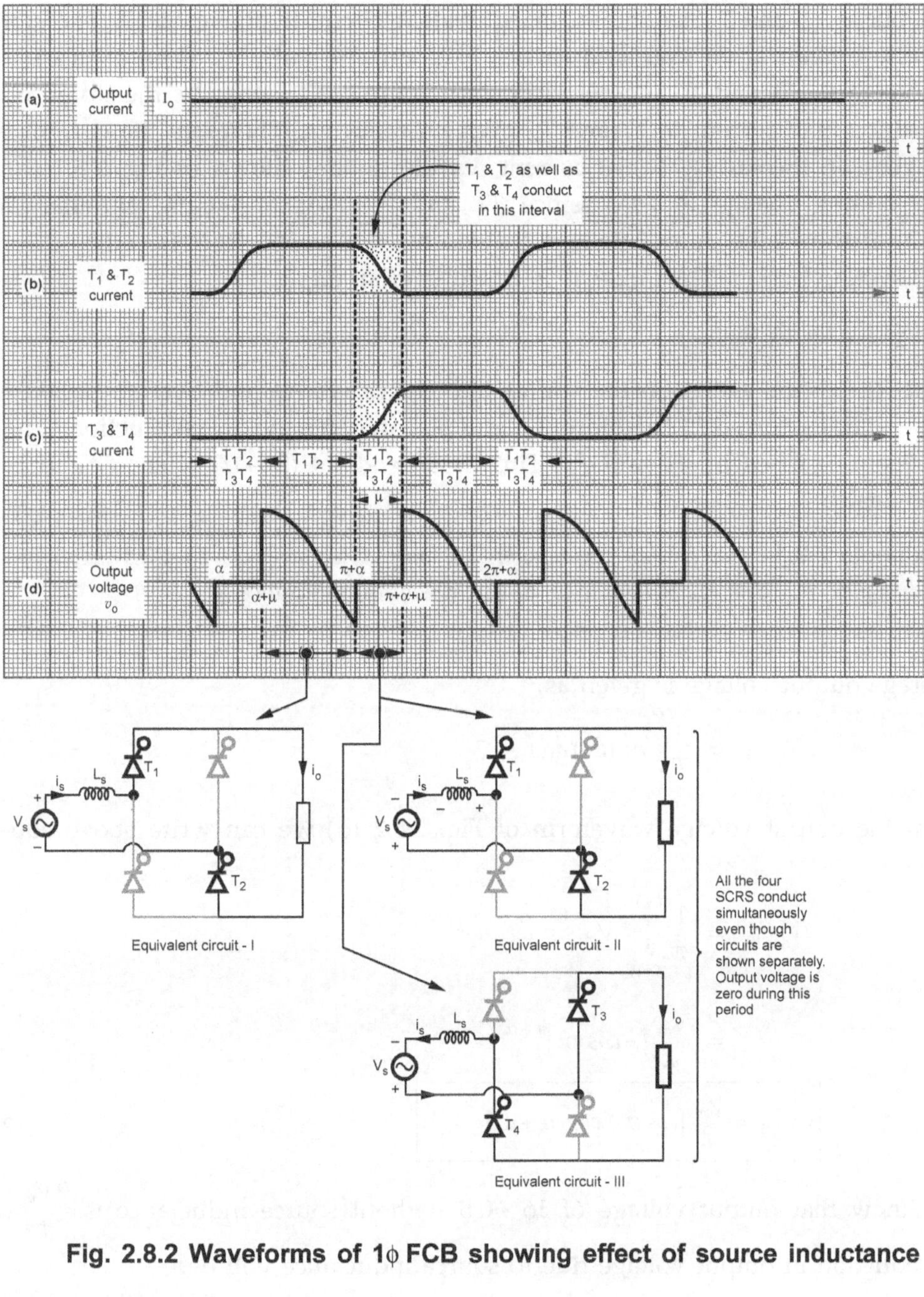

Fig. 2.8.2 Waveforms of 1ϕ FCB showing effect of source inductance

Mode-II ($\pi+\alpha$ to $\pi+\alpha+\mu$)

At $\pi+\alpha$, the supply voltage is negative. SCRs T_3 and T_4 are triggered at $\pi+\alpha$. Since T_3 and T_4 are forward biased they turn-on and load current starts flowing through them. Hence the current through 'L_s' also tries to reverse as shown in equivalent circuit-III of Fig. 2.8.2. But the energy stored in 'L_s' induces a voltage that forward biases T_1 and T_2 as shown in equivalent circuit-II of Fig. 2.8.2. Therefore T_1 and T_2 also remain in on state and carries load current partially. Thus in this interval all the four SCRs, T_1 and T_2 as well as T_3 and T_4 are conducting. As the stored energy in 'L_s' reduces, the current in T_1 and T_2 goes on reducing. Therefore the current in T_3 and T_4 goes on increasing. These current waveforms are shown in Fig. 2.8.2 (b) and (c). The output voltage is zero, since T_3 and T_2 are conducting simultaneously. Fig. 2.8.2 (d) shows the output voltage waveform. The period from $(\pi+\alpha)$ to $(\pi+\alpha+\mu)$ is called overlap period. The angle 'μ' is called overlap angle.

Mode-III ($\pi+\alpha+\mu$ to $2\pi+\alpha$)

At $(\pi+\alpha+\mu)$, the energy in the source inductance is zero and overlap period ends. Hence T_1 and T_2 turn-off. Therefore T_3 and T_4 only remain in conduction. The full load current is carried by T_3 and T_4. This mode ends at $(2\pi+\alpha)$ when T_1 and T_2 are triggered again.

Example 2.8.1 : *Derive an expression for output voltage of 1φ FCB having source inductance 'L_s'. The overlap angle is 'μ' due to source inductance.*

Solution :

Average output voltage is given as,

$$V_{o(av)} = \frac{1}{T}\int_0^T v_o(\omega t)\, d\omega t$$

From the output voltage waveform of Fig. 2.8.2 (d) we can write above equation as,

$$V_{o(av)} = \frac{1}{\pi}\int_{\alpha+\mu}^{\pi+\alpha} V_m \sin \omega t\, d\omega t$$

$$= \frac{V_m}{\pi}[-\cos \omega t]_{\alpha+\mu}^{\pi+\alpha}$$

$$\therefore \quad \boxed{V_{o(av)} = \frac{V_m}{\pi}[\cos\alpha + \cos(\alpha+\mu)]} \quad \text{... (2.8.1)}$$

We know that output voltage of 1φ FCB without source inductance is $\frac{2V_m}{\pi}\cos\alpha$.

Hence reduction in output voltage due to source inductance will be,

$$V_{o(reduced)} = V_{o(av)} - V_{o(av)} \text{ with } L_s$$

$$= \frac{2V_m}{\pi}\cos\alpha - \frac{V_m}{\pi}[\cos\alpha + \cos(\alpha+\mu)]$$

$$= \frac{V_m}{\pi}[\cos\alpha - \cos(\alpha + \mu)] \quad \text{...(2.8.2)}$$

Average output voltage in terms of L_s

When all the four SCRs are conducting in overlap period, the output voltage is zero. Therefore the complete source voltage appears across 'L_s'. Fig. 2.8.3 shows the loop, that shows supply voltage appears across L_s through T_1 and T_2. By KVL to this loop,

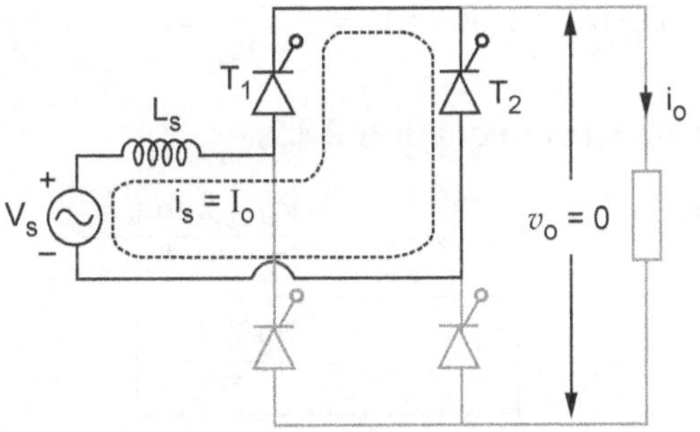

Fig. 2.8.3 Source voltage appears across 'L_s' during overlap

$$v_s = L_s \frac{d\,i_s(t)}{dt}$$

Since $v_s = V_m \sin\omega t$, above equation becomes,

$$V_m \sin\omega t = L_s \frac{d\,i_s(t)}{dt}$$

$$\therefore \quad V_m \sin\omega t\, dt = L_s\, d\,i_s(t)$$

Integrating both the sides,

$$\int V_m \sin\omega t\, dt = \int L_s\, d\,i_s(t)$$

Above integration is for the period $(\pi+\alpha)$ to $(\pi+\alpha+\mu)$. Hence,

$$\int_{\pi+\alpha}^{\pi+\alpha+\mu} V_m \sin\omega t\, dt = \int L_s\, d\,i_s(t)$$

Note that at $(\pi+\alpha)$, $i_s = I_o$ and at $(\pi+\alpha+\mu)$, $i_s = -I_o$. Hence above equation becomes,

$$\int_{\pi+\alpha}^{\pi+\alpha+\mu} V_m \sin\omega t\, dt = \int_{I_O}^{-I_O} L_s\, d\,i_s(t)$$

Above integration is with respect to 't'. Hence 'ω' will be treated as constant. i.e.,

$$V_m\left[\frac{-\cos\omega t}{\omega}\right]_{\pi+\alpha}^{\pi+\alpha+\mu} = L_s[i_s(t)]_{I_o}^{-I_o}$$

$$\therefore \quad \frac{V_m}{\omega}[\cos(\alpha+\mu)-\cos\alpha] = -2L_s I_o$$

$$\cos(\alpha+\mu)-\cos\alpha = -\frac{2\omega L_s I_o}{V_m}$$

i.e.

$$\cos\alpha-\cos(\alpha+\mu) = \frac{2\omega L_s I_o}{V_m} \qquad \text{... (2.8.3)}$$

Putting for $\cos\alpha-\cos(\alpha+\mu)$ in equation 2.8.2,

$$V_{o(reduced)} = \frac{V_m}{\pi}\times\frac{2\omega L_s I_o}{V_m}$$

$$\therefore \quad \boxed{V_{o(reduced)} = \frac{2\omega L_s I_o}{\pi}} \qquad \text{... (2.8.4)}$$

Hence output voltage is given as,

$$V_{o(av)} = V_{o(av)} \text{ without } L_s - V_{o(reduced)}$$

$$= \frac{2V_m}{\pi}\cos\alpha-\frac{2\omega L_s I_o}{\pi} \qquad \text{... (2.8.5)}$$

This is the formula for output voltage in terms of L_s.

To obtain an expression for overlap angle 'μ'

Consider equation 2.8.3,

$$\cos\alpha-\cos(\alpha+\mu) = \frac{2\omega L_s I_o}{V_m}$$

$$\therefore \quad \cos(\alpha+\mu) = \cos\alpha-\frac{2\omega L_s I_o}{V_m}$$

$$\therefore \quad \alpha+\mu = \cos^{-1}\left[\cos\alpha-\frac{2\omega L_s I_o}{V_m}\right]$$

$$\therefore \quad \boxed{\mu = \cos^{-1}\left[\cos\alpha-\frac{2\omega L_s I_o}{V_m}\right]-\alpha} \qquad \text{... (2.8.6)}$$

This is an expression for overlap angle 'μ'.

Relations of overlap angle and reduced output voltage for 3ϕ full converter

Because of overlap, two line voltages are connected simultaneously to load. Hence,

$$V_{o(av)} = \frac{1}{\pi/6}\left[\int_{\frac{\pi}{6}+\alpha}^{\frac{\pi}{6}+\alpha+\mu} \frac{V_{RY}(\omega t)+V_{BY}(\omega t)}{2}\, d\omega t + \int_{\frac{\pi}{6}+\alpha+\mu}^{\frac{\pi}{2}+\alpha} V_{RY}(\omega t)\, d\omega t\right]$$

Putting expressions in above equation and simplifying we get,

$$V_{o(av)} = \frac{3\sqrt{3}\,V_m}{2\pi}[\cos\alpha + \cos(\alpha+\mu)] \quad \text{...(2.8.7)}$$

Here V_m is peak value of the phase voltage. The reduction in output voltage due to source inductance will be,

$$V_{o(reduced)} = V_{o(av)} - V_{o(av)} \text{ with } L_s$$

$$= \frac{3\sqrt{3}\,V_m}{\pi}\cos\alpha - \frac{3\sqrt{3}\,V_m}{2\pi}[\cos\alpha + \cos(\alpha+\mu)]$$

$$= \frac{3\sqrt{3}\,V_m}{2\pi}[\cos\alpha - \cos(\alpha+\mu)] \quad \text{...(2.8.8)}$$

If the same inductance is L_s and output current is I_o then reduction in output voltage is given as,

$$V_{o(reduced)} = \frac{3\omega L_s I_o}{\pi} \quad \text{...(2.8.9)}$$

And output voltage with source inductance is,

$$V_{o(av)} = V_{o(av)} \text{ without } L_s - V_{o(reduced)}$$

$$= \frac{3\sqrt{3}\,V_m}{\pi}\cos\alpha - \frac{3\omega L_s I_o}{\pi} \quad \text{...(2.8.10)}$$

We know that

$$V_{o(reduced)} = \frac{3\sqrt{3}\,V_m}{2\pi}[\cos\alpha - \cos(\alpha+\mu)]$$

Equating above equation with $V_{o(reduced)} = \dfrac{3\omega L_s I_o}{\pi}$,

$$\frac{3\omega L_s I_o}{\pi} = \frac{3\sqrt{3}\,V_m}{2\pi}[\cos\alpha - \cos(\alpha+\mu)]$$

$$\omega L_s I_o = \frac{\sqrt{3}\,V_m}{2}[\cos\alpha - \cos(\alpha+\mu)]$$

$$\therefore \quad \mu = \cos^{-1}\left[\cos\alpha - \frac{2\omega L_s I_o}{\sqrt{3}\,V_m}\right] - \alpha \quad \text{...(2.8.11)}$$

Example 2.8.2 : *A three phase bridge converter is supplied from a star connected 208 V, 50 Hz supply. The average load current is 60 A and has negligible ripple. Calculate the percentage reduction of output voltage due to commutation if the line inductance for phase is 0.5 mH.*

Solution :

Here $V_{line} = 208$ V. Hence $V_{ph} = \frac{V_{line}}{\sqrt{3}} = \frac{208}{\sqrt{3}} = 120$ V

$$V_m = \sqrt{2}\,V_{ph} = \sqrt{2}\times 120 = 169.7 \text{ V}$$

$$\omega = 2\pi f = 2\pi\times 50 = 100\,\pi$$

$$I_o = 60 \text{ A}$$

L_s = 0.5 mH + 0.5 mH = 1 mH, since line inductance is 0.5 mH for one phase.

$$\therefore \quad V_{o(reduced)} = \frac{3\omega L_s I_o}{\pi}$$

$$= \frac{3\times 100\pi\times 1\times 10^{-3}\times 60}{\pi} = 18 \text{ V}$$

Output voltage without source inductance is,

$$V_{o(av)} = \frac{3\sqrt{3}\,V_m}{\pi}\cos\alpha$$

$$= \frac{3\sqrt{3}\times 169.7}{\pi}\cos 0 \quad \text{since } \alpha = 0$$

$$= 280.68 \text{ V}$$

Thus there is a reduction of 18 V in 280.68 V. Hence percentage reduction is $\frac{18}{280.68}\times 100 = 6.41$ %.

Review Questions

1. *Explain the effect of soure inductance on the output of converters.*
2. *Derive an expression for reduction in output voltage and overlop angle due to source inductance.*

2.9 Single Phase Dual Converters

We have seen that the single phase full converter operates in two quadrants for inductive load. If two such single phase full converters are connected back to back, then it is possible to get four quadrant operation. Such converters are called as dual converters. They operate in all the four quadrants. Fig. 2.9.1 shows the circuit diagram of 1ϕ dual converter. The P-converter has the output V_{o1} and it operates in rectification mode (i.e. $\alpha < 90°$). The N-converter has the output V_{o2} and it operates in the inversion mode (i.e. $\alpha > 90°$). For $\alpha > 90$, the average output V_{o2} of N-converter is negative. The two outputs V_{o1} and V_{o2} appear across the load. The two converters operate on inductive load. Hence their average values are given as,

$$\left.\begin{aligned} V_{o1(av)} &= \frac{2V_m}{\pi} \cos \alpha_1 \\ \text{and} \quad V_{o2(av)} &= \frac{2V_m}{\pi} \cos \alpha_2 \end{aligned}\right\} \quad \text{... (2.9.1)}$$

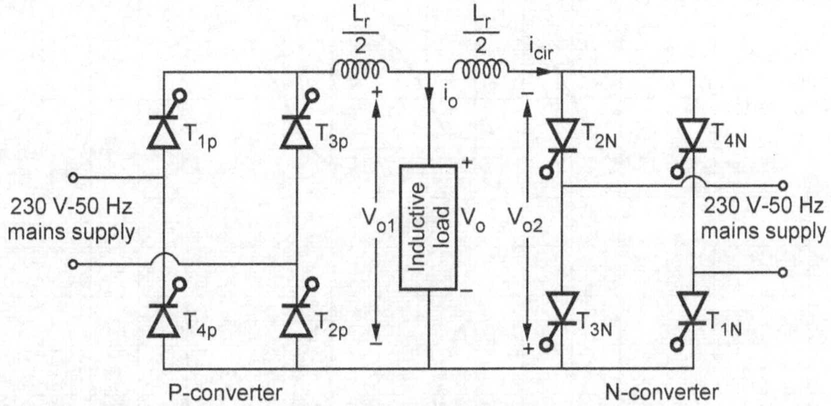

Fig. 2.9.1 1ϕ dual converter

Since one converter operates in rectifying mode and other converter operates in inversion mode, their average values must be equal and opposite in sign. i.e.,

$$V_{o1(av)} = -V_{o2(av)} \quad \text{... (2.9.2)}$$

Putting for $V_{o1(av)}$ and $V_{o2(av)}$ from equation 2.9.1 in above equation,

$$\frac{2V_m}{\pi} \cos \alpha_1 = -\frac{2V_m}{\pi} \cos \alpha_2$$

$$\therefore \quad \cos \alpha_1 = -\cos \alpha_2$$

$$\therefore \quad \alpha_2 = \pi - \alpha_1$$

$$\text{or} \quad \alpha_1 + \alpha_2 = \pi \quad \text{... (2.9.3)}$$

Thus if $\alpha_1 < 90°$, then $\alpha_2 > 90°$ and their sum is 180° or π. Fig. 2.9.2 shows the waveforms of 1ϕ dual converter.

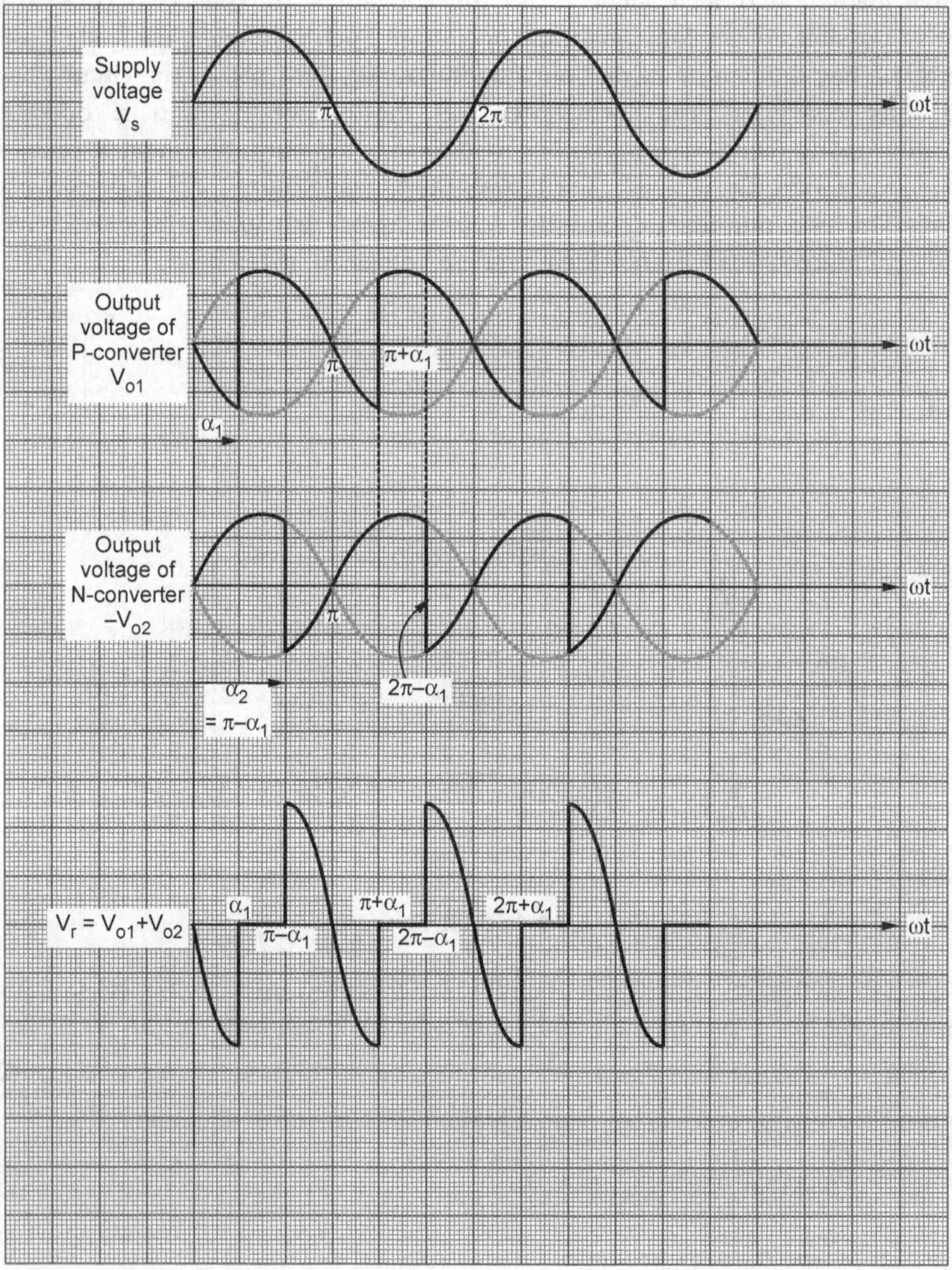

Fig. 2.9.2 Waveforms of 1ϕ dual converter

Observe that P-converter is triggered at α_1 which is less than 90°. The N-converter is triggered at α_2 which is greater than 90. Hence the average values of V_{01} and V_{02} are equal but opposite in sign. In Fig. 2.9.2 observe that the instantaneous values of V_{01} and V_{02} are not same. Hence a small amount of circulating current flows between the two converters. This current does not flow through the load. The circulating current is limited by the circulating current reactor L_r. The circulating current reactor is connected between the two converters as shown in Fig. 2.9.1.

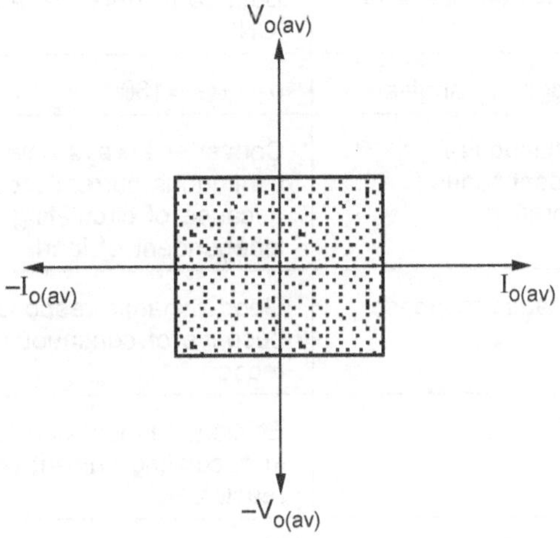

Fig. 2.9.3 4-Quadrant operation of 1ϕ dual converter

Four quadrant operation

By controlling the firing angles α_1 and α_2 it is possible to change the polarity of output voltage completely. Consider that V_0 is positive when $\alpha_1 < 90°$ and $\alpha_2 > 90°$. Then V_o is negative when $\alpha_1 > 90°$ and $\alpha_2 < 90°$. Similarly direction of output current i_o will also be changed. Thus 1ϕ dual converter operates in 4 quadrants of V_o and i_o.

2.9.1 Circulating Current and Non-circulating Current Dual Converters

The 1ϕ dual converter discussed in this section is *circulating current mode*. In such dual converter, both the P and N converters are 'ON' simultaneously. One converter operates in rectification mode and other converter operates in inversion mode. The circulating current flows between the two converters because both are 'ON' simultaneously. In *non-circulating current mode,* only one converter is 'ON' at a time. The other converter is completely 'OFF' (i.e. no gate drives are applied). Hence circulating current doesnot flow in such operation. The non-circulating current mode dual converters also provide 4-quadrant operation.

2.9.2 Comparison between Circulating and Non-circulating Current Modes

The following Table 2.9.1 shows the comparison between circulating current and noncirculating current mode dual converters.

Sr. No.	Parameter	Circulating current mode	Non-circulating current mode
1	Converter operation	Both converters are always 'ON'.	Only one converter is 'ON' and other is 'OFF'.
2	Triggering angles	$\alpha_1 + \alpha_2 = 180$	α_1 and α_2 are less than 90°.
3	Continuous / discontinuous operation	Converter always operate in continuous current mode because of circulating current independent of load.	The operation can be continuous or discontinuous depending upon the load.
4	Dynamic response	Good dynamic response because of continuous current mode.	Slow dynamic response due to deadband during changeover.
5	Efficiency	Efficiency is low due to losses in circulating current limiting reactor.	Efficiency is higher since there are no losses.
6	Cost and complexity	Higher due to circulating current and its limiting reactor.	Lower since only one converter is to be controlled at a time.

Table 2.9.1 Comparison between circulating and non-circulating current modes

Example 2.9.1 : *For a single phase dual converter, obtain an equation for circulating current i_{cir}.*

Solution : We have seen that the average values of v_{o1} and v_{o2} are same but their instantaneous values are not always same. Hence circulating current flows between the two converters. The circulating current can be obtained by integrating the instantaneous voltage difference. i.e.,

$$i_{cir} = \frac{1}{\omega L_r} \int v_r \, d\omega t$$

In Fig. 2.9.2 observe that v_{o1} and v_{o2} are same in the interval from $\pi+\alpha_1$ to $2\pi-\alpha_1$. Hence there is no instantaneous voltage difference. But from $2\pi-\alpha_1$, v_{o1} and v_{o2} are different. Hence above equation will be,

$$i_{cir} = \frac{1}{\omega L_r} \int_{2\pi-\alpha_1}^{\omega t} v_r \, d\omega t \qquad \text{... (2.9.4)}$$

Here we can write,

$$v_r = v_{o1} + v_{o2}$$

In Fig. 2.9.2 observe that $v_{o1} = -V_m \sin \omega t$ and $v_{o2} = -V_m \sin \omega t$ from $2\pi-\alpha_1$ to $2\pi+\alpha_1$. Hence above equation becomes,

$$v_r = -V_m \sin \omega t - V_m \sin \omega t$$

$$= -2 V_m \sin \omega t$$

Hence equation 2.9.4 becomes,

$$i_{cir} = \frac{1}{\omega L_r} \int_{2\pi-\alpha 1}^{\omega t} -2 V_m \sin \omega t \, d\omega t$$

$$= \frac{2 V_m}{\omega L_r} \int_{2\pi-\alpha 1}^{\omega t} (-\sin \omega t) \, d\omega t$$

$$= \frac{2 V_m}{\omega L_r} [\cos \omega t]_{2\pi-\alpha 1}^{\omega t}$$

$$= \frac{2 V_m}{\omega L_r} [\cos \omega t - \cos (2\pi - \alpha_1)]$$

$$= \frac{2 V_m}{\omega L_r} [\cos \omega t - \cos \alpha_1] \quad \text{... (2.9.5)}$$

This is the required expression for circulating current in the 1 ϕ dual converter.

2.10 Three Phase Dual Converters

We have discussed 1 ϕ dual converters earlier. We know that dual converter operates in all the four quadrants of $v_o - i_o$ plane. Dual converters are used in reversible speed drives. The 3 ϕ dual converters supply more power compared to 1ϕ dual converters.

2.10.1 Circulating Current Mode

Fig. 2.10.1 shows the circuit diagram of 3 ϕ dual converter. It uses two 3 ϕ full converters. We know that the average output voltage of 3 ϕ full converter is given for highly inductive as,

$$V_{o(av)} = \frac{3\sqrt{3} V_m}{\pi} \cos \alpha \quad \text{... (2.10.1)}$$

Hence the outputs of the two converters will be,

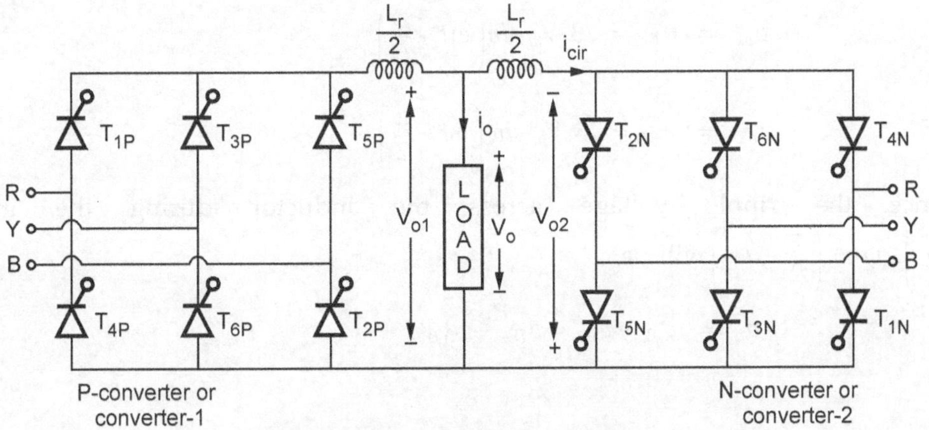

Fig. 2.10.1 3 ϕ dual converter

$$V_{o1\,(av)} = \frac{3\sqrt{3}\,V_m}{\pi} \cos\alpha_1$$

and $$V_{o2\,(av)} = \frac{3\sqrt{3}\,V_m}{\pi} \cos\alpha_2$$

From Fig. 2.10.1 it is clear that,

$$V_{o\,(av)} = V_{o1\,(av)} = -V_{o2\,(av)} \qquad \text{... (2.10.2)}$$

From equation 2.10.1 and above equation we have,

$$\frac{3\sqrt{3}\,V_m}{\pi} \cos\alpha_1 = -\frac{3\sqrt{3}\,V_m}{\pi} \cos\alpha_2$$

$$\cos\alpha_1 = -\cos\alpha_2$$

or $$\alpha_2 = \pi - \alpha_1 \qquad \text{... (2.10.3)}$$

This is the relationship between triggering angles of the two converters. The converter having $\alpha < 90°$ operates in rectifying mode. And the converter having $\alpha > 90°$ operates in inverting mode. Fig. 2.10.2 shows the waveforms of 3ϕ dual converter. Fig. 2.10.2 (c) shows the output voltage waveform of converter 1 for $\alpha_1 = 60°$. It operates in rectification mode. Fig. 2.10.2 (d) shows the output voltage waveform of converter-2 (inverted output i.e. $-v_{o2}$) for $\alpha_2 = 120°$. It operates in the inverting mode.

The two outputs v_{o1} and v_{o2} are same in magnitude. But their instantaneous values are different. Hence small amount of circulating current flows from one converter to other converter. This current can be limited by the circulating current reactor L_r connected between the two converters.

Example 2.10.1 : *Derive an expression for circulating current in 3 ϕ dual converter.*

Solution : Consider the interval from $\left(\frac{\pi}{6}+\alpha_1\right)$ to $\left(\frac{\pi}{2}+\alpha_1\right)$ in Fig. 2.10.2. The voltage $v_{o1} = V_{RY}$ and $v_{o2} = V_{YB}$ in this interval. From Fig. 2.10.2 (b) we can write,

$$v_{o1} = v_{RY} = \sqrt{3}\,V_m \sin\left(\omega t + \frac{\pi}{6}\right)$$

and $$v_{o2} = v_{RB} = \sqrt{3}\,V_m \sin\left(\omega t - \frac{\pi}{2}\right)$$

Hence the ripple voltage across the inductor during the interval $\left(\frac{\pi}{6}+\alpha_1\right) \le \omega t \le \left(\frac{\pi}{2}+\alpha_1\right)$ will be,

$$v_r = v_{o1} - v_{o2} = v_{RY} - v_{YB}$$

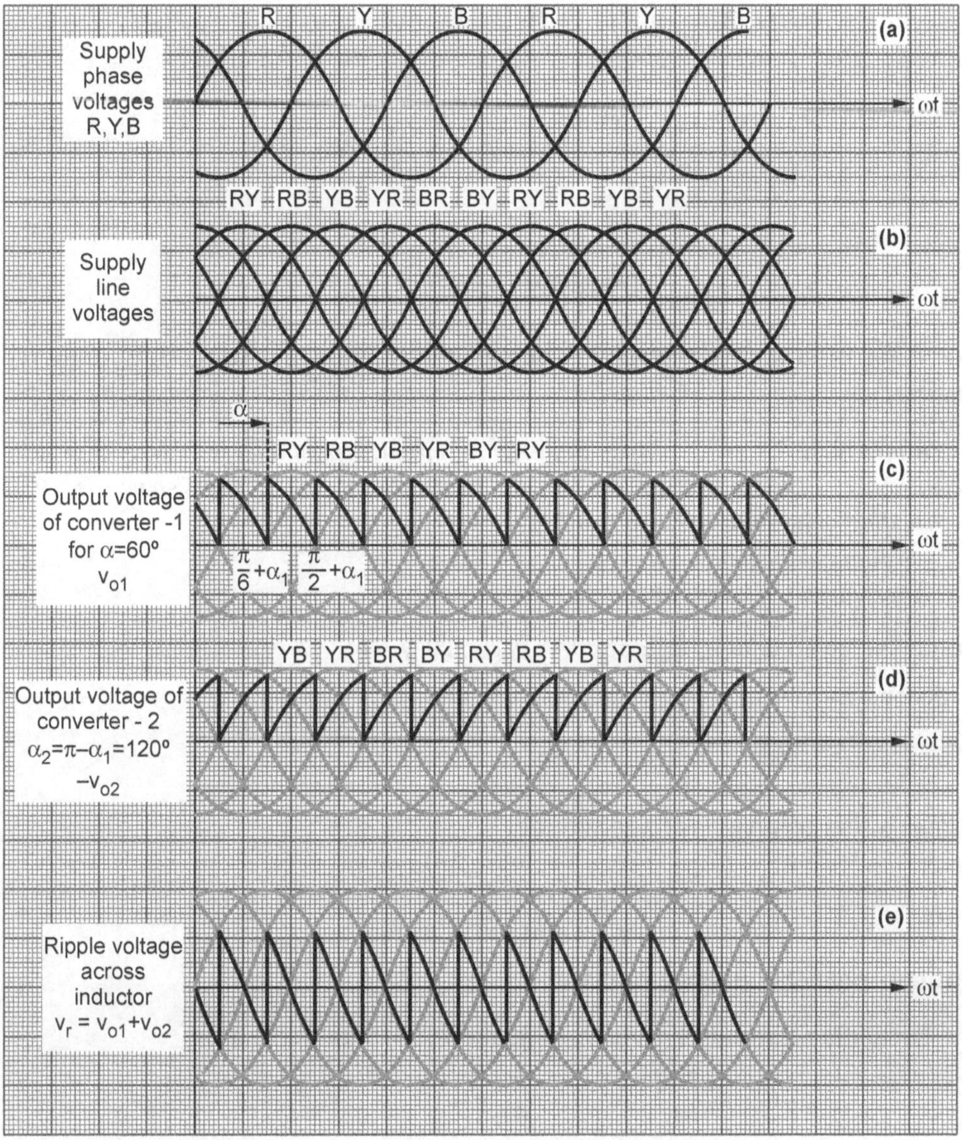

Fig. 2.10.2 Waveforms of 3 φ dual converter

$$= \sqrt{3}\, V_m \sin\left(\omega t+\frac{\pi}{6}\right)-\sqrt{3}\, V_m \sin\left(\omega t-\frac{\pi}{2}\right)$$

$$= 3V_m \cos\left(\omega t-\frac{\pi}{6}\right) \qquad \text{... (2.10.4)}$$

The waveform of this voltage is shown in Fig. 2.10.2 (e). Now the circulating current can be obtained as,

$$i_{cir}(t) = \frac{1}{\omega L_r} \int_{\frac{\pi}{6}+\alpha 1}^{\omega t} v_r(\omega t)\, d\omega t$$

$$= \frac{1}{\omega L_r} \int_{\frac{\pi}{6}+\alpha 1}^{\omega t} 3\, V_m \cos\left(\omega t - \frac{\pi}{6}\right) d\omega t$$

$$= \frac{3 V_m}{\omega L_r}\left[\sin\left(\omega t - \frac{\pi}{6}\right) - \sin\alpha_1\right] \qquad \text{... (2.10.5)}$$

This expression shows that circulating depends upon the triggering angle α_1 and inductance L_r.

Example 2.10.2 : *A 3 ϕ dual converter is operated from the secondary of a delta star connected transformer of 220 volts, 50 Hz supply. If the load resistance is 10 Ω, the circulating inductance is 7.5 mH and α_1 =50°, find,*

i. Peak circulating current

ii. Peak current of converter-I

Solution : The given data is, line input voltage 220 V

∴ Secondary phase voltage will be

$$V_{ph} = \frac{220}{\sqrt{3}} = 127 \text{ V}$$

∴ Peak value of phase voltage will be,

$$V_m = \sqrt{2}\, V_{ph} = \sqrt{2} \times 127 = 179.63 \text{ V}$$

$$\omega = 2\pi f = 2\pi \times 50 = 314.16 \text{ rad/sec.}$$

i) To obtain peak circulating current

The circulating current is given by equation 2.10.5 as

$$i_{cir}(t) = \frac{3V_m}{\omega L_r}\left[\sin\left(\omega t - \frac{\pi}{6}\right) - \sin\alpha_1\right]$$

This current will be maximum when $\sin\left(\omega t - \frac{\pi}{6}\right) = 1$ i.e. $\omega t = \frac{2\pi}{3}$. Putting values in above equation,

$$I_{cir(peak)} = \frac{3 \times 179.63\, V}{314.16 \times 7.5 \times 10^{-3}}\left[\sin\left(\frac{2\pi}{3} - \frac{\pi}{6}\right) - \sin 50°\right] = 53.5 \text{ A}$$

ii) To obtain peak current of converter - I

The load resistance is 10 Ω. Hence the peak load current will be,

$$\text{Peak load current} = \frac{3V_m}{R}$$

$$= \frac{3 \times 179.63}{10} = 53.89 \text{ A}$$

∴ Peak current of converter - 1 will be,

$$I_{peak}(\text{Conv1}) = \text{Peak load current} + \text{peak circulating current}$$

$$= 53.5 + 53.89$$

$$= 107.39 \text{ A}$$

2.10.2 Non-circulating Current Mode

We know that both the converters are 'ON' (i.e. conducting) in circulating current mode. The directions of voltage and current both can be reversed to operate it into all the four quadrants. In the noncirculating current mode, only one converter operates at a time. Hence circulating current doesnot flow. Hence circulating current inductor L_r is not required. When converter-1 is 'ON' output voltage is positive. When converter-2 is 'ON' output voltage is negative. The output current direction will also be reversed. But the response of noncirculating mode dual converter is slow. Table 2.9.1 given earlier for comparison of circulating and noncirculating current modes is applicable for 3ϕ dual converters also.

Review Questions

1. *Compare circulating and non circulating current modes.*
2. *Derive an equation for circulating current in dual converter.*

Summary

Serial No.	Type of converter	Load	Parameter
1.	1 ϕ half wave	R	$V_{o(av)} = \frac{2V_m}{\pi}(1+\cos\alpha)$
		RL	$V_{o(av)} = \frac{V_m}{2\pi}(\cos\alpha - \cos\beta)$
		RL with freewheeling diode	$V_{o(av)} = \frac{V_m}{2\pi}(1+\cos\alpha)$

2.	1 ϕ semi converter	R	$V_{o(av)} = \frac{V_m}{\pi}(1+\cos\alpha)$ $V_{o(rms)} = \left\{\frac{V_m^2}{2\pi}\left[\pi-\alpha+\frac{1}{2}\sin 2\alpha\right]\right\}^{\frac{1}{2}}$
		RL	$V_{o(av)} = \frac{V_m}{\pi}(1+\cos\alpha)$ $V_{o(rms)} = \left\{\frac{V_m^2}{2\pi}\left[\pi-\alpha+\frac{1}{2}\sin 2\alpha\right]\right\}^{\frac{1}{2}}$
		RL	$DF = \cos\frac{\alpha}{2}$ $PF = \sqrt{\frac{8}{\pi(\pi-\alpha)}}\cos^2\frac{\alpha}{2}$ $HF = \sqrt{\frac{\pi(\pi-\alpha)}{8\cos^2\frac{\alpha}{2}}-1}$
3.	1 ϕ full converter	R	$V_{o(av)} = \frac{V_m}{\pi}(1+\cos\alpha)$ $V_{o(rms)} = \left\{\frac{V_m^2}{2\pi}\left[\pi-\alpha+\frac{1}{2}\sin 2\alpha\right]\right\}^{\frac{1}{2}}$
		RL	$V_{o(av)} = \frac{2V_m}{\pi}\cos\alpha$ $V_{o(rms)} = \frac{V_m}{\sqrt{2}}$
		RL with freewheeling diode	$V_{o(av)} = \frac{V_m}{\pi}(1+\cos\alpha)$ $V_{o(rms)} = \left\{\frac{V_m^2}{2\pi}\left[\pi-\alpha+\frac{1}{2}\sin 2\alpha\right]\right\}^{\frac{1}{2}}$
		RL	$DF = \cos\alpha$ $PF = \frac{2\sqrt{2}}{\pi}\cos\alpha$ HF = 0.4834 or 48.34%
4.	3 ϕ halfwave converter	R	$V_{o(av)} = \frac{3\sqrt{3}V_m}{2\pi}\cos\alpha$ *for* $\alpha \le 30^o$ $= \frac{3V_m}{2\pi}\left[1+\cos\left(\frac{\pi}{6}+\alpha\right)\right]$ *for* $\alpha > 30^o$
		RL	$V_{o(av)} = \frac{3\sqrt{3}V_m}{2\pi}\cos\alpha$

5.	3 ϕ semiconverter	R, RL	$V_{o(av)} = \frac{3\sqrt{3}V_m}{2\pi}(1+\cos\alpha)$
6.	3 ϕ fullconverter	R RL RL - with freewheeling diode	$V_{o(av)} = \frac{3\sqrt{3}V_m}{\pi}\cos\alpha$ *for* $\alpha \le 60^\circ$ $V_{o(av)} = \frac{3\sqrt{3}V_m}{\pi}\left[1+\cos\left(\frac{\pi}{3}+\alpha\right)\right]$ *for* $\alpha > 60^\circ$ $V_{o(av)} = \frac{3\sqrt{3}V_m}{\pi}\cos\alpha$ $V_{o(av)}$ is same as that with R-load.
7.	1 ϕ dual converter	RL	$V_{o1(av)} = \frac{2V_m}{\pi}\cos\alpha_1$ $V_{o2(av)} = \frac{2V_m}{\pi}\cos\alpha_2$ $\alpha_1+\alpha_2 = 180^\circ$ $i_{cir} = \frac{2V_m}{\omega L_r}[\cos\omega t - \cos\alpha_1]$

❑❑❑

3 AC Voltage Controllers and Cyclo Converters

Objectives

- *Principle of operation of AC voltage controllers.*
- *Types of AC controllers* → *on / off control*, *phase control*
- *Phase controlled AC voltage controllers*
  - *(i) Bidirectional (full wave) controllers*
  - *(ii) Unidirectional (half wave) controllers*
- *Controllers with resistive load and inductive load.*

3.1 Introduction

- Most of the loads need variable or regulated AC power. For example induction motors, heating furnaces, blowers, pumps, lights etc need variable AC supply.
- Then AC voltage controllers are used to drive such loads. The AC voltage controller consists of SCRs as switches. It normally takes mains supply as the input and provides variable AC to the load.
- The SCRs are used to control the rms value of the load voltage. If the input AC is varying, then same applications need regulated AC. The AC voltage controllers are also used in such applications. Then they are also called AC voltage regulators.
- The AC voltage controllers use either on-off control or phase-angle control to transfer required power to the load.
- In the on-off control, the SCRs conduct fully (like diodes) for few cycles and they are fully off for some cycles. Thus power is

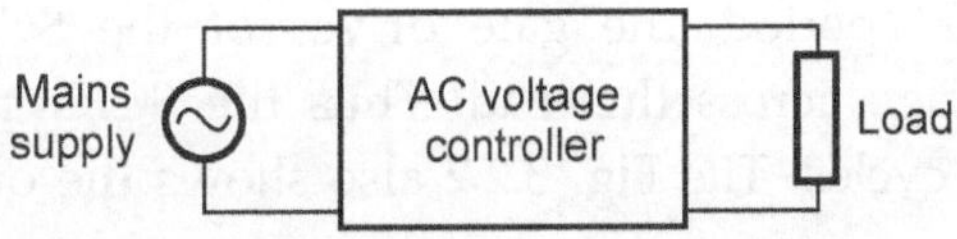

Fig. 3.1.1 The AC voltage controller

supplied to the load intermittently. The number of 'ON' and 'OFF' cycles are controlled such that required power is transformed to the load.

- In the phase-angle control, the SCRs are triggered in each half cycle of the supply with some delay. This delay is such that it transfers required power to the load.
- The AC voltage controllers have mains supply as the input. The SCRs in these controllers are turned on by applying gate drives. The SCRs are turned off by natural (line) commutation. Hence extra commutation components are not required. Therefore AC voltage controllers are simple and easy to implement if SCRs are used. Some applications also use Triacs instead of using antiparallel SCRs.
- The AC voltage controllers can be single phase or three phase. Each of these types can be unidirectional (i.e. half wave control) and bidirectional (i.e. full wave control). In this chapter we will study these basic types of AC voltage controllers.

3.2 Principle of Integral Cycle (On-off) Control

3.2.1 Circuit Diagram and Waveforms

Fig. 3.2.1 shows the circuit diagram of single phase full wave controller. This controller can be operated in on-off control mode.

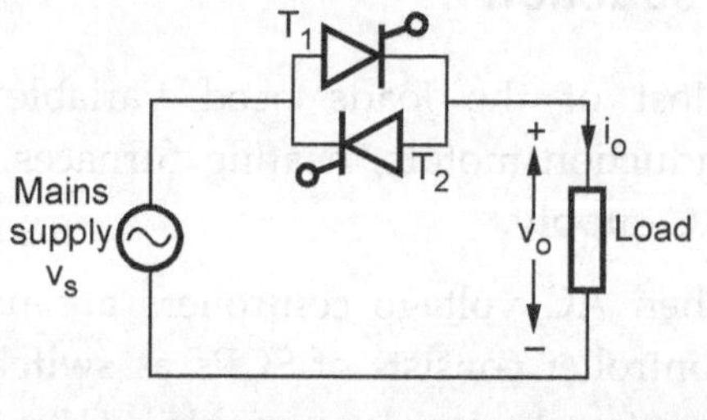

Fig. 3.2.1 1ϕ full wave controller for on-off control

The SCR T_1 conducts in the positive half cycle of the supply and T_2 conducts in the negative half cycle. The SCRs are triggered at $\alpha = 0$; i.e. zero triggering delay for few cycles. The mains supply is connected fully to the load during these cycles. Fig. 3.2.2 shows the waveforms of this AC controller for resistive load.

Observe that T_1 is triggered in positive half cycles with $\alpha = 0$ (i.e. at zero crossings). Similarly T_2 is triggered in negative half cycles with $\alpha = 0$ (i.e. at zero crossings). Hence full cycle of the mains supply is applied across the load. During the 'OFF' period, the gate drives of the SCRs are totally removed. Hence there is no voltage across the load. Thus the SCRs conduct for 'n' cycles and they are 'OFF' for 'm' cycles. The Fig. 3.2.2 also shows the output current waveform for resistive load.

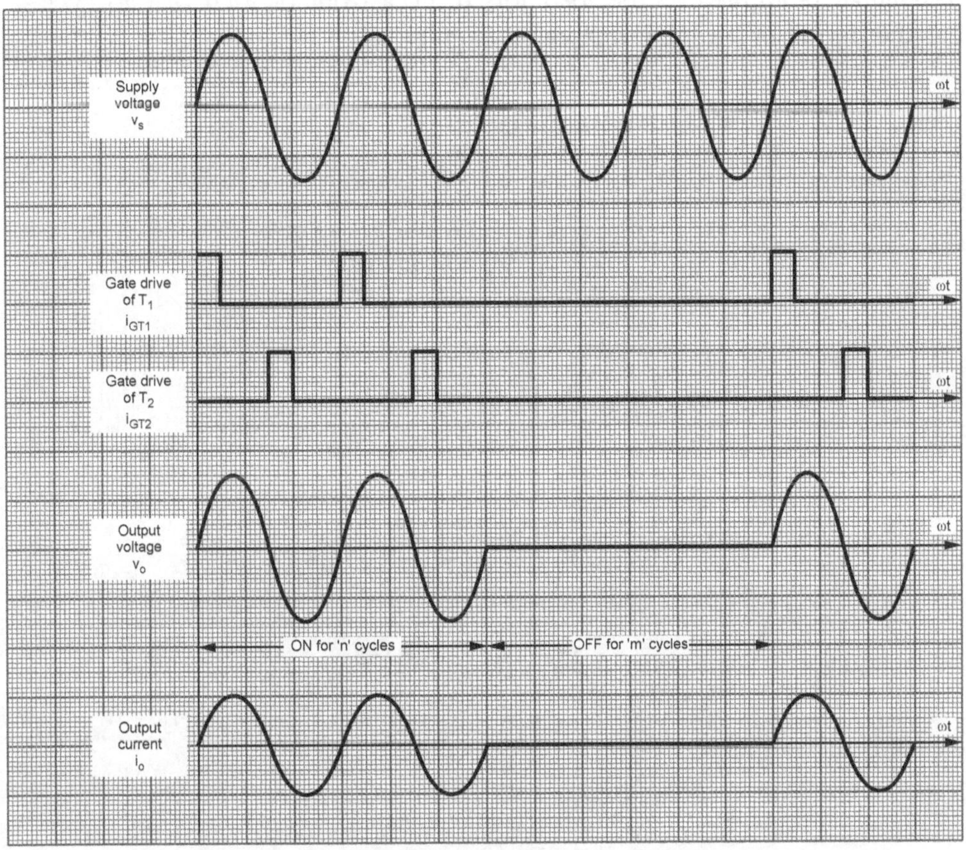

Fig. 3.2.2 Waveforms of on-off type AC voltage controller

3.2.2 Advantages and Disadvantages

Advantages

The SCRs are switch on at zero crossings. Hence the harmonics due to switching actions are reduced.

Disadvantages

A full supply voltage is applied across the load during 'ON' period and load voltage is zero during the 'OFF' period. Hence the load voltage is not smooth rather it is intermittent. The load has to sustain these variations.

Example 3.2.1 : *Obtain an expression for rms value of load voltage in on-off AC voltage controller.*

Solution : The supply voltage is,

$$v_s = V_m \sin \omega t$$

This voltage is applied to the load for 'n' cycles out of total $(n+m)$ cycles. Each cycle has the period of 2π. The rms value is given as,

$$V_{o\,(rms)} = \left[\frac{1}{T}\int_0^T v_s^2\,(\omega t)\,d\,\omega t\right]^{\frac{1}{2}}$$

$$= \left[\frac{n}{n+m}\cdot\frac{1}{2\pi}\int_0^{2\pi} V_m^2\,sin^2\,(\omega t)\,d\,\omega t\right]^{\frac{1}{2}}$$

$$= \left[\frac{n}{n+m}\cdot\frac{V_m^2}{2\pi}\int_0^{2\pi} sin^2\,\omega t\;d\,\omega t\right]^{1/2}$$

$$= \left[\frac{n}{m+n}\cdot\frac{V_m^2}{2\pi}\int_0^{2\pi}\frac{1-cos\,2\omega t}{2}\,d\,\omega t\right]^{\frac{1}{2}}$$

$$= \left[\frac{n}{m+n}\cdot\frac{V_m^2}{2\pi}\left(\int_0^{2\pi}\frac{1}{2}\,d\,\omega t-\frac{1}{2}\int_0^{2\pi} cos\,2\omega t\;d\,\omega t\right)\right]^{\frac{1}{2}}$$

$$= \left[\frac{n}{m+n}\cdot\frac{V_m^2}{2\pi}\cdot\pi\right]^{\frac{1}{2}}$$

$$= \frac{V_m}{\sqrt{2}}\cdot\sqrt{\frac{n}{m+n}} \quad \text{... (3.2.1)}$$

Normally $\frac{n}{m+n} = k$ is call the duty cycle of the on-off control. Hence above equation becomes,

$$V_{o(rms)} = \frac{V_m}{\sqrt{2}}\cdot\sqrt{k}$$

Here $\frac{V_m}{\sqrt{2}} = V_{s(rms)}$ i.e. rms value of the supply. Hence above equation becomes,

$$V_{o(rms)} = V_{s\,(rms)}\cdot\sqrt{k} \quad \text{... (3.2.2)}$$

Example 3.2.2 : *In an ON-OFF control circuit using 1ϕ, 230 V, 50 Hz supply the ON time is 10 cycles and OFF time is 4 cycles. Calculate the RMS value of the output voltage.*

Solution : The given data is,

$$V_{s\,(rms)} = 230 \text{ V}$$

$$\text{ON cycles n} = 10$$

$$\text{OFF cycles m} = 4$$

$$\therefore \quad \text{Duty cycle k} = \frac{n}{m+n}$$

$$= \frac{10}{10+4} = \frac{10}{14}$$

From equation 3.2.2 the rms value of output voltage is given as,

$$V_{o(rms)} = V_{s(rms)} \cdot \sqrt{k}$$

Putting the values in above equation,

$$V_{o(rms)} = 230 \cdot \sqrt{\frac{10}{14}}$$

$$= 194.38 \text{ volts.}$$

Example 3.2.3 : *An AC voltage controller has a resistive load of $R = 10\ \Omega$ and rms input voltage is V_s = 120 V, 60 Hz. The thyristor switch is on for n = 25 cycles and off for m = 75 cycles.*
Determine
i) the rms output voltage
ii) input power factor
iii) average and rms current of thyristors

Solution : The given data is

$$R = 10\ \Omega$$

$$V_s = 120 \text{ V}$$

$$m = 75 \text{ and } n = 25$$

$$\therefore \quad \text{Duty cycle } k = \frac{n}{m+n}$$

$$= \frac{25}{75+25} = 0.25$$

i) To obtain rms output voltage

The rms value of output voltage is given by equation 3.2.2 as,

$$V_{o(rms)} = V_s \sqrt{k}$$

$$= 120 \sqrt{0.25}$$

$$= 60 \text{ V}$$

ii) To obtain input power factor

The rms output current can be calculated as,

$$I_{o\,(rms)} = \frac{V_{o(rms)}}{R}$$

$$= \frac{60}{10} = 6 \text{ A}$$

The load is resistive. Hence power assumed by the load is active. Hence,

$$\text{Active load power} = I_{o(rms)}^2 R$$

$$= 6^2 \times 10 = 360 \text{ W}$$

In Fig. 3.2.2 observe that the supply current is same as output current. Hence rms value of supply current is,

$$I_{s\,(rms)} = I_{o\,(rms)} = 6 \text{ A}$$

Hence total rms input power (total input VA) is,

$$\text{Total rms power} = V_{s\,(rms)} \times I_{s(rms)}$$

$$= 120 \times 6 = 720 \text{ VA}$$

Therefore supply power factor is

$$\text{PF} = \frac{\text{Active load power}}{\text{Total rms input power}}$$

$$= \frac{360}{720} = 0.5 \text{ (lagging)}$$

iii) To obtain the rms and average currents of thyristors

rms thyristor current

In on/off controller, there are two antiparallel SCRs as shown in Fig. 3.2.1. The output current is shared equally by these thyristors. The rms value of load current is calculated in part (ii) as,

$$I_{o(rms)} = 6 \text{ A}$$

Let the rms current of T_1 be I_{T1} and that of T_2 be I_{T2}. Then we can write,

$$I_{o(rms)}^2 = I_{T1}^2 + I_{T2}^2$$

Since T_1 and T_2 carry equal amount of currents, $I_{T1} = I_{T2} = I_T$. Hence above equation becomes,

$$I_{o(rms)}^2 = I_T^2 + I_T^2$$

$$\text{or} \qquad \frac{I_{o(rms)}^2}{2} = I_T^2$$

$$\therefore \qquad I_T = \frac{I_{o(rms)}}{\sqrt{2}}$$

$$= \frac{6}{\sqrt{2}} = 4.24 \text{ A}$$

Average thyristor current

The average current of both the thyristors will be same. Since they share the current equally. Fig. 3.2.3 shows the thyristor current waveform.

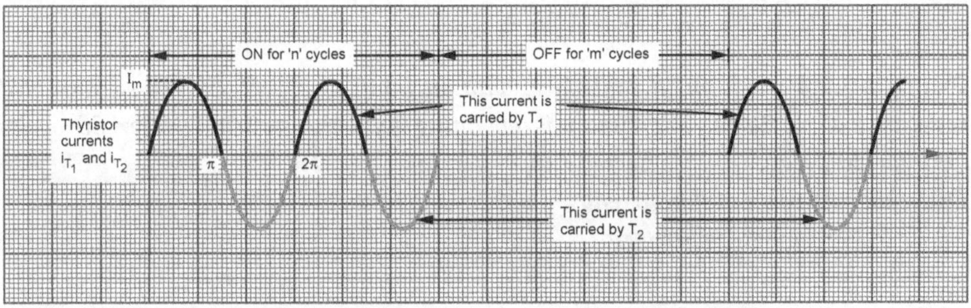

Fig. 3.2.3 Thyristor currents in on/off control

The peak value of the current will be

$$I_m = \frac{V_m}{R} = \frac{\sqrt{2}V_s}{R}$$

Hence the current can be expressed as,

$$i(t) = I_m \sin \omega t$$

From above waveform we can write the expression for average value of current in a thyristor T_1 as,

$$I_{T1(av)} = \frac{n}{m+n}\left\{\frac{1}{2\pi}\int_0^{\pi} i(t)\,dt\right\} = \frac{n}{m+n}\left\{\frac{1}{2\pi}\int_0^{\pi} I_m \sin(\omega t)\,d\omega t\right\}$$

$$= \frac{n}{m+n}\left\{\frac{I_m}{2\pi}[-\cos\omega t]_0^{\pi}\right\}$$

$$= \frac{n}{m+n}\left\{\frac{I_m}{2\pi}[-\cos\pi + \cos 0]\right\}$$

$$= \frac{n}{m+n} \cdot \frac{I_m}{\pi}$$

Putting for I_m in above equation,

$$I_{T1(av)} = \frac{n}{m+n} \cdot \frac{\sqrt{2}V_s}{\pi R}$$

Putting values in above equation,

$$I_{T1(av)} = \frac{25}{75+25} \cdot \frac{\sqrt{2}\times 120}{\pi \times 10}$$

$$= 1.35 \text{ A}$$

Since both the thyristors share the load current equally, their average currents will be same. i.e.,

$$I_T = I_{T1} = I_{T2} = 1.35 \text{ A}$$

Example 3.2.4 : *An on/off controller, with an input of 230 V, 50 Hz is connected to a resistive load of 20 Ω. The circuit is operating with the switch ON for 30 cycles and OFF for 30 cycles.*

Determine i) rms output current

ii) input power factor

Solution :

ON cycles, $n = 30$

OFF cycles, $m = 30$

Load, $R = 20\ \Omega$

$$V_{s(rms)} = 230 \text{ V}$$

The duty cycle, $k = \frac{n}{m+n} = \frac{30}{30+60} = 0.5$

i) To obtain rms output current

RMS output voltages will be,

$$V_{0(rms)} = V_s\sqrt{k}$$

$$= 230\sqrt{0.5} = 162.6 \text{ V}$$

ii) To obtain input power factor

Since the load is resistive, the active load power will be,

$$\text{Active load power} = \frac{V_{0(rms)}^2}{R}$$

$$= \frac{(162.6)^2}{20} = 1322.5 \text{ W}$$

Since supply current is same as output current, their rms values will be same.

Hence

$$I_{s(rms)} = I_{o(rms)}$$

$$= \frac{V_{o(rms)}}{R}$$

$$= \frac{162.6}{20}$$

$$= 8.13 \text{ A}$$

Hence total rms input power will be,

$$\text{Total rms power} = V_{s(rms)} \times I_{s(rms)}$$

$$= 230 \times 8.13 = 1870 \text{ VA}$$

Therefore input power factor is given as,

$$PF = \frac{\textit{Active load power}}{\textit{Total rms power}}$$

$$= \frac{1322.5}{1870} = 0.707$$

Review Question

1. *Explain the principle of ON-OFF control. Derive the expression for rms value of output voltage.*

3.3 Principle of Phase Control

3.3.1 Circuit Diagram and Waveforms

Fig. 3.3.1 shows the circuit diagram of single phase half wave controller employing phase angle control. Observe that there is one SCR T_1 and antiparallel diode D_1. Hence only positive half cycle of the supply is controlled. The negative half cycle is not controlled since diode D_1 conducts fully. Fig. 3.3.2 shows the waveforms of this circuit for resistive load. Observe that the output voltage waveform is not symmetric. Hence it will have a dc component. This may saturate the loads like induction motors, pumps etc. Observe that the output current waveform is not symmetric. The supply

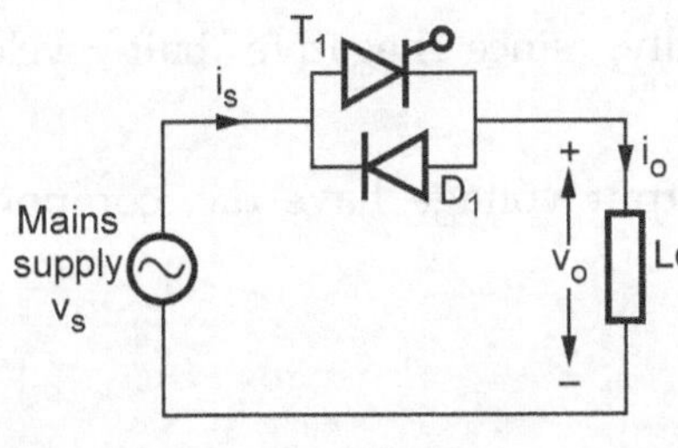

Fig. 3.3.1 1 ϕ half wave controller (unidirectional controller)

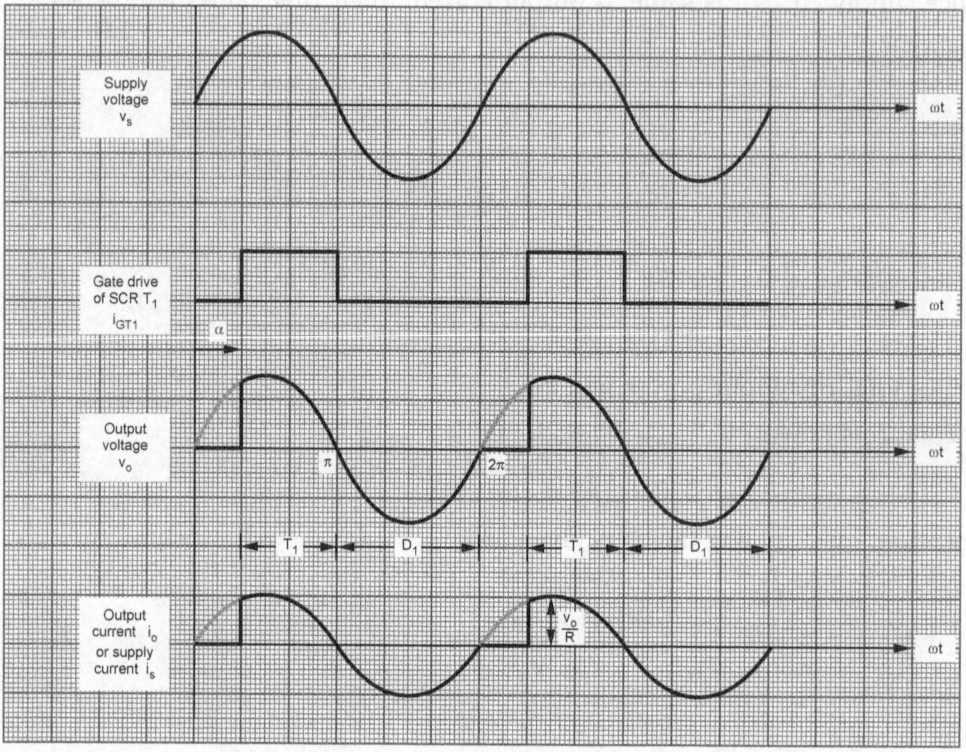

Fig. 3.3.2 Waveforms of 1ϕ half wave controller

current waveform is same as output current waveform. Hence the supply current i_s is not symmetric. The supply current also contains d.c. component. If there is input transformer, then it will be saturated due to d.c. component of supply current.

3.3.2 Advantages, Disadvantages and Applications

Advantages

Since there is only one SCR, the control is simple.

Disadvantages

1) The output voltage is not controlled fully, since negative half cycle is uncontrolled due to diode D_1.

2) The supply current, output current and output voltage have d.c. component. This is a problem for inductive loads.

Applications

Due to dc component in the output current and voltage this is used for heating and lighting (mainly resistive) applications.

Example 3.3.1 : *Derive an expression for rms and average value of output voltage for 1ϕ half wave controller.*

Solution : The supply voltage is,

$$v_{s(\omega t)} = V_m \sin \omega t \quad \text{... (3.3.1)}$$

The rms value is given as,

$$V_{o(rms)} = \left[\frac{1}{T}\int_0^T v_o^2(\omega t)\, d\omega t\right]^{\frac{1}{2}}$$

From output voltage waveform of Fig. 3.3.2 we can write above equation as,

$$V_{o(rms)} = \left\{\frac{1}{2\pi}\left[\int_\alpha^\pi v_o^2(\omega t)\, d\omega t + \int_\pi^{2\pi} v_o^2(\omega t)\, d\omega t\right]\right\}^{\frac{1}{2}}$$

In Fig. 3.3.2, observe that whenever SCR or diode conducts, the supply voltage v_s is applied to the load. Hence $v_{o(\omega t)} = v_s(\omega t)$. Hence above equation can be written as,

$$V_{o(rms)} = \left\{\frac{1}{2\pi}\left[\int_\alpha^\pi V_m^2 \sin^2 \omega t\, d\omega t + \int_\pi^{2\pi} V_m^2 \sin^2 \omega t\, d\omega t\right]\right\}^{\frac{1}{2}}$$

$$= \left\{\frac{V_m^2}{2\pi}\left[\int_\alpha^\pi \sin^2 \omega t\, d\omega t + \int_\pi^{2\pi} \sin^2 \omega t\, d\omega t\right]\right\}^{\frac{1}{2}}$$

$$= \left\{\frac{V_m^2}{2\pi}\left[\int_\alpha^\pi \frac{1-\cos 2\omega t}{2}\, d\omega t + \int_\pi^{2\pi} \frac{1-\cos 2\omega t}{2}\, d\omega t\right]\right\}^{\frac{1}{2}}$$

Solving the above integration,

$$V_{o(rms)} = \frac{V_m}{2}\sqrt{\frac{2\pi - \alpha + \frac{\sin 2\alpha}{2}}{\pi}} \quad \text{... (3.3.2)}$$

And rms value of current will be,

$$I_{o(rms)} = \frac{V_{o(rms)}}{R}$$

We know that the average value is given as,

$$V_{o(av)} = \frac{1}{T}\int_0^T v_o\,(\omega t)\,d\omega t$$

We know that $v_o\,(\omega t) = v_s\,(\omega t)$ whenever SCR or diode conducts. Hence above equation can be written as,

$$V_{o(av)} = \frac{1}{2\pi}\left[\int_\alpha^\pi V_m \sin \omega t\, d\omega t + \int_\pi^{2\pi} V_m \sin \omega t\, d\omega t\right]$$

$$= \frac{V_m}{2\pi}\left[\int_\alpha^\pi \sin \omega t\, d\omega t + \int_\pi^{2\pi} \sin \omega t\, d\omega t\right]$$

Solving the above integration,

$$V_{o(av)} = \frac{V_m}{2\pi}(\cos\alpha - 1) \qquad \text{... (3.3.3)}$$

Thus the d.c. component in the output depends upon firing angle α.

➠ **Example 3.3.2 :** *A single phase half wave a.c. voltage controller has a resistance load of $R = 5\ \Omega$ and input voltage V_s = 120 V, 60 Hz. The delay angle of thyristor is $\alpha = \frac{\pi}{3}$. Determine :*

i) rms output voltage

ii) input power factor

iii) average input current

Solution : The given data is,

$$V_s = 120\text{ V}$$

$$R = 5\ \Omega$$

$$\alpha = \frac{\pi}{3}$$

i) rms output voltage

The rms output voltage of 1ϕ half wave controller is given by equation 3.3.2 as,

$$V_{o(rms)} = \frac{V_m}{2}\sqrt{\frac{2\pi - \alpha + \frac{\sin 2\alpha}{2}}{\pi}}$$

Putting values with $\qquad V_m = \sqrt{2}\,V_s = \sqrt{2}\times 120,$

$$V_{o(rms)} = \frac{\sqrt{2}\times 120}{2}\sqrt{\frac{2\pi - \frac{\pi}{3} + \frac{\sin\left(2\cdot\frac{\pi}{3}\right)}{2}}{\pi}}$$

$$= 113.984 \approx 114 \text{ V.}$$

ii) Input power factor

The rms load current can be calculated as

$$I_{o(rms)} = \frac{V_{(rms)}}{R}$$

$$= \frac{113.984}{5} = 22.797 \text{ A}$$

Since the load is resistive, the power consumed by load is active. Hence,

$$\text{Active load power} = I^2_{o(rms)} R$$

$$= (22.797)^2 \times 5$$

$$= 2598.478 \text{ watts}$$

In Fig. 3.3.2, observe that input current is same as output current. Hence the rms value of input current is,

$$I_{s(rms)} = I_{o(rms)} = 22.797 \text{ A}$$

Hence total rms input power (total input VA) is,

$$\text{Total rms power} = V_{s(rms)}\, I_{s(rms)}$$

$$= 120\times 22.797$$

$$= 2735.64 \text{ VA}$$

Therefore supply power factor is,

$$\text{PF} = \frac{\text{Active load power}}{\text{Total rms input power}}$$

$$= \frac{2598.478}{2735.64} = 0.9498 \text{ (lagging)}$$

iii) Average input current

The average output voltage is given by equation 3.3.3 as,

$$V_{o(av)} = \frac{V_m}{2\pi}(\cos\alpha - 1)$$

$$= \frac{\sqrt{2} \cdot V_s}{2\pi}(\cos\alpha - 1)$$

Putting the values in above equation,

$$V_{o(av)} = \frac{\sqrt{2} \times 120}{2\pi}\left(\cos\frac{\pi}{3} - 1\right)$$

$$= -13.5 \text{ V}$$

Hence the average input current will be,

$$I_{o(av)} = \frac{V_{o(av)}}{R} = \frac{-13.5}{5} = -2.7 \text{ A}$$

Example 3.3.3 : *A 230 V, 1.5 kW, heater is connected as shown in Fig. 3.3.3. Find the conduction angle of the SCR if the power dissipated in the resistor is 500 W.*

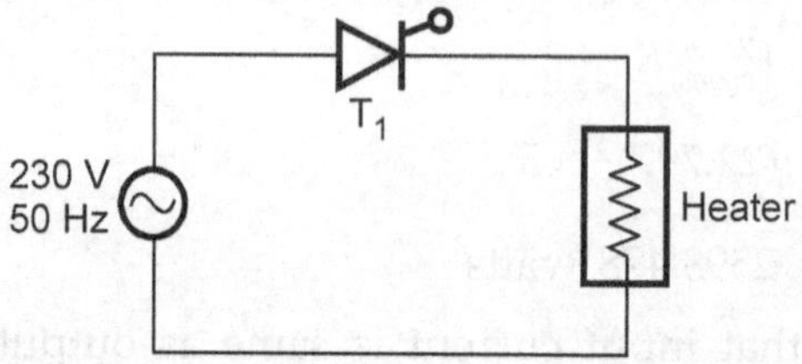

Fig. 3.3.3 AC controller of example 3.3.3

Solution : The SCR will conduct only in positive half cycle. The SCR will be triggered at α and conduct upto π. The output voltage waveform is shown in Fig. 3.3.4.

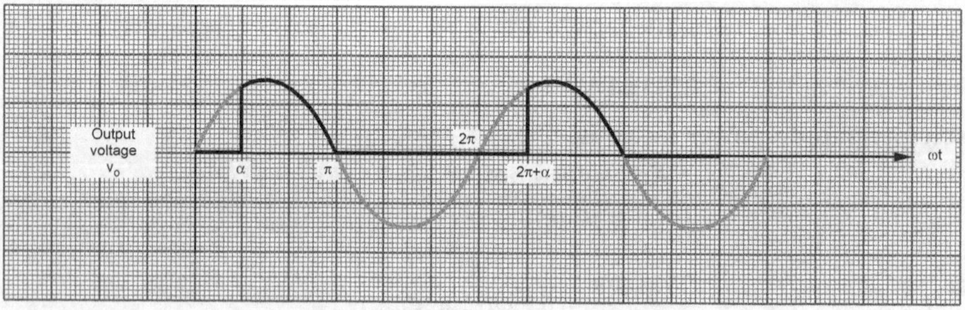

Fig. 3.3.4 Output voltage waveform of AC controller of Fig. 3.3.3

Hence load voltage is equal to supply voltage from α to π. At π, the supply voltage becomes zero. Hence SCR turns-off by natural commutation. The rms value of output can be calculated as,

$$V_{o\,(rms)} = \left[\frac{1}{T}\int_0^T v_o^2(\omega t)\,d\omega t\right]^{\frac{1}{2}}$$

$$= \left[\frac{1}{2\pi}\int_\alpha^\pi V_m^2 \sin^2 \omega t \;\; d\omega t\right]^{\frac{1}{2}}$$

$$= \left[\frac{V_m^2}{2\pi}\int_\alpha^\pi \frac{1-\cos 2\omega t}{2}\,d\omega t\right]^{\frac{1}{2}}$$

$$= \left\{\frac{V_m^2}{4\pi}\left[\int_\alpha^\pi d\omega t - \int_\alpha^\pi \cos 2\omega t \;\, d\omega t\right]\right\}^{\frac{1}{2}}$$

$$= \frac{V_m}{2}\sqrt{1-\frac{\alpha}{\pi}+\frac{\sin 2\alpha}{2\pi}} \qquad \text{... (3.3.4)}$$

The heater dissipates 1.5 kW at 230 V. Hence heater resistance is,

$$R = \frac{V_{o(rated)}^2}{P_{o(rated)}}$$

Here $V_{o(rated)} = 230$ V and $P_{o(rated)} = 1.5$ kW

$$R = \frac{230^2}{1500}$$

$$= 35.267\ \Omega$$

The power dissipated in the load is

$$P_o = 500\text{ W}$$

And $$P_o = \frac{V_{o\,(rms)}^2}{R}$$

or $$V_{o(rms)}^2 = P_o R$$

$$= 500 \times 35.267$$

$$= 17633.33$$

The supply voltage is $V_s = 230$ V. Hence $V_m = 230\sqrt{2}$. Putting values in equation 3.3.4.

$$V_{o(rms)}^2 = \left(\frac{V_m}{2}\right)^2\left[1-\frac{\alpha}{\pi}+\frac{\sin 2\alpha}{2\pi}\right]$$

$$17633.33 = \left(\frac{230\sqrt{2}}{2}\right)^2 \left[1 - \frac{\alpha}{\pi} + \frac{\sin 2\alpha}{2\pi}\right]$$

$$\therefore \quad \sin 2\alpha - 2\alpha + 2.0944 = 0$$

This equation can be solved using numerical methods. Using bisection method, the value of α at the end of 5th iteration is 1.3 radians.

Hence $\alpha = 1.3$ radians $= 74.48°$

Thus at the triggering angle of 74.48°, the AC controller will dissipate 500 W in the heater load.

3.4 Single Phase Controllers with Resistive Loads (Bidirectional Controllers)

3.4.1 Circuit Diagram and Waveforms

Fig. 3.4.1 shows the circuit diagram of 1ϕ full wave controller. It has two SCRs, T_1 and T_2. In the positive half cycle of the supply T_1 controls the power flow to the load. And in the negative half cycle of the supply T_2 controls the power flow to the load. The waveforms of this circuit are shown in Fig. 3.4.2 for resistive load.

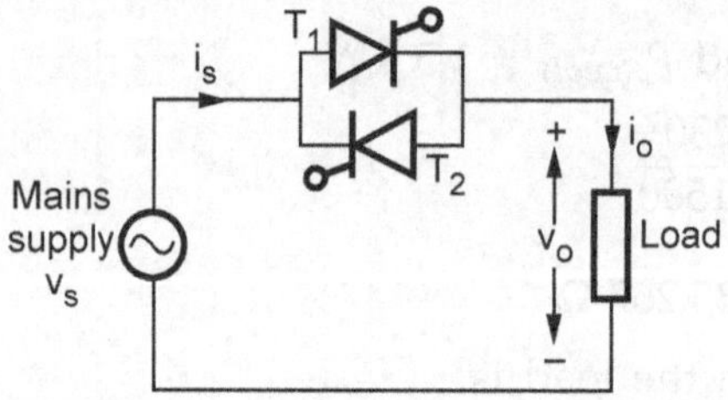

Fig. 3.4.1 1ϕ full wave (bidirectional) controller

The output current waveform is shown for resistive load. It is similar to the voltage waveform. The output current and the supply current flow in the same loop. Hence $i_o = i_s$. Observe that the voltage and current waveforms are symmetric. Hence there is no dc component in v_o, i_o and i_s. Also, it is possible to control the output fully from zero to maximum value. The output is controlled in positive as well as negative half cycles due to two SCRs.

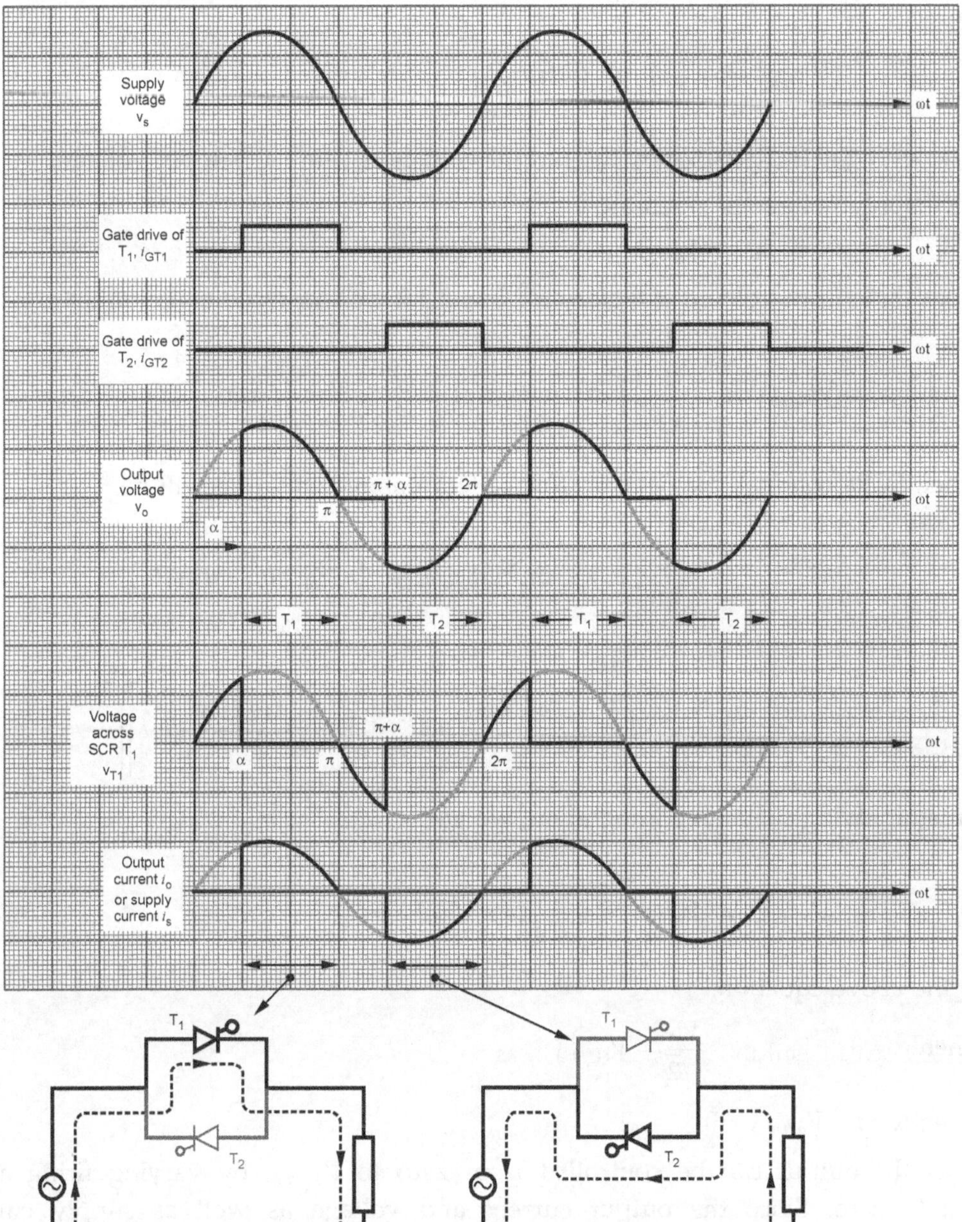

Fig. 3.4.2 Waveforms of 1ϕ full wave controller for resistive load

Example 3.4.1 : *Derive an expression for rms value of the output voltage for 1ϕ full wave (bidirectional) controller.*

Solution : We know that the rms value is given as,

$$V_{o(rms)} = \left[\frac{1}{T}\int_0^T v_o^2(\omega t)\,d\omega t\right]^{\frac{1}{2}} \qquad \text{... (3.4.1)}$$

Whenever the SCRs conduct, the output voltage is same as supply voltage.

The supply voltage is given as

$$v_s(\omega t) = V_m \sin \omega t$$

From the output voltage waveform of Fig. 3.4.2 we can write equation 3.4.1 as,

$$V_{o\,(rms)} = \left\{\frac{1}{2\pi}\left[\int_{\alpha}^{\pi} v_o^2(\omega t)\,d\omega t + \int_{\pi+\alpha}^{2\pi} v_o^2(\omega t)\,d\omega t\right]\right\}^{\frac{1}{2}}$$

When SCRs conduct, $v_o\ (\omega t) = v_s(\omega t)$. Hence above equation will be,

$$V_{o(rms)} = \left\{\frac{1}{2\pi}\left[\int_{\alpha}^{\pi} V_m^2 \sin^2 \omega t\,d\omega t + \int_{\pi+\alpha}^{2\pi} V_m^2 \sin^2 \omega t\,d\omega t\right]\right\}^{\frac{1}{2}}$$

$$= \left\{\frac{V_m^2}{2\pi}\left[\int_{\alpha}^{\pi} \frac{1-\cos 2\omega t}{2}\,d\omega t + \int_{\pi+\alpha}^{2\pi} \frac{1-\cos 2\omega t}{2}\,d\omega t\right]\right\}^{\frac{1}{2}}$$

On simplifying the above integration, we get,

$$V_{o(rms)} = V_m\sqrt{\frac{\pi-\alpha+\dfrac{\sin 2\alpha}{2}}{2\pi}} \qquad \text{... (3.4.2)}$$

In the above equation,

when $\alpha = 0$ $\quad V_{o\,(rms)} = \dfrac{V_m}{\sqrt{2}} = V_{s(rms)}$

when $\alpha = \pi$ $\quad V_{o(rms)} = 0$

Thus the output can be controlled from zero to $V_{s\,(rms)}$ by varying firing angle from π to zero. Since the output current and voltage as well as supply current waveforms are symmetric, their dc/average values are zero. Hence transformer saturation problems are absent.

Example 3.4.2 : *A single phase bidirectional regulator is feeding resistive load of 10 Ω. The supply voltage is 230 V-50 Hz. If the firing angle is 45 degrees, calculate the power absorbed by the load. Derive necessary equations.*

Solution : Given data is,

$$R = 10\ \Omega$$

$$V_{s(rms)} = 230\ \text{V}$$

$$\alpha = 45° \text{ or } \frac{\pi}{4}$$

The rms value of output voltage is given by equation 3.4.2 as,

$$V_{o\,(rms)} = V_m \sqrt{\frac{\pi - \alpha + \dfrac{\sin 2\alpha}{2}}{2\pi}}$$

Here $V_m = \sqrt{2}\ V_{s(rms)} = \sqrt{2} \times 230 = 325.26$ V.

Hence above equation becomes,

$$V_{o\,(rms)} = 325.26 \sqrt{\frac{\pi - \dfrac{\pi}{4} + \dfrac{\sin \dfrac{\pi}{2}}{2}}{2\pi}}$$

$$= 219.3 \text{ volts}$$

The power absorbed by the load is given as,

$$P_o = \frac{V_{o(rms)}^2}{R}$$

$$= \frac{(219.3)^2}{10}$$

$$= 4.8 \text{ kW.}$$

Example 3.4.3 : *A voltage source $v_s = 100 \sin 377t$ supplies a resistive load of 100 Ω through a pair of back to back connected thyristors (ac regulator). Calculate the average power in the load, if the thyristor's firing angle is fixed at 45° with respect to the supply voltage.*

Solution : This is a 1φ full wave controller. The supply voltage is $v_s(\omega t) = 100 \sin 377t$.

Hence,

$$V_m = 100 \text{ volts}$$

$$\text{Load } R = 100\ \Omega$$

$$\alpha = 45° \text{ or } \frac{\pi}{4}$$

The rms value of output is given by equation 3.4.2 as,

$$V_{o\,(rms)} = V_m \sqrt{\frac{\pi - \alpha + \frac{\sin 2\alpha}{2}}{2\pi}}$$

Putting values in this equation,

$$V_{o\,(rms)} = 100 \sqrt{\frac{\pi - \frac{\pi}{4} + \frac{\sin \frac{\pi}{2}}{2}}{2\pi}}$$

$$= 67.42 \text{ volts}$$

The load power can be obtained as

$$P_o = \frac{V_{o(rms)}^2}{R}$$

$$= \frac{(67.42)^2}{100}$$

$$= 45.45 \text{ watts}$$

Example 3.4.4 : *Find the power consumed in the heater element shown in Fig. 3.4.3, if both SCRs are triggered with delay angle of 45°.*

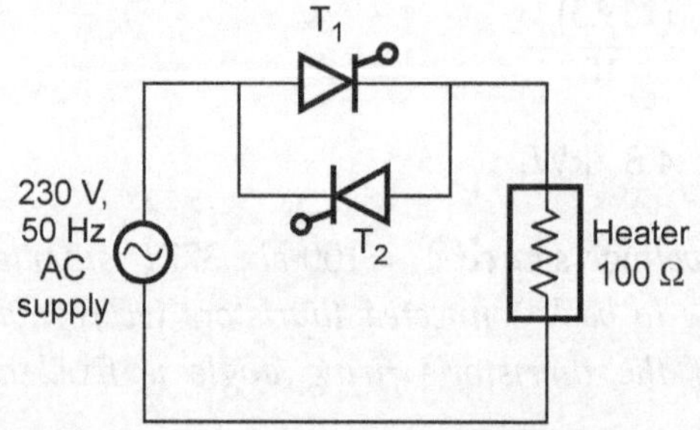

Fig. 3.4.3 AC controller of example 3.4.4

In the circuit of Fig. 3.4.3, if the load is 2 kW, 230 V heater and V_s = 230 V, 50 Hz, Calculate - (i) $V_{o(rms)}$ and

(ii) Power dissipated in heater for α = 45°

Solution : The given data is,

$$\alpha = 45° = \frac{\pi}{4}$$

$$V_s = 230 \text{ V}, \qquad \therefore V_m = 230\sqrt{2}$$

Load resistance, R = 100 Ω

The rms value of output in bidirectional regulator is given by equation 3.4.2 as,

$$V_{o\,(rms)} = V_m \sqrt{\frac{\pi - \alpha + \frac{\sin 2\alpha}{2}}{2\pi}}$$

Putting values in above equation,

$$V_{o\,(rms)} = 230\sqrt{2}\sqrt{\frac{\pi - \frac{\pi}{4} + \frac{\sin\left(2\times\frac{\pi}{4}\right)}{2}}{2\pi}}$$

$$= 219.3 \text{ V}$$

The power absorbed in the load can be calculated as

$$P_o = \frac{V_{o(rms)}^2}{R}$$

$$= \frac{219.3^2}{100}$$

$$= 480.92 \text{ W}$$

Solution to the second question :

RMS valve of output voltage is calculated as,

$$V_{o(rms)} = 219.3 \text{ V}$$

A heater is 2 kW, 230 V. This means it dissipates 2 kW at 230 V. Therefore we can write,

$$P_{rated} = \frac{V_{rated}^2}{R}$$

Here $P_{rated} = 2$ kW and $V_{rated} = 230$ V. Hence,

$$2 \times 10^3 = \frac{(230)^2}{R}$$

$$\therefore \quad R = 26.45 \ \Omega$$

Hence power dissipation at $V_{o(rms)} = 219.3$ V will be,

$$P_o = \frac{V_{o(rms)}^2}{R} = \frac{(219.3)^2}{26.45}$$

$$= 1.818 \text{ kW}$$

Example 3.4.5 : *For the AC voltage controller shown in Fig. 3.4.4, the delay angles of the thyristors T_1 and T_2 are equal, $\alpha_1 = \alpha_2 = \frac{2\pi}{3}$. Determine -*

(i) rms output voltage

(ii) input power factor

(iii) average current of thyristors and

(iv) rms current of thyristors

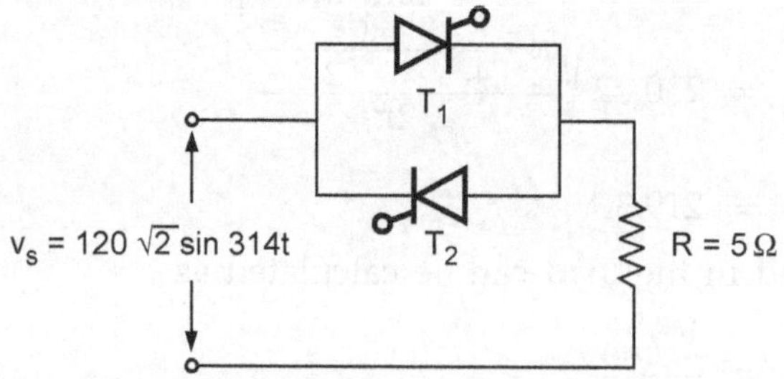

Fig. 3.4.4 Controller of example 3.4.5

Solution : Given data

$$v_s = 120\sqrt{2}\sin 314\, t$$

$$\therefore \quad V_m = 120\sqrt{2} = 169.7 \text{ V}$$

Firing angle $\quad \alpha = \alpha_1 = \alpha_2 = \frac{2\pi}{3}$

Load resistance, $R = 5\ \Omega$

(i) To obtain rms output voltage ($V_{o(rms)}$)

RMS valve of output is given by equation 3.4.2 as,

$$(V_{o(rms)}) = V_m \sqrt{\frac{\pi - \alpha + \frac{\sin 2\alpha}{2}}{2\pi}}$$

$$= 169.7 \sqrt{\frac{\pi - \frac{2\pi}{3} + \frac{\sin\left(2 \cdot \frac{2\pi}{3}\right)}{2}}{2\pi}}$$

$$= 53 \text{ V}$$

(ii) To obtain input power factor

RMS value of load current can be calculated as,

$$I_{o(rms)} = \frac{V_{o(rmo)}}{R}$$

$$= \frac{53}{5} = 10.6 \text{ A}$$

The active load power becomes,

$$\text{Active load power} = I^2_{o(rms)} R$$

$$= (10.6)^2 \times 5 = 561.8 \text{ W}$$

In Fig. 3.4.2 observe that the supply current is same as the output current.

Hence rms value of supply current will be,

$$I_{s(rms)} = I_{o(rms)} = 10.6 \text{ A}$$

Thesefore total rms input power will be,

$$\text{Total rms input power} = V_{s(rms)} \times I_{s(rms)}$$

Here the $V_m = 120\sqrt{2}$. Hence $V_{rms} = 120$ V.

Therefore above equation becomes,

$$\text{Total rms input power} = 120 \times 10.6 = 1272 \text{ VA}$$

Therefore input power factor is given as,

$$PF = \frac{\text{Active load power}}{\text{Total rms input power}} = \frac{561.8}{1272} = 0.4416 \text{ or } 44.16 \text{ \%}$$

This is lagging power factor since load current lags behind the supply voltage.

(iii) To obtain average current of thyristors

Fig. 3.4.5 shows the current waveform of thyristor T_1. This waveform is prepared from waveforms of Fig. 3.4.2.

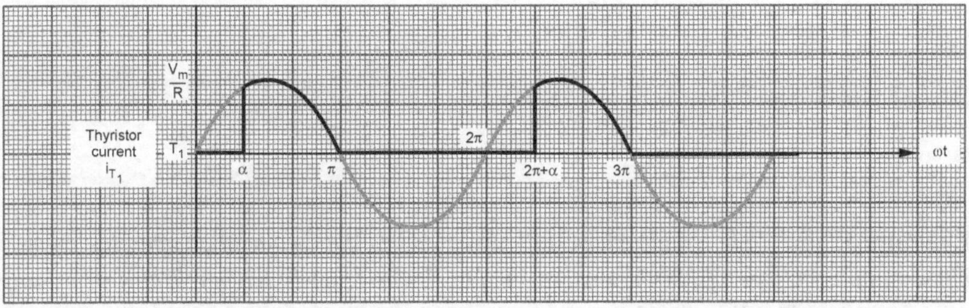

Fig. 3.4.5 Current waveform of thyristor in full wave phase control

The current waveform can be expressed as,

$$i(t) = I_m \sin \omega t$$

Peak value of current is, $I_m = \frac{V_m}{R}$. Average value of current in above figure can be given as,

$$I_{T(av)} = \frac{1}{2\pi}\int_{\alpha}^{\pi} I_m \sin(\omega t)\, d\omega t$$

$$= \frac{1}{2\pi}\int_{\alpha}^{\pi} \frac{V_m}{R} \sin(\omega t)\, d\omega t$$

$$= \frac{V_m}{2\pi R}[-\cos\omega t]_{\alpha}^{\pi}$$

$$= \frac{V_m}{2\pi R}(1+\cos\alpha)$$

Putting values in above equation

$$I_{T(av)} = \frac{169.7}{2\pi\times 5}\left(1+\cos\frac{2\pi}{3}\right)$$

$$= 2.7 \text{ A}$$

(iv) To obtain rms thyristor current

Let the rms currents of thyristors be $I_{T1(rms)}$ and $I_{T2(rms)}$. Since output rms current is made up of thyristor currents, we can write,

$$I_{o(rms)}^2 = I_{T1(rms)}^2 + I_{T2(rms)}^2$$

Both the thyristors conduct for same interval and they carry equal amount of current. Hence rms currents of both the thyristors will be same. i.e.,

$$I_{T1(rms)} = I_{T2(rms)} = I_{T(rms)}$$

$$\therefore \quad I_{o(rms)}^2 = I_{T(rms)}^2 + I_{T(rms)}^2$$

$$\therefore \quad I_{T(rms)} = \frac{I_{o(rms)}}{\sqrt{2}} = \frac{10.6}{\sqrt{2}}$$

$$= 7.49 \text{ A}$$

Example 3.4.6 : *The single phase full wave AC voltage controller operates on a single phase supply voltage of 230 V rms, at 50 Hz. If the triac is triggered at a delay angle of 45°, during each half cycle of input supply,*

Calculate,

i) RMS value of output voltage

ii) RMS value of current through heater

iii) Averge value of triac current

iv) RMS value of triac current

v) Input PF

Derive any expressions used.

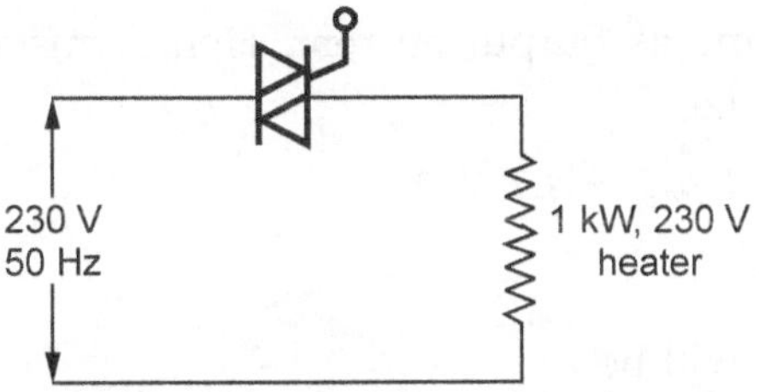

Fig. 3.4.6 AC controller of example 3.4.6

Solution :

Given data

$$V_S = 230 \text{ V} \quad \therefore \; V_m = 230\sqrt{2}$$

$$\alpha = 45° \text{ i.e. } \frac{\pi}{4}$$

i) To obtain $V_{o\,(rms)}$

The rms value of output voltage is given by equation 3.4.2 as,

$$V_{o\,(rms)} = V_m \sqrt{\frac{\pi - \alpha + \frac{\sin 2\alpha}{2}}{2\pi}}$$

$$= 230\sqrt{2} \cdot \sqrt{\frac{\pi - \frac{\pi}{4} + \frac{\sin\left(2 \cdot \frac{\pi}{4}\right)}{2}}{2\pi}}$$

$$= 219.3 \text{ volts}$$

ii) To obtain rms value of output current

The resistance of the heater will be,

$$R_L = \frac{(230)^2}{1000} = 53\ \Omega$$

Hence rms current through heater will be,

$$I_{o(rms)} = \frac{V_{o(rms)}}{R_L} = \frac{219.3}{53} = 4.137 \text{ A}$$

iii) Average value of triac current

The output current, triac current and supply current waveforms are same. It is shown in Fig. 3.4.2. Since the positive and negative half cycles of current flowing through triac are symmetric, the average value is zero.

iv) RMS value of triac current

The triac current is same as output current. Hence rms value of triac current will be same as output current i.e.,

$$I_{T(rms)} = I_{o(rms)} = 4.137 \text{ A}$$

v) To obtain input PF

The active load power will be,

$$\text{Active load power} = I_{o\,(rms)}^2 \; R_L$$

$$= (4.137)^2 \times 53$$

$$= 907 \text{ W}$$

The supply current is same as output current (see Fig. 3.4.2). Hence rms value of supply current will be same as that of output current i.e.

$$I_{s(rms)} = I_{o(rms)} = 4.137 \text{ A}$$

Therefore total rms input power will be,

$$\text{Total rms input power} = V_{s(rms)} \times I_{s(rms)}$$

$$= 230 \times 4.137$$

$$= 951.5 \text{ VA}$$

Then input power factor will be,

$$\text{PF} = \frac{\text{Active load power}}{\text{Total input power}} = \frac{907}{951.5} = 0.953$$

3.4.2 Other Circuit Configurations of 1ϕ Full Wave Controller

In the circuit of Fig. 3.4.1, the gate-cathode drives of the two SCRs must be isolated. Observe that (Fig. 3.4.1) the cathode of T_2 is connected to anode of T_1. Hence the gates as well as cathodes of T_1 and T_2 must be isolated. Hence other circuit configurations are developed which eliminate necessity of isolation.

3.4.2.1 1ϕ Full Wave Controller with Common Cathode

Fig. 3.4.7 shows the circuit diagram of 1ϕ full wave controller having common cathode connection of SCRs. In this figure observe that the cathodes of two SCRs T_1 and T_2 are connected together. In the positive half cycle of the supply SCR T_1 and diode D_1 conducts. And in the negative half cycle SCR T_2 and diode D_2 conducts. In this circuit, the gate cathode drives need not be isolated. But isolation is normally provided between the control and power circuits. Four devices are required in the above circuit. The efficiency of this circuit is slightly reduced due to increased power dissipation in the devices. The waveforms of this circuit are similar to those shown in Fig. 3.4.2.

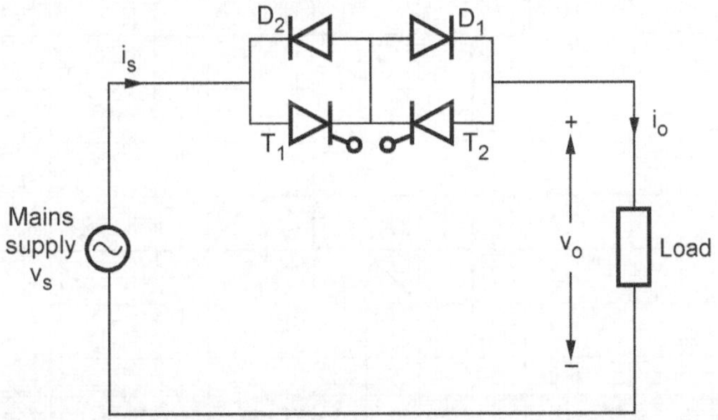

Fig. 3.4.7 1ϕ full wave controller having common cathode

3.4.2.2 1ϕ Full Wave Controller with One SCR

Fig. 3.4.8 shows another configuration of 1ϕ full wave controller having one SCR T_1. There is a diode bridge having four diodes D_1 , D_2 , D_3 and D_4. The waveforms of this circuit are shown in Fig. 3.4.9.

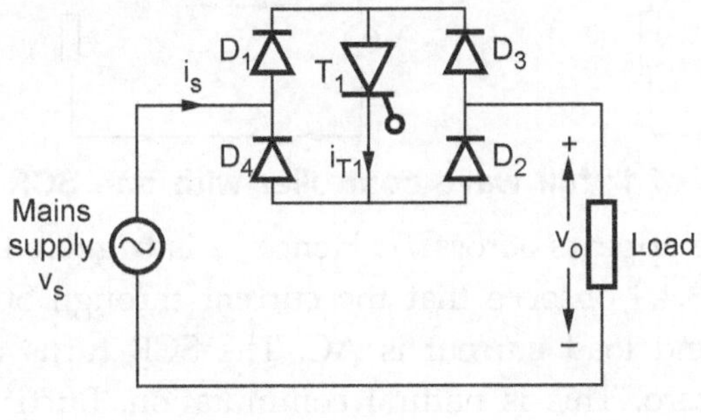

Fig. 3.4.8 1ϕ full wave controller with diode bridge and one SCR

In the waveforms, observe that T_1 is triggered in every half cycle. In the positive half cycle, current flows through $D_1 - T_1 - D_2$ and load. In the negative half cycle current flows through $D_3 - T_1 - D_4$ and load. Thus the diodes form a rectifier bridge

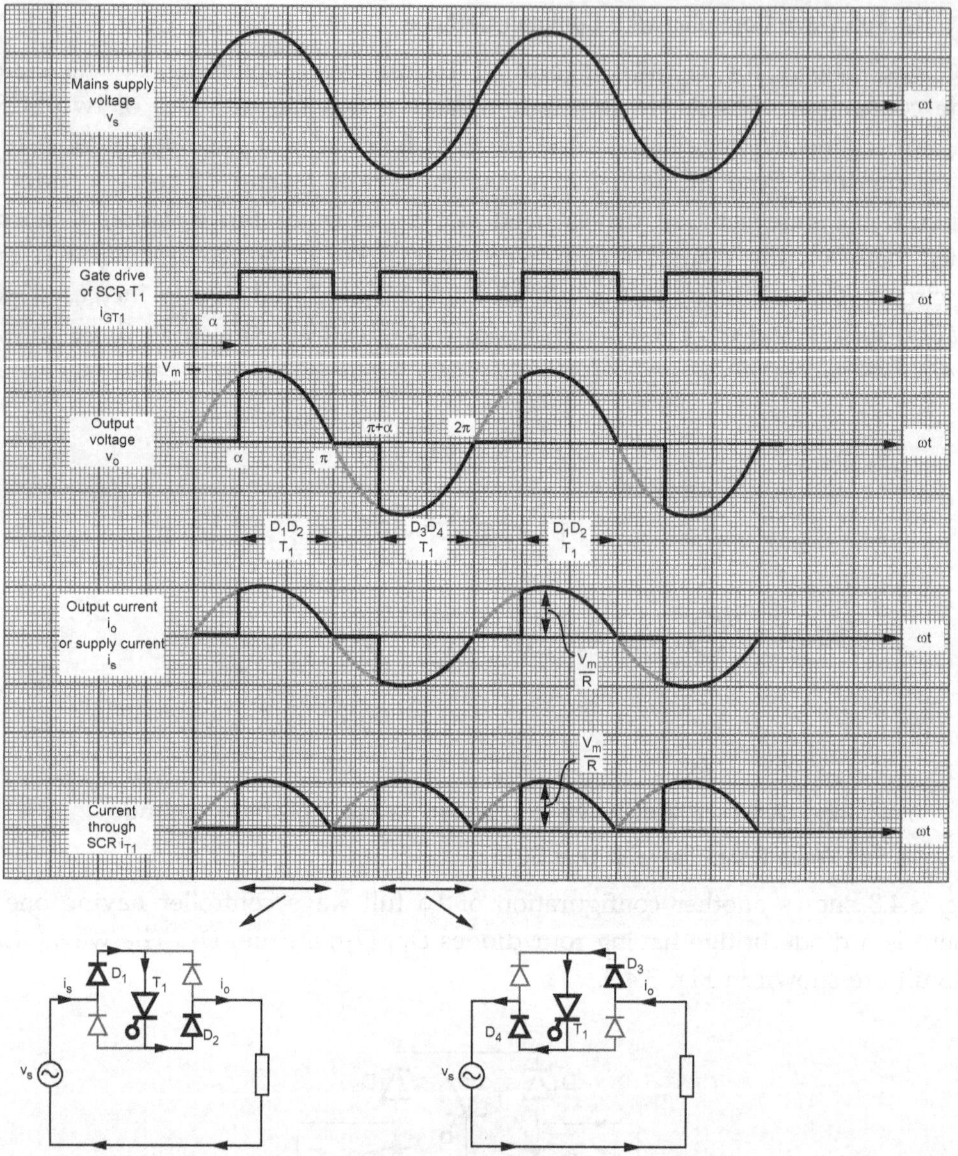

Fig. 3.4.9 Waveforms of 1ϕ full wave controller with one SCR and rectifier bridge

and the positive voltage appears across T_1. Hence T_1 is triggered in every half cycle. In the waveforms of Fig. 3.4.9 observe that the current through SCR flows in every half cycle. But the supply and load current is AC. The SCR turns off in every half cycle when current goes to zero. This is natural commutation. But if the load is inductive, then current will not go to zero at π, $2\pi, 3\pi$, ... etc. Hence SCR will not turn off and control will be lost. This is the drawback of the circuit. In this circuit, three devices (i.e. two diodes and one SCR) conduct at any time. Hence losses are increased. Hence efficiency is slightly reduced.

The rms output voltage of this circuit is same as that given by equation 3.4.2.

3.4.3 Advantages, Disadvantages and Applications

Advantages

1. The input, output currents and voltage are symmetric. Hence there is no dc component.
2. Transformers and motors saturation problems are absent.

Disadvantage

1. Control circuit is little complex.

Applications

Full wave controllers are used extensively for induction motors, pumps, fans etc.

Review Question

1. *Explain the operation of 1ϕ full wave controller with the help of circuit diagram and waveforms. What are the different configurations of this circuit ?*

Unsolved Examples

1. *Calculate the rms current flowing through the heater element shown in Fig. 3.4.10. Both the SCRs are triggered with a delay angle of 30°.* **[April - 98]**

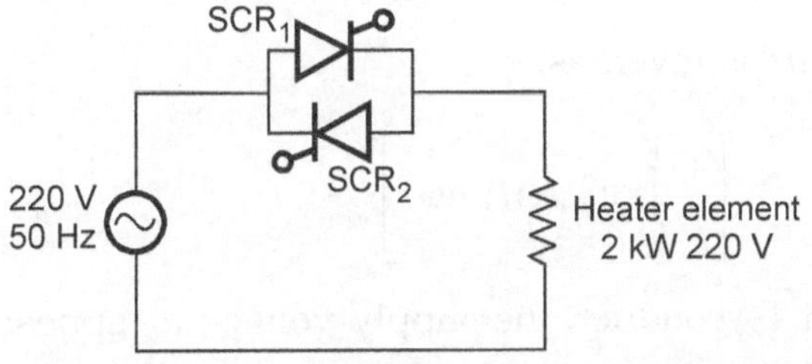

Fig. 3.4.10 AC regulator of example 3.4.10

[Ans. : $I_{o(rms)} = 9.23\ A$]

2. *A single phase AC voltage controller has the resistive load of $R = 10\ \Omega$. It has the supply voltage of 230 V, 50 Hz. The triggering angle is $\alpha = \frac{\pi}{2}$. Determine (i) rms output voltage (ii) input power factor.* **[Ans. : $V_{o\,(rms)} = 162.63\ V$, PF = 0.707]**

3.5 Single Phase Controllers with Inductive Loads

3.5.1 Circuit Diagram

In the previous section we discussed 1ϕ controllers with resistive loads. Most of the times the load on the controllers is inductive. Such loads are induction motors, water pumps, fans, blowers etc. The output current waveform is

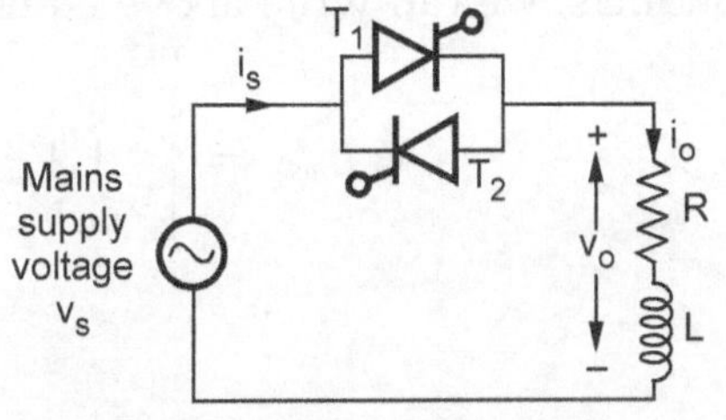

Fig. 3.5.1 1ϕ full wave controller with inductive load

different in case of inductive load. Fig. 3.5.1 shows the circuit diagram of 1ϕ full wave controller having inductive load.

3.5.2 Discontinuous Output Current

The SCR T_1 is triggered in positive half cycle with delay angle of 'α'. The output current starts increasing from zero. The waveforms are shown in Fig. 3.5.2. At π, the supply voltage is zero, but output current is not zero. The load inductance tries to maintain the current in the same direction. Hence SCR T_1 keeps on conducting. Refer to circuit-I in Fig. 3.5.2. The SCR T_1 conducts from π to β due to energy stored in the load inductance. At β the output current becomes zero. Hence T_1 turns off by natural commutation.

In the waveforms of Fig. 3.5.2 observe that T_2 is triggered at $\pi+\alpha$. The output current starts increasing in the negative direction from zero. And negative supply voltage appears across the load. Thus negative cycle of v_o and i_o starts. The supply current is same as output current. These currents are nonsinusoidal due to inductive load.

Example 3.5.1 : *Derive an expression for rms output voltage of 1ϕ full wave controller having inductive load for discontinuous load current.*

Solution : The rms value is given as,

$$V_{o(rms)} = \left[\frac{1}{T}\int_0^T v_o^2\,(\omega t)\;d\omega t\right]^{\frac{1}{2}} \qquad \text{... (3.5.1)}$$

When SCRs T_1 and T_2 conduct, the supply voltage v_s appears across the load. The supply voltage can be expressed as,

$$v_s\,(\omega t) = V_m\; sin\;\omega t \qquad \text{... (3.5.2)}$$

From the output voltage waveform of Fig. 3.5.2 we can write equation 3.5.1 as follows :

$$V_{o\,(rms)} = \left[\frac{1}{\pi}\int_\alpha^\beta v_o^2\,(\omega t)\;d\omega t\right]^{\frac{1}{2}}$$

Here observe that we have taken $T=\pi$, since rms values of positive and negative half cycles are same for symmetric waveforms. Since $v_o\,(\omega t) = v_s\,(\omega t)$ when SCR conducts, we can write above equation as,

$$V_{o\,(rms)} = \left[\frac{1}{\pi}\int_\alpha^\beta V_m^2\; sin^2\,\omega t\; d\omega t\right]^{\frac{1}{2}}$$

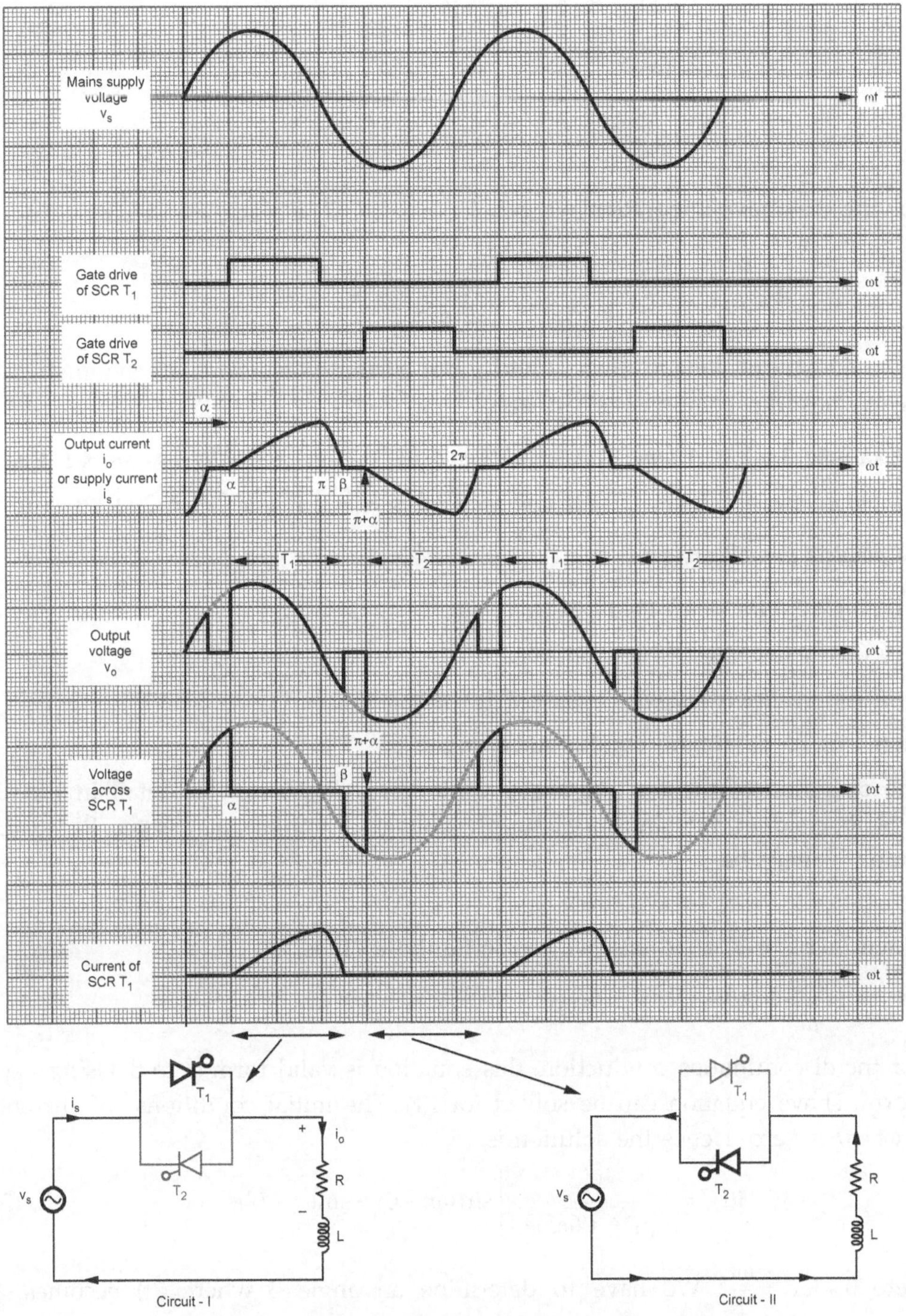

Fig. 3.5.2 Waveforms of 1ϕ full wave controller for inductive load (Discontinuous output current)

$$= \left[\frac{V_m^2}{\pi} \int_{\alpha}^{\beta} \frac{1-\cos 2\,\omega t}{2}\, d\omega t \right]^{\frac{1}{2}}$$

$$= \left[\frac{V_m^2}{2\pi} \int_{\alpha}^{\beta} (1-\cos 2\omega t)\, d\omega t \right]^{\frac{1}{2}}$$

Solving the above integration we get,

$$V_{o\,(rms)} = V_m \sqrt{\frac{\beta - \alpha + \frac{\sin 2\alpha}{2} - \frac{\sin 2\beta}{2}}{2\pi}} \quad \text{... (3.5.3)}$$

Since the output voltage, current and input current waveforms are symmetric, their average values (d.c. components) are zero.

Example 3.5.2 : *A single phase full wave AC voltage controller has an R-L load. The input voltage is 230 V, 50 Hz and the load is R = 2 Ω and X_L = 2 Ω, $\alpha_1 = \alpha_2 = \frac{\pi}{2}$.*

Calculate the following :

i) Angle β until which the thyristor conducts.

ii) Conduction angle of the thyristor.

iii) rms voltage of output. Derive the formulae you use.

Solution : i) To determine conductance angle β

The operation of the 1 ϕ full wave controller is explained in detail in Fig. 3.5.2. when SCRs conduct, the sinusoidal supply voltage is applied across the RL load. Let the supply voltage be,

$$v_s = V_m \sin \omega t$$

When thyristor T_1 conducts, we can write the KVL as,

$$L\frac{di(t)}{dt} + Ri(t) = V_m \sin \omega t$$

For the discontinuous conduction, this equation is valid form α to β. Using Laplace transform, above equation can be solved for i(t). The initial conditions of current are i(t) at $\omega t = \alpha$ is zero. Hence the solution is,

$$\text{i(t)} = \frac{V_m}{\sqrt{R^2 + (\omega L)^2}} \left\{ \sin(\omega t - \theta) - \sin(\alpha - \theta) e^{\frac{R(\alpha - \omega t)}{\omega L}} \right\} \quad \text{... (3.5.4)}$$

Here $\theta = \tan^{-1}\frac{\omega L}{R}$. We have to determine an angle β where $i(t)$ becomes zero. (observe output current waveform of Fig. 3.5.2). Therefore at $\omega t = \beta$, $i(t) = 0$. Putting this condition in above equation,

$$0 = \frac{V_m}{\sqrt{R^2+(\omega L)^2}}\left\{\sin(\beta-\theta)-\sin(\alpha-\theta)e^{\frac{R(\alpha-\beta)}{\omega L}}\right\}$$

$$\sin(\beta-\theta)-\sin(\alpha-\theta)e^{\frac{R(\alpha-\beta)}{\omega L}} = 0 \quad \text{... (3.5.5)}$$

If α and θ are known, solution of above equation gives β.

In this example the given data is,

$$R = 2\,\Omega$$

$$X_L = \omega L = 2\,\Omega$$

$$\alpha = \frac{\pi}{2}$$

$$\therefore \quad \theta = \tan^{-1}\left(\frac{\omega L}{R}\right)$$

$$= \tan^{-1}\left(\frac{X_L}{R}\right)$$

$$= \tan^{-1}\left(\frac{2}{2}\right) = 0.7853981$$

Putting the values of θ, α, R and ωL in equation 3.5.5,

$$\sin(\beta-0.7853981)-\sin\left(\frac{\pi}{2}-0.7853981\right)e^{\frac{2\left(\frac{\pi}{2}-\beta\right)}{2}} = 0$$

$$\therefore \quad \sin(\beta-0.7853981)-0.7071068\,e^{\left(\frac{\pi}{2}-\beta\right)} = 0$$

This is transcendental equation and it can be solved using numerical method. Using bisection method, the solution of above equation at the end of 11th iteration is,

$$\beta = 3.8552772 \text{ or } 220.89^\circ$$

Thus thyristor conducts till 220.89° due to inductive load.

ii) To determine conduction angle

The SCR conducts from α to β. Hence conduction angle (δ) is,

$$\delta = \beta-\alpha = 3.8552772-\frac{\pi}{2} = 2.2844809 \text{ or } 130.89^\circ$$

iii) rms output voltage

The rms value of output voltage is given by equation 3.5.3 as,

$$V_o(rms) = V_m\sqrt{\frac{\beta-\alpha+\frac{\sin 2\alpha}{2}-\frac{\sin 2\beta}{2}}{2\pi}}$$

Putting value in above equation,

$$V_o(rms) = \sqrt{2}\times 230\left\{\frac{2.2844809+\frac{\sin\left(2\frac{\pi}{2}\right)}{2}-\frac{\sin(2\times 3.8552772)}{2}}{2\pi}\right\}^{\frac{1}{2}}$$

$$= 173.59 \text{ volts}$$

3.5.3 Continuous Output Current

We discussed the operation of 1ϕ full wave controller for inductive load for discontinuous load current. In this mode the current through T_1 goes to zero before T_2 is triggered. Let us consider the case of large inductive loads. The waveforms are shown in Fig. 3.5.3. The SCR T_2 is given the drive at $\pi+\alpha$. But T_1 is conducting and its current is not zero (see Fig. 3.5.3 (d) and (f)). Hence T_1 continues to conduct. This current is maintained by the load inductance. After some time, output current goes to zero. Then T_2 starts conducting and negative cycle of i_o starts as shown in Fig. 3.5.3(d).

Here note that gate drive of T_2 must be present when i_o falls to zero. Otherwise T_2 will not be triggered. Hence gate pulses with wide duration are always required as shown in Fig. 3.5.3 (b) and (c). Observe that the output current waveform is continuous. It's shape depends upon the triggering angle and load inductance. The output voltage waveform is shown in Fig. 3.5.3 (e). The SCRs T_1 and T_2 conduct continuously. Hence the supply v_s is connected to load always. Hence output voltage waveform is same as supply voltage. In other words, the control on output voltage is lost due to load inductance.

3.5.4 Why Short Duration Gating Pulses are not Suitable with Inductive Loads ?

Observe the waveforms of Fig. 3.5.3. These are the waveforms of bidirectional AC controller for inductive load. Gating pulse is applied to T_2 at $\pi+\alpha$. But T_1 continues to conduct beyond $(\pi+\alpha)$ due to inductive load. Hence T_2 doesnot turn-on at $(\pi+\alpha)$ even if gate pulse is applied. T_2 starts conducting when current of T_1 goes to zero.

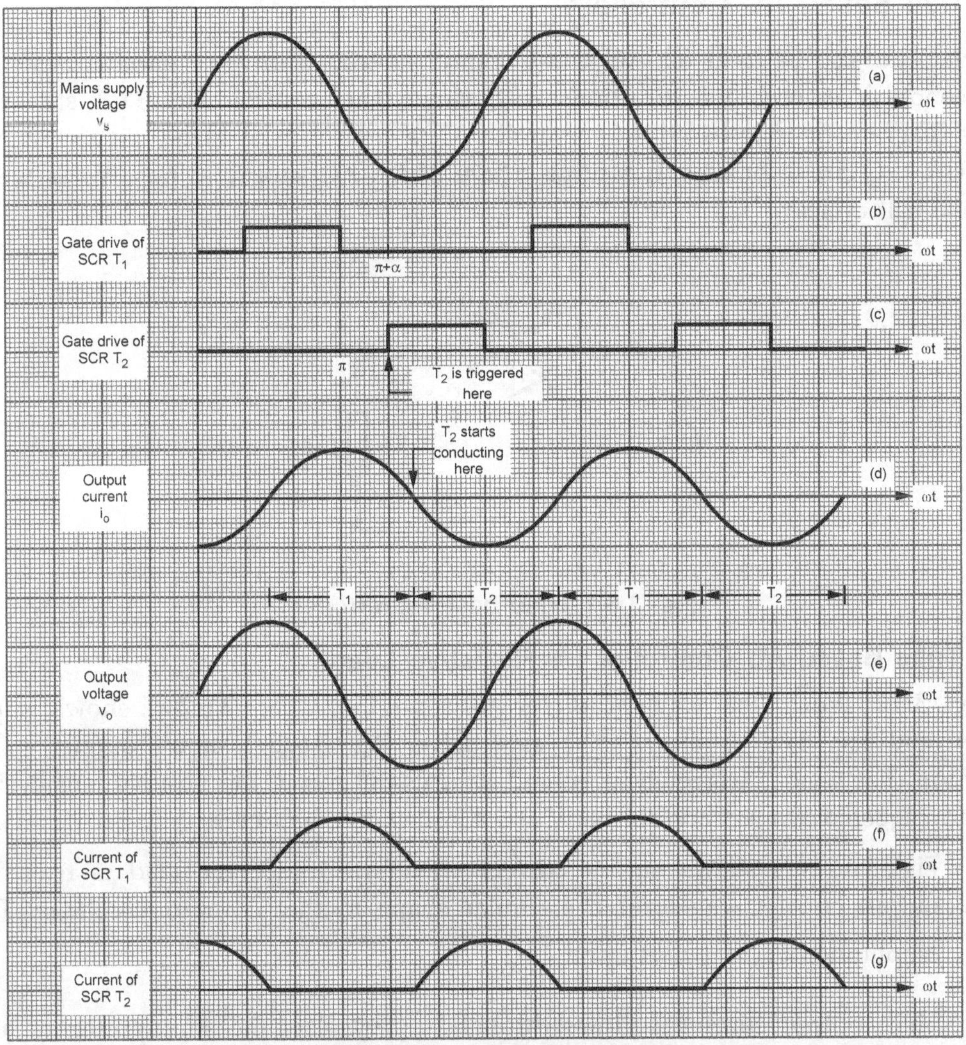

Fig. 3.5.3 Waveforms of 1φ full wave controller for inductive load (continuous output current)

Hence gate pulse of T_2 must be wide enough so that it is present at the time when T_1 current goes to zero. Then only T_2 will turn-on. If gate pulse is of short duration, then T_2 will not conduct, since there is not gating pulse when it is forward biased. Therefore short duration gating pulses are not suitable for inductive loads.

Review Question

1. *Explain the operation of 1φ full wave (bidirectional) controller for inductive load with the help of circuit diagram and waveforms. What care must be taken in triggering of SCRs in such controller?*

3.6 Triac Based AC Controllers

Triac can also be used for AC phase control instead of inverse parallel SCRs. The triac conducts in both the directions with single gate control. Fig. 3.6.1 shows the circuit diagram of triac based AC controller. The triac can be given triggering pulse from UJT, RC circuit or diac. It controls phase of the AC supply in positive as well as negative half cycles.

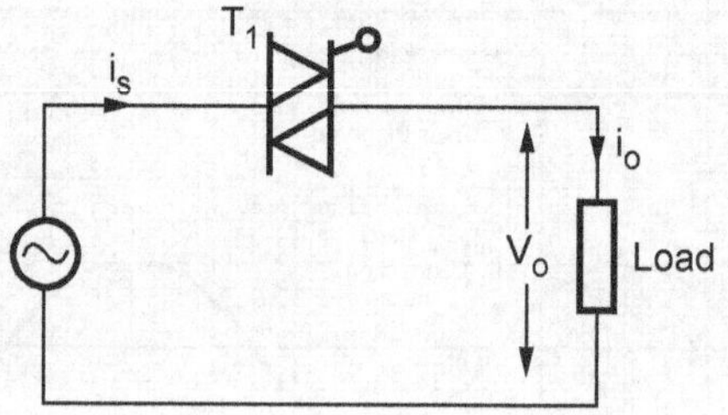

Fig. 3.6.1 Triac based controller

3.6.1 Triac Based AC Controller for Resistive Load

Fig. 3.6.2 shows the waveforms for resistive load. Fig. 3.6.2(a) shows supply voltage and Fig. 3.6.2(b) shows triac gate drive. Fig. 3.6.2(c) shows output voltage waveform. The triac is triggered at 'α' and it turns-off at π in the positive half cycle. It is again triggered at $\pi+\alpha$ and turns-off at 2π in negative half cycle.

- Fig. 3.6.2 (d) shows output current waveform. Since the load is resistive, output current waveform has similar shape to that of output voltage. The amplitude of i_o depends upon $\frac{v_o}{R}$.
- Since output current i_o flows through supply also, the supply current and output current is same. Hence waveform Fig. 3.6.2 (d) also represents supply current i_s.

RMS value of output voltage for RL load

For the waveform of Fig. 3.6.2 (c) we have obtained rms value of output voltage in example 3.4.1. It is given as,

$$V_{o(rms)} = V_m \sqrt{\frac{\pi - \alpha + \frac{\sin 2\alpha}{2}}{2\pi}}$$

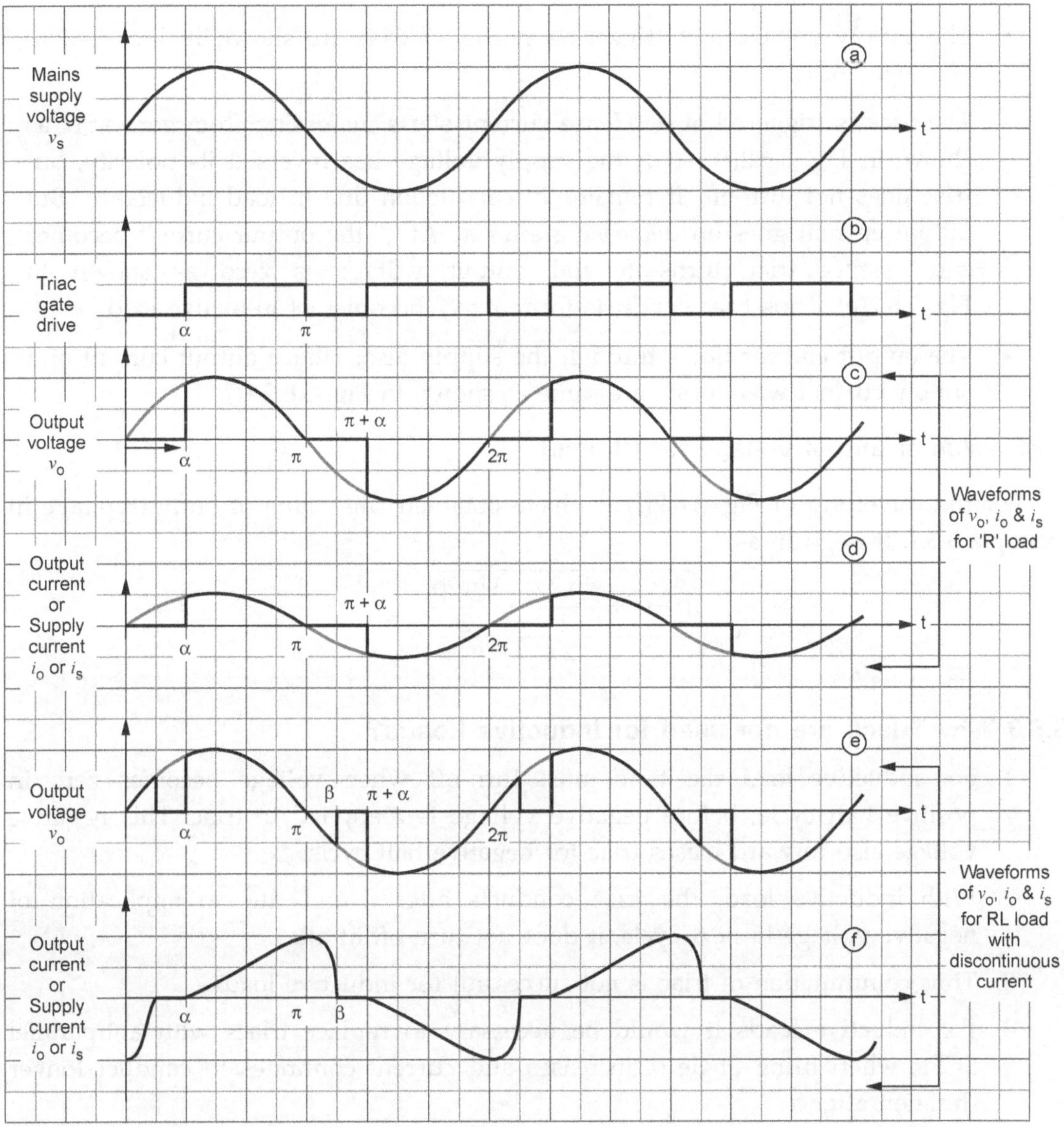

Fig. 3.6.2 Waveforms of triac based AC controller for R & RL loads

3.6.2 Triac Based AC Controller for Inductive Load

Fig. 3.6.2 (e) and (f) shows the waveforms for inductive load.

- The supply voltage and triggering pulses of triac are shown in Fig. 3.6.2(a) and Fig. 3.6.2(d).
- The triac is triggered at α. Hence current starts increasing from zero at α as shown in Fig. 3.6.2(f). At π, the supply voltage has reversed its polarity, but triac does not turn-off. It remains in conduction due to load inductance. But output current goes on decreasing after π. At 'β' the output current becomes zero. Hence triac turns-off and output voltage is zero as shown in Fig. 3.6.2(e). Thus triac conducts form α to β becomes of inductive load.
- The output current flows through the supply also. Hence output current and supply current waveforms are same as shown in Fig. 3.6.2 (f).

RMS value of output voltage for RL load

For the waveform of Fig. 3.6.2(f) we have obtained rms value of output voltage in example 3.5.1. It is given as,

$$V_{o(rms)} = V_m \sqrt{\frac{\beta - \alpha + \frac{\sin 2\alpha}{2} - \frac{\sin 2\beta}{2}}{2\pi}}$$

3.6.3 Why Triacs are not used for Inductive Loads?

1. For inductive load the triac must turn-off when voltage becomes zero in positive half cycle, before negative voltage is reapplied to triac. This negative voltage also forward biases triac for negative half cycle.
2. With inductive load, the triac conducts after π and due to application of negative voltage in next cycle, it does not turn-off at all.
3. Thus commutation of triac is not successful for inductive loads.
4. For inductive loads it would be necessary to replace triacs with antiparallel SCRs, when firing angle α increases and current continues to conduct longer duration after π.

3.6.4 Triac Based Single Phase Control using Diac Triggering

The diac can be used as a triggering device for triac. Fig. 3.6.3 shows the circuit diagram.

Operation

- In this figure the diac starts conducting when there is enough voltage across the capacitor.

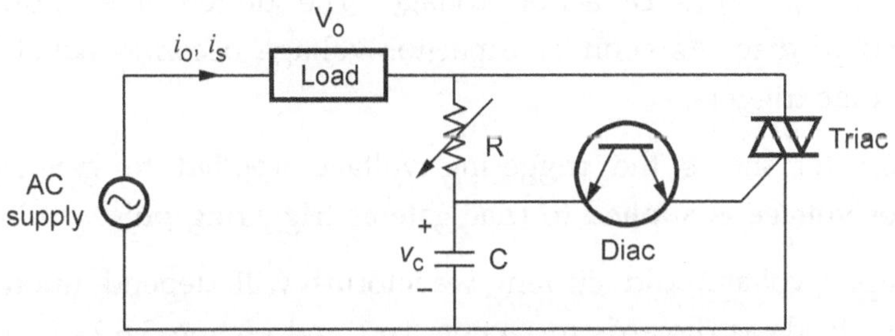

Fig. 3.6.3 Diac triggering for triac based single phase control

- The capacitor voltage is applied as gate voltage to the triac.
- The firing angle 'α' can be controlled by varying the resistance R. This resistance controls the charging of capacitor C.
- The diac triggers the triac in positive as well as negative half cycles.

Waveforms

- Fig. 3.6.4 shows the waveforms of above circuit.
- Fig. 3.6.4 (a) shows supply voltage waveform.

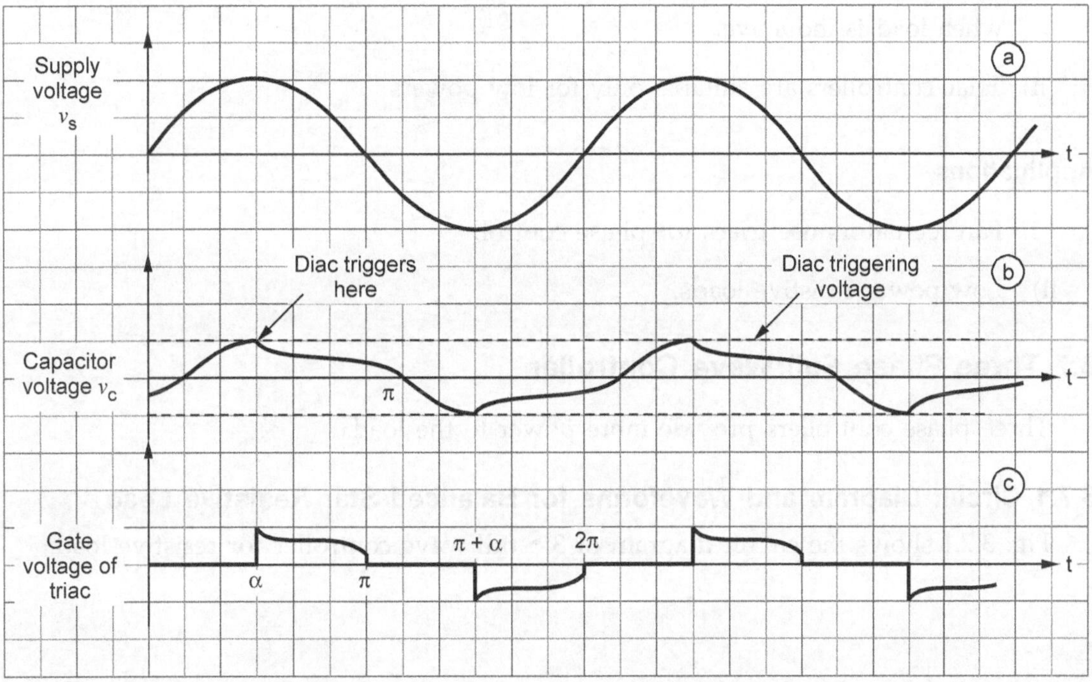

Fig. 3.6.4 Waveforms of diac triggering for triac

- Fig. 3.6.4 (b) shows capacitor voltage. The dotted line shows triggering threshold of diac. As soon as capacitor voltage becomes equal to threshold voltage diac triggers.
- Fig. 3.6.4 (c) shows the triggering voltage applied to gate of triac. The capacitor voltage is applied to triac gate as triggering pulse.
- The output voltage and current waveforms will depend upon resistive or inductive loads as discards in section 3.6.1 and section 3.6.2.

3.6.5 Advantages, Disadvantages and Applications of Triac Controllers

Advantages of triac controller

i) Triac controller is simple, since there is only one drive.

ii) It is best device for resistive low power loads.

iii) Triggering of triac controllers is simpler using devices such as diacs.

Disadvantages

i) Triac is not suitable for inductive load, since it does not turn-off properly when load is inductive.

ii) Triac controllers are suitable only for low powers.

Applications

i) Fan regulators use triacs for phase control.

ii) Low power resistive loads.

3.7 Three Phase Full Wave Controller

Three phase controllers provide more power to the load.

3.7.1 Circuit Diagram and Waveforms for Balanced Star Resistive Load

Fig. 3.7.1 shows the circuit diagram of 3 ϕ full wave controller for resistive load.

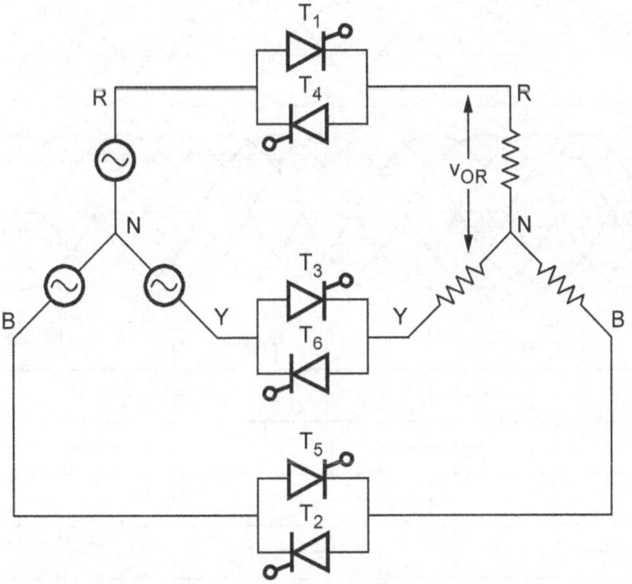

Fig. 3.7.1 Circuit diagram of 3ϕ controller for star load

In the above diagram observe that the star connected 3 ϕ supply is applied to the star connected resistive load through antiparallel SCRs in each phase. Triacs can also be used instead of antiparallel SCRs. At any time current flows in two phases. Third phase is open. This is because two SCRs from different phases conduct at a time.

Supply phase and line voltage

In Fig. 3.7.2 observe that supply phase and line voltages are shown first. The supply phase voltages are R, Y and B. The line voltages are RY, RB, YB, YR, BR, BY.

Fig. 3.7.3 shows the vector diagram for line and phase voltages. In this figure observe that the phase shift between phase voltages R, Y and B is 120°. Line voltage RY is obtained by adding R and –Y. Observe that line voltage RY leads R phase

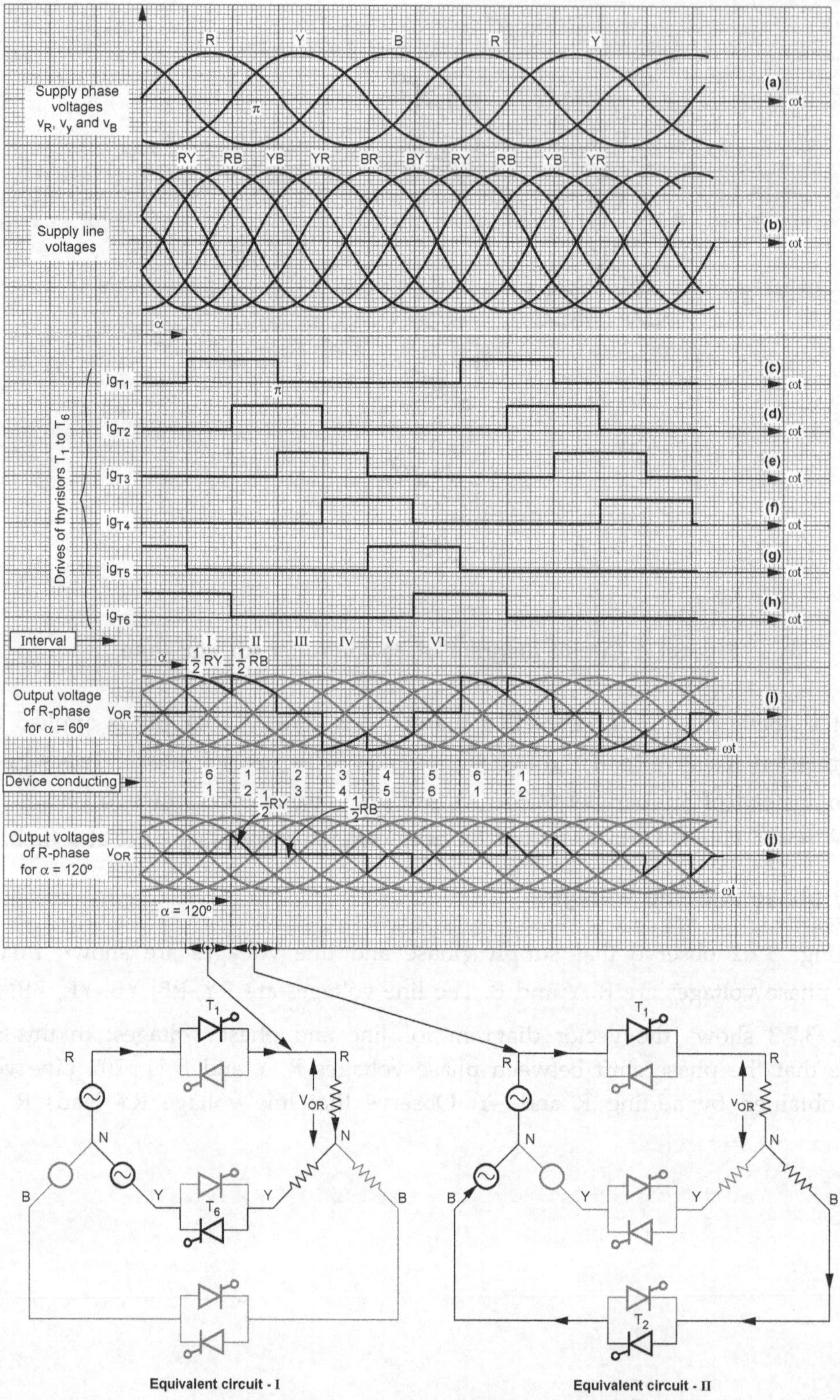

3.7.2 Waveforms of 3ϕ full wave controller

voltage by 30°. This is shown in the waveforms of Fig. 3.7.2 (a) and (b). Similarly line voltage RB lags R-phase voltage by 30°. The phase shift between individual line voltages is 60°.

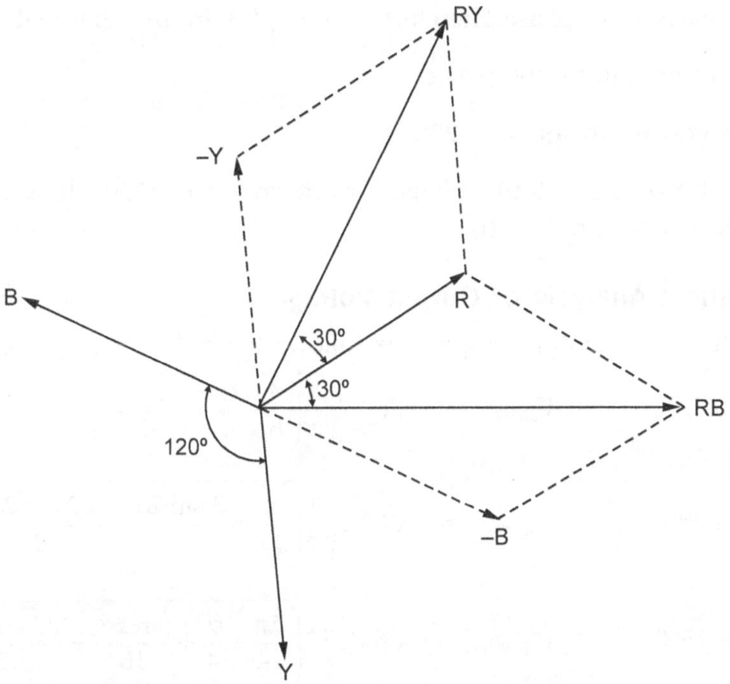

Fig. 3.7.3 Supply phase and line voltages

Gate drives of thyristors T_1 - T_6

The waveforms of Fig. 3.7.2 (c) to (h) shows the gate drives of all the six thyristors. Note that T_1 is applied the drive at $\alpha = 60°$. Note that 'α' is counted from zero crossing of the supply phase voltage. The firing pulse of T_1 ends at π. Similarly other firing pulses are generated.

Note that in the first interval thyristor T_1 and T_6 are applied the gate drive. In the second interval thyristor T_1 and T_2 are applied the gate drive. Like this the cycle repeats after VIth interval.

Output voltage waveform

Consider an equivalent circuit-I of Fig. 3.7.2 when thyristors T_1 and T_6 are conducting. Observe that the line voltage RY appears across load. Since load is resistive and balanced, half of the line voltage appears across each of the load phase. Therefore $v_{OR} = \frac{1}{2} RY$. The waveform of Fig. 3.7.2(i) shows line voltages of half amplitude. And voltage $\frac{1}{2}$ RY appears as output in interval. Similarly in interval-II, thyristors T_1 and T_2 conduct. The line voltage RB appears across the load as shown in equivalent circuit-II of Fig. 3.7.2. Since the load is balanced, half of the line voltage appears across each load phase.Therefore $v_{OR} = \frac{1}{2} RB$ in IInd interval. Like this the remaining waveforms can be completed.

Output voltage waveform for $\alpha = 120°$

Fig. 3.7.2(j) shows the output voltage waveform for $\alpha = 120°$. It is constructed on the similar lines that of Fig. 3.7.2 (i).

3.7.2 Mathematical Analysis of Output Voltage

The rms value of the output is given as follows for different ranges of α,

For $0 \leq \alpha \leq 60°$,

$$V_{o(rms)} = \sqrt{6}\, V_s \sqrt{\frac{1}{\pi}\left(\frac{\pi}{6} - \frac{\alpha}{4} + \frac{\sin 2\alpha}{8}\right)} \qquad \ldots(3.7.1)$$

For $60° \leq \alpha \leq 90°$,

$$V_{o(rms)} = \sqrt{6}\, V_s \sqrt{\frac{1}{\pi}\left(\frac{\pi}{12} + \frac{3 \sin 2\alpha}{16} + \frac{\sqrt{3} \cos 2\alpha}{8}\right)} \qquad \ldots(3.7.2)$$

For $90° \leq \alpha \leq 150°$,

$$V_{o(rms)} = \sqrt{6}\, V_s \sqrt{\frac{1}{\pi}\left(\frac{5\pi}{24} - \frac{\alpha}{4} + \frac{\sin 2\alpha}{16} + \frac{\sqrt{3} \cos 2\alpha}{16}\right)} \qquad \ldots(3.7.3)$$

3.7.3 Advantages, Disadvantages and Applications of 3 ϕ Controllers

Advantages

i) 3 ϕ controllers can deliver more power.

ii) 3 ϕ controllers contain less hormonics in output for high powers.

Disadvantages

i) The control of 3 ϕ controllers is complex.

ii) 3 ϕ controllers are not much suitable for resistive loads.

Applications

i) Induction motor control.

ii) Large fans control.

Review Questions

2. *With the help of circuit diagram and waveforms explain the operation of 3ϕ full wave controller with resistive load.*
3. *What are the advantages and disadvantages of 3 ϕ voltage controllers ?*

3.8 AC Phase Control using TCA785

The TCA785 IC is used for triggering thyristors, triacs and transistors in phase control applications. The range of phase control is from 0° to 180°.

3.8.1 Pin Definition of TCA785

Fig. 3.8.1 shows the pin definitions of TCA785. It is 16-pin IC. It requires synchronization input. Ramp resistance and ramp capacitance at pins 9 and 10. The triggering pulses are generated.

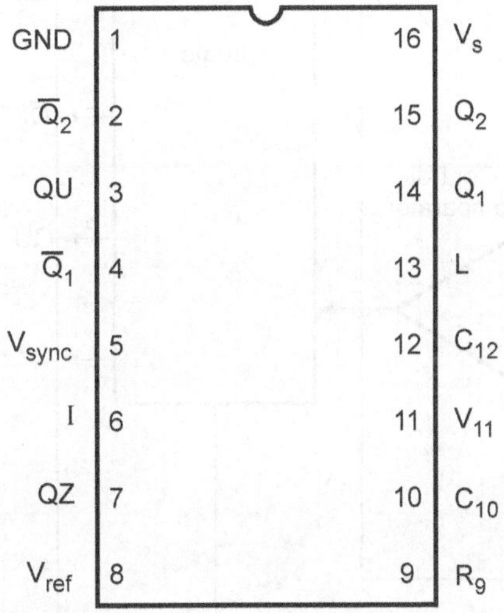

Pin no.	Symbol	Function
1	GND	Ground
2	$\overline{Q}_2$	Output 2 inverted
3	QU	Output U
4	$\overline{Q}_1$	Output 1 inverted
5	V_{sync}	Synchronizing voltage
6	I	Inhibit
7	QZ	Output Z
8	V_{ref}	Stabilized voltage
9	R_9	Ramp resistance
10	C_{10}	Ramp capacitance
11	V_{11}	Control voltage
12	C_{12}	Pulse extension
13	L	Long pulse
14	Q_1	Output 1
15	Q_2	Output 2
16	V_s	Supply voltage

Fig. 3.8.1 Pin definitions of TCA785

3.8.2 Internal Diagram and Waveforms of TCA785

Fig. 3.8.2 shows the internal block diagram of the IC TCA785. The synchronization signal from line voltage is given to V_{sync} input of the IC. The internal zero voltage detector passes ZCD(Zero Crossing Detection) pulses to the synchronization register. The synchronization register controls the changing of external ramp capacitor C_{10}.

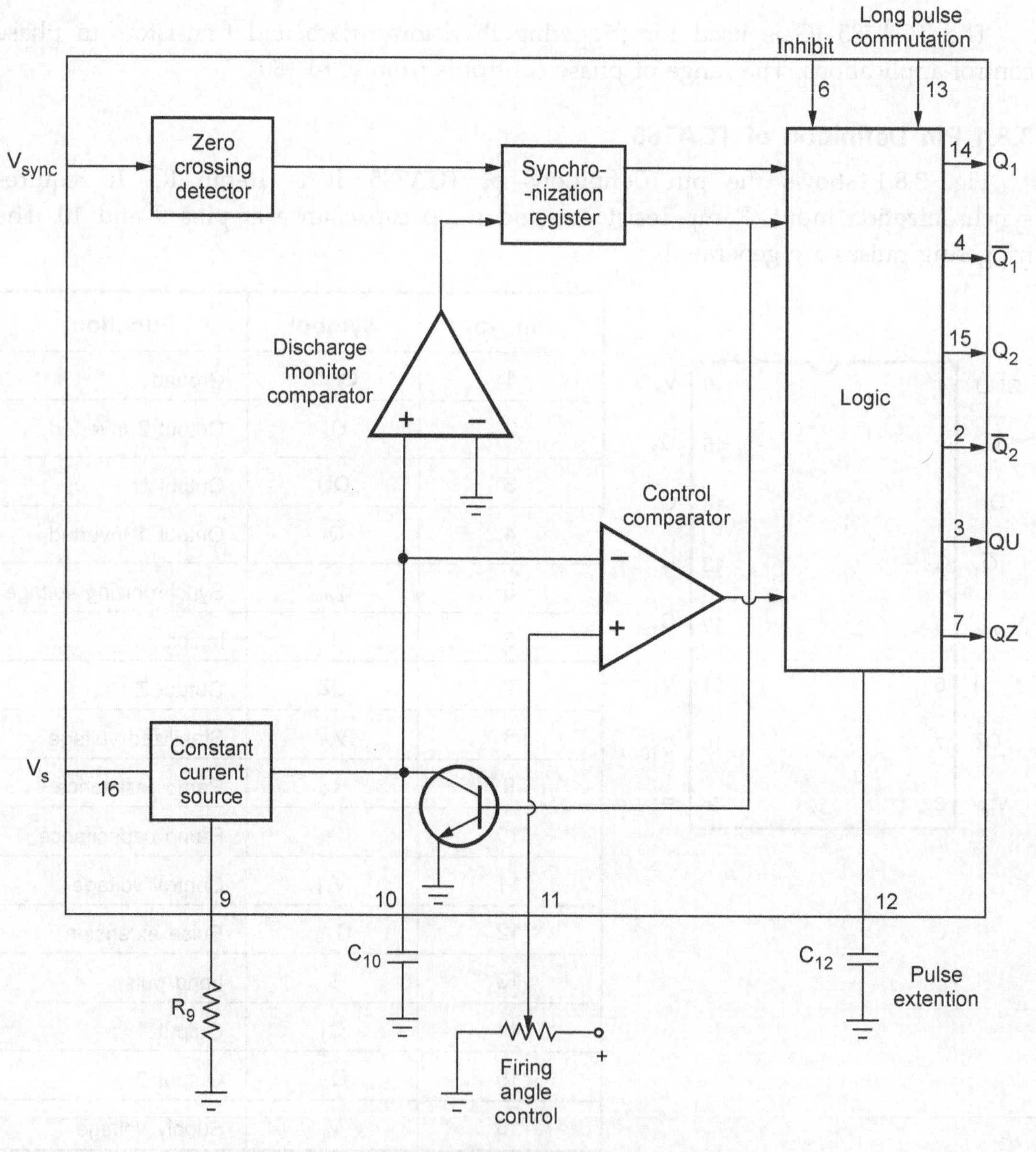

Fig. 3.8.2 Internal block diagram of TCA785

The ramp capacitor is charged by constant current determined by external ramp resistor R_9. The control voltage is applied at pin 11. When ramp voltage exceeds the control voltage, the control comparator passes a signal to the logic. The logic generates a short pulse of duration 30 μs at the output Q_2 on pin 15. This is shown in the waveforms of Fig. 3.8.3(c).

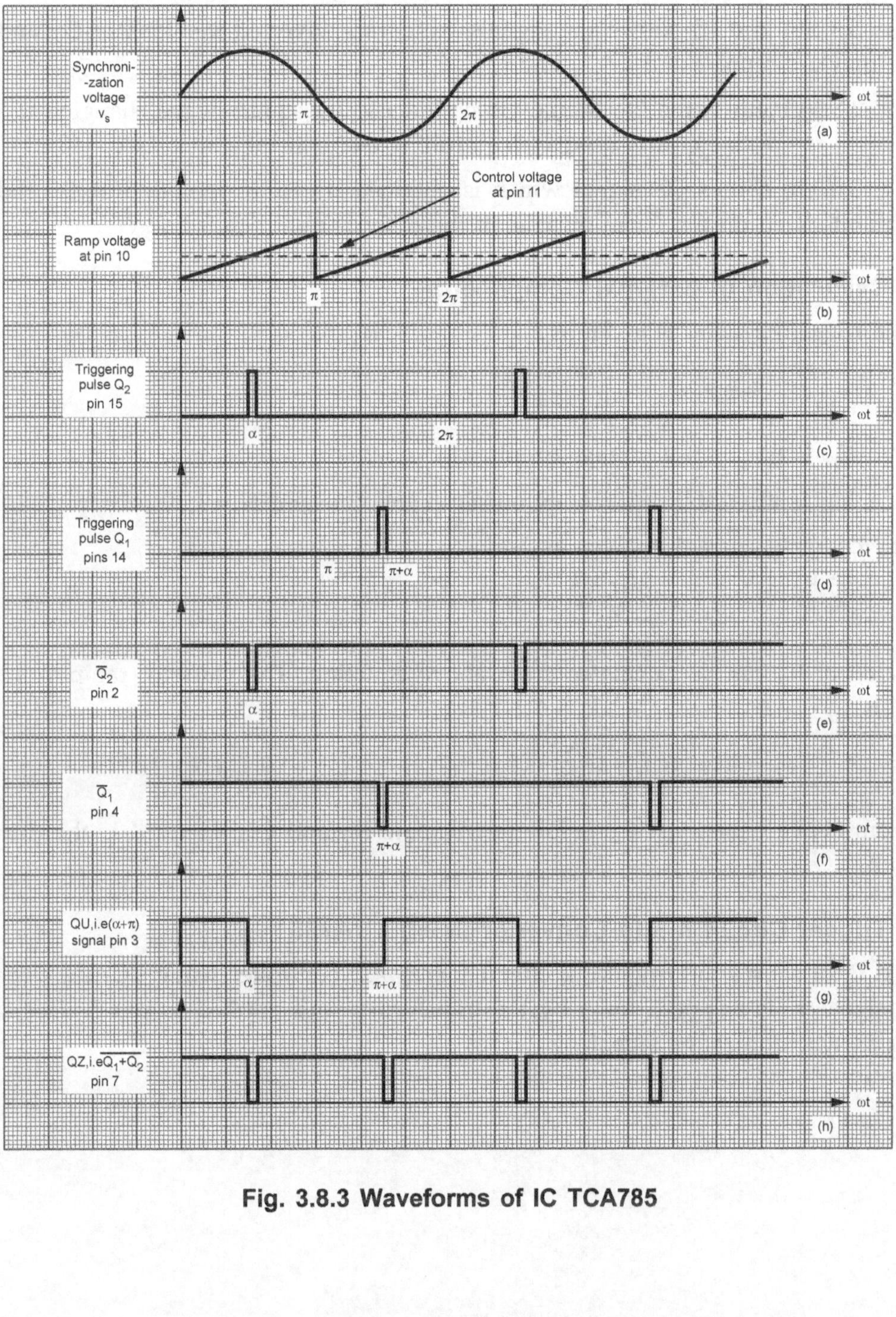

Fig. 3.8.3 Waveforms of IC TCA785

Fig. 3.8.3(d) shows that output Q_1 goes high when ramp voltage exceeds control voltage in negative half cycle. The firing (α) angle can be varied from 0 to 180° by changing the control voltage at pin 11. The discharge monitor comparator gives discharge signal of ramp capacitor to synchronization register. The width of the firing pulse at Q_2 and Q_1 can be adjusted by connecting one capacitor on pin 12. If this pin is connected to ground, then firing pulse from α to π duration is generated.

Fig. 3.8.3(e) and (f) shows the signals which are inverted versions of Q_1 and Q_2. Hence they are represented as $\overline{Q_1}$ and $\overline{Q_2}$.

Fig. 3.8.3(g) shows the QU signal which is basically square wave changing its level at $\alpha, \pi+\alpha, 2\pi+\alpha, ...$ etc. It is used for controlling external logic.

Fig. 3.8.3(h) shows QZ signal which is obtained by NOR operation of Q_1 and Q_2. The inhibit signal on pin 6 can disable all the outputs $Q_1, Q_2, \overline{Q_1}$ and $\overline{Q_2}$.

3.8.3 Calculation of External Components

The triggering point is given as,

$$t_{trigger} = \frac{V_{11} \times R_9 \times C_{10}}{V_{ref} \times K}$$

Here V_{11} is the control voltage on pin 11

V_{ref} is the stabilized voltage on pin 8 and constant K = 1.10 ± 20 %.

The charging current is given as,

$$I_{10} = \frac{V_{ref} \times K}{R_9}$$

Value of R_9 can be calculated from above equation. The value of ramp voltage is given as,

$$V_{10} = \frac{V_{ref} \times K \times t}{R_9 \times C_{10}}$$

From above equations it possible to calculate R_9 and C_{10}. Here 't' is the duration of ramp.

3.8.4 Trigger Circuit for 1ϕ Converter using TCA785

Fig. 3.8.4 shows the triggering circuit for 1ϕ AC controller using TCA785. The AC input signal v_s is stepped down by the transformer and applied to pin 5. This signal serves as synchronization signal.

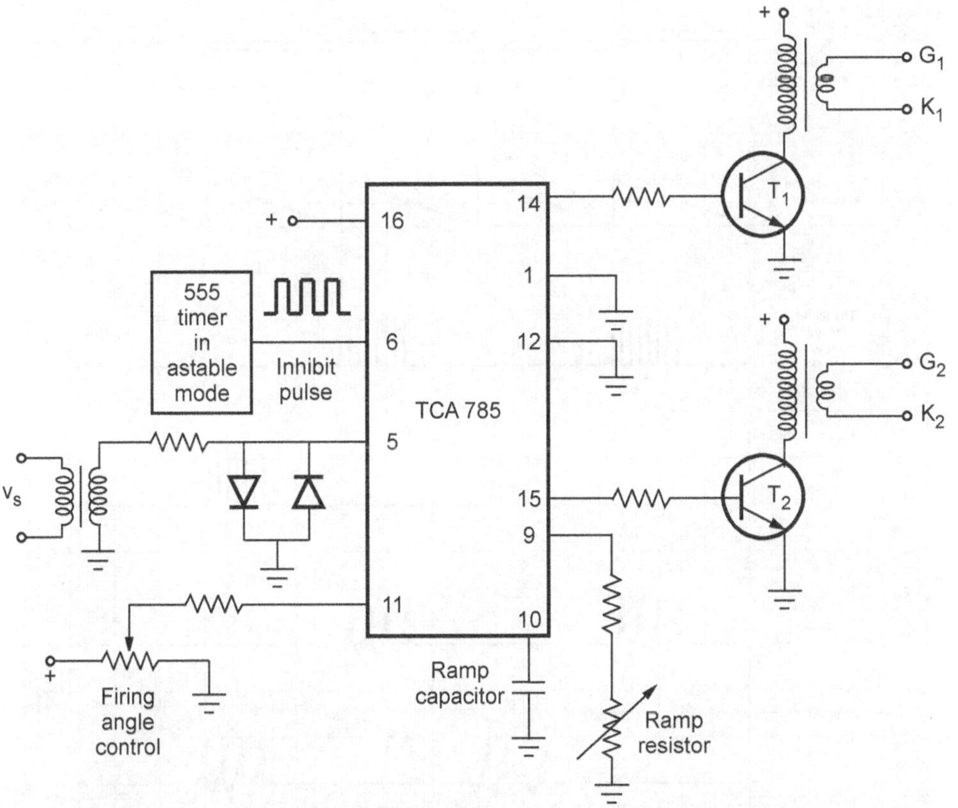

Fig. 3.8.4 AC phase control using TCA785 in 1ϕ AC controller

At pin-5 there is a square wave of ± 1V as shown in waveforms of Fig. 3.8.5.

Fig. 3.8.5(b) shows the ramp voltage at pin 10 and the control voltage at pin 11. In Fig. 3.8.4 note that pin '12' is connected to ground. Hence the firing pulse from α to π is generated. Pin '6' is fed with inhibit pulses from 555 timer in astable mode. These pulses are applied at about 2 kHz. Therefore the triggering signals generated on pins 14 and 15 are pulsed. The pulse pin 14 (i.e. Q_1) and pin 15 (i.e. Q_2) are passed only when 555 timer output is high. This results in pulsed base drives for pulse amplifiers T_1 and T_2. The pulse amplifier T_1 generates firing pulses for SCR 1. The pulse amplifier T_2 generates firing pulses for SCR 2. Fig. 3.8.5(e) and (f) shows these firing pulses. The last waveform in Fig. 3.8.5 shows the output voltage of the 1ϕ AC controller. The ramp resistor on pin 9 can be varied to vary the firing angle (α).

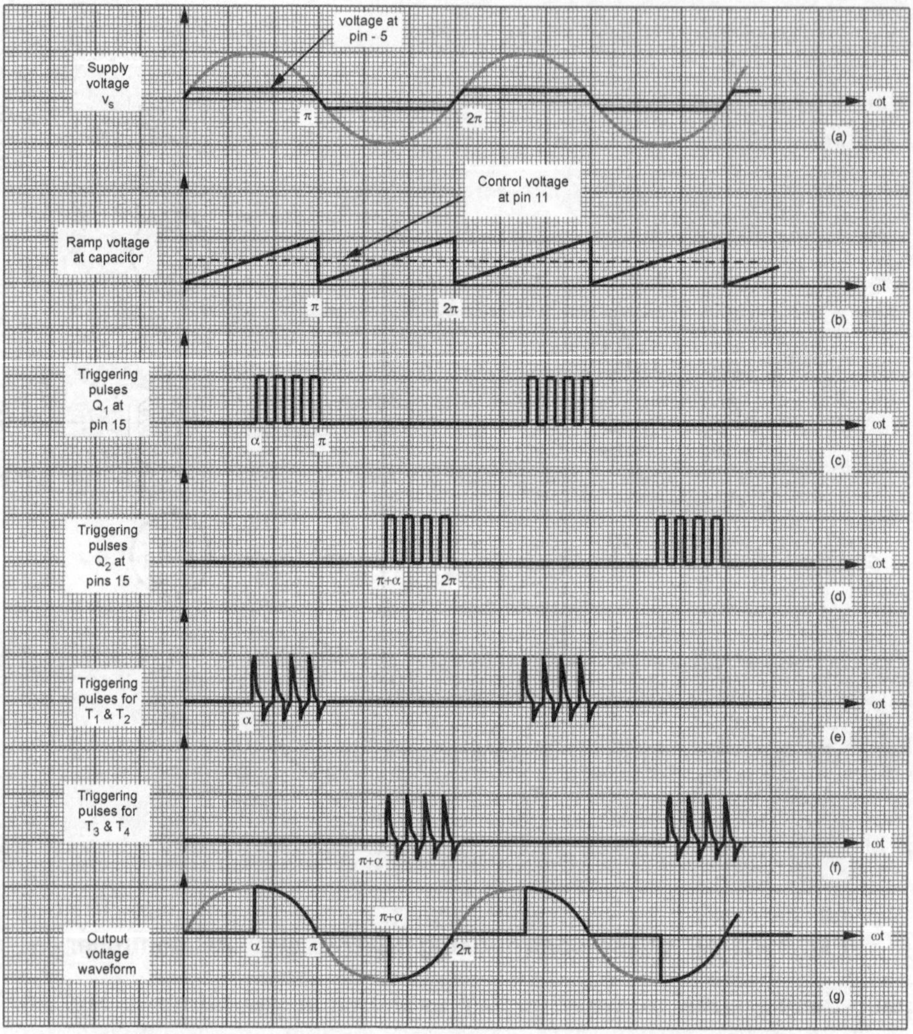

Fig. 3.8.5 Waveforms of AC phase control using TCA 785

Review Question

1. *Draw the diagram of triggering circuit for 1ϕ AC voltage controller using TCA 785. Explain its operation with the help of waveforms.*

3.9 Cycloconverters

Cycloconverters is another class of AC to AC converters. Output frequency can also be varied. It is possible to obtain 3 phase to 1 phase regulator. In this section we will study some of these aspects.

3.9.1 Step-up Cycloconverter

When the frequency of the output is higher than the frequency of input, then it is called step-up cycloconverter. Fig. 3.9.1 shows the mid point type step-up cycloconverter.

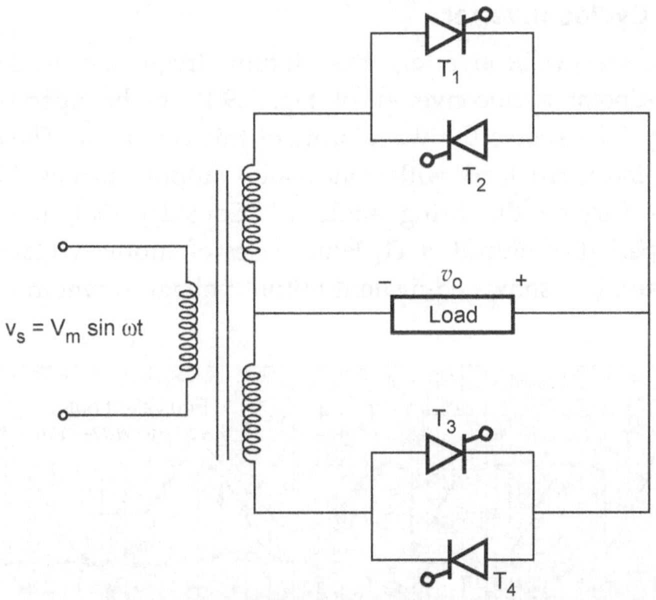

Fig. 3.9.1 Midpoint type cycloconverter

The waveforms of this converter are shown in Fig. 3.9.2. In the positive half cycle, T_1 conducts from 0 to t_1, hence output voltage is positive. At t_1, T_1 is forced commutated and T_4 is triggered. Hence load voltage becomes negative as shown in above waveform. Then at t_2, T_4 is forced commutated and T_1 is turned-on again. Therefore output voltage is again positive. At t_3, T_4 is turned-on and T_1 is forced commutated. Hence output voltage is negative. At π , T_3 is turned on. Therefore output voltage is positive. At t_4, T_3 is forced commutated and T_2 is triggered. Hence the load voltage is negative. This sequence continues. Observe that output voltage waveform has the frequency of,

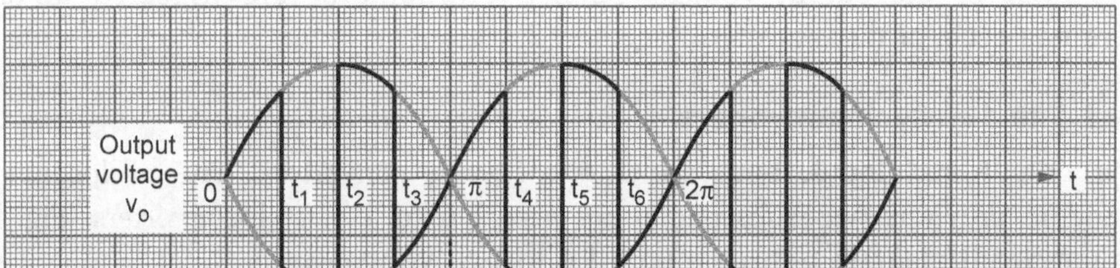

Fig. 3.9.2 Waveforms of step-up cycloconverter

$$f_o = \frac{1}{t_2 - t_1}$$

This frequency is higher than supply frequency.

3.9.2 Step-down Cycloconverter

In the step-down cycloconverter, the output frequency is less than supply frequency. The midpoint cycloconverter of Fig. 3.9.1 can be operated as step-down cycloconverter. Fig. 3.9.3 shows the waveforms of this converter. These waveforms are shown for highly inductive load with continuous output current. Output voltage is also controlled by varying the firing angle 'α'. Consider that, the period of input supply is 'T'. The period of output is 4T. Four cycles of supply voltage make one cycle of output. The dotted line shows equivalent output voltage waveform.

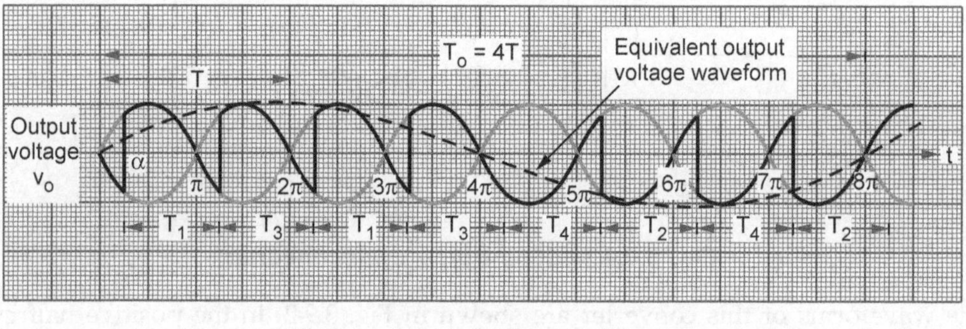

Fig. 3.9.3 Step-down cycloconverter

Output frequency will be,

$$f_o = \frac{1}{T_o}$$

$$= \frac{1}{4T} = \frac{f}{4}$$

Thus the supply frequency is divided by four. T_1 and T_3 are triggered to obtain positive cycle of the output voltage. Similarly, T_2 and T_4 are triggered to obtain negative half cycle of the output voltage.

3.9.3 1 ϕ to 1 ϕ Cycloconverters

The output and input, both are 1 ϕ for such converters. The cycloconverters discussed in previous two subsections are also 1 ϕ to 1 ϕ cycloconverters. Fig. 3.9.4 shows the circuit diagram of bridge type 1 ϕ to 1 ϕ cycloconverter.

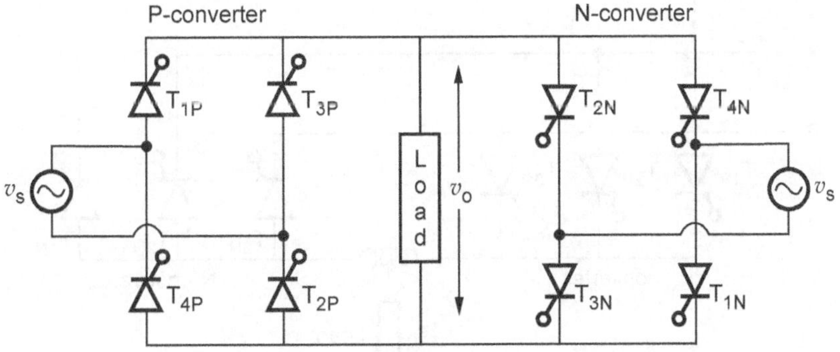

Fig. 3.9.4 1 φ to 1 φ bridge type cycloconverter

Output waveform of the above cycloconverter will be similar to that of Fig. 3.9.3 for inductive load and continuous output current. Fig. 3.9.5 shows the waveform of above converter for resistive load.

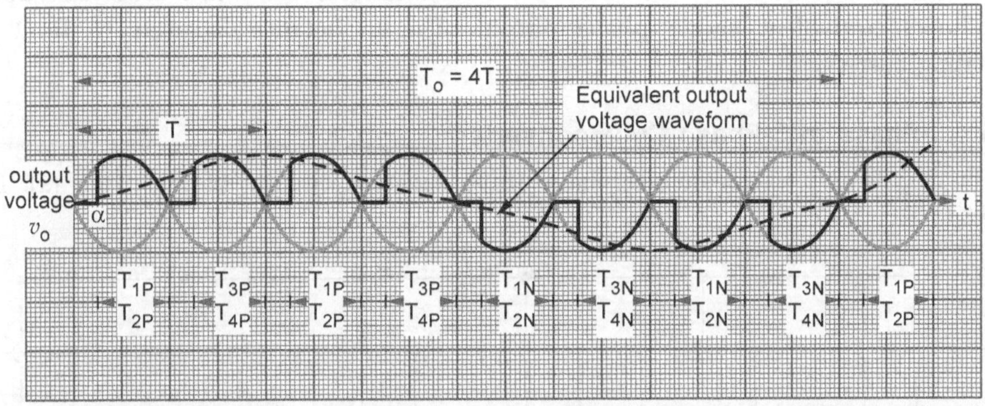

Fig. 3.9.5 Waveforms of 1 φ to 1 φ cycloconverter for resistive load

Output voltage is positive, when SCRs of P-converter are conducting. Similarly the output voltage is negative when SCRs of N-converter are conducting. The firing angle of the SCRs is varied to control the output voltage. Above waveforms are shown for step down operation. Step-up operation is also possible in this converter. Four cycles of input make one cycle of output. Hence output frequency is,

$$f_o = \frac{1}{T_o} = \frac{1}{4T} = \frac{f}{4}$$

3.9.4 3-Phase to 1-Phase Cycloconverter

The 3 φ to 1 φ cycloconverter converts 3 φ input supply to 1 φ output. Fig. 3.9.6 (a) shows the circuit diagram of 3 φ to 1 φ cycloconverter.

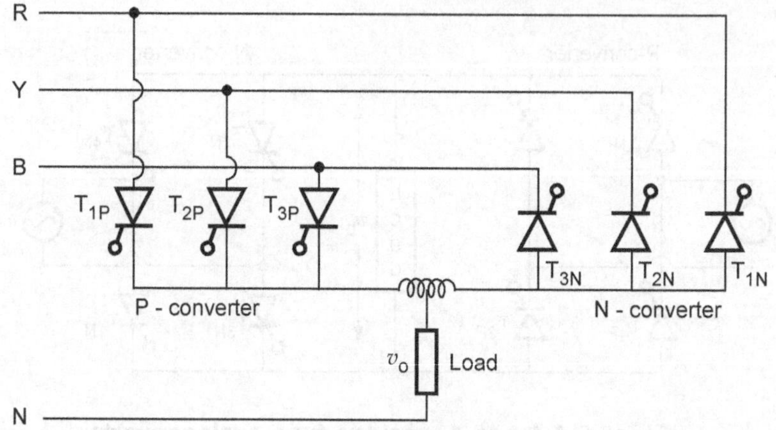

Fig. 3.9.6 (a) Circuit diagram of 3 - ϕ converter

Output voltage is positive, when SCRs of P-converter are conducting. And output voltage is negative when N-converter is operating. Thyristor T_{1p} is triggered when R-phase is dominant. Similarly T_{2p} is triggered when Y-phase is dominant, and so on. Fig. 3.9.6 (b) shows the waveforms of this converter. It is 3 - ϕ converter. The dotted line shows equivalent single phase output. The inductance is used to limit the circulating current.

3.9.5 3-Phase to 3-Phase Cycloconverter

The 3 ϕ to 3 ϕ cycloconverter have three separate circuits of Fig. 3.9.6 (a). It is shown in Fig. 3.9.7.

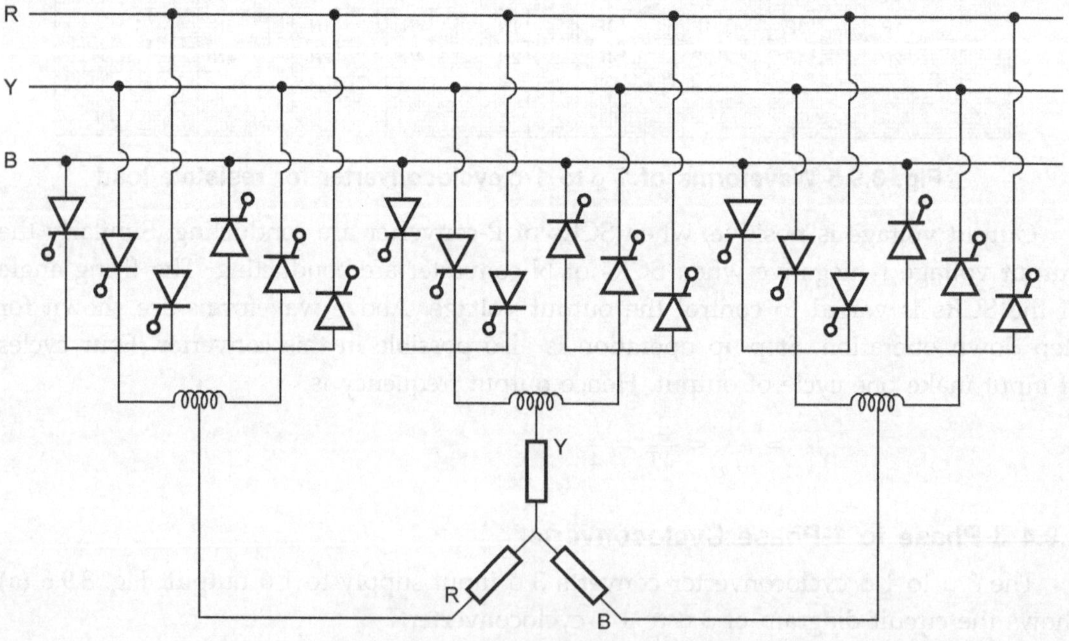

Fig. 3.9.7 3ϕ to 3ϕ cycloconverter

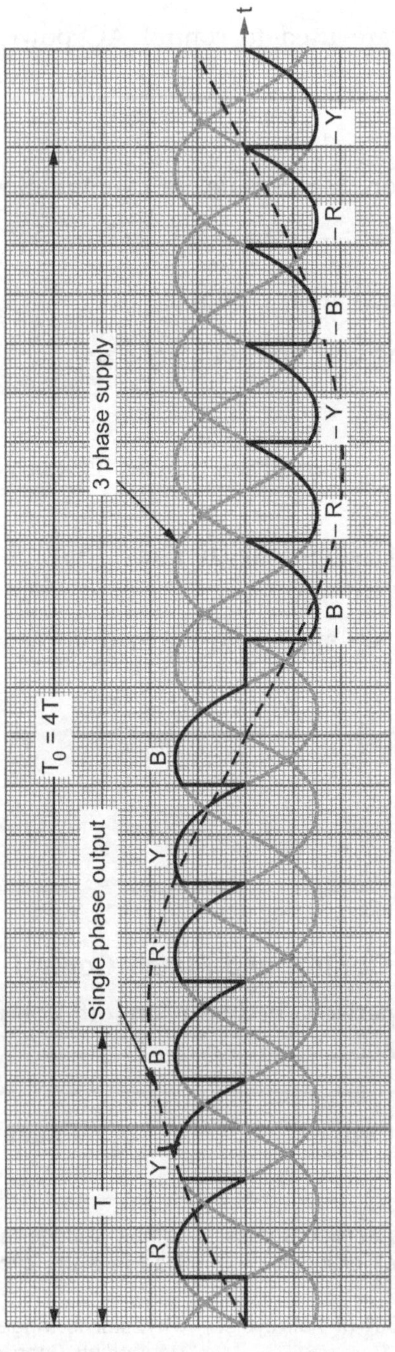

Fig. 3.9.6 (b) Waveforms of 3 ϕ to 1 ϕ cycloconverter

As shown in Fig. 3.9.7, each phase is generated with the help of P-converter and N-converter. The two converters of each phase are connected through inductor. This is also 3 - ϕ converter.

3.10 Merits, Demerits and Applications of AC Voltage Controllers

The AC voltage controllers are used to control AC power delivered to the load from AC supply.

Merits of AC controllers

i. They use line commutation. Hence no extra commutation circuit is required.

ii. They have high efficiencies, since devices losses are reduced.

iii. Circuits and their implementation is simple.

iv. Less maintenance, running cost and compact size.

Demerits of AC controllers

i. In case of on-off control intermittent power is delivered to the load. This affects load performance.

ii. The load current is not sinusoidal in phase control. Hence harmonics are introduced in supply current also.

iii. Sharply varying voltages are applied to the load in case of phase angle control.

Applications of AC controllers

i. Fans and lighting controls

ii. Speed control of AC motors

iii. Saturable core reactors

iv. Industrial heating furnaces.

3.10.1 Comparison of On/Off Control and Phase Control

Table 3.10.1 lists the comparison between on/off control and phase control techniques of AC voltage controllers.

Serial No.	On/off control	Phase control
1.	Full supply voltage is applied across the load for few cycles of input.	Supply voltage is partially applied to load in every cycle of input.
2.	Thyristor conducts 180° or half cycle. It is triggered at $\alpha = 0$.	Thyristor is triggered at some angle depending upon output power.
3.	Since thyristors are switched on at $\alpha = 0$, harmonics due to switching are reduced.	Thyristors are switched on at angles other than $\alpha = 0$, hence switching harmonics are present.

4.	Load voltage will be equal to full supply voltage or zero. Hence output waveform has high voltage fluctuations.	Supply voltage is applied partially to load. Hence there are less voltage fluctuations.
5.	Power control is achieved by varying the on and off times for which supply voltage is applied to the load.	Power control is achieved by controlling the conduction period (phase) of thyristor in every half cycle.

Table 3.10.1 Comparison between on/off control and phase control

3.10.2 Comparison between Full Wave (Bidirectional) and Half Wave (Unidirectional) AC Controllers

Table 3.10.2 lists the comparison between full wave and half wave AC controllers.

Serial No.	Full wave controller	Half wave controller
1.	Both the half cycles of supply are controlled.	Only half cycle of the supply controlled.
2.	These controllers generate symmetric voltage and currents across load.	These controllers generate asymmetric voltage and currents across load.
3.	Supply current is symmetric.	Supply current is asymmetric.
4.	Complete range of supply voltage can be controlled.	Complete range of supply voltage cannot be controlled.
5.	Full wave controllers are preferred for large loads.	Half wave controllers are preferred for small loads.

Table 3.10.2 Full wave and half wave controllers

Review Question

1. *Compare the full wave and half wave AC controllers.*

Summary

1. On/off control

Output rms voltage $$V_{o(rms)} = V_{s(rms)} \cdot \sqrt{k}$$

Here $$k = \frac{n}{m+n}$$

2. Half wave controller

RMS output voltage, $V_{o(rms)} = \frac{V_m}{2}\sqrt{\frac{2\pi - \alpha + \frac{\sin 2\alpha}{2}}{\pi}}$

Average output voltage, $V_{o(av)} = \frac{V_m}{2\pi}(\cos\alpha - 1)$

3. Full wave controller (R load)

RMS output voltage, $V_{o(rms)} = V_m\sqrt{\frac{\pi - \alpha + \frac{\sin 2\alpha}{2}}{2\pi}}$

RMS thyristor current, $I_{T(rms)} = \frac{I_{o(rms)}}{\sqrt{2}}$

Average thyristor current, $I_{T(av)} = \frac{V_m}{2\pi R}(1 + \cos\alpha)$

4. Full wave controller (RL load)

For discontinuous conduction, RMS valve of output is given as,

$$V_{o(rms)} = V_m\sqrt{\frac{\beta - \alpha + \frac{\sin 2\alpha}{2} - \frac{\sin 2\beta}{2}}{2\pi}}$$

$V_{o(rms)} = V_{s(rms)}$ for continous conduction

❑❑❑

4

Choppers

Objectives

- *Principle of DC to DC conversion in power electronics.*
- *Step down and step up principle in choppers.*
- *Control strategies of choppers.*
- *Choppers with RL and RLE loads, their current analysis.*
- *Performance parameters of choppers.*
- *Classification of choppers according to their V_o - I_o quadrant of operation.*
- *Impulse commutated thyristor chopper.*
- *Applications of choppers.*
- *Study of SMPS (flyback converter)*
- *Resonant converters.*

4.1 Introduction

The DC choppers convert the input DC voltage into fixed or variable DC output. Fig. 4.1.1 shows the basic block diagram of the chopper.

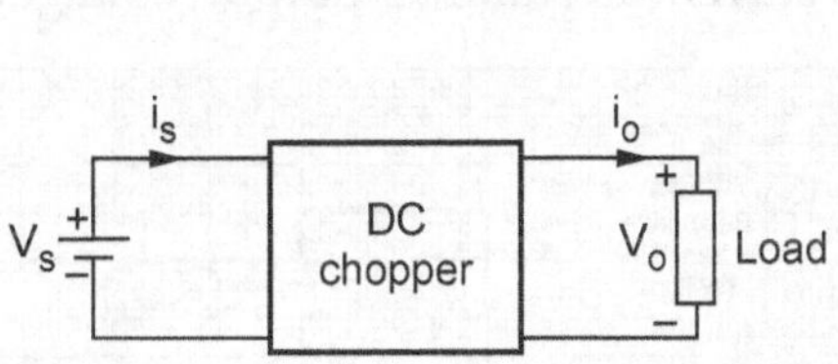

Fig. 4.1.1 Basic inputs/outputs of DC chopper

- The chopper has fixed or variable DC input, V_S. And the output V_0 is also fixed or variable. Hence DC chopper is also called as dc to dc converter.
- The output V_0 can be greater or less than the input. Hence the choppers can be step down or step up type.

- Choppers are used in dc traction drives, trolley cars, marine lifts etc. The dc choppers use switching principle. Hence they have high efficiency. The choppers are also used in dc voltage regulators for high efficiency.
- The dynamic response of choppers is fast due to switching nature of the devices.

4.2 Step down Chopper

Fig. 4.2.1 shows the circuit diagram of the basic step down chopper. The switch (sw) can be a power transistor, SCR, GTO, power MOSFET, IGBT or similar switching device. Normally the drop in the switch is very small and it is neglected. Fig. 4.2.2 shows the waveforms of the step down chopper with resistive load.

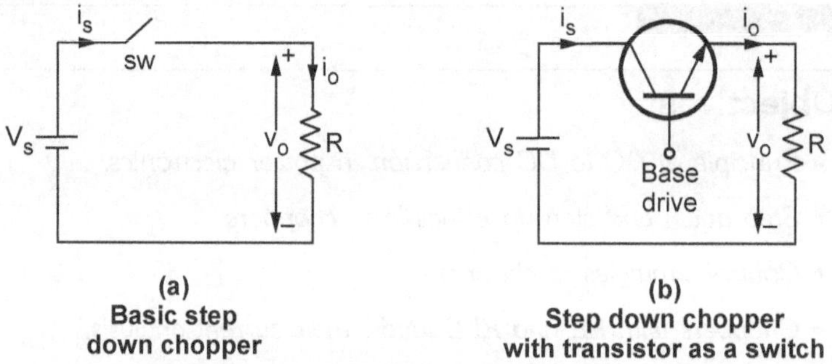

Fig. 4.2.1

Fig. 4.2.2 (a) shows the drive of the switch. In case of power transistor, it will be a base drive. The drive is applied for the period 0 to δT. Hence the switch turns ON for this period and connects supply V_s to the load. Hence $v_o = V_s$ in this period. From δT to T the drive of the switch is removed, hence it turns off. Hence the load voltage is zero. Since the load is resistive, output current will be, $i_o = \frac{v_o}{R}$. Hence the nature of output current is same as output voltage.

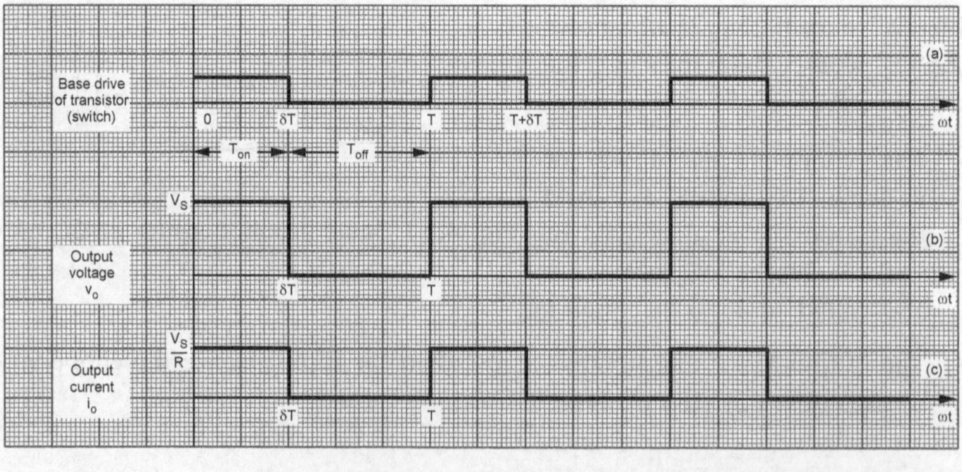

Fig. 4.2.2 Waveforms of the step down chopper with resistive load

Example 4.2.1 : *For the step down chopper having resistive load derive an expression for the following -*

(i) Average output voltage $V_{o\,(av)}$

(ii) rms output voltage $V_{o(rms)}$

Solution : (i) To obtain $V_{o(av)}$

The average value is given as,

$$V_{o\,(av)} = \frac{1}{T}\int_0^T v_o(t)\,dt$$

In the output voltage waveform of Fig. 4.2.2 observe that $v_o = V_s$ from 0 to δT, rest of the time v_o is zero. Hence above equation can be written as,

$$V_{o(av)} = \frac{1}{T}\int_0^{\delta T} V_s\,dt = \frac{V_s}{T}\int_0^{\delta T} dt = \frac{V_s}{T}\cdot\delta T$$

$$\therefore \quad V_{o(av)} = \delta V_s \qquad \text{... (4.2.1)}$$

Here $\delta = \frac{T_{on}}{T}$ is called the *duty cycle of the chopper. The value of duty cycle lies between* $0 \le \delta \le 1$.

(ii) To obtain $V_{o(rms)}$

The rms value of output is given as,

$$V_{o(rms)} = \left[\frac{1}{T}\int_0^T v_o^2(t)\,dt\right]^{\frac{1}{2}}$$

We know from Fig. 4.2.2 that $v_o = V_s$ from 0 to δT (i.e. when the transistor switch is on). Hence above equation becomes,

$$V_{o\,(rms)} = \left[\frac{1}{T}\int_0^{\delta T} V_s^2\,dt\right]^{\frac{1}{2}} = \left[\frac{V_s^2}{T}\int_0^{\delta T} dt\right]^{\frac{1}{2}} = \left[\frac{V_s^2}{T}\cdot\delta T\right]^{\frac{1}{2}}$$

$$\therefore \quad V_{o(rms)} = \sqrt{\delta}\,V_s \qquad \text{... (4.2.2)}$$

Example 4.2.2 : *Explain the basic principle of step down chopper and write down the expressions for -*

(i) Average output voltage

(ii) Output power

(iii) Effective input resistance interms of chopper duty cycle.

Solution : (i) Average output voltage

It is derived in previous example. i.e.,

$$V_{o(av)} = \delta V_s$$

Here $\delta = \frac{T_{on}}{T}$ is called the duty cycle of the chopper.

(ii) Output power

The load power can be calculated as,

$$P_o = \frac{1}{T}\int_0^{\delta T} \frac{v_o^2}{R}\,dt = \frac{1}{T}\int_0^{\delta T} \frac{(V_s - V_{ch})^2}{R}$$

$$= \frac{1}{T}\cdot\frac{(V_s - V_{ch})^2}{R}\int_0^{\delta T} dt$$

$$= \frac{1}{T}\cdot\frac{(V_s - V_{ch})^2}{R}\cdot\delta T$$

$$= \frac{\delta (V_s - V_{ch})^2}{R}$$

If the chopper is lossless, then $V_{ch} = 0$ and output power will be,

$$P_o = \frac{\delta V_s^2}{R}$$

(iii) Effective input resistance

Fig. 4.2.3 shows that the average current flowing from the input is basically output average current, i.e. $I_{o(av)}$. The input voltage is supply voltage V_s. Hence effective input resistance will be,

$$R_{in} = \frac{V_s}{I_{o(av)}}$$

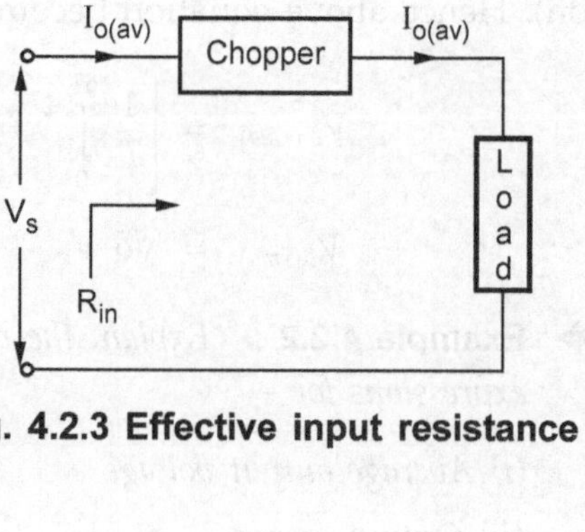

Fig. 4.2.3 Effective input resistance

Putting for $I_{o(av)} = \frac{V_{o(av)}}{R}$ in above equation,

$$R_{in} = \frac{V_s}{\frac{V_{o(av)}}{R}} = R\cdot\frac{V_s}{V_{o(av)}}$$

Since $\delta = \frac{V_{o(av)}}{V_s}$, above equation becomes,

$$R_{in} = \frac{1}{\delta} R$$

This is an expression for effective input resistance.

Example 4.2.3 : *For a chopper shown below, DC source voltage is 230 V, load resistance is 10Ω. Consider the voltage drop of 2 V across chopper when it is on. For a duty cycle of 0.4 calculate,*
i) average and rms value of output voltage
ii) chopper efficiency

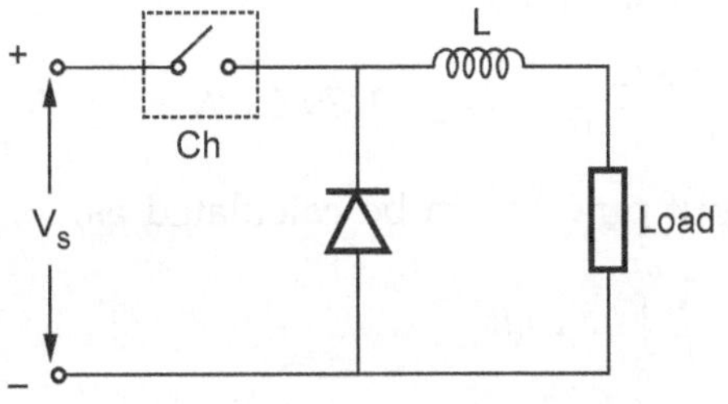

Fig. 4.2.4 Chopper circuit of example 4.2.3

Solution : The given data is,

$$V_s = 230\text{ V}, R = 10\ \Omega, V_{ch} = 2\text{ V}, \delta = 0.4$$

It is a step down chopper. The inductance filter is present. We will consider that the load is fairly resistive.

i) To obtain average and rms values

The average value is given by equation 4.2.1 as,

$$V_{o(av)} = \delta V_s$$

This equation neglects the drop in chopper. Considering the drop, above equation will be,

$$V_{o(av)} = \delta(V_s - V_{ch})$$

$$= 0.4\ (230 - 2) = 91.2\text{ V}$$

The rms value is given by equation 4.2.2 as,

$$V_{o(rms)} = \sqrt{\delta}\, V_s$$

Considering the voltage drop across chopper,

$$V_{o(rms)} = \sqrt{\delta}\,(V_s - V_{ch})$$

$$= \sqrt{0.4}\,(230-2) = 144.2 \text{ V}$$

ii) To obtain chopper efficiency

The load power can be calculated as,

$$P_o = \frac{1}{T}\int_0^{\delta T} \frac{v_o^2}{R}\,dt = \frac{1}{T}\int_0^{\delta T} \frac{(V_s - V_{ch})^2}{R}\,dt$$

$$= \frac{1}{T}\frac{(V_s - V_{ch})^2}{R}\int_0^{\delta T} dt = \frac{1}{T}\cdot\frac{(V_s - V_{ch})^2}{R}\cdot\delta T$$

$$= \frac{\delta\,(V_s - V_{ch})^2}{R} \qquad \text{... (4.2.3)}$$

Putting values in above equation,

$$P_o = \frac{0.4(230-2)^2}{10} = 2079.36 \text{ W}$$

The supply power (input power) can be calculated as,

$$P_s = \frac{1}{T}\int_0^{\delta T} v_s\, i_s\, dt$$

In the Fig. 4.2.2, observe that $i_s = i_o$ and $v_s = V_s$. Hence above equation becomes,

$$P_s = \frac{1}{T}\int_0^{\delta T} V_s\, i_o\, dt = \frac{1}{T}\int_0^{\delta T} V_s\left(\frac{V_s - V_{ch}}{R}\right)dt$$

$$= \frac{1}{T}\cdot V_s\cdot\frac{(V_s - V_{ch})}{R}\int_0^{\delta T} dt = \frac{1}{T}\cdot V_s\cdot\frac{V_s - V_{ch}}{R}\cdot\delta T$$

$$= \delta\, V_s\cdot\frac{V_s - V_{ch}}{R} \qquad \text{... (4.2.4)}$$

Putting the values,

$$P_s = 0.4\times 230\times\frac{230-2}{10} = 2097.6 \text{ W}$$

Therefore efficiency of the chopper is,

$$\eta = \frac{\text{Load power (output power)}}{\text{Supply power(input power)}}$$

$$= \frac{2079.36}{2097.6} = 0.9913 \text{ or } 99.13\ \%$$

Here note that efficiency is very high since voltage drop in the chopper is very small.

4.2.1 Chopper Control Techniques

The chopper can be operated as constant frequency or variable frequency. These are also called as Time Ratio Control (TRC) techniques.

1. Constant frequency operation :

The chopper frequency is kept constant. Hence total period T remains constant. T_{on} and T_{off} both are varied to vary the duty cycle. The advantage is that the filter components are easy to design.

2. Variable frequency operation :

The frequency of the chopper varies when duty cycle is to be varied. When T_{on} is varied T_{off} is kept constant and vice versa. The filter design is difficult for this type of chopper.

Example 4.2.4 : *A chopper circuit is operating on TRC at a frequency of 2 kHz on a 460 V supply of the load voltage of 350V. Calculate the conduction period of the thyristor in each cycle.*

Solution : The given data is,

$$f = 2 \text{ kHz} \qquad \therefore \; T = \frac{1}{f} = \frac{1}{2000} = 5\times10^{-4} \text{ sec}$$

$$V_s = 460 \text{ V}$$

$$V_{o(av)} = 350 \text{ V}$$

For the step down chopper, the average value of output voltage is given as,

$$V_{o(av)} = \delta V_s$$

and duty cycle $\delta = \frac{T_{on}}{T}$. Hence above equation will be,

$$V_{o(av)} = \frac{T_{on}}{T} V_s$$

$$\therefore \quad 350 = \frac{T_{on}}{5\times10^{-4}} \times 460$$

$$\therefore \quad T_{on} = 3.8\times10^{-4} = 380 \text{ } \mu\text{s}$$

Thus the conduction period of thyristor is 380 μs.

Example 4.2.5 : *Considering the switch to be ideal in the circuit of Fig. 4.2.5 determine :*

i) The duty cycle, for which the output average dc voltage and rms voltage are equal.

ii) The chopper efficiency

Solution : i) To obtain duty cycle

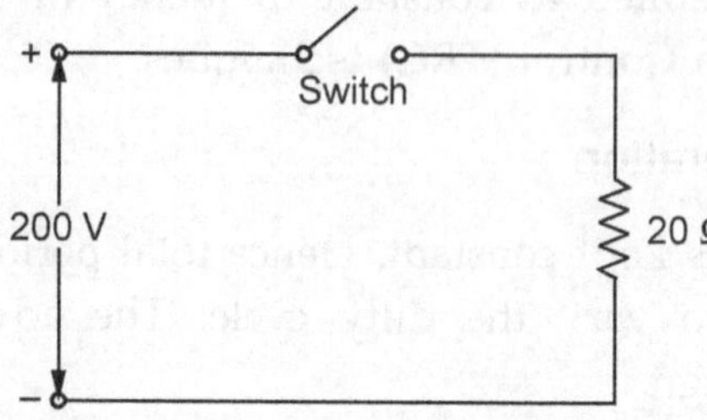

Fig. 4.2.5 Chopper of example 4.2.5

Average voltage is given as,

$$V_{o(av)} = \delta V_s$$

And rms voltage is given as,

$$V_{o(rms)} = \sqrt{\delta}\, V_s$$

Here not that $V_{o(av)} = V_{o(rms)}$ if $\sqrt{\delta} = \delta$. This is possible only for

$$\delta = 1$$

ii) Chopper efficiency

The switch is ideal. Hence there are no losses in the chopper. Therefore efficiency of the chopper is 100%.

Example 4.2.6 : *A transistor chopper is shown in Fig. 4.2.6. If the DC supply of 120 V is applied and the transistor is switched by applying periodic base pulses of voltage varying between 0 to 1 V, sketch the load voltage with reference to the input base pulses. Calculate i) Average/DC output voltage, and ii) Average load current assuming the chopper frequency as 200 Hz, duty cycle of operation as 33.33 %, and load resistance of 10 Ω.*

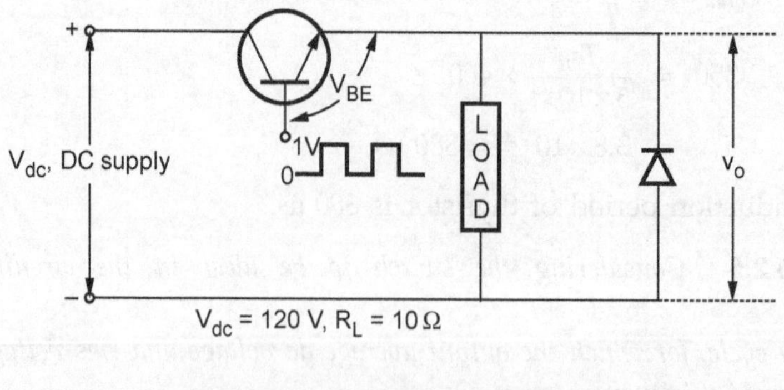

Fig. 4.2.6 Chopper circuit of example 4.2.6

Solution : This is a step down chopper. The base drive and load voltage waveforms are shown in the Fig. 4.2.2. The average value of output is derived in equation 4.2.1 as,

$$V_{o\,(av)} = \delta V_s$$

i) Average load current $I_{o(av)}$

Here the given data is,

$$V_s = V_{dc} = 120 \text{ V}$$

$$\delta = 33.33\ \% = 0.3333 = \frac{1}{3}$$

$$R = 10\ \Omega$$

Hence average output voltage will be,

$$V_{o(av)} = \delta V_s$$

$$= \frac{1}{3} \times 120 = 40 \text{ V}$$

Since the load is resistive, averge load current will be,

$$I_{o(av)} = \frac{V_{o(av)}}{R}$$

$$= \frac{40}{10} = 4 \text{ A}$$

Example 4.2.7 : *With the aid of basic circuit and waveform explain the basic principles of operation of a step down chopper with resistive load. Obtain expressions for i) DC output voltage ii) Output power and iii) Chopper efficiency*

Solution : i) DC output voltage

The DC output voltage is given by equation 4.2.1 as,

$$V_{o(av)} = \delta V_s$$

ii) Output power

Since the drop in chopper switch is not given, we will consider ideal switch. Hence output power as calculated in example 4.2.2 will be,

$$P_o = \frac{\delta V_s^2}{R}$$

iii) Chopper efficiency

The supply power to the chopper is given by equation 4.2.4.

$$P_s = \delta V_s \cdot \frac{V_s - V_{ch}}{R}$$

Since chopper is lossless,

$$P_s = \delta V_s \cdot \frac{V_s}{R} = \frac{\delta V_s^2}{R}$$

Chopper efficiency is, $\eta = \frac{P_o}{P_s} = \frac{\delta V_s^2 / R}{\delta V_s^2 / R} = 1$

The efficiency is 100% since chopper is lossless.

Review Question

1. *Draw the circuit diagram and explain the operation of step down chopper with the help of waveforms.*

Unsolved Examples

1. *In a DC chopper the average load current is 30 A. The chopping frequency is 250 Hz and supply voltage is 110 volts. Calculate ON and OFF periods, if the load resistance is 2 Ω.*

   [Ans. : T_{on} = 2.18 msec, T_{off} = 1.818 msec]

2. *A chopper has supply voltage of 220 V, load resistance R = 10 Ω and operating frequency of 1 kHz. The duty cycle is 50% Determine,*

   (i) Average and rms output voltage

   (ii) Output power.

   [Ans. : $V_{o(av)}$ = 110 V, $V_{o(rms)}$ = 155.56 V, P_o = 2420 W]

4.3 Step Down Chopper with RL Load

Normally the choppers are used to drive the DC motors. These motors are considered as RL (inductive) load. Fig. 4.3.1 shows the circuit diagram of step down chopper having inductive load.

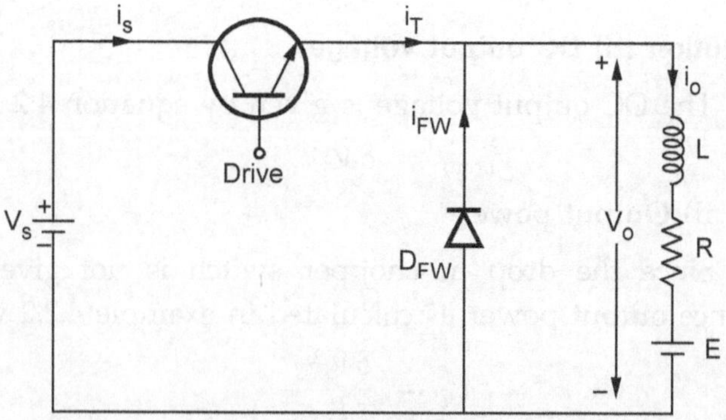

Fig. 4.3.1 Step down chopper with RL load

Transistor is used as a switch. It can be MOSFET, GTO, IGBT or SCR also. Since the load is inductive, freewheeling diode D_{FW} is used in the circuit. Normally the inductive loads are motors. Hence back emf 'E' is also shown in the circuit diagram as a part of load. Such loads are also called as 'RLE' load.

4.3.1 Continuous Load Current

Fig. 4.3.2 shows the waveforms of this circuit. The switch is turned 'on' for 0 to δT. Here 'δ' is the duty cycle of the chopper. The output voltage is equal to supply voltage ($v_o = v_s$) when the switch is 'ON'. The equivalent circuit-I in Fig. 4.3.2 shows the current flow when the switch is on. The output current is assumed continuous. At δT, the output current reaches to maximum value I_{max}.

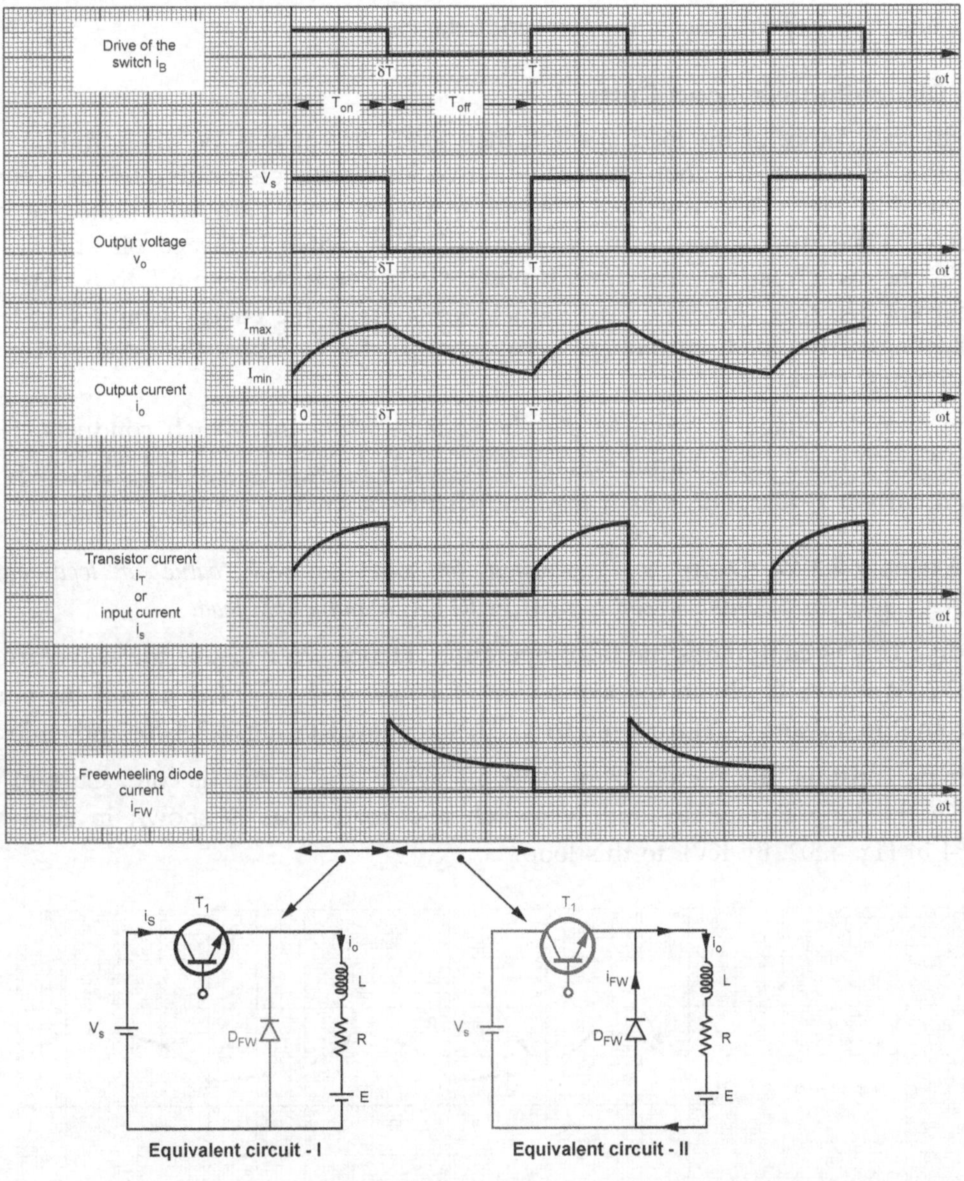

Fig. 4.3.2 Waveforms of step down chopper for inductive load (continuous load current)

From δT to T the switch is 'OFF'. At δT, the output current is at I_{max}. When the switch is turned off, the load inductance tries to maintain the output current in the same direction. This current flows through the freewheeling diode D_{FW}. The equivalent circuit-II in Fig. 4.3.2 shows this situation. The freewheeling diode is forward biased due to load inductance voltage $L\frac{di_o}{dt}$. Output voltage is zero when freewheeling diode conducts. In Fig. 4.3.2, observe that input current i_s flows only when switch (transistor conducts). Hence i_s and switch current i_T are same. The values of I_{max} and I_{min} depends upon the load inductance.

4.3.2 Discontinuous Load Current

If the inductance of the load is small then load current may be discontinuous. The waveforms of step down chopper for inductive load with discontinuous load current are shown in Fig. 4.3.3. The load current increases from zero when the switch is turned 'on'. The current reaches to 'I_{max}' at δT. The switch is turned off at δT. From δT to T, the switch is 'off'. The freewheeling diode conducts from δT to t_1. At t_1, the load current becomes zero. The load inductance supplies the energy from δT to t_1 (i.e. freewheeling). The load current is zero from t_1 to T. Thus the load current is discontinuous.

The output voltage is equal to supply voltage when the switch conducts. Hence $v_o = V_s$ from 0 to δT. The output voltage is zero when the switch is off. Here note that the load current is discontinuous if load inductance is small.

Example 4.3.1 : *Derive an expression for peak to peak value of load current (i.e. ripple) for continuous load current in RLE (i.e. inductive) load.*

Solution : This is continuous mode of operation. Hence we will use the waveforms of Fig. 4.3.2. We have to obtain the expressions for I_{max} and I_{min}. Let us call the current $i_1(t)$ from 0 to δT and $i_2(t)$ from δT to T. This is shown in Fig. 4.3.4 for convenience.

Let us consider the period from 0 to δT when the switch is 'on'. Current $i_1(t)$ flows in the loop formed by supply switch and load. This is shown in equivalent circuit-I in Fig. 4.3.2. By KVL to this loop,

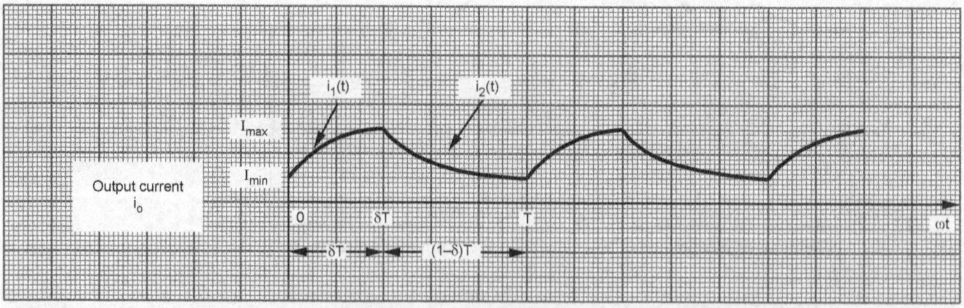

Fig. 4.3.4 Output current waveform for inductive load

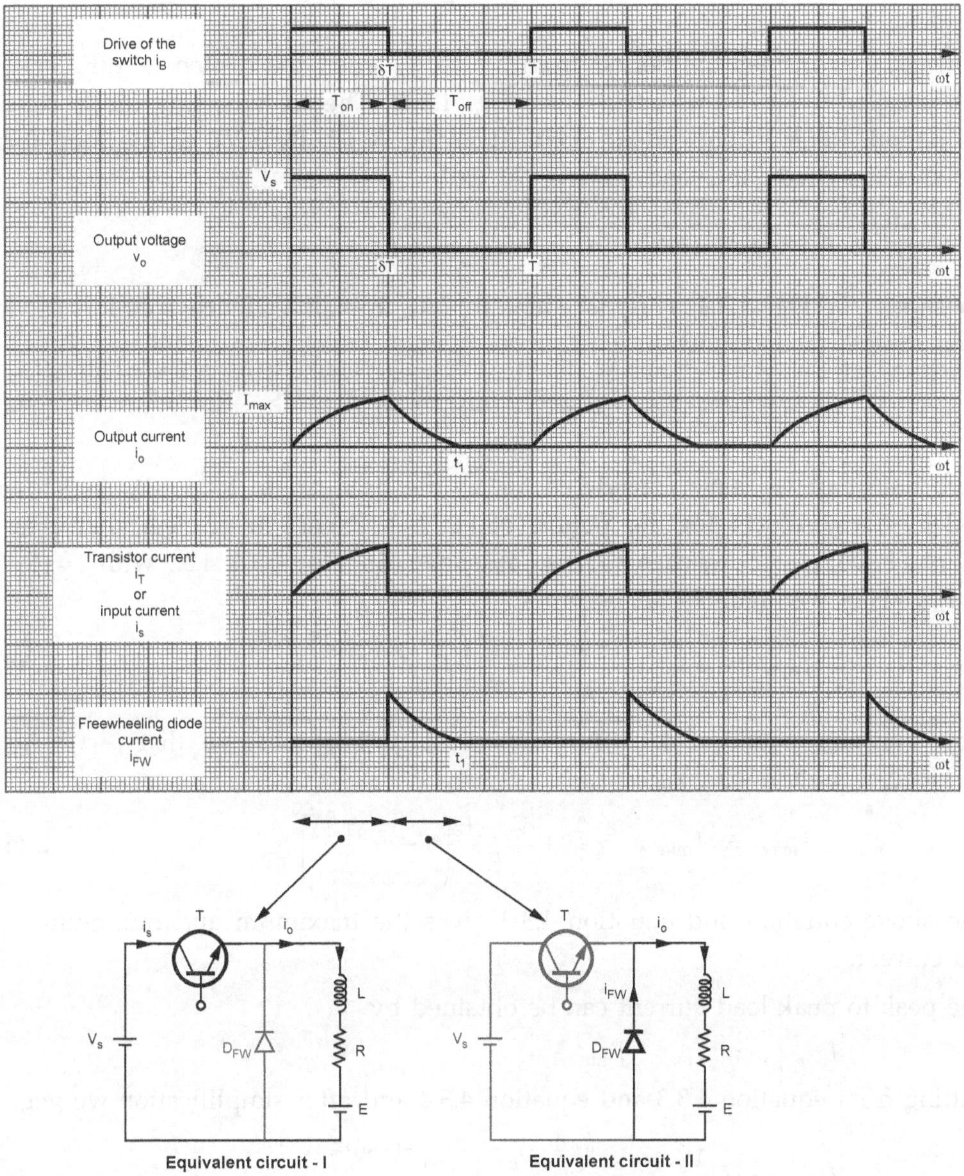

Fig. 4.3.3 Waveforms of step down chopper for inductive load (discontinuous load current)

$$V_s = L \frac{d i_1(t)}{dt} + R\, i_1(t) + E$$

Here we have neglected the drop in the switch (transistor). At $t = 0$, $i_1(t) = I_{min}$. Hence solution of above equation will be,

$$i_1(t) = I_{min}\, e^{-t\frac{R}{L}} + \frac{V_s - E}{R}\left(1 - e^{-t\frac{R}{L}}\right) \qquad \text{... (4.3.1)}$$

Now let us consider the period from δT to T, when the switch is 'off'. The load current is called $i_2(t)$ in this period (see Fig. 4.3.4). The freewheeling diode conducts in this period. Refer to equivalent circuit-II in Fig. 4.3.2. By KVL to the loop formed by freewheeling diode and load,

$$L\frac{d i_2(t)}{dt} + R\, i_2(t) + E = 0$$

As shown in Fig. 4.3.4, $i_2(t)$ is equal to I_{max} initially. Hence solution of above equation gives $i_2(t)$ as follows :

$$i_2(t) = I_{max}\, r^{-t\frac{R}{L}} - \frac{E}{R}\left(1 - e^{-t\frac{R}{L}}\right) \qquad \text{... (4.3.2)}$$

Here note that we have considered δT as the time origin for $i_2(t)$. As shown in Fig. 4.3.4 $i_2(t)$ conducts for the period of $(1-\delta)T$. Hence it is clear that after the period of $(1-\delta)T$, $i_2(t)$ reaches to I_{min} as shown in Fig. 4.3.4. i.e., with $t = (1-\delta)T$, $i_2(t) = I_{min}$. Hence equation 4.3.2 becomes,

$$I_{min} = I_{max}\, e^{-(1-\delta)T\frac{R}{L}} - \frac{E}{R}\left(1 - e^{-(1-\delta)T\frac{R}{L}}\right) \qquad \text{... (4.3.3)}$$

Now consider equation 4.3.1 for $i_1(t)$. In Fig. 4.3.4 observe that $i_1(t) = I_{max}$ at $t = \delta T$. Hence equation 4.3.1 becomes,

$$I_{max} = I_{min}\, e^{-\delta T\frac{R}{L}} + \frac{V_s - E}{R}\left(1 - e^{-\delta T\frac{R}{L}}\right) \qquad \text{... (4.3.4)}$$

The above equation and equation 4.3.3 gives the maximum and minimum values of load current.

The peak to peak load current can be obtained by,

$$I_{o(p-p)} = I_{max} - I_{min}$$

Putting from equation 4.3.3 and equation 4.3.4 and after simplification we get,

$$I_{o(p-p)} = \frac{V_s}{R}\,\frac{1 - e^{-\delta T\frac{R}{L}} + e^{-T\frac{R}{L}} - e^{-(1-\delta)T\frac{R}{L}}}{1 - e^{-T\frac{R}{L}}} \qquad \text{... (4.3.5)}$$

This equation gives peak to peak ripple in the load current.

Example 4.3.2 : *Derive an expression for average and rms value of the load voltage for the step down chopper having inductive load.*

Solution : The waveform of output voltage is shown in Fig. 4.3.2 for inductive load when the load current is continuous. Fig. 4.3.3 shows the output voltage waveform for inductive load when the load current is discontinuous. And Fig. 4.2.2 shows the output voltage waveform for the step down chopper for resistive load. Note that the output voltage waveform is exactly same in all these cases. Hence their rms and average values must be same. Hence average value of step down chopper for inductive load can be given by equation 4.2.1 as,

$$V_{o(av)} = \delta\ V_s \qquad \text{... (4.3.6)}$$

Here $\delta = \frac{T_{on}}{T}$ is called duty cycle of the chopper. Similarly rms value of the output voltage for inductive load can be obtained from equation 4.2.2 as,

$$V_{o(rms)} = \sqrt{\delta}\ V_s \qquad \text{... (4.3.7)}$$

Example 4.3.3 : *Prove that maximum ripple occurs in the output current of step down chopper when duty cycle is 0.5. And the value of maximum ripple current is given as,*

$$I_{o\max(p-p)} = \frac{V_s}{R} \tanh \frac{TR}{4L}$$

Solution : The peak to peak ripple in the load current is given by equation 4.3.5 as,

$$I_{o(p-p)} = \frac{V_s}{R} \frac{1-e^{-\delta T\frac{R}{L}} + e^{-T\frac{R}{L}} - e^{-(1-\delta)T\frac{R}{L}}}{1-e^{-T\frac{R}{L}}} \qquad \text{... (4.3.8)}$$

For the maximum value of $I_{o(p-p)}$ with respect to duty cycle δ, we have to set the following condition,

$$\frac{d}{d\delta} I_{o(p-p)} = 0$$

That is differentiate peak to peak ripple current with respect to duty cycle δ. Then differentiating equation 4.3.8 with respect to δ and equating to zero we get,

$$e^{-\delta T\frac{R}{L}} - e^{-(1-\delta)T\frac{R}{L}} = 0$$

$$\therefore \quad e^{-\delta T\frac{R}{L}} = e^{-(1-\delta)T\frac{R}{L}}$$

$$\therefore \quad \delta T \frac{R}{L} = (1-\delta)\ T\frac{R}{L}$$

$$\therefore \quad \delta = \frac{1}{2} = 0.5$$

This shows that at duty cycle $\delta = 0.5$, the peak to peak value of ripple current is maximum. Hence by putting $\delta = 0.5$ equation 4.3.8 becomes,

$$I_{o\,max\,(p-p)} = \frac{V_s}{R}\frac{1-e^{-0.5T\frac{R}{L}}+e^{-T\frac{R}{L}}-e^{-(1-0.5)\,T\frac{R}{L}}}{1-e^{-T\frac{R}{L}}}$$

On simplifying the above equation,

$$I_{o\,max\,(p-p)} = \frac{V_s}{R}\tanh\left(\frac{TR}{4L}\right) \quad \text{... (4.3.9)}$$

Note that the above equation is derived only for continuous load current.

Example 4.3.4 : *Derive an expression for the load current of the step down chopper with inductive load having discontinuous load current. Also obtain the time at which the load current becomes zero.*

Solution : The waveforms of step down chopper for inductive load (discontinuous mode) are shown in Fig. 4.3.3 The equivalent circuits are similar to those in Fig. 4.3.2. Hence the equations derived in example 4.3.1 are also applicable to discontinuous mode. Here $I_{min} = 0$, since load current is discontinuous. Hence equation 4.3.1 will be,

$$i_1(t) = \frac{V_s - E}{R}\left(1-e^{-t\frac{R}{L}}\right) \quad \text{... (4.3.10)}$$

This is the equation for the load current from 0 to δT, i.e. when the switch is 'on'. Equation 4.3.2 gives the load current when the chopper switch is 'off'. i.e.,

$$i_2(t) = I_{max}\,e^{-t\frac{R}{L}} - \frac{E}{R}\left(1-e^{-t\frac{R}{L}}\right) \quad \text{... (4.3.11)}$$

This current goes to at $t = t_1$ (see Fig. 4.3.3), hence above equation becomes,

$$0 = I_{max}\,e^{-t_1\frac{R}{L}} - \frac{E}{R}\left(1-e^{-t_1\frac{R}{L}}\right)$$

$$\therefore \quad I_{max}\,e^{-t_1\frac{R}{L}} = \frac{E}{R}\left(1-e^{-t_1\frac{R}{L}}\right)$$

$$\therefore \quad \left(I_{max}+\frac{E}{R}\right)e^{-t_1\frac{R}{L}} = \frac{E}{R}$$

$$\therefore \quad e^{-t_1\frac{R}{L}} = \frac{\frac{E}{R}}{I_{max}+\frac{E}{R}}$$

$$\therefore \quad -t_1\frac{R}{L} = \ln\left[\frac{\frac{E}{R}}{I_{max}+\frac{E}{R}}\right]$$

$$\therefore \qquad t_1 = \frac{L}{R}\, ln \left[\frac{I_{\max} + \frac{E}{R}}{\frac{E}{R}} \right]$$

$$\therefore \qquad t_1 = \frac{L}{R}\, ln \left(1 + \frac{RI_{\max}}{E} \right) \qquad \text{... (4.3.12)}$$

This equation gives the time at which the load current will be zero or discontinuous.

Example 4.3.5 : *A chopper circuit drives an inductive load from 200 V DC supply. Given the load resistance as 4 Ω, the average load current as 30 A and operating frequency is 400 Hz. Compute the ON period and OFF period of the chopper. Also determine the duty cycle of the chopper.*

Solution : Given data :

Supply voltage $V_s = 200\ V$, Load resistance $R = 4\ \Omega$

Output current $I_{o\,(av)} = 30\ A$, Frequency $f = 400\ Hz$

(i) To obtain $V_{o\,(av)}$

Output average voltage is given as

$$V_{o\,(av)} = R \times I_{o\,(av)} = 4 \times 30 = 120\ V$$

(ii) To obtain δ

We know that

$$V_{o\,(av)} = \delta\, V_s$$

$$\therefore \qquad \delta = \frac{V_{o(av)}}{V_s} = \frac{120\,V}{200\,V} = 0.6$$

(iii) To obtain T_{on} and T_{off}

We know that,

$$\delta = \frac{T_{on}}{T} = T_{on} \times f$$

$$\therefore \qquad T_{on} = \frac{\delta}{f} = \frac{0.6}{400} = 1.5 \text{ ms}$$

and

$$T = T_{on} + T_{off}$$

$$\therefore \qquad T_{off} = T - T_{on} = \frac{1}{f} - T_{on}$$

$$= \frac{1}{400} - 1.5 \times 10^{-3} = 1 \text{ ms}$$

Example 4.3.6 : *A DC chopper has a resistive load of 20 Ω and input voltage $V_s = 220$ V. When the chopper is on, its voltage drop is 1.5 V and chopping frequency is 10 kHz. If duty cycle is 80% determine the average output voltage, rms output voltage and chopper on time.*

Solution : Given data :

Supply voltage $V_s = 220$, Drop in switch $V_T = 1.5$ V

Frequency f = 10 kHz, Duty cycle $\delta = 0.8$

Load resistance $R = 20\ \Omega$

(i) To determine $V_{o\,(av)}$

$V_{o\,(av)}$ is given as,

$$V_{o(av)} = \delta\, V_s$$

Considering the drop in the switch,

$$V_{o\,(av)} = \delta\,(V_s - V_T)$$

$$= 0.8\,(220 - 1.5) = 174.8$$

(ii) To determine $V_{o(rms)}$

$V_{o(rms)}$ is given as,

$$V_{o(rms)} = \sqrt{\delta}\, V_s$$

Considering drop in the switch,

$$V_{o(rms)} = \sqrt{\delta}\,(V_s - V_T)$$

$$= \sqrt{0.8}\,(220 - 1.5) = 195.4 \text{ V}$$

(iii) To determine T_{on}

Duty cycle is given as,

$$\delta = \frac{T_{on}}{T} = f\, T_{on}$$

$$\therefore \quad T_{on} = \frac{\delta}{f} = \frac{0.8}{10000}$$

$$= 80\ \mu\, sec$$

Example 4.3.7 : *A chopper is supplying an inductive load with a freewheeling diode. The load inductance is 5 H and resistance is 10 Ω. The input to the chopper is 200 V and the chopper frequency is 100 Hz. The ON to OFF time ratio is 2 : 3 calculate*

(i) Average load current

(ii) Limits between which current fluctuates.

Solution : Given data :

supply voltage, V_s = 200 V

frequency, f = 100 Hz

ON/OFF ratio = $\frac{2}{3}$

$\therefore$ duty cycle, $\delta = \frac{2}{2+3} = 0.4$

resistance, R = 10 Ω

inductance, L = 5 H

(i) To obtain load current

The average value of output voltage is given as,

$$V_{o(av)} = \delta V_s$$

$$= 0.4 \times 200$$

$$= 80\ V$$

Average load current will be,

$$I_{o(av)} = \frac{V_{o(av)}}{R}$$

$$= \frac{80}{10}$$

$$= 8\ A$$

(ii) To obtain I_{min} and I_{max}

I_{min} is given by equation 4.3.3 as,

$$I_{min} = I_{max}\, e^{-(1-\delta)T\frac{R}{L}} - \frac{E}{R}\left(1 - e^{-(1-\delta)T\frac{R}{L}}\right)$$

Here $E = 0$, hence above equation will be,

$$I_{min} = I_{max}\, e^{-(1-\delta)T\frac{R}{L}}$$

Putting the values with $T = \frac{1}{f} = \frac{1}{100}$,

$$I_{min} = I_{max}\, e^{-(1-0.4)\frac{10}{100\times 5}}$$

$$= 0.98807\ I_{max} \qquad \text{... (4.3.13)}$$

I_{max} is given by equation 4.3.4 as,

$$I_{max} = I_{min}\ e^{-\delta T \frac{R}{L}} + \frac{V_s - E}{R}\left(1 - e^{-\delta T \frac{R}{L}}\right)$$

Since $E = 0$,
$$I_{max} = I_{min}\ e^{-\delta T \frac{R}{L}} + \frac{V_s}{R}\left(1 - e^{-\delta T \frac{R}{L}}\right)$$

$$= I_{min}\ e^{-0.4 \frac{10}{100 \times 5}} + \frac{200}{10}\left(1 - e^{-0.4 \times \frac{10}{100 \times 5}}\right)$$

$$= I_{min} \times 0.99203 + 0.15936$$

Putting the value of I_{min} (equation 4.3.13) in above equation,

$$I_{max} = 0.98807\ I_{max} \times 0.99203 + 0.15936$$

$$\therefore \quad I_{max} = 8.0465 \text{ A}$$

And I_{min} will be (from equation 4.3.13),

$$I_{min} = 0.98807\ I_{max}$$

$$= 0.98807 \times 8.0465$$

$$= 7.9505 \text{ A}$$

$I_{o(av)}$ can also be calculated as,

$$I_{o(av)} = \frac{I_{max} + I_{min}}{2}$$

$$= \frac{8.0465 + 7.9505}{2} = 7.9985 \text{ A}$$

Example 4.3.8 : *A chopper is feeding an R-L load as shown in the Fig. 4.3.5. $V_s = 220\ V$, $R = 5\ \Omega$, $L = 7.5\ mH$, $f = 1\ kHz$, $\delta = 0.5$ and $E = 0$ volts. Calculate -*

i. *Minimum instantaneous load current, I_{min}*
ii. *Peak instantaneous load current, I_{max}*
iii. *Maximum peak to peak load ripple current.*
iv. *Average value of load current, $I_{o(av)}$*
v. *rms load current $I_{o(rms)}$*
vi. *Effective input resistance R_i seen by the source.*
vii. *rms chopper current, $I_{T(rms)}$*

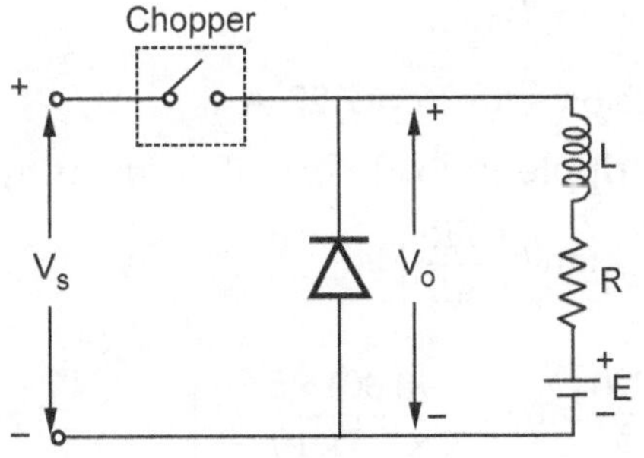

Fig. 4.3.5

Solution : The given data is

$$V_s = 220 \text{ V}$$

$$R = 5\ \Omega$$

$$L = 7.5 \text{ mH}$$

$$f = 1 \text{ kHz} \qquad \therefore \ T = \frac{1}{f} = 0.001$$

$$\delta = 0.5$$

$$E = 0$$

(i) and (ii) To obtain I_{min} and I_{max}

The expressions for I_{min} and I_{max} are given by equation 4.3.3 and equation 4.3.4. Putting values in equation 4.3.3,

$$I_{min} = I_{max}\, e^{-(1-0.5)\times 0.001 \times \frac{5}{7.5\times 10^{-3}}} - \frac{0}{R}\left[1 - e^{-(1-0.5)\times 0.001\times\frac{5}{7.5\times 10^{-3}}}\right]$$

$$\therefore \quad I_{min} = 0.7165\, I_{max} + 0 \qquad \text{.... (4.3.14)}$$

Similarly putting values in equation 4.3.4 we get,

$$I_{max} = I_{min}\, e^{-0.5\times 0.001\times\frac{5}{7.5\times 10^{-3}}} + \frac{220-0}{5}\left[1 - e^{-0.5\times 0.001\times\frac{5}{7.5\times 10^{-3}}}\right]$$

$$= 0.7165\, I_{min} + 12.4726 \qquad \text{... (4.3.15)}$$

Solving equation 4.3.14 and above equation we get,

$$I_{max} = 25.63 \text{ A and } I_{min} = 18.36 \text{ A}$$

(iii) Maximum peak to peak ripple in load current

The ripple in load current is,

$$I_{o(p-p)} = I_{max} - I_{min}$$

$$= 25.63 - 18.36 = 7.27 \text{ A}$$

Maximum peak to peak ripple in load current is given by equation 4.3.9,

$$I_{o\,max(p-p)} = \frac{V_s}{R} \tan h\left(\frac{TR}{4L}\right)$$

$$= \frac{220}{5} \tan h\left(\frac{0.001 \times 5}{4 \times 7.5 \times 10^{-3}}\right)$$

$$= 7.266 \text{ A}$$

Thus the maximum ripple is 7.266 ≈ 7.27 A since δ = 0.5.

iv) Average value of load current

The average value of load current is approximately given as,

$$I_{o(av)} = \frac{I_{max} + I_{min}}{2} = \frac{25.63 + 18.36}{2}$$

$$= 21.995 \approx 22 \text{ A.}$$

v) rms load current

The rms load current is given as,

$$I_{o(rms)} = \left\{ I_{min}^2 + \frac{I_{o(p-p)}^2}{3} + I_{min}\, I_{o(p-p)} \right\}^{1/2} \quad \text{... (4.3.16)}$$

Putting values,

$$I_{o(rms)} = \left\{ 18.36^2 + \frac{7.27^2}{3} + 18.36 \times 7.27 \right\}^{1/2}$$

$$= 22.095 \text{ A}$$

vi) Effective input resistance

The average source current can be calculated from average load current as

$$I_{s(av)} = \delta\, I_{o(av)}$$

$$= 0.5 \times 22 = 11 \text{ A.}$$

$$\therefore \text{ Input resistance } R_i = \frac{V_s}{I_{s(av)}}$$

$$= \frac{220}{11} = 20\ \Omega$$

vii) rms chopper current

The rms chopper current can be calculated as (see Fig. 4.3.3),

$$I_{T(rms)} = \left\{ \frac{1}{T} \int_0^{\delta T} i^2(t)\, dt \right\}^{1/2}$$

$$= \sqrt{\delta}\left\{I_{min}^2 + \frac{I_{o(p-p)}^2}{3} + I_{min} I_{o(p-p)}\right\}^{1/2}$$

$$= \sqrt{0.5}\left\{18.36^2 + \frac{7.27^2}{3} + 18.36 \times 7.27\right\}^{1/2}$$

$$= 15.62 \text{ A}$$

This value can also be calculated as,

$$I_{T(rms)} = \sqrt{\delta}\ I_{o(rms)} = \sqrt{0.5} \times 22.095$$

$$= 15.62 \text{ A}$$

Example 4.3.9 : *An SCR chopper is supplying an inductive load with R = 10 Ω and L = 2 H. The DC supply to the chopper is 200 V. If the chopper is operating at a frequency of 200 Hz and ON/OFF time ratio of chopper is 2 : 3, calculate,*

(i) Maximum and minimum values of load current in one cycle of chopper operation under steady state conditions.

(ii) Average load current

Solution : Given data

Load resistance, R = 10 Ω

Load inductance, L = 2 H

Supply voltage, V_s = 200 V

Chopper frequency, f = 200 Hz, hence $T = \frac{1}{200}$

ON/OFF ratio = 2:3

$$\therefore \quad \delta = \frac{T_{on}}{T_{on} + T_{off}} = \frac{2}{2+3} = \frac{2}{5} = 0.4$$

(i) To obtain I_{min} and I_{max}

From equation 4.3.3,

$$I_{min} = I_{max} e^{-(1-\delta) T \frac{R}{L}} - \frac{E}{R}\left(1 - e^{-(1-\delta) T \frac{R}{L}}\right)$$

Here, E = 0, hence above equation becomes,

$$I_{min} = I_{max} e^{-(1-0.4) \cdot \frac{1}{200} \cdot \frac{10}{2}}$$

$$= 0.98511\, I_{max} \qquad \text{... (4.3.17)}$$

Similarly from equation 4.3.4,

$$I_{max} = I_{min}\, e^{-\delta T \frac{R}{L}} + \frac{V_s - E}{R}\left(1 - e^{-\delta T \frac{R}{L}}\right)$$

$$= I_{min}\, e^{-0.4 \cdot \frac{1}{200} \cdot \frac{10}{5}} + \frac{200 - 0}{10}\left(1 - e^{-0.4 \cdot \frac{1}{200} \cdot \frac{10}{5}}\right)$$

$$= 0.996\, I_{min} + 0.07984 \qquad \text{... (4.3.18)}$$

Putting for I_{min} from equation 4.3.17 in above equation,

$$I_{max} = 0.996\,(0.98511\, I_{max}) + 0.07984$$

$$\therefore \quad I_{max} = 4.24 \text{ A}$$

Similarly from equation 4.3.17 we can get I_{min} as,

$$I_{min} = 0.98511\, I_{max}$$

$$= 0.98511 \times 4.24 = 4.1768 \text{ A}$$

(ii) To obtain average load current

The average load current is given as,

$$I_{o(av)} = \frac{I_{max} + I_{min}}{2} = \frac{4.24 + 4.1768}{2}$$

$$= 4.2 \text{ A}$$

Example 4.3.10 : *For the chopper circuit shown in Fig. 4.3.6, the duty cycle, k = 0.5 and chopping frequency f = 5 kHz. Determine*

(i) Minimum instantaneous load current

(ii) Peak instantaneous load current

(iii) Maximum peak to peak current in the load

(iv) Average load current

(v) rms load current

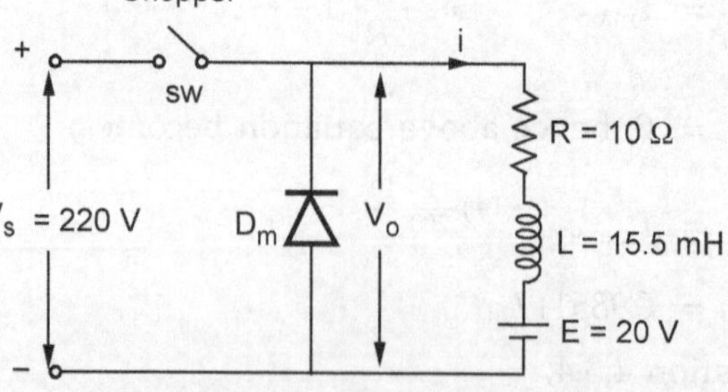

Fig. 4.3.6 Chopper of example 4.3.10

Solution : The given data is,

Supply voltage, $V_s = 220$ V

Duty cycle, $\delta = 0.5$

Chopper frequency, $f = 5000$ Hz. Hence $T = \frac{1}{5000}$

Load resistance, $R = 10\ \Omega$

Load inductance, $L = 15.5$ mH

and $E = 20$ V

(i) and (ii) To obtain I_{max} and I_{min}

From equation 4.3.3 I_{min} is given as,

$$I_{min} = I_{max} e^{-(1-\delta)T\frac{R}{L}} - \frac{E}{R}\left[1 - e^{-(1-\delta)T\frac{R}{L}}\right]$$

Putting values in above equation,

$$I_{min} = I_{max} e^{-(1-0.5)\cdot\frac{1}{5000}\cdot\frac{10}{15.5\times10^{-3}}} - \frac{20}{10}\left[1 - e^{-(1-0.5)\cdot\frac{1}{5000}\cdot\frac{10}{15.5\times10^{-3}}}\right]$$

$$= 0.9375\ I_{max} - 0.12496 \qquad \text{... (4.3.19)}$$

From equation 4.3.4 I_{max} is given as,

$$I_{max} = I_{min}\, e^{-\delta T\frac{R}{L}} + \frac{V_s - E}{R}\left(1 - e^{-\delta T\frac{R}{L}}\right)$$

Putting values in above equation,

$$I_{max} = I_{min}\, e^{-0.5\cdot\frac{1}{5000}\cdot\frac{10}{15.5\times10^{-3}}} + \frac{220-20}{10}\left(1 - e^{-0.5\cdot\frac{1}{5000}\cdot\frac{10}{15.5\times10^{-3}}}\right)$$

$$= 0.9375\ I_{min} + 1.2495$$

Putting for I_{min} from equation 4.3.19 in above equation,

$$I_{max} = 0.9375\ (0.9375\ I_{max} - 0.12496) + 1.2495$$

$$= 9.351\ \text{A}$$

Putting this value of I_{max} in equation 4.3.19 we get,

$$I_{min} = 0.9375\ (9.351) - 0.12496$$

$$= 8.642\ \text{A}$$

(iii) Maximum peak to peak current in the load

Maximum peak to peak current in the load is given as,

$$I_{o(p-p)} = I_{max} - I_{min}$$

$$= 9.351 - 8.642 = 0.709 \text{ A}$$

(iv) To obtain average load current

Average load current is given as,

$$I_{o(av)} = \frac{I_{max} + I_{min}}{2} = \frac{9.351 + 8.642}{2} = 8.996 \approx 9 \text{ A}$$

Average load current can also be obtained by following equation,

$$I_{o(av)} = \frac{V_{o(av)}}{R} = \frac{\delta V_s - E}{R} = \frac{0.5 \times 220 - 20}{10}$$

$$= 9 \text{ A}$$

(v) RMS load current

The rms value of load current is given by equation 4.3.16 as,

$$I_{o(rms)} = \left[I_{min}^2 + \frac{I_{o(p-p)}^2}{3} + I_{min} I_{o(p-p)} \right]^{\frac{1}{2}}$$

Putting values in above equation,

$$I_{o(rms)} = \left[(8.642)^2 + \frac{(0.709)^2}{3} + 8.642 \times 0.709 \right]^{\frac{1}{2}} = 8.9988 \approx 9 \text{ A}$$

Review Question

1. *Explain the operation of step down chopper for inductive load with the help of waveforms.*

4.4 Step Up Chopper

In the previous section we studied step down chopper. The average value of output voltage $V_{o(av)}$ is always less than or equal to supply voltage V_s in step down chopper. In the step-up chopper $V_{o\,(av)} \geq V_s$.

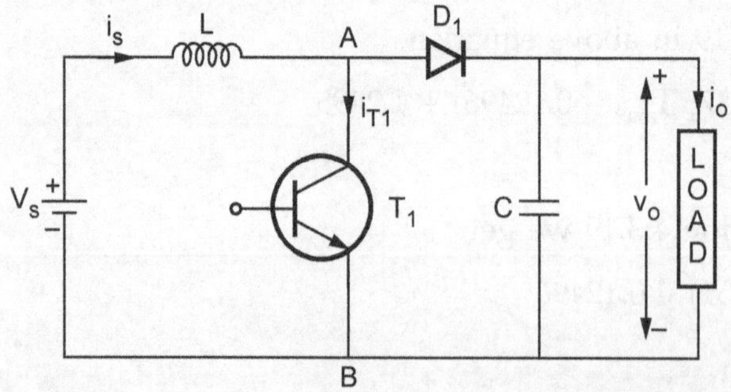

Fig. 4.4.1 Step up chopper having transistor as a switch

Fig. 4.4.1 shows the circuit diagram of the step up chopper. Observe that there is an inductance in series with the supply V_s. A switch (transistor GTO, MOSFET etc.) is connected across inductance and supply. A filter capacitor C is used across the load to make v_o smooth. The diode D_1 blocks the reverse flow of output current when switch is turned 'on'. Fig. 4.4.2 shows the waveforms of this step up chopper. The drive of the switch is shown at the begining. The transistor is turned on from 0 to δT. Hence current flows through the inductance from the supply. The inductance current rises and inductance stores the energy from the supply. The equivalent circuit-I in Fig. 4.4.2 shows this operation. Note that the drop in the inductance is $L\frac{di_L}{dt}$ with the polarity shown. The voltage $v_{AB} = 0$ since transistor is

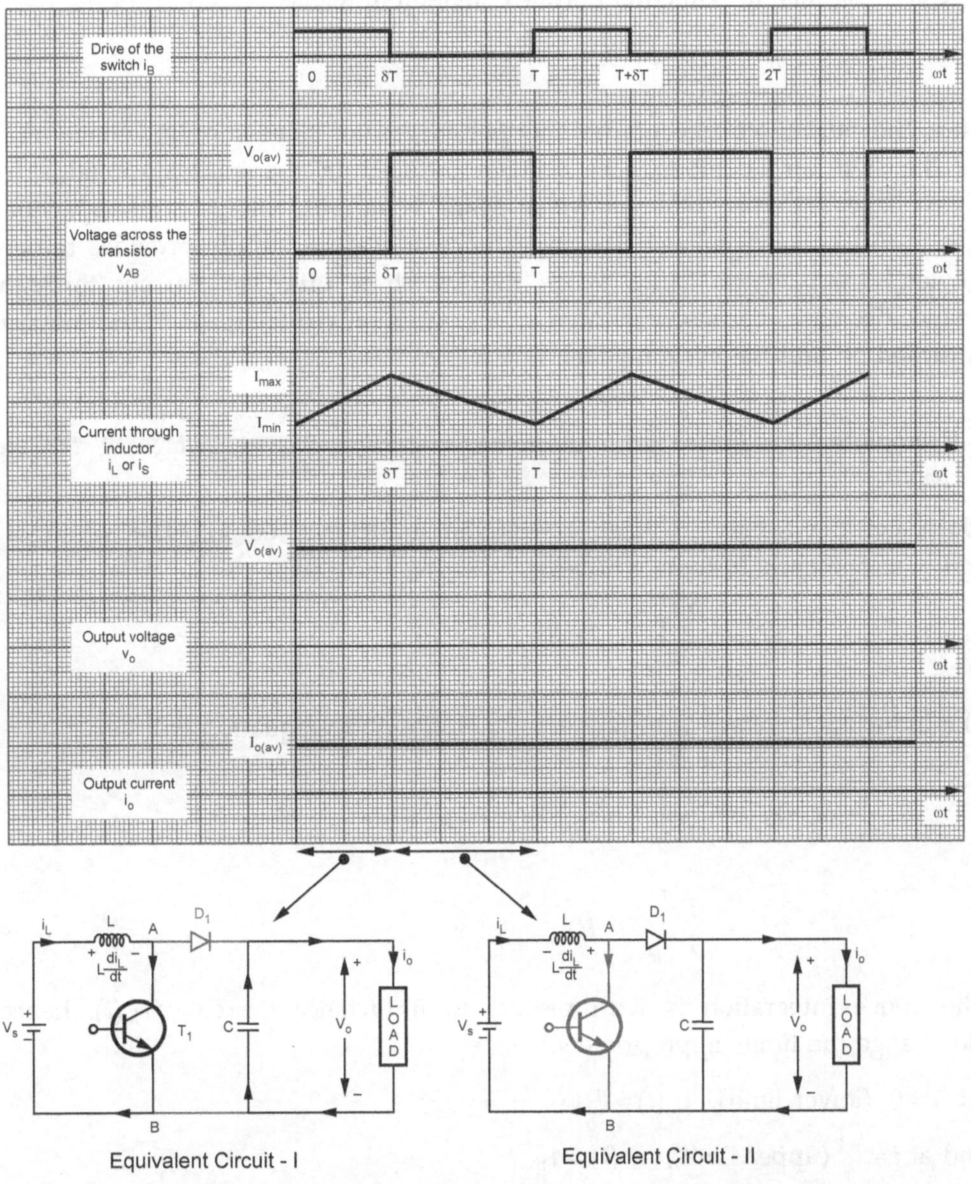

Fig. 4.4.2 Waveforms of the step up chopper for ripple-free output voltage and currents

conducting. Fig. 4.4.2 also show the output voltage and current waveforms. Here we have assumed that the output voltage and current are continuous and ripple-free. The capacitor maintains the voltage 'v_o' and supplies the current 'i_o' when transistor is 'on'. Hence the diode D_1 is reverse biased and it does not conduct.

At δT transistor (switch) is turned off. Hence the inductance generates a large voltage $L \frac{di_L}{dt}$ to maintain the current i_L in the same direction. Note the polarity of inductance voltage in equivalent circuit-II of Fig. 4.4.2. The diode D_1 is forward biased and it starts conducting. Thus the output voltage will be

$$v_o = V_s + L \frac{di_L}{dt}$$

Thus the output voltage of the chopper is greater than supply voltage V_s. This shows the step up operation. The voltage induced in the inductance adds to the supply voltage and this total voltage appears as output voltage. The capacitor also charges to this boosted voltage. The inductance as well as supply provides the energy to the load from δT to T (i.e. when the switch is off). The current through the inductance decreases because its stored energy goes on reducing. At T, the transistor is again turned on and the cycle repeats.

Example 4.4.1 : *Derive an expression for average output voltage of the step-up chopper.*

Solution : Let the average voltage across the inductance be V_L. The value of this voltage over one cycle (0 to T) will be,

$$V_L = \frac{1}{T} \int_0^T v_L(t)\, dt$$

The inductance voltage is $v_L(t) = L \frac{di_L}{dt}$. Hence above equation will be,

$$V_L = \frac{1}{T} \int_0^T L \frac{di_L(t)}{dt} dt$$

$$= \frac{L}{T} \int_0^T di_L(t) \qquad \text{... (4.4.1)}$$

The above integration is with respect to inductance current $di_L(t)$, hence we should change the limits appropriately.

At $t = 0$ (lower limit), $i_L(t) = I_{min}$

and at $t = T$ (upper limit), $i_L(t) = I_{min}$

Hence equation 4.4.1 will be,

$$V_L = \frac{L}{T}\int_{I_{min}}^{I_{min}} d\, i_L(t) = \frac{L}{T}\left[i_L(t)\right]_{I_{min}}^{I_{min}}$$

$$= \frac{L}{T}\left[I_{min} - I_{min}\right] = 0 \qquad \text{... (4.4.2)}$$

Thus the average voltage across the inductance is zero. The inductance stores the energy when the switch is 'ON' and supplies the energy to the load when the switch is 'OFF'.

Now let us consider the loop formed by supply voltage V_s, inductance and switch. This loop is shown in Fig. 4.4.3. The voltage across the switch is v_{AB}. The waveform of v_{AB} is shown in Fig. 4.4.2. The average value of v_{AB} can be obtained as follows :

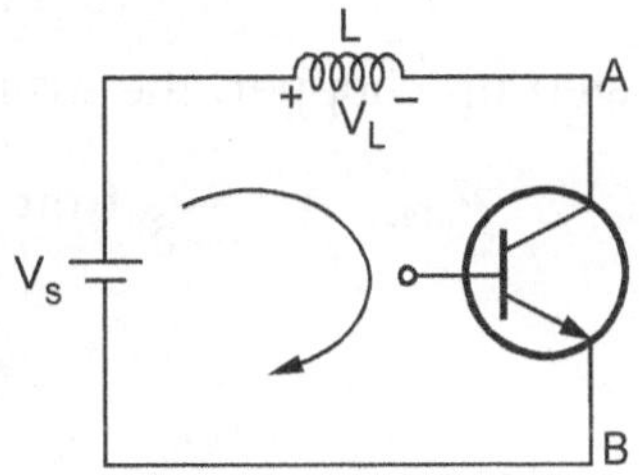

Fig. 4.4.3 Loop formed by supply, inductance and switch in step up chopper

$$V_{AB} = \frac{1}{T}\int_0^T v_{AB}(t)\, dt$$

In Fig. 4.4.2 observe that $v_{AB} = V_{o(av)}$ from δT to T and rest of the period it is zero. Hence above equation becomes,

$$V_{AB} = \frac{1}{T}\int_{\delta T}^{T} V_{o(av)}\, dt = \frac{V_{o(av)}}{T}(T - \delta T)$$

$$= V_{o(av)}(1-\delta) \qquad \text{... (4.4.3)}$$

By KVL to the loop shown in Fig. 4.4.3,

$$V_s = V_L + V_{AB}$$

The above equation holds for steady state. Putting values from equation 4.4.2 and 4.4.3 in above equation,

$$V_s = 0 + V_{o(av)}(1-\delta)$$

$$\therefore \quad V_{o(av)} = \frac{V_s}{1-\delta} \qquad \text{... (4.4.4)}$$

This equation gives the value of average output voltage. When $\delta = 0$, $V_{o(av)} = V_s$ and $V_{o(av)} \rightarrow \infty$ as δ approaches to unity. The value of duty cycle 'δ' lies between 0 and 1.

Example 4.4.2 : *Explain the principle of operation of a step up chopper. A step up DC chopper has an input of 200 volts and an output of 250 volts. The blocking period in each cycle of operation is* 0.6×10^{-3} *seconds. Find the period of conduction in each cycle.*

Solution : Given $V_{o(av)} = 250$ volts

$$V_s = 200 \text{ volts}$$

$$T_{off} = 0.6\times10^{-3} \text{ sec}$$

For the step up chopper, the average value of output voltage is given as,

$$V_{o(av)} = \frac{V_s}{1-\delta} \text{ from equation 4.4.4}$$

Here $\delta = \dfrac{T_{on}}{T_{on}+T_{off}}$ is the duty cycle.

Let us calculate δ first. Putting $V_{o(av)}$ and V_s,

$$250 = \frac{200}{1-\delta}$$

$$\therefore \quad \delta = 0.2$$

And $$\delta = \frac{T_{on}}{T_{on}+T_{off}}$$

Putting for δ and T_{off},

$$0.2 = \frac{T_{on}}{T_{on}+0.6\times10^{-3}}$$

$$\therefore \quad T_{on} = 0.15\times10^{-3}$$

Thus the period of conduction in each cycle is 0.15×10^{-3} seconds.

Example 4.4.3 : *Input to the step up chopper is 200 V. The output required is 600 V. If the conducting time of thyristor is 200 μs, compute-*
i) Chopping frequency (ii) If pulse width is halved for constant frequency of operation, find the new output voltage.

Solution : The given data is,

$$V_s = 200 \text{ V}$$

$$V_{o(av)} = 600 \text{ V}$$

$$T_{on} = 200\ \mu s$$

i) To obtain chopping frequency

For the step up chopper, the average value of output voltage is given as,

$$V_{o(av)} = \frac{V_s}{1-\delta}$$

$$\therefore \quad 600 = \frac{200}{1-\delta}$$

$$\therefore \quad \delta = 0.6667$$

The duty cycle δ is given as,

$$\delta = \frac{T_{on}}{T_{on}+T_{off}}$$

$$= \frac{T_{on}}{T}$$

$$= f T_{on} \qquad \text{since } f = \frac{1}{T}$$

$$\therefore \quad f = \frac{\delta}{T_{on}}$$

$$= \frac{0.6667}{200 \times 10^{-6}}$$

$$= 3333.33 \text{ Hz} \approx 3.3 \text{ kHz}$$

This is the chopping frequency of the chopper.

(ii) To obtain new output voltage

The chopping frequency is constant. Hence

$$T = \frac{1}{f} = \frac{1}{3333.33} = 3 \times 10^{-4} = 300 \ \mu s$$

The T_{on} pulse width is halved. Hence new T_{on} will be $\frac{200\mu s}{2}$ = 1000 μs.

Hence duty cycle will be,

$$\delta = \frac{T_{on}}{T} = \frac{100 \times 10^{-6}}{300 \times 10^{-6}} = 0.3333$$

The new output voltage of chopper will be,

$$V_{o(av)} = \frac{V_s}{1-\delta} = \frac{200}{1-0.3333}$$

$$= 300 \text{ V}$$

Thus the output is also reduced by half.

Example 4.4.4 : *A step up chopper has input voltage of 220 V and output voltage of 660 volts. If the nonconducting time of thyristor chopper is 100 μsec, compute the pulse width of output voltage. In case pulsewidth is halved for constant frequency operation, find new output voltage.*

Solution : Given data

Supply voltage, $V_s = 220\text{ V}$

Output voltage, $V_{o(av)} = 660\text{ V}$

$T_{off} = 100\ \mu\text{sec}$

(i) To calculate T_{on}

We know that,

$$V_{o(av)} = \frac{V_s}{1-\delta}$$

$$\therefore \quad 660 = \frac{220}{1-\delta}$$

$$\therefore \quad \delta = 0.6666 \text{ or } \frac{2}{3}$$

Duty cycle is given as,

$$\delta = \frac{T_{on}}{T_{on} + T_{off}}$$

$$\frac{2}{3} = \frac{T_{on}}{T_{on} + 100 \times 10^{-6}}$$

$$\therefore \quad T_{on} = 200\ \mu\text{sec}$$

(ii) To obtain output voltage if T_{on} is halved

Chopper frequency is,

$$f = \frac{1}{T} = \frac{1}{T_{on} + T_{off}}$$

$$= \frac{1}{200\ \mu s + 100\ \mu s} = \frac{1}{300 \times 10^{-6}}$$

Hence chopper period is,

$$T = \frac{1}{f} = 300 \times 10^{-6} \text{ sec}$$

Since T_{on} is halved, it will be,

$$T_{on} = \frac{200 \ \mu s}{2} = 100 \ \mu\text{sec}$$

$$\therefore \text{ duty cycle,} \quad \delta = \frac{T_{on}}{T} = \frac{100 \ \mu\text{sec}}{300 \ \mu\text{sec}} = \frac{1}{3}$$

Therefore new output voltage will be,

$$V_{o(av)} = \frac{V_s}{1-\delta} = \frac{220}{1-\frac{1}{3}} = 330 \text{ V}$$

Thus the output voltage is also reduced by half.

4.4.1 Use of Step Up Operation for Energy Transfer

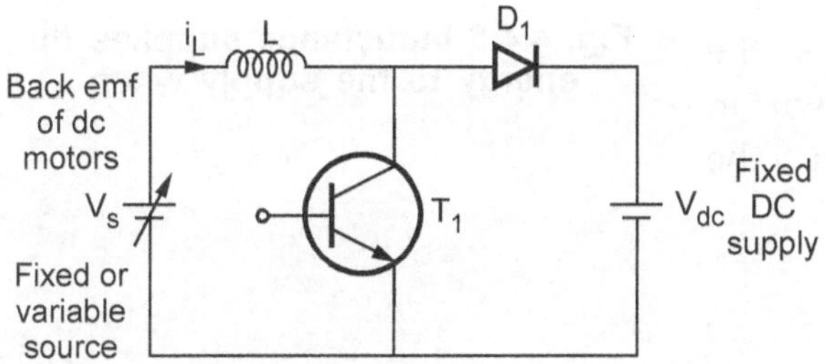

Fig. 4.4.4 Use of step up operation

Normally step up operation is used in chopper based dc motor drives for energy transfer. The energy is transferred from fixed or variable source (normally dc motors) to fixed dc source (i.e. supply). Fig. 4.4.4 shows the circuit diagram for such operation. The waveforms and functioning of this circuit is same as shown in Fig. 4.4.2.

The step up operation is normally used in chopper based dc drives. Let the voltage V_s be due to back emf of the motor. It will depend upon the speed of the motor. Normally during the braking the speed is decreased from the large values. The energy associated with the back emf (speed) of the motor is fedback to the dc supply (V_{dc}).

This operation is called *regenerative braking*. In regenerative braking, the braking energy is not wasted, rather it is fed back to the supply. This increases efficiency of the chopper drives.

When the switch is 'on', the current flows through the inductance. The loop formed is given in Fig. 4.4.5. The inductance stores the energy from the source V_s, when switch is 'on'. The KVL to the above loop gives,

$$V_s = \frac{d\,i_1(t)}{dt} \qquad \text{... (4.4.5)}$$

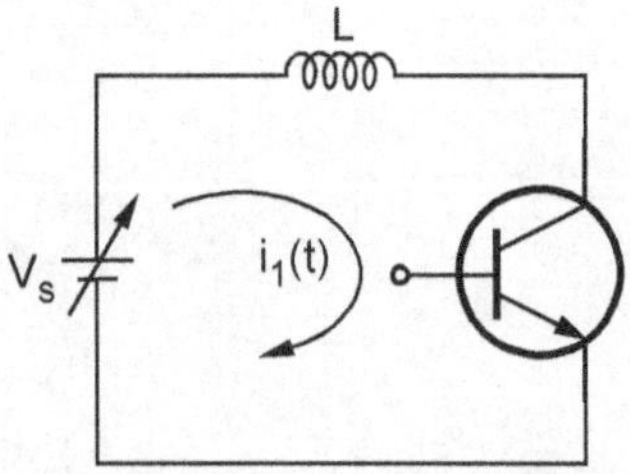

Fig. 4.4.5 Inductance stores the energy from V_s, when the switch is 'on'

If the inductance stores the energy, then it goes on saturating. Hence $i_1(t)$ goes on increasing. Hence $\frac{d\,i_1(t)}{dt}$ must be positive. i.e. from equation 4.4.5 we can write,

$$V_s = \frac{d\,i_1(t)}{dt} > 0$$

$$\therefore\ V_s > 0 \qquad \text{... (4.4.6)}$$

Now consider the situation when chopper switch is turned off. From Fig. 4.4.4 the equivalent circuit will be as shown in Fig. 4.4.6 below.

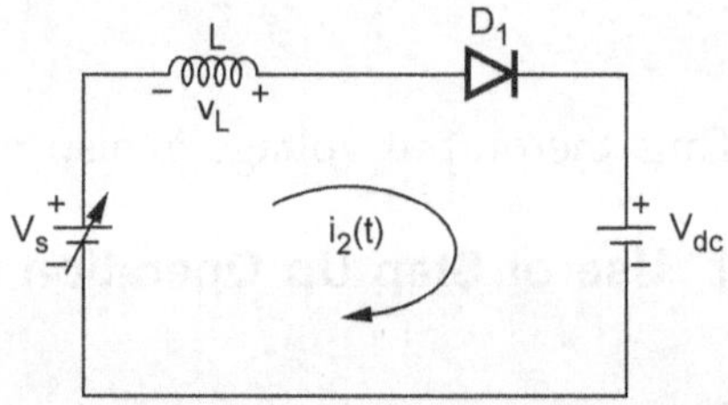

Fig. 4.4.6 Inductance supplies the energy to the supply when

As shown in the above figure, the current flows through inductance and diode D_1 to the supply V_{dc}. The inductance generates the voltage v_L with the polarity as shown in Fig. 4.4.6. The energy stored in the inductance is supplied to V_{dc}.

We can write KVL in Fig. 4.4.6 as,

$$V_s = L\frac{d\,i_2(t)}{dt} + V_{dc} \qquad \text{... (4.4.7)}$$

$$\therefore \qquad V_s - V_{dc} = L\frac{d\,i_2(t)}{dt} \qquad \text{... (4.4.8)}$$

Since the energy in the inductance is transferred to V_{dc}, current $i_2(t)$ goes on reducing. Hence $\frac{d\,i_2(t)}{dt}$ must be negative. i.e. above equation becomes,

$$V_s - V_{dc} = L\frac{d\,i_2(t)}{dt} < 0$$

$$\therefore \qquad V_s - V_{dc} < 0$$

$$\text{or} \qquad V_s < V_{dc} \qquad \text{... (4.4.9)}$$

Thus energy is transferred to V_{dc} when above condition is satisfied. We can combine above equation and equation 4.4.6, i.e.,

$$0 < V_s < V_{dc} \qquad \text{... (4.4.10)}$$

This is the condition for power transfer from fixed or variable source (V_s) to the fixed dc supply (V_{dc}).

Here note that step up operation is used for regenerative braking or to transfer power from motors to the dc supply. Condition of equation 4.4.10 must be satisfied to implement such operation.

Review Question

1. *Explain the operation of step up chopper with the help of waveforms.*

4.5 Chopper Classification

In the previous sections we discussed the basic principles of chopper operation. The choppers are classified depending upon the directions of current and voltage flows. These choppers operate in different quadrants of $v_o - i_o$ plane.

There are broadly following types of choppers :

(i) Class A chopper

(ii) Class B chopper

(iii) Class C chopper

(iv) Class D chopper

(v) Class E chopper

4.5.1 Class A Chopper

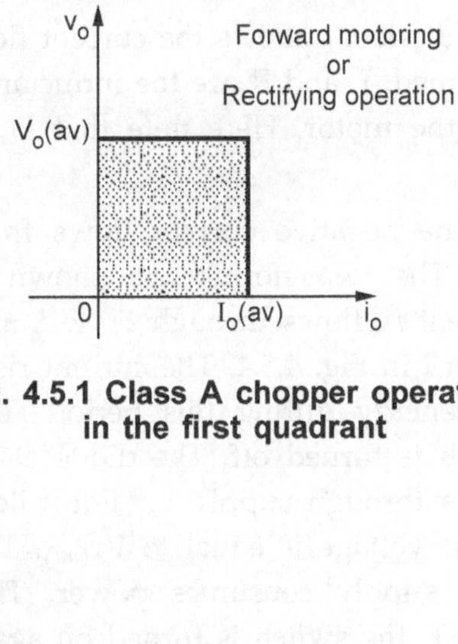

Fig. 4.5.1 Class A chopper operates in the first quadrant

The class A chopper operates in the first quadrant of $v_o - i_o$ plane as shown in Fig. 4.5.1. The output current and output voltage both are positive. These values never go negative. This type of chopper operates as the rectifier. The energy always flows from source to the load. Hence it is also called as forward motoring. The step down chopper discussed earlier is basically class A chopper.

4.5.2 Class B Chopper

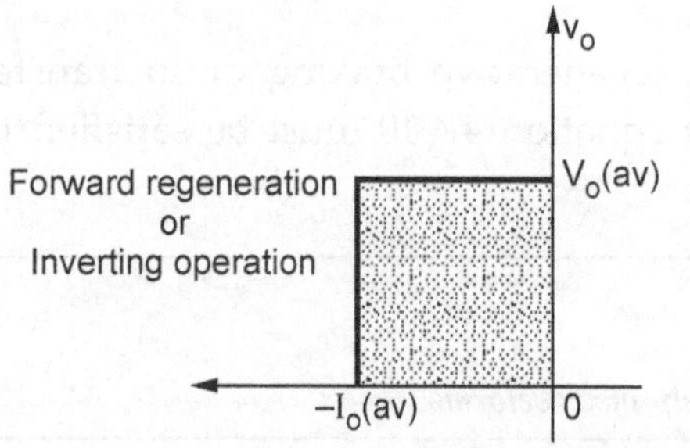

Fig. 4.5.2 Class B chopper operates in the second quadrant

The class B chopper operates in the second quadrant of the $v_o - i_o$ plane as shown in Fig. 4.5.2. The load voltage is positive and load current is negative. The load current flows out of the load. Since the current flows from load to the source, the energy is transferred from load to the source. This is also called as inverting operation. Such situation occurs during the braking of dc motor. The energy associated with back emf of the motor is fed back to the source. Hence current becomes negative. Since motor rotates in the same direction, it is called forward regenerative braking. The word *regeneration* means energy is fed from load to the source during braking. Fig. 4.5.3 shows the circuit diagram of class B chopper.

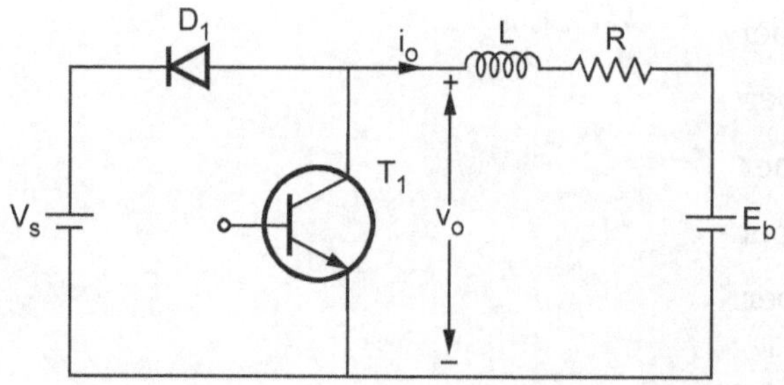

Fig. 4.5.3 Circuit diagram of class B chopper

In the above circuit, the supply voltage is V_s. The diode D_1 allows the current flow only from load to the source. Here motor load is assumed. L and R are the inductance and resistance of the motor. E_b is the back emf of the motor. Here note that E_b is responsible for negative current flow.

When the transistor (i.e. switch) is turned on, the negative current flows from E_b, R, L and T_1. This current keeps on increasing. The waveforms are shown in Fig. 4.5.4. The switch is 'on' from 0 to δT. The current i_o flows through E_b, R, L and switch in negative direction. Refer to equivalent circuit-I in Fig. 4.5.4. The current rises and reaches to I_{max} at δT. The inductance stores energy during this period. The energy is supplied by back emf E_b. At δT, the switch is turned off. The diode D_1 is forward biased and the negative output current flows through supply V_s. Equivalent circuit-II in Fig. 4.5.4 shows this situation. The output voltage is equal to $V_{o(av)}$. The current is forced through supply voltage V_s. Thus supply consumes power. This power is transferred from load inductance and E_b. At T, the switch is turned on again and the cycle repeats.

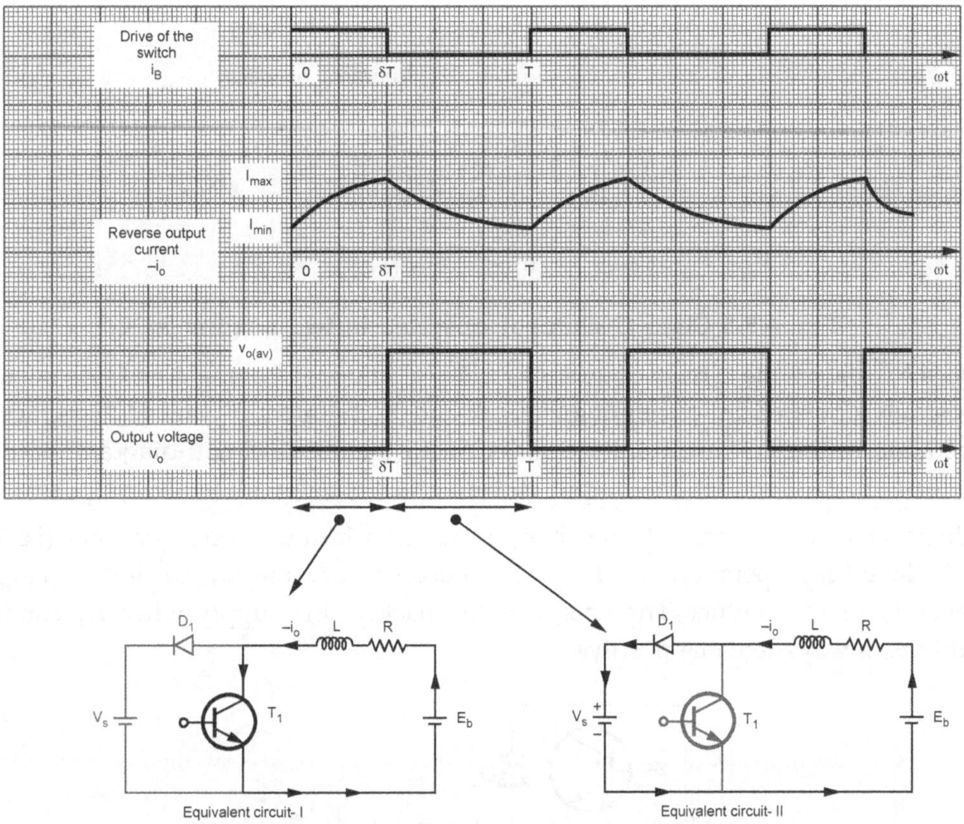

Fig. 4.5.4 Waveforms and operation of class B chopper

The equation for the output current can be obtained on the similar lines as done in example 4.3.1. The expressions for output current are given as follows :

$$i_o(t) = I_{\min} e^{-t\frac{R}{L}} - \frac{E_b}{R}\left(1 - e^{-t\frac{R}{L}}\right) \quad \text{for } 0 \le t \le \delta T \qquad \text{... (4.5.1)}$$

$$i_o(t) = I_{\max} e^{-t\frac{R}{L}} + \frac{V_s - E_b}{R}\left(1 - e^{-t\frac{R}{L}}\right) \text{for } \delta T \le t \le T \qquad \text{... (4.5.2)}$$

Here note that the class B operation is never implemented independently. It takes place in two and four quadrant chopper having motor loads.

4.5.3 Class C Chopper

The class C chopper operates in two quadrants. It is the combination of class A and B choppers. Fig. 4.5.5 shows the quadrants of operation of this chopper. It operates as a rectifier as well as invertor. In the first quadrant forward motoring takes place and in the second quadrant forward regenerative braking takes place.

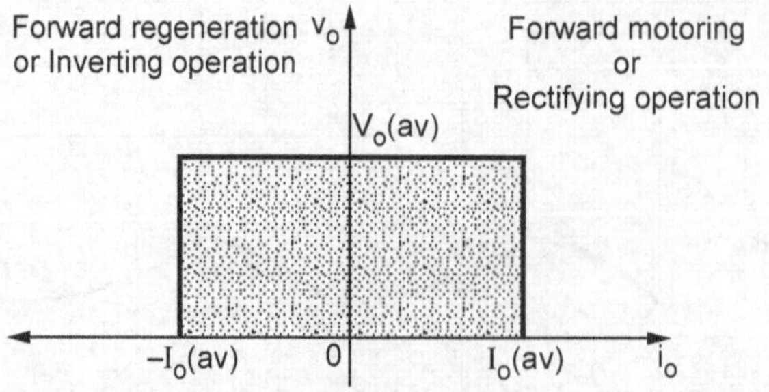

Fig. 4.5.5 Class C chopper operates in two quadrants

Fig. 4.5.6 shows the circuit diagram of class C chopper having transistor switches. It is basically obtained by combining class A and B choppers. T_1 and D_2 conducts for the operation in the first quadrant (i.e. class A). In the above circuit diagram, note that whenever T_1 or D_2 conduct, the output current and voltage will be always positive.

Whenever T_2 or D_1 conduct, the chopper operates in the second quadrant (i.e. class B). It is inverting operation. In Fig. 4.5.6 observe that output current is negative whenever T_2 or D_1 conduct. The energy is fed back to the supply when D_1 conducts. Note that v_o always remains positive.

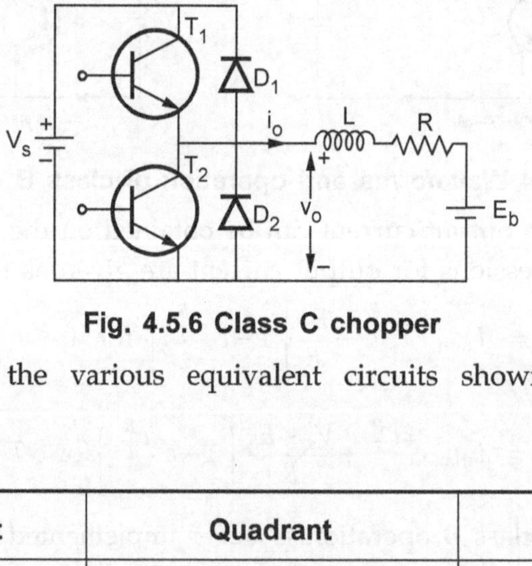

Fig. 4.5.6 Class C chopper

Table 4.5.1 shows the various equivalent circuits showing operation of this chopper.

Equivalent circuit	Quadrant	Description
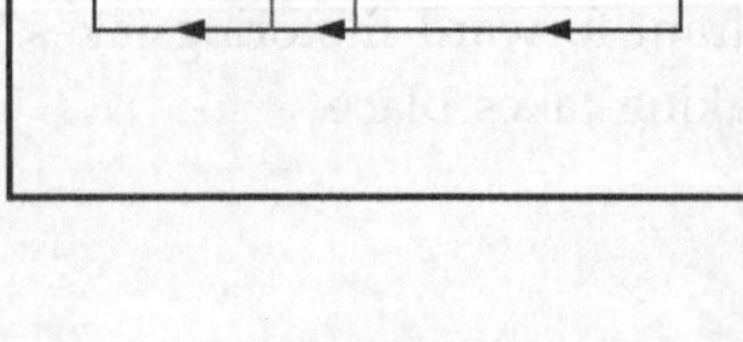	I Forward motoring or rectifying	T_1 conducts and energy flows from source to load. v_o and i_o are positive.

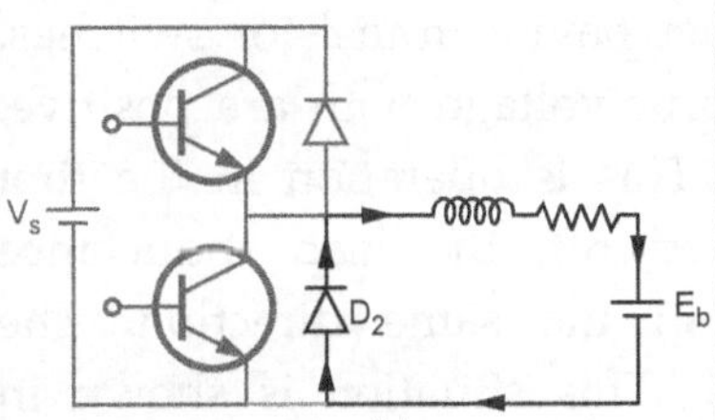 (b)	I Forward motoring or rectifying	D_2 conducts i_o is positive and v_o is zero. Freewheeling takes place. Inductance supplies energy to the load.
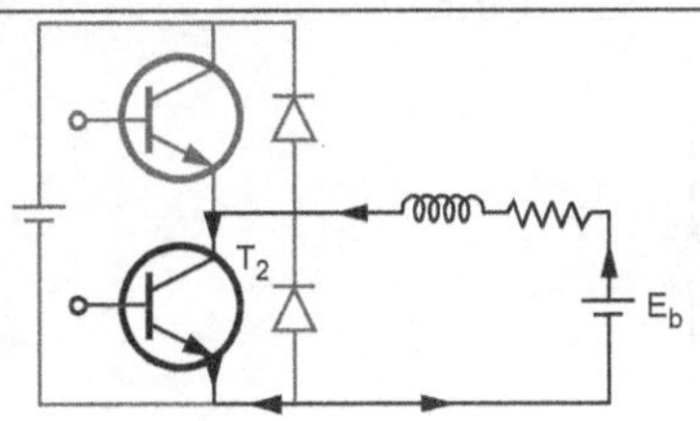 (c)	II Forward regenerative braking	T_2 conducts i_o is negative. E_b supplies energy to the inductance v_o is zero.
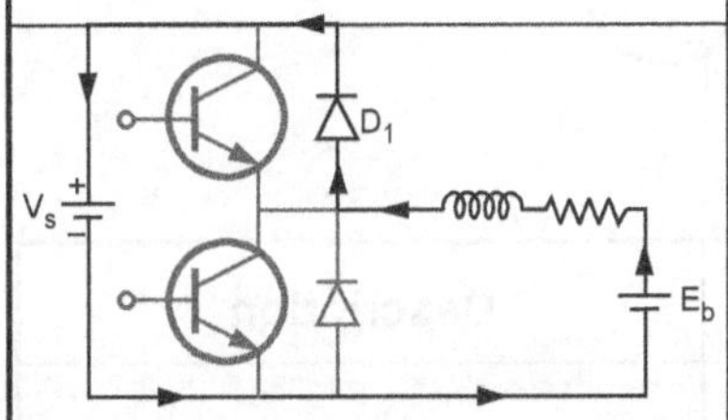 (d)	II Forward regenerative braking or inverting	D_1 conducts inductance supplies energy to the source v_o is positive and i_o is negative.

Table 4.5.1 Operation of class C chopper

4.5.4 Class D Chopper

The class D chopper also operates in two quadrants. Fig. 4.5.7 shows the quadrants of operation. The output current is always positive. The output voltage can be positive or negative. When v_o and i_o both are positive, the rectifying operation takes place. It is also called forward motoring. When the voltage is reversed, inverting operation takes place. The energy is fed from load to the source. The IVth quadrant operation is also called reverse regeneration.

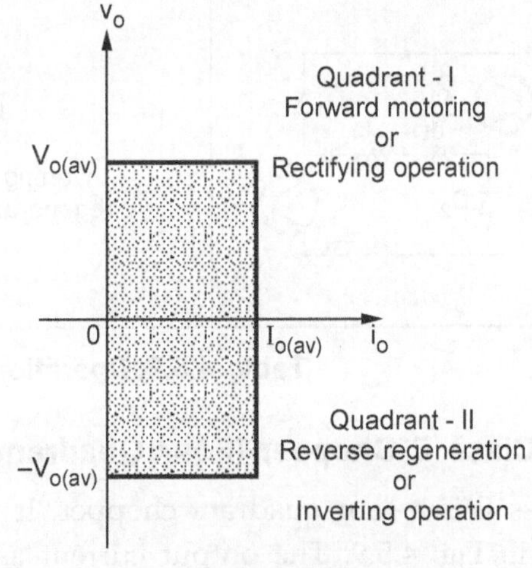

Fig. 4.5.7 Class D chopper operates in I and IV quadrants

Fig. 4.5.8 shows the circuit diagram of class D chopper having transistor switches. When T_1 and T_2 are conducting, output current and output voltage both are positive. Power is taken from the source and given to the load. This is operation in the first quadrant, i.e. rectifying. When T_1 and T_2 are switched off, the load inductance generates the large voltage to maintain the current in the same direction. The inductance voltage forward biases diodes D_1 and D_2. This situation is shown in Table 4.5.2. The diodes conduct and supply energy from load to the source. The output voltage is negative. Hence chopper operates in IVth quadrant.

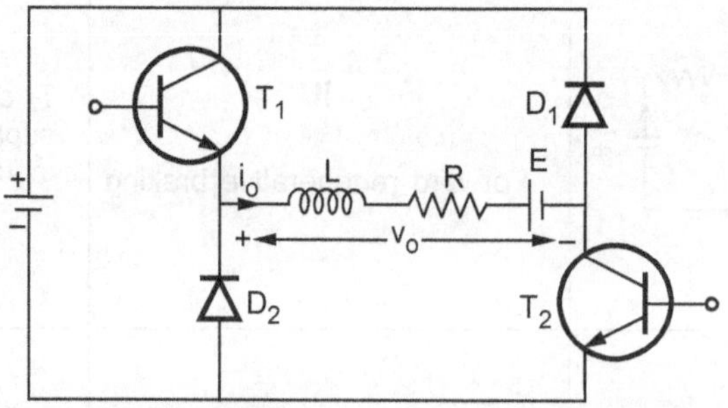

Fig. 4.5.8 Class D chopper

Equivalent circuit	Quadrant	Description
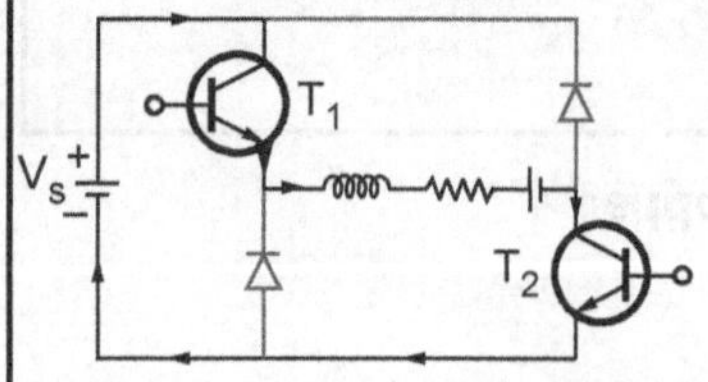	**I** Rectifying or forward motoring	T_1, T_2 conduct power flows from source to the load
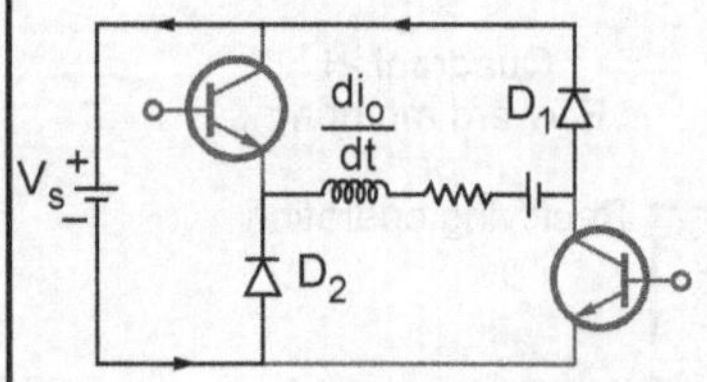	**IV** Inverting or reverse regenerative braking	i_o is maintained in the same direction by inductance voltage $L\frac{di_o}{dt}$. Diodes D_1 and D_2 are forward biased and conduct.

Table 4.5.2 Operation of class D chopper

4.5.5 Class E Chopper (Four Quadrant Chopper)

Class E is a four quadrant chopper. It operates in the four quadrants of $v_o - i_o$ as shown in Fig. 4.5.9. The output current as well as voltage both can take positive or negative values. The first quadrant is forward motoring. The output voltage and current both are positive. The IIIrd quadrant is reverse motoring. This means the motor rotates in the opposite direction compared to first quadrant. Since the power flows from source to the load in IIIrd quadrant, it is called rectifying operation. In IInd and IVth quadrant, the power flows from load to the source hence it is called inverting operation.

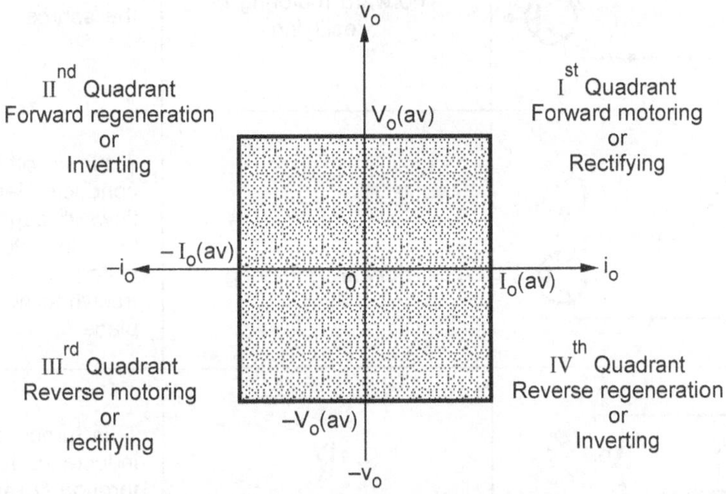

Fig. 4.5.9 Four quadrant operation

Fig. 4.5.10 shows the circuit diagram of the four quadrant chopper having transistor switches. Whenever T_1 and T_2 conducts the chopper operates in the I^{st} quadrant. v_o and i_o both are positive. When T_3 and T_4 conduct v_o and i_o both are negative. The chopper operates in the third quadrant. Table 4.5.3 shows the operation of this chopper with equivalent circuits.

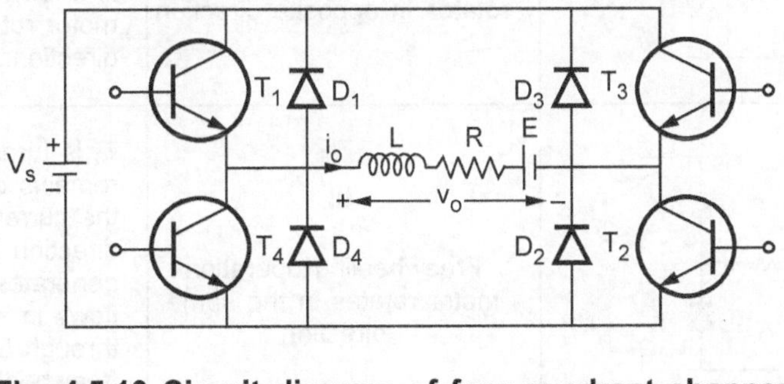

Fig. 4.5.10 Circuit diagram of four quadrant chopper

Equivalent circuit	Quadrant	Description
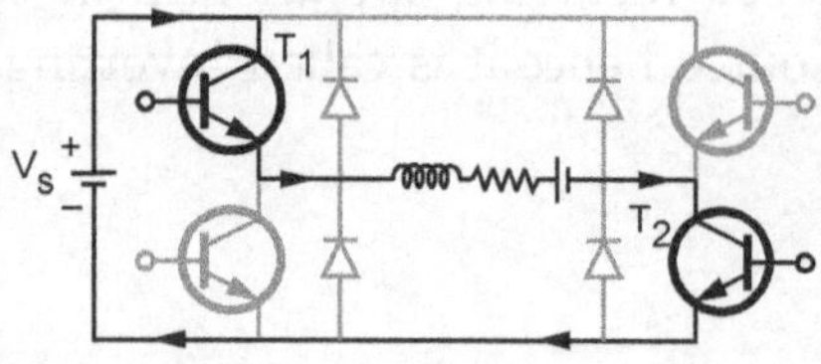	I Forward motoring or rectifying	T_1 and T_2 conducts load consumes the power from the source
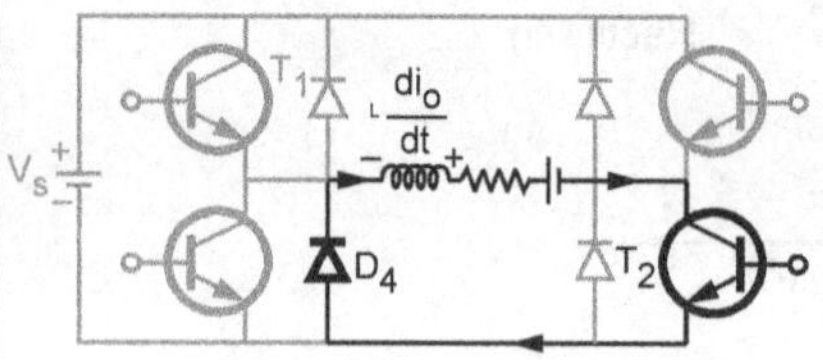	I Forward motoring	T_1 turned off but T_2 conducts Hence current i_o flows through T_2 and D_4. Load inductor induces voltage as shown freewheeling action takes place.
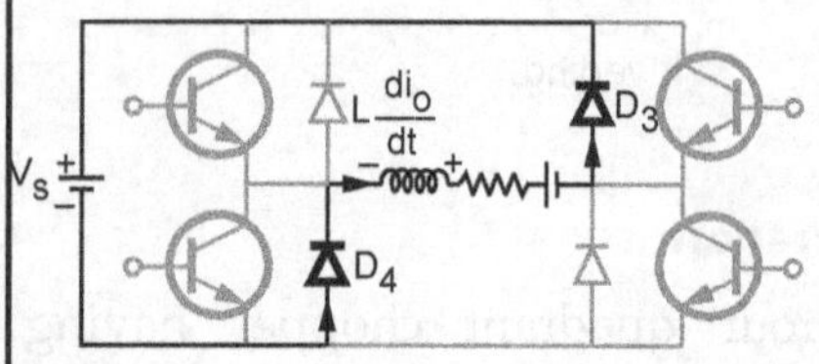	IV Inverting operation i_o positive v_o negative.	T_2 is turned off. Hence inductance forces current through D_3 and D_4. This current flows through supply. load energy is fed to the supply.
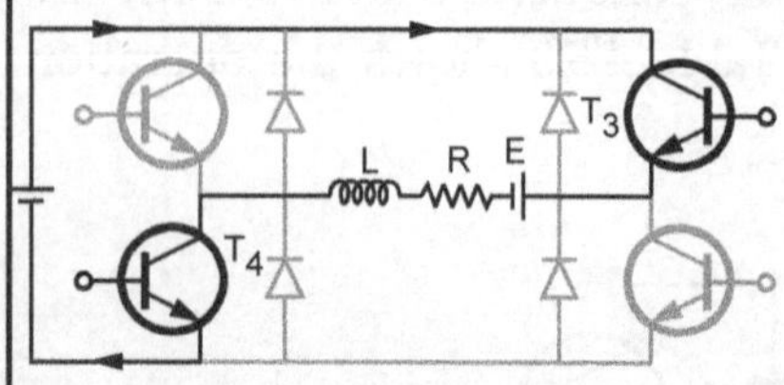	III Rectifying operation motor rotates in opposite direction	To reverse the direction of rotation of the motor, T_3 and T_4 are turned on. v_o and i_o both are negative. E is shown negative since motor rotates in opposite direction.
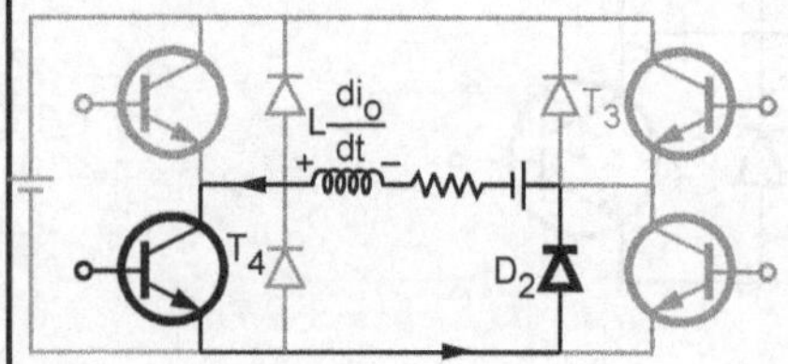	III Freewheeling operation motor rotates in the same direction	T_3 is turned off, but T_4 remains on. To maintain the current in the same direction inductance generates voltage and i_o flows in same direction through D_2 and T_4 This is freewheeling action.

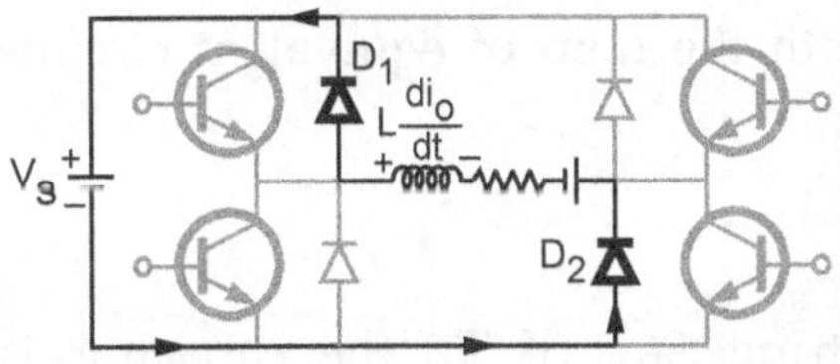	II Inverting operation i_o negative v_o positive	T_4 is turned off. Hence inductance forces current through D_1 and D_2. This current flows through supply. load energy is fed to the supply.

Table 4.5.3 Operation of four quadrant chopper

Four quadrant chopper has the capability to operate in all the four quadrants. Hence it is used in reversible dc drives. The braking is regenerative. Hence Four quadrant chopper drives are highly efficient. Their dynamic response is also fast.

Review Questions

1. *How choppers are classified ?*
2. *Explain two quadrant and four quadrant choppers.*

4.6 Voltage Commutated Thyristor Chopper

We know that thyristors have DC supply. In chopper circuits the switches are turned on and off to get the required output voltage. When thyristors are used in the chopper circuits, they are to be turned off by commutation circuits. Hence chopper must include commutation circuit also. Fig. 4.6.1 shows the circuit diagram of the thyristor chopper which uses voltage commutation. As shown in this circuit, T_1 is the main current carrying thyristor. T_2 is the auxiliary thyristor. It is used to turn-off main thyristor. C, L_1 and D_1 are the commutating components. L_s is the inductance of the DC source. D_{FW} is the freewheeling diode.

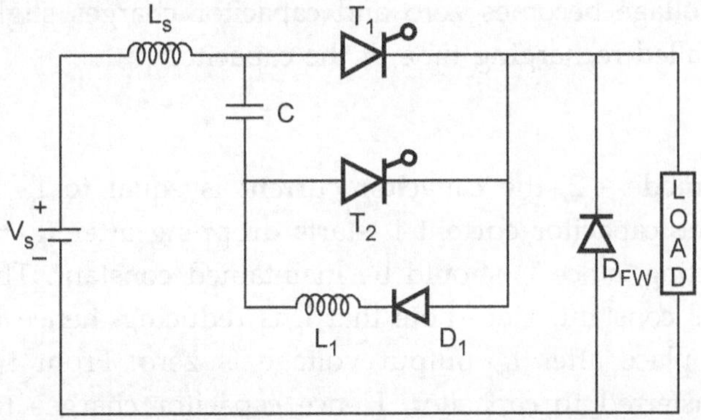

Fig. 4.6.1 Impulse commutated thyristor chopper

4.6.1 Operation of the Circuit

The operation of the chopper can be explained with the help of equivalent circuits and waveforms shown in Fig. 4.6.2.

Mode - 1

Thyristor T_1 is triggered at t_1. At the time of triggering of T_1, the capacitor is charged to supply voltage V_s. The voltage polarities as shown in equivalent circuit 1 in Fig. 4.6.2. That is, upper plate is positive with respect to lower plate. When T_1 is triggered, it starts conducting and load current flows through it. It is assumed that load current I_o is constant. The capacitor now starts discharging through T_1-D_1-L_1-C. Observe that if we neglect the drops between T_1 and D_1 it becomes an LC circuit. Hence the capacitor charges to opposite polarity. At t_2, the capacitor current becomes zero. The capacitor does not discharge again, since its current is blocked by diode D_1. The time period $t_2 - t_1 = t_r$ is called commutation readiness time of the chopper. Note that commutation of T_1 cannot be initiated before t_2. The load current then continues to flow through T_1.

Mode - 2

At t_3, an auxiliary thyristor T_2 is triggered to commutate T_1. The capacitor voltage is applied across T_1 and it is reverse biased. Therefore T_1 current becomes zero at t_3. The load current starts flowing through L_s-C-T_2. This current is constant. In the waveforms, observe that the output voltage becomes $v_o + v_c$ due to capacitor voltage. The output voltage linearly starts decreasing since capacitor charges in opposite direction with constant I_o. This mode is shown in Fig. 4.6.2 by equivalent circuit-2. The capacitor discharges through the load from $-V_s$ to 0. The time of this discharge is $t_4 - t_3 = t_{off}$. It is called circuit turn-off time. A reverse bias is placed across T_1 during this time. Hence T_1 must be turned off before t_4. For successful commutation $t_q > t_{off}$.

From t_4 to t_5 the capacitor charges towards positive voltage (upper plate positive). At t_5, the output voltage becomes zero and capacitor charges slightly above V_s. The time $t_5 - t_4 = t_d$ is called recharging time of the capacitor.

Mode - 3

At the end of mode - 2, the capacitor current is equal to I_o. Due to rise in the capacitor voltage, the capacitor current i_1 starts dropping after t_5. Hence freewheeling diode starts conducting, since I_o should be maintained constant. Thus i_c and i_{FW} both flows such that I_o is constant. Here note that i_c is reducing, hence i_{FW} increases. Since freewheeling takes place after t_5, output voltage is zero. From t_5 to t_6, the energy stored in L_s is transferred to capacitor. Hence capacitor charges to a voltage higher than supply. At t_6, the capacitor current becomes zero, and I_o flows completely through freewheeling diode. Since capacitor current becomes zero, T_2 turns-off.

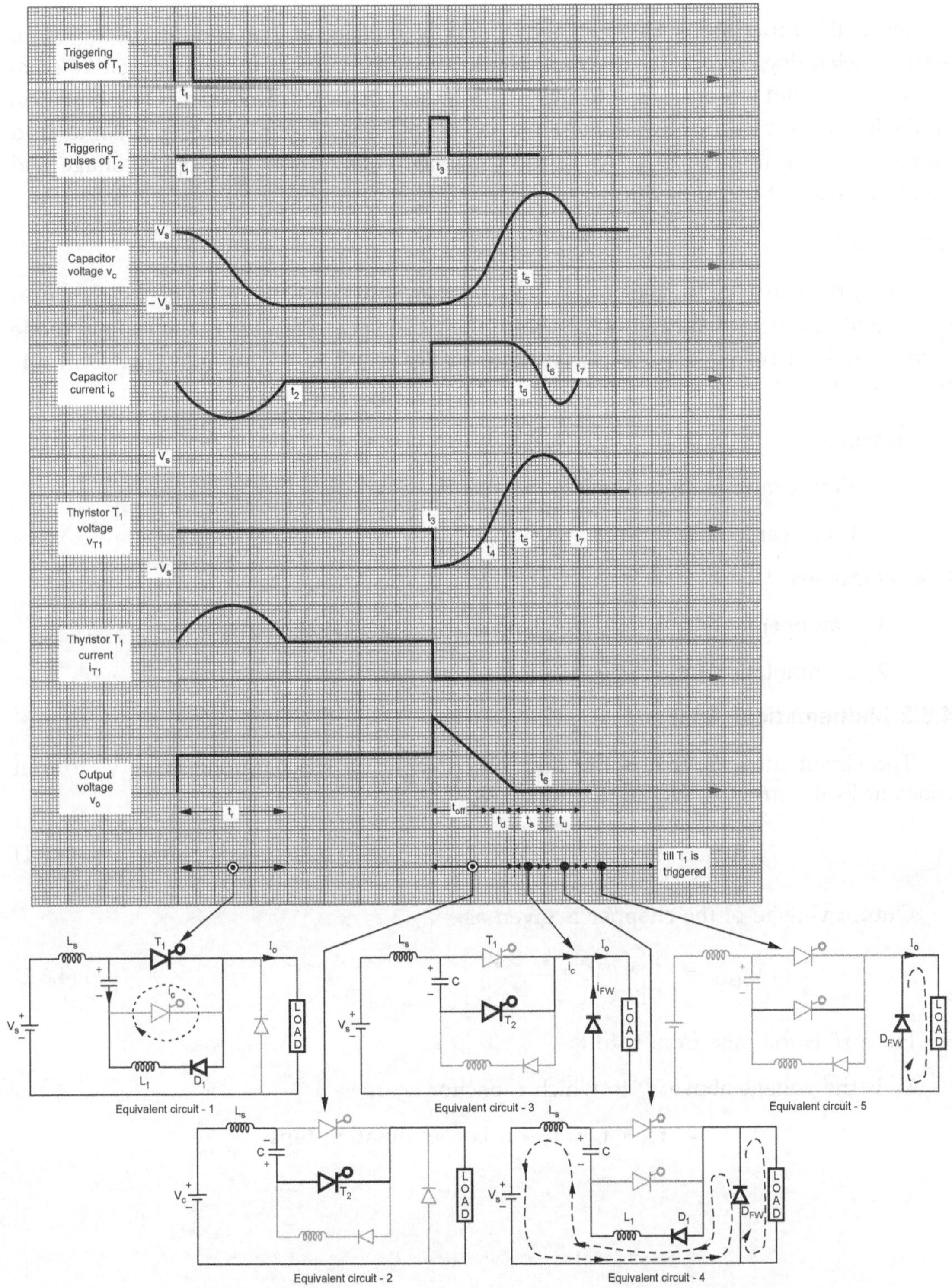

Fig. 4.6.2 Operation of impulse commutated thyristor chopper

Mode - 4

Since the capacitor is charged above supply voltage at the end of mode - 4, it starts discharging in reverse direction. Hence capacitor current becomes negative from t_6. The capacitor discharges through L_S-V_S-D_{FW}-D_1-L_1-C. This path is shown in equivalent circuit - 4 in Fig. 4.6.2. At the same time freewheeling takes place and load current I_o flows through D_{FW}. At t_7, the capacitor voltage falls to supply voltage and discharge current becomes zero.

Mode - 5

At t_7, the capacitor current is zero. Its voltage is V_s with the polarity as shown in equivalent circuit - 5 in Fig. 4.6.2. Both the SCRs are off. The freewheeling diode carries the load current till SCR T_1 is triggered again. Mode - 1 repeats when SCR - T_1 is triggered.

Advantages

1. Very simple in operation.
2. It can carry high powers.

Disadvantages

1. Can operate only at low frequencies.
2. Commutation time depends on load current.

4.6.2 Mathematical Analysis

The circuit turn-off time is the time when capacitor discharges from $-V_s$ to 0 at constant load current I_o. Hence it is given as,

$$t_{off} = \frac{V_S C}{I_o} \quad \text{... (4.6.1)}$$

Output voltage of the chopper is given as,

$$V_{o(av)} = \frac{1}{T}\left[\delta T\, V_S + \frac{V_C + V_S}{2t_C}\right] \quad \text{... (4.6.2)}$$

Here δT is the time from t_1 to t_3.

V_c is the voltage above V_s to which capacitor charges.

$$t_c = t_{off} + t_d = t_5 - t_3 \text{ is commutation time.}$$

4.6.3 Jones Chopper

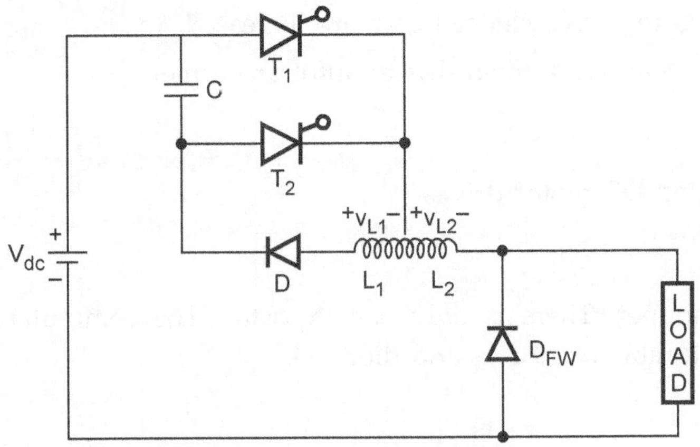

Fig. 4.6.3 Jones chopper

Observe that the above circuit is similar to that of Fig. 4.6.1. The only difference is that of inductances L_1 and L_2. Actually it is a single coil having one center tap. We know that capacitor changes to a negative voltage due to L_1 and D_1. The energy stored in L_1 is used to overcharge C in opposite direction.

In the above circuit whenever load current flows through 'L_2', the voltage is also induced in 'L_1'. The polarity of these voltages is as shown in Fig. 4.6.3. The polarity of voltage across L_1 (i.e. v_{L1}) is such that it charges the capacitor 'C' to higher negative voltage. This current path is shown below.

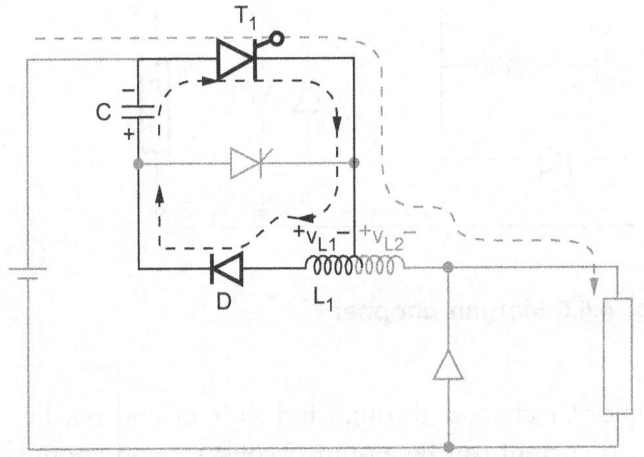

Fig. 4.6.4 Overcharging of capacitor due to autotransformer L_1-L_2

The overcharge on the capacitor increases the negative voltage. The discharge of this voltage is blocked by diode D. This capacitor voltage provides reliable turn-off of the main SCR (T_1). The waveforms of this circuit are same as that given in the Fig. 4.6.2. The only difference will be that, negative capacitor volatge will be higher than $-V_s$ between t_2 and t_3.

Advantages

(i) Reliable turn-off of main SCR due to auto transformer.

(ii) The auto transformer becomes the part of load. This provides filtering action.

Disadvantages

(i) Bulky autotransformer is required. Hence losses are increased.

(ii) Higher operating frequencies are limited due to autotransformer.

Application

It is used for separately excited DC motor drives.

4.6.4 Morgan Chopper

Fig. 4.6.5 shows Morgan chopper. There is only one thyristor. The commutating components are capacitor C_1, saturable reactor L_1 and diode D.

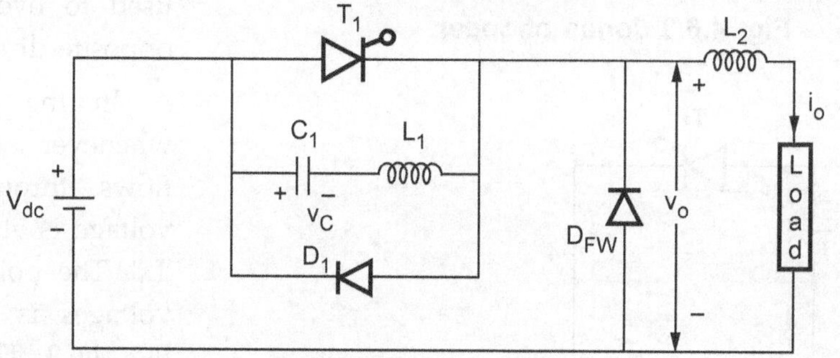

Fig. 4.6.5 Morgan chopper

Operation and Waveforms

Initially, when T_1 is off, capacitor C_1 charges through inductor L_1 and reaches to a steady state voltage of V_{dc}. The load current can be assumed constant and continues to flow through freewheeling diode D_{FW}

At t_1 SCR T_1 is triggered. Hence capacitor starts discharging through $C_1 - T_1 - L_1$. This forms L_1C_1 tank circuit. At t_2, the capacitor charges to negative peak value and capacitor current tries to flow in flow in opposite direction. This puts reverse voltage across T_1 and turns it off. The capacitor current then flow through D_1.

At t_3 the capacitor again charges to + V_{dc} and formed biases T_1. The output voltage remains positive from t_1 to t_2. The cycle repeats when SCR T_1 is again triggered at t_4.

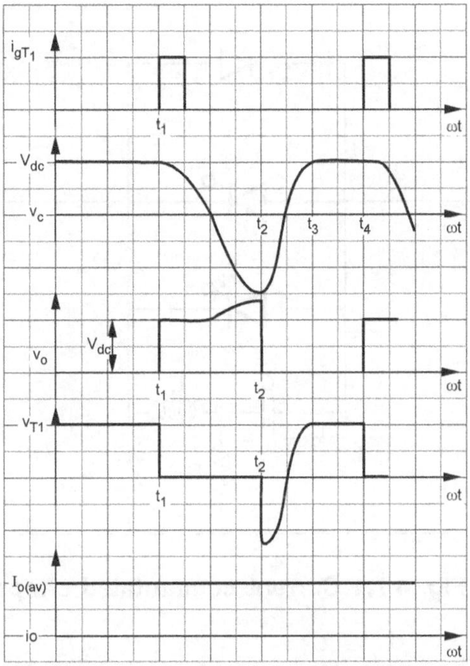

Fig. 4.6.6 Waveforms of Morgan chopper

Advantages

i) Low cost circuit since only one used.

ii) Operation of the circuit is simplified.

Disadvantages

i) The conduction of T_1 depends upon L_1 and C_1

ii) Freqeuncy of the chopper is to be adjusted to control output voltage.

4.7 Current Commutated Chopper

Fig. 4.7.1 shows the circuit diagram of current commutated chopper. The main current carrying thyristor is turned off by injecting reverse current pulse. T_1 is main load current carrying SCR and T_2 is auxiliary SCR. D_f ,D_s, C and L are commutating components.

The operation of the circuit can be explained through following modes :

Mode I

At the begining of this mode, it is assumed capacitor is charged to $+V_{dc}$ with polarities as shown in Fig. 4.7.1. The load current is being carried by T_1. The thyristor T_2 is triggered at the instant 't_1'. This shorts the LC circuit as shown in equivalent

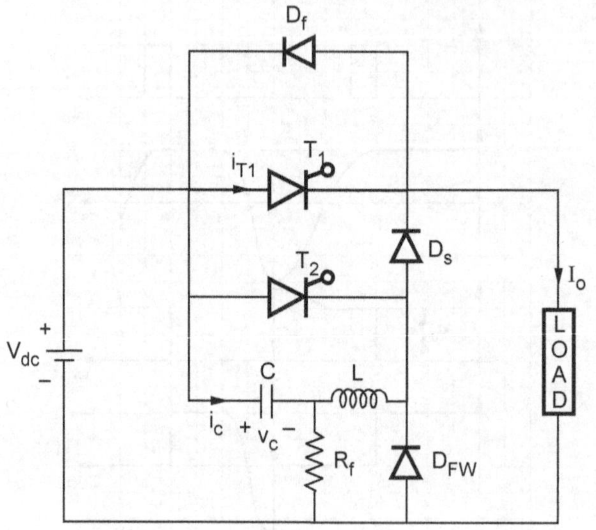

Fig. 4.7.1 Current commutated chopper

circuit-I of Fig. 4.7.2. The capacitor current starts flowing in opposite direction. Therefore capacitor voltage also starts falling. At 't_2' the capacitor current is zero, hence T_2 turns off. The capacitor voltage reaches to negative maximum as shown in waveform of Fig. 4.7.2 (c). This mode ends at 't_2' when SCR T_2 turns-off.

Mode II

This mode begins when capacitor current reverses its direction (positive) at 't_2'. This capacitor current flows in reverse direction through T_2 as shown in equivalent circuit-II of Fig. 4.7.2. When $i_c = I_0$, no current flows through T_1 and it turns-off. The load current flows through C-L-D_s-load-V_{dc}. This mode ends at 't_3' when SCR T_1 turns-off.

Mode III

When SCR T_1 turns-off, diode D_f is forward biased and it starts conducting. The constant load current continues to flow through C-L-D_s-load-V_{dc}. And excess capacitor current flows through the loop formed by C-L-D_s-D_f as shown in equivalent circuit-III of Fig. 4.7.2. This mode ends at t_3 when capacitor current is again equal to load current I_o.

Mode IV

When capacitor current is equal to I_o, diode D_f stops conducting. The complete load current now flows through C-L-load-V_{dc} loop as shown in equivalent circuit-IV of Fig. 4.7.2. The capacitor starts charging above supply voltage. Hence capacitor current starts falling. Therefore freewheeling diode D_{FW} is forward biased and starts carrying load current after t_4. This mode ends at t_5 when capacitor current is zero.

Fig. 4.7.2 (d) shows the output voltage waveform of the chopper. After the turn-off period, T_1 is triggered again and the cycle repeats. The capacitor discharges to V_{dc} level through R_f-C-V_{dc}. Therefore the capacitor voltage is equal to + V_{dc} as assumed at the begining of mode-I.

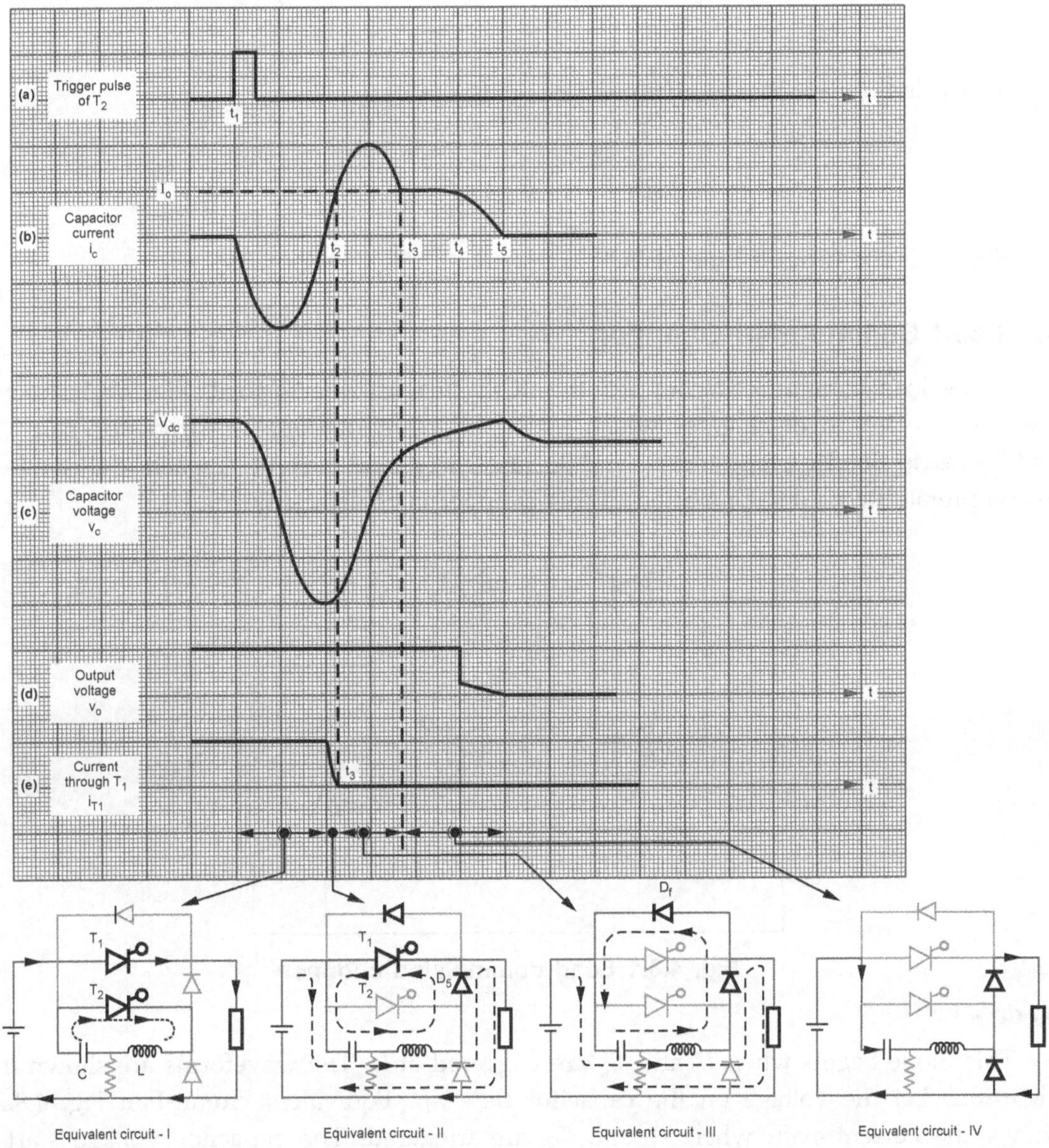

Fig. 4.7.2 Waveforms of current commutated chopper

Mathematical Analysis

Let 'x' denote the ratio of peak capacitor current to peak output current. i.e.,

$$x = \frac{I_{c(peak)}}{I_{o(peak)}}$$

Then values of inductance is given as,

$$L = \frac{V_{in}\, t_q}{I_c\left(\pi - 2\sin^{-1}\frac{1}{x}\right)}$$

And value of capacitor is given as,

$$C = \frac{I_c\, t_q}{\left(\pi - 2\sin^{-1}\frac{1}{x}\right)V_{dc}}$$

Here t_q is the turn-off time of SCR T_1.

4.8 Load Commutated Chopper

In the load commutated chopper, the SCRs turn-off due to load. Fig. 4.8.1 shows the circuit diagram of chopper which uses load commutation. The thyristor pair T_1 and T_2 carries load current and T_3 and T_4 also carry load current. Capacitor C is used to commutate thyristors. Operation of this chopper is explained next.

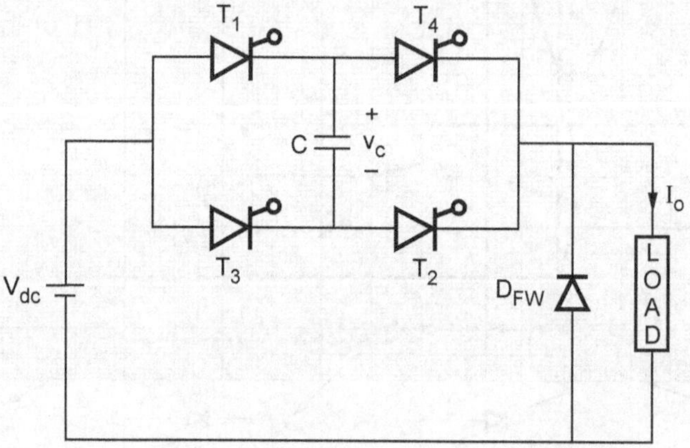

Fig. 4.8.1 Load commulated chopper

Mode - I

This mode begins when T_1 and T_2 are triggered at t_1. The waveforms are shown in Fig. 4.8.2. Let the voltage on the capacitor be $-V_{dc}$. Equivalent circuit-I in Fig. 4.8.2 shows the current path when T_1 and T_2 are triggered. The capacitor voltage starts rising towards $+V_{dc}$. The load voltage is 2 V_{dc} when T_1 and T_2 are triggered. The load voltage goes on reducing as capacitor charges. When capacitor voltage becomes equal to V_{dc}, current becomes zero and T_1 and T_2 turn-off.

Mode-II

When T_1 and T_2 turn-off at 't_2', the freewheeling diode is forward biased and it starts conducting. Equivalent circuit-II shows freewheeling action.

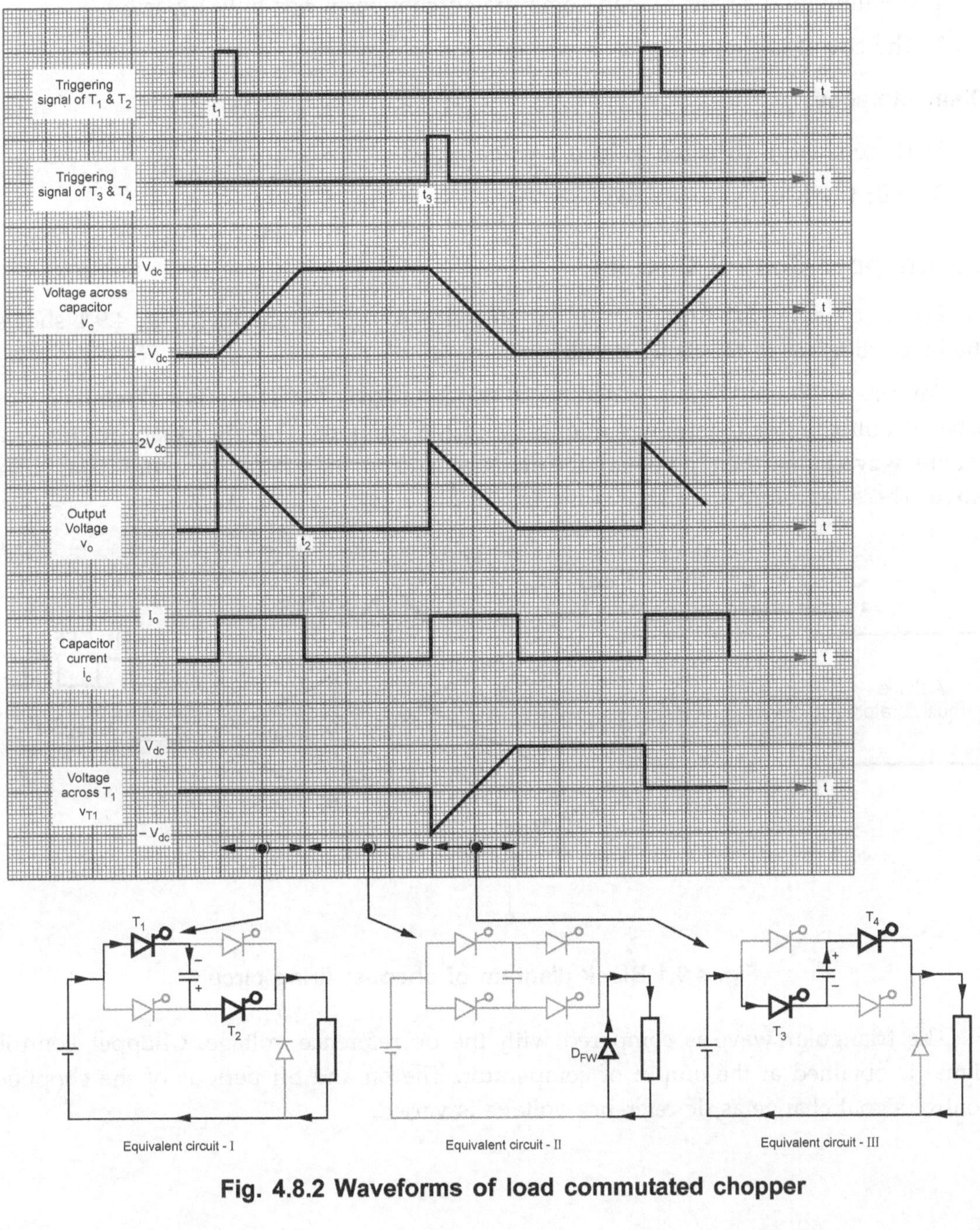

Fig. 4.8.2 Waveforms of load commutated chopper

Mode-III

T_3 and T_4 are triggered at t_3. Hence capacitor voltage adds to supply voltage and load voltage becomes 2 V_{dc}. The load current is carried by T_3 and T_4. Equivalent circuit-III in Fig. 4.8.2 shows the conduction in this mode.

Advantages

1. Commutation circuit of load commutated chopper needs only capacitor.
2. The circuit is simple.

Disadvantages

1. There is large variation in load voltage.
2. Four thyristors are required.

4.9 Chopper Control Circuits

Let us consider the firing circuit for step down, type A chopper. Fig. 4.9.1 shows the block diagram of the firing circuit.

An astable multivibrator generates a trigger signal for D-flipflop. Frequency of astable multivibrator determines frequency of the chopper. The D-flipflop output is a square wave. It is then given to the integrator. The integrator generates triangular wave. These waveforms are shown in Fig. 4.9.2.

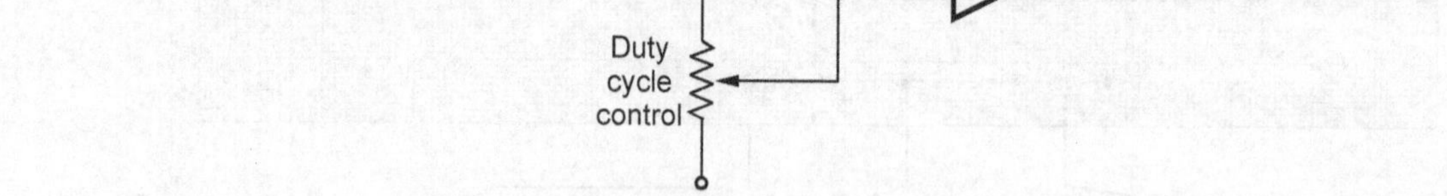

Fig. 4.9.1 Block diagram of chopper firing circuit

The triangular wave is compared with the dc reference voltage. Chopper control signal is obtained at the output of comparator. The on and off periods of the chopper control signal change as dc reference voltage is varied.

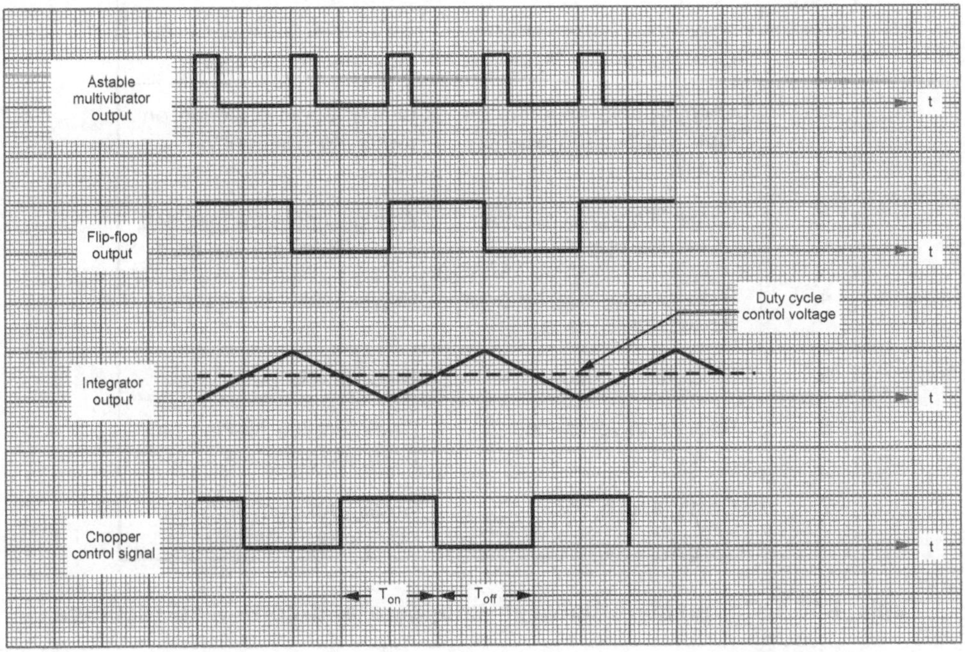

Fig. 4.9.2 Waveforms of chopper firing circuit

4.10 Applications of Choppers

Choppers are used in following applications :

i. DC motor drives when the DC supply is available.

ii. Battery operated vehicles.

iii. Switched mode power supplies.

iv. Battery charges where uncontrolled rectifiers give DC to choppers.

v. Traction drives use four quadrant choppers for energy saving.

vi. Lighting and lamp controls also prefer choppers.

Table 4.10.1 lists the comparison between chopper and controlled rectifiers :

Sr. No.	Controlled rectifiers	Choppers
1.	Operate on AC input	Operate on DC input
2.	Output ripple depends on input supply and type of rectifier	Output ripple depends on switching frequency.
3.	Large filters are required to reduce ripple	At high ripple frequencies, small filters are required.
4.	Power factor can be poor for inductive loads.	Power factor is better.
5.	Efficiency is poor	Efficiency is better.
6.	Regenerative braking is slow	Regenerative braking can be made effective.
7.	Costly and bulky	Small and compact, but circuits are complex.

Table 4.10.1 Comparison between choppers and controlled rectifiers

Summary

1. Step down chopper with resistive load

Average output voltage, $V_{o(av)} = \delta V_s$

RMS output voltage, $V_{o(rms)} = \sqrt{\delta}\, V_s$

Load power, $P_o = \dfrac{\delta V_s^2}{R}$ } Neglecting drop in chopper.

Supply power $P_s = \dfrac{\delta V_s^2}{R}$

Output current, $I_{o(av)} = \dfrac{V_{o(av)}}{R}$

2. Step down chopper with RLE load

$$I_{min} = I_{max} e^{-(1-\delta)\frac{TR}{L}} - \frac{E}{R}\left(1 - e^{-(1-\delta)T\frac{R}{L}}\right)$$

$$I_{max} = I_{min} e^{-\delta T\frac{R}{L}} + \frac{V_s - E}{R}\left(1 - e^{-\delta T\frac{R}{L}}\right)$$

$$I_{o(p\text{-}p)} = I_{max} - I_{min}$$

$$V_{o(av)} = \delta V_s, \quad I_{o(av)} = \frac{V_{o(av)}}{R}$$

$$I_{o(rms)} = \left\{ I_{min}^2 + \frac{I_{o(p-p)}^2}{3} + I_{min} I_{o(p-p)} \right\}^{\frac{1}{2}}$$

$$I_{o(av)} = \frac{I_{max} + I_{min}}{2}$$

$$I_{T(rms)} = \sqrt{\delta}\, I_{o(rms)}$$

3. Step up chopper

Output average voltage, $V_{o(av)} = \dfrac{V_s}{1-\delta}$

❑❑❑

5

Inverters

Objectives

- *Principle of DC to AC conversion.*
- *1ϕ and 3ϕ bridge inverters.*
- *Half bridge and full bridge inverters with resistive and inductive loads.*
- *120° and 180° modes of operation of 3ϕ inverters.*
- *Harmonic analysis of output waveforms of inverters.*
- *Voltage control in inverters.*
- *Performance parameters for inverters.*

5.1 Introduction

The inverters are DC to AC converters. The DC source is normally a battery or output of the controlled rectifier. Inverters are used in induction motor drives, UPS, standby power supplies, induction heating etc.

- The output voltage waveform of the inverter can be square wave, quasi-square wave or low distorted sine wave.
- The output voltage can be controlled (i.e. adjustable) with the help of drives of the switches.
- The pulse width modulation (PWM) techniques are most commonly used to control the output voltage of inverters. Such inverters are called PWM inverters. The output voltage of the inverter contain harmonics whenever it is non sinusoidal. These harmonics can be reduced by using proper control schemes.
- The inverters can be classified as voltage source inverters or current source inverters. When input DC voltage remains constant, then it is called *voltage source inverter* (VSI) or *voltage fed inverter* (VFI). When input current is

maintained constant, then it is called *current source inverter* (CSI) or *current fed inverter* (CFI).

- Some times, the DC input voltage to the inverter is controlled to adjust the output. Such inverters are called variable DC link inverters.
- The inverters can have single phase or three phase output.

5.2 Half Bridge Inverter

5.2.1 Operation with Resistive Load

Fig. 5.2.1 shows the circuit diagram of single phase half bridge inverter. The two transistors T_1 and T_2 are used as switching devices. They can be MOSFET, GTO, SCR, IGBT etc. Fig. 5.2.2 shows the waveforms of the half bridge inverter having resistive load.

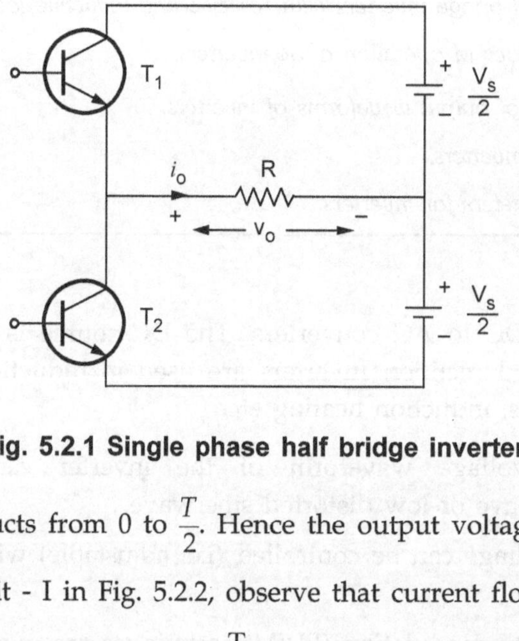

Fig. 5.2.1 Single phase half bridge inverter

Transistor T_1 conducts from 0 to $\frac{T}{2}$. Hence the output voltage is positive and it is $\frac{V_s}{2}$. In equivalent circuit - I in Fig. 5.2.2, observe that current flows from point A to B in the load. Transistor T_2 conducts from $\frac{T}{2}$ to T and T_1 is off. Equivalent circuit - II in Fig. 5.2.2 shows the situation when T_2 conducts. Current flows from point B to A in the load. The output voltage is $-\frac{V_s}{2}$. This is the negative half cycle of output. Since the load is resistive, output current waveform is same as voltage waveform. Fig. 5.2.2 also shows the currents through the transistors. The output of this inverter is a square wave. Since there are two transistors in the bridge, it is called half bridge inverter.

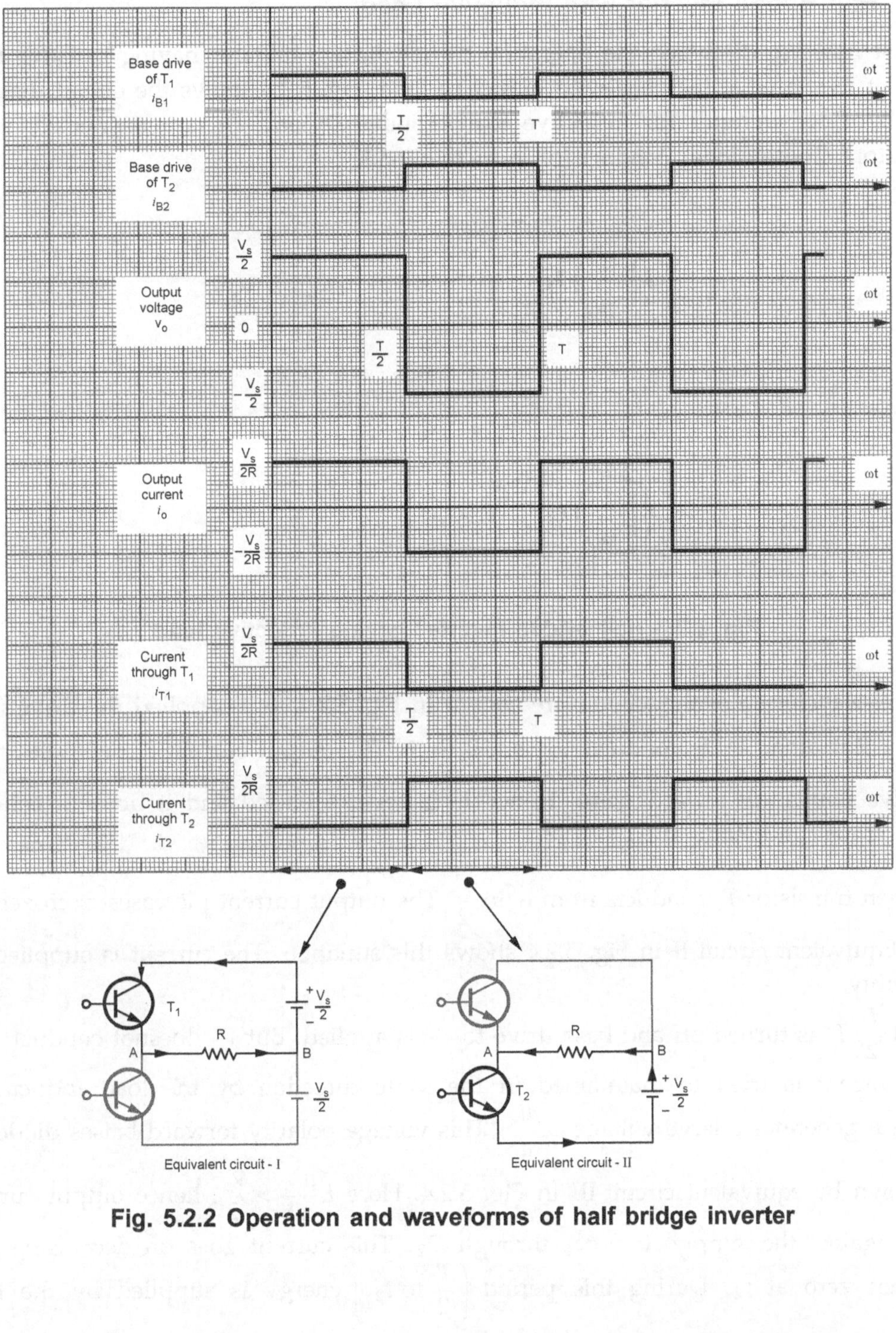

Fig. 5.2.2 Operation and waveforms of half bridge inverter

5.2.2 Half Bridge Inverter with Inductive Load

Just now we discussed the operation of half bridge inverter having resistive load. Now let us see what happens with inductive load. Fig. 5.2.3 shows the circuit diagram of half bridge inverter for inductive load. Diodes D_1 and D_2 are connected across transistors. These diodes conduct for inductive load.

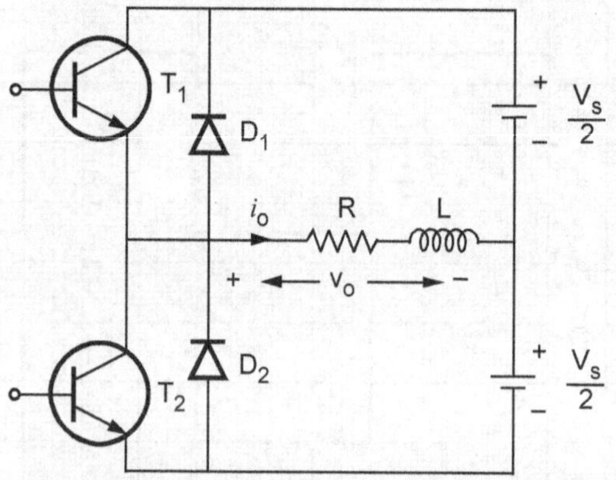

Fig. 5.2.3 Half bridge inverter having inductive load

The waveforms of this circuit are shown in Fig. 5.2.4. T_1 is applied the drive from 0 to $\frac{T}{2}$. But diode D_1 is conducting from 0 to t_1. The output current i_0 decreases from negative maximum towards zero. Hence T_1 is reverse biased and it doesnot conduct till D_1 stops conducting at t_1.

Then transistor T_1 conducts from t_1 to $\frac{T}{2}$. The output current increases from zero to I_{max}. Equivalent circuit-II in Fig. 5.2.4 shows this situation. The current is supplied by dc supply.

At $\frac{T}{2}$, T_1 is turned off and base drive to T_2 is applied. But T_2 doesnot conduct. The load current is tried to maintained in the same direction by the load inductance. Hence it generates a large voltage $L\frac{di_o}{dt}$. This voltage polarity forward biases diode D_2 as shown by equivalent circuit-III in Fig. 5.2.4. Here $L\frac{di_o}{dt} > \frac{V_s}{2}$, hence output current flows against the supply. It flows through D_2. This current goes on decreasing and becomes zero at t_2. During this period $\left(\frac{T}{2} \text{ to } t_2\right)$ energy is supplied by the load inductance to the DC supply. Hence it is also called feedback operation.

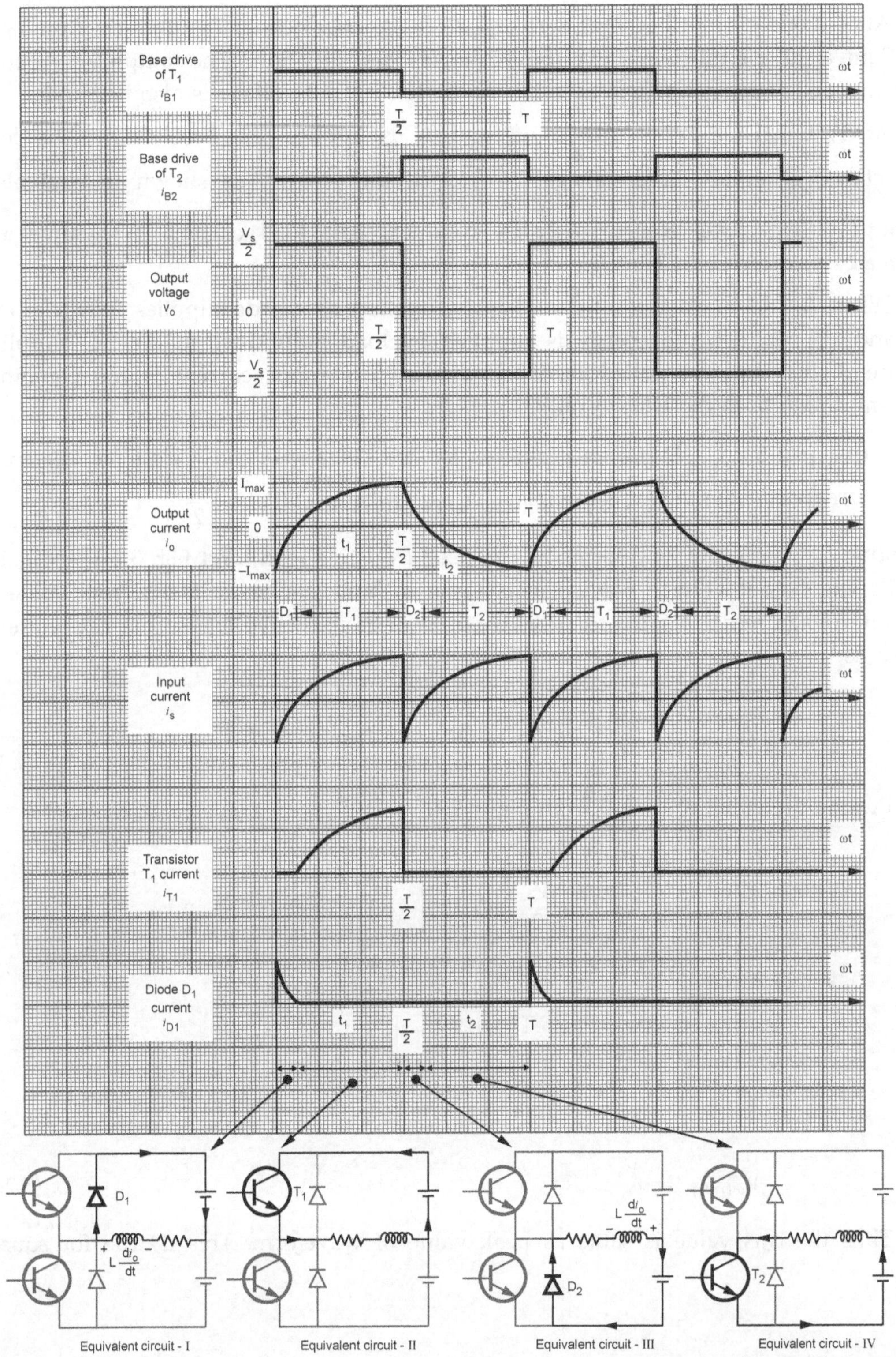

Fig. 5.2.4 Operation and waveforms of half bridge inverter for inductive load

At t_2 diode current becomes zero and T_2 starts conducting. T_2 conducts from t_2 to T. Equivalent circuit-IV in Fig. 5.2.4 shows this situation. The output current is negative and it increases from zero to $-I_{max}$. The output voltage is also negative.

At T, transistor T_2 is turned off and T_1 is applied the base drive. The load inductance generates large voltage $L\frac{di_o}{dt}$ with the polarity as shown in equivalent circuit-I of Fig. 5.2.4. Hence diode D_1 conducts. The load current is negative and decreases towards zero. Thus the cycle repeats.

When T_1 and T_2 conduct, the energy is supplied by the dc supplies. When diodes D_1 and D_2 conduct, the energy is supplied by load inductance to the DC supplies. Hence diodes are also called feedback diodes. The input current is the combined current due to both the DC supplies.

Example 5.2.1 : *Derive an expression for rms value of output voltage for half bridge inverter having square wave output. The peak value of output is* $\frac{V_s}{2}$.

Solution : Observe the waveforms of Fig. 5.2.2 (resistive load) and Fig. 3.2.4 (inductive load). The output voltage waveform is same in both the cases. Hence rms value of output voltage remains same for resistive as well as inductive loads. The rms value is given as,

$$V_{o\,(rms)} = \left[\frac{1}{T}\int_0^T v_0^2\,(\omega t)\,d\omega t\right]^{\frac{1}{2}}$$

Putting the value of $v_o\,(\omega t)$ from Fig. 5.2.4,

$$V_{o\,(rms)} = \left\{\frac{1}{T}\left[\int_0^{\frac{T}{2}}\left(\frac{V_s}{2}\right)^2 dt + \int_{\frac{T}{2}}^{T}\left(-\frac{V_s}{2}\right)^2 dt\right]\right\}^{\frac{1}{2}}$$

$$= \left\{\frac{1}{T}\left(\frac{V_s}{2}\right)^2\left[\int_0^{\frac{T}{2}} dt + \int_{\frac{T}{2}}^{T} dt\right]\right\}^{\frac{1}{2}}$$

$$V_{o\,(rms)} = \frac{V_s}{2} \qquad \text{... (5.2.1)}$$

Thus the rms value is same as peak value of waveform. This is true for square wave.

Example 5.2.2 : *Obtain Fourier series for the output voltage waveform of half bridge inverter. Determine the rms value of the fundamental component of output voltage.*

Solution : i) To obtain Fourier series

Fig. 5.2.5 shows the waveform of output voltage observe that it is square wave of amplitude $\pm\frac{V_s}{2}$. The period of this square wave is T.

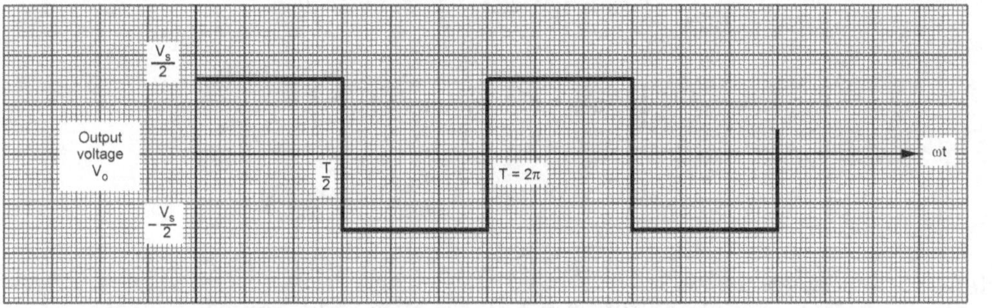

Fig. 5.2.5 Output voltage waveform of half bridge inverter

The general expression for Fourier series is given as,

$$v_o(\omega t) = V_{o(av)} + \sum_{i=1}^{\infty} c_n \sin(n\omega t + \phi_n) \quad \text{... (5.2.2)}$$

where

$$c_n = \sqrt{a_n^2 + b_n^2}$$

$$\phi_n = \tan^{-1}\left(\frac{a_n}{b_n}\right)$$

Here

$$a_n = \frac{2}{T}\int_0^T v_o(\omega t)\cos n\omega t \; d\omega t$$

From the waveform of Fig. 5.2.5 we can write,

$$a_n = \frac{2}{2\pi}\left[\int_0^{\pi}\frac{V_s}{2}\cos n\omega t \; d\omega t + \int_{\pi}^{2\pi}\left(-\frac{V_s}{2}\right)\cos n\omega t \; d\omega t\right]$$

$$= \frac{V_s}{2\pi}\left\{\int_0^{\pi}\cos n\omega t \; d\omega t - \int_{\pi}^{2\pi}\cos n\omega t \; d\omega t\right\}$$

$$= \frac{V_s}{2n\pi}\left\{[\sin n\omega t]_0^{\pi} - [\sin n\omega t]_{\pi}^{2\pi}\right\}$$

$$= 0 \quad \text{for all values of n.} \quad \text{... (5.2.3)}$$

Now the value of b_n can be calculated as,

$$b_n = \frac{2}{T}\int_0^T v_o(\omega t)\sin n\omega t\, d\omega t$$

From the waveform of Fig. 5.2.5, we can write,

$$= \frac{2}{2\pi}\left[\int_0^{\pi}\frac{V_s}{2}\sin n\omega t\, d\omega t + \int_{\pi}^{2\pi}\left(-\frac{V_s}{2}\right)\sin n\omega t\, d\omega t\right]$$

$$= \frac{V_s}{2\pi}\left\{\int_0^{\pi}\sin n\omega t\, d\omega t - \int_{\pi}^{2\pi}\sin n\omega t\, d\omega t\right\}$$

$$= \frac{V_s}{2n\pi}\left\{[-\cos\omega t]_0^{\pi} - [-\cos\omega t]_{\pi}^{2\pi}\right\}$$

$$= \frac{V_s}{2n\pi}\{2(1-\cos n\pi)\} = \frac{V_s}{n\pi}(1-\cos n\pi)$$

Here $\cos n\pi = \begin{cases} 1 & \text{for } n = 2, 4, 6, \text{....i.e. even values} \\ -1 & \text{for } n = 1, 3, 5, \text{....i.e. odd values} \end{cases}$

Hence b_n becomes,

$$b_n = \begin{cases} \dfrac{2V_s}{n\pi} & \text{for odd values of n} \\ 0 & \text{for even values of n} \end{cases} \qquad \text{... (5.2.4)}$$

$$c_n = \sqrt{a_n^2 + b_n^2} = \sqrt{0 + b_n^2} = b_n$$

$$\therefore \quad c_n = \frac{2V_s}{n\pi} \quad \text{for odd n} \qquad \text{..... (5.2.5)}$$

And $\phi_n = \tan^{-1}\dfrac{a_n}{b_n} = \tan^{-1} 0 = 0$

The waveform of Fig. 5.2.5 has symmetric positive and negative half cycles. Hence average value of such waveform is zero i.e.,

$$V_{o(av)} = 0$$

Therefore fourier series of equation 5.2.2 can be written as,

$$v_o(\omega t) = \sum_{n=1,3,5,\ldots}^{\infty}\frac{2V_s}{n\pi}\sin n\omega t \qquad \text{... (5.2.6)}$$

Thus output voltage contains only odd harmonics.

ii) To obtain rms value of the fundamental

From equation 5.2.6, we can write an equation for the fundamental component of output voltage by taking n = 1 i.e.,

$$v_1(\omega t) = \frac{2V_s}{\pi}\sin\omega t \qquad .... (5.2.7)$$

This has same frequency (ωt) as that of square wave. The peak value of fundamental is,

$$c_1 = \frac{2V_s}{\pi}$$

rms value of the fundamental is,

$$V_{1(rms)} = \frac{c_1}{\sqrt{2}} = \frac{2V_s}{\sqrt{2}\,\pi} = 0.45\ V_s \qquad ... (5.2.8)$$

Similarly rms values of other harmonics can be calculated as,

$$V_{n(rms)} = \frac{2V_s}{\sqrt{2}\,n\pi} = \frac{\sqrt{2}\ V_s}{n\,\pi} = \frac{0.45\ V_s}{n} \qquad ... (5.2.9)$$

Example 5.2.3 : *For a half bridge inverter feeding RL load, derive an expression for output current. Determine maximum and minimum values of load current.*

Solution : Fig. 5.2.6 shows the output current waveform for half bridge inverter. Let the current in positive half cycle be $i_1(t)$. During this period $\left(0 \text{ to} \frac{T}{2}\right)$ the load voltage is $\frac{V_s}{2}$. Fig. 5.2.7 shows an equivalent circuit. Writing KVL to this loop we get,

$$L\frac{di_1(t)}{dt} + Ri_1(t) = \frac{V_s}{2}$$

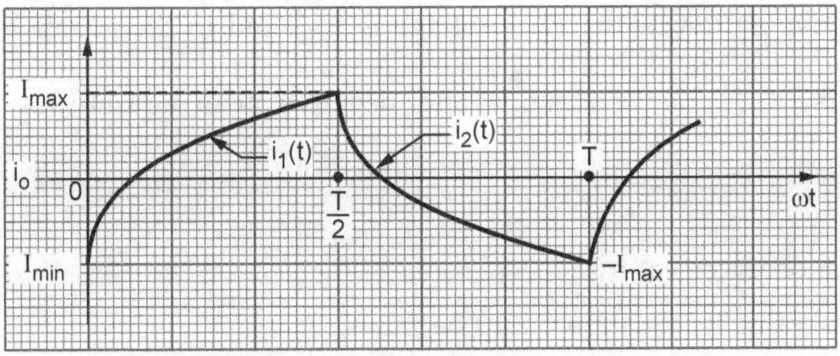

Fig. 5.2.6 Output current waveform

Taking Laplace transform of above equation,

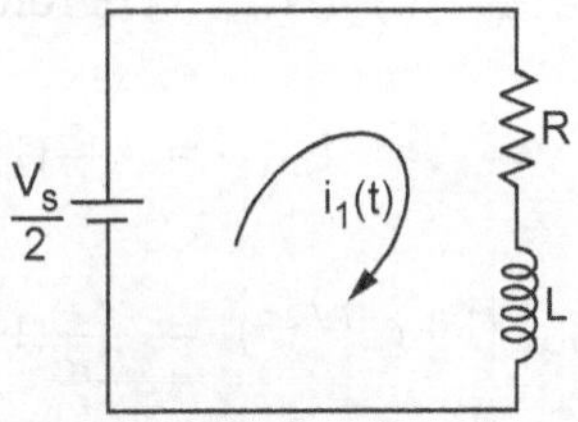

Fig. 5.2.7

$$L[sI_1(s) - I_1(0)] + RI_1(s) = \frac{V_s}{2s}$$

$$\therefore \quad I_1(s)[R + sL] = \frac{V_s}{2s} + LI_1(0)$$

$$\therefore \quad I_1(s) = \frac{V_s}{2s(R+sL)} + \frac{LI_1(0)}{(R+sL)}$$

$$= \frac{V_s}{2L} \cdot \frac{1}{s\left(s + \frac{R}{L}\right)} + \frac{I_1(0)}{s + \frac{R}{L}}$$

$$= \frac{V_s}{2R}\left(\frac{1}{s} - \frac{1}{s + \frac{R}{L}}\right) + I_1(0) \cdot \frac{1}{s + \frac{R}{L}}$$

Taking inverse Laplace transform of above equation,

$$i_i(t) = \frac{V_s}{2R}(1 - e^{-t \cdot R/L}) + i_1(0)e^{-t \cdot R/L}$$

Let $\tau = \frac{L}{R}$ be the load time constant. Then above equation will be,

$$i_i(t) = \frac{V_s}{2R}(1 - e^{-t/\tau}) + i_1(0)e^{-t/\tau} \quad \text{...(5.2.10)}$$

In the positive half cycle the load voltage is $+\frac{V_s}{2}$ and in the negative half cycle the load voltage is $-\frac{V_s}{2}$. Hence positive and negative peak amplitudes will be same. Therefore,

$$I_{min} = -I_{max} \quad \text{...(5.2.11)}$$

In Fig. 5.2.6 observe that $i_1(0) = I_{min} = -I_{max}$.

Putting this value in equation 5.2.10 becomes,

$$i_1(t) = \frac{V_s}{2R}(1 - e^{-t/\tau}) - I_{max}e^{-t/\tau} \quad \text{...(5.2.12)}$$

At $t = \frac{T}{2}$, $i_1(t) = I_{max}$. Therefore above equation will be,

$$I_{max} = \frac{V_s}{2R}(1 - e^{-T/2\tau}) - I_{max}e^{-T/2\tau}$$

$$\therefore \quad I_{max}(1 + e^{-T/2\tau}) = \frac{V_s}{2R}(1 - e^{-T/2\tau})$$

$$\therefore \qquad I_{max} = \frac{V_s}{2R}\frac{1-e^{-T/2\tau}}{1+e^{-T/2\tau}} \qquad ...(5.2.13)$$

and hence,
$$I_{min} = -\frac{V_s}{2R}\frac{1-e^{-T/2\tau}}{1+e^{-T/2\tau}} \qquad ...(5.2.14)$$

From equation 5.2.12 we can write the equation for $i_2(t)$ keeping in mind that (i) $i_2(t)$ is delayed by $\frac{T}{2}$ and (ii) polarities are reversed. i.e.,

$$i_2(t) = -\frac{V_s}{2R}(1-e^{-\left(t-\frac{T}{2}\right)/\tau}) + I_{max}e^{-\left(t-\frac{T}{2}\right)/\tau} \qquad ...(5.2.15)$$

Review Question

1. *Explain the working of half bridge inverter with the help of waveforms.*

5.3 Performance Parameters

The inverters generate the AC output. But the output voltage waveform can be square wave, quasi-square wave or low distorted sine wave. These waveforms contain harmonics (i.e. distortion). The performance parameters represent the amount of distortion in inverter output.

1. Harmonic factor of nth harmonic (HF_n)

The harmonic factor of the nth harmonic is defined as the ratio of rms value of nth harmonic to rms value of the fundamental. i.e.,

$$HF_n = \frac{V_n}{V_1} \qquad ...\ (5.3.1)$$

Here V_n is rms value of nth harmonic

V_1 is rms value of fundamental

The harmonic factor (HF_n) indicates contribution of each harmonic to the harmonic distortion. This factor is evaluated to check the dominant harmonics in the output.

2. Total harmonic distortion (THD)

The total harmonic distortion is the ratio of rms values of all the harmonics to rms value of the fundamental component. i.e.,

$$\text{THD} = \frac{\left(\sum_{n=2,3,...}^{\infty} V_n^2\right)^{\frac{1}{2}}}{V_1} \qquad ...\ (5.3.2)$$

The total harmonic distortion indicates the distortion in the waveform. It is the measure of closeness of the waveform to sine wave.

3. Distortion factor (DF)

The distortion factor is defined as,

$$DF = \frac{1}{V_1}\left[\sum_{n=2,3,\ldots}^{\infty}\left(\frac{V_n}{n^2}\right)^2\right]^{\frac{1}{2}} \quad \text{... (5.3.3)}$$

The distortion factor is the measure of harmonic distortion that remains in the particular waveform after filtering. Due to filtering higher order harmonics are eliminated. The distortion factor of nth harmonic component is given as,

$$DF_n = \frac{V_n}{V_1\, n^2} \quad \text{... (5.3.4)}$$

4. Lowest order harmonic (LOH)

The lowest order harmonic is the harmonic component which has nearest frequency to the fundamental and its amplitude is within 3% of the fundamental component. The lowest order harmonic should have maximum possible frequency, so that the harmonic distortion is less.

Example 5.3.1 : *The single phase half bridge inverter has the DC input of 48 V. The load resistance is 4.8 Ω. Determine*

(i) RMS value of output voltage

(ii) RMS value of fundamental component

(iii) Total harmonic distortion.

Solution : Given $V_s = 48\ V$ and $R = 4.8\ \Omega$

(i) The rms value of output voltage is $\frac{V_s}{2}$. i.e.,

$$V_{o\,(rms)} = \frac{V_s}{2} \quad \text{... (5.3.5)}$$

$$= \frac{48}{2} = 24\ V$$

(ii) The output waveform of half bridge inverter is square wave of amplitude $\frac{V_s}{2}$. The fundamental component is given for such square wave is given by equation 5.2.8 as,

$$V_{1\,(rms)} = \frac{2\,V_s}{\sqrt{2}\,\pi} = 0.45\,V_s \quad \text{... (5.3.6)}$$

$$= 0.45 \times 48 = 21.6\ V$$

(iii) The rms voltage due to all the harmonics can be obtained as,

$$V_{(harmonics)} = \left(\sum_{n=3,5,7}^{\infty} V_n^2\right)^{\frac{1}{2}} \quad \text{... (5.3.7)}$$

The square wave contains only odd harmonics. This value can be obtained as,

$$V_{(harmonics)} = \left(V_{o\,(rms)}^2 - V_1^2\right)^{\frac{1}{2}} \quad \text{... (5.3.8)}$$

$$= \left[(24)^2 - (21.6)^2\right]^{\frac{1}{2}}$$

$$= 10.46$$

The THD is obtained as the ratio of rms value due to harmonics to rms value of fundamental i.e.,

$$THD = \frac{10.46}{21.6} = 0.484 \text{ or } 48.4\%$$

From equation 5.3.2, equation 5.3.7 and equation 5.3.8 we can obtain an expression for THD as,

$$THD = \frac{\sqrt{V_{o(rms)}^2 - V_1^2}}{V_1} = \sqrt{\left(\frac{V_{o\,(rms)}}{V_1}\right)^2 - 1} \quad \text{... (5.3.9)}$$

The square wave contains 48.4% of harmonic distortion.

Review Question

1. *State the various performance parameters used for inverters.*

5.4 Full Phase Bridge Inverter

We studied half bridge inverter earlier. Fig. 5.4.1 shows the circuit diagram of full bridge inverter. Observe that there are four transistors and four diodes. The diodes are required for feedback when the load is inductive.

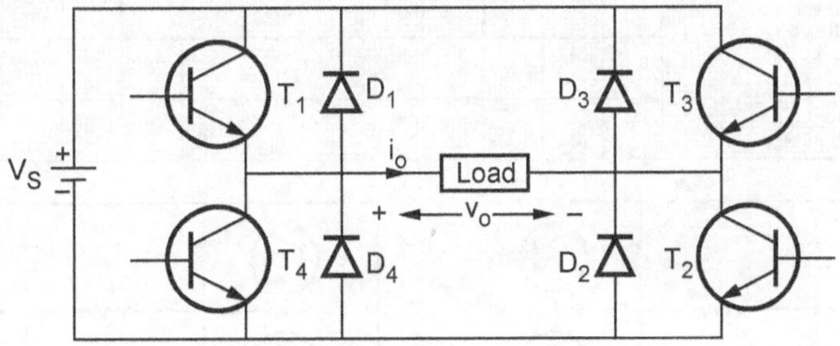

Fig. 5.4.1 Single phase bridge inverter

5.4.1 Operation with Resistive Load

When the load is resistive, the diodes doesnot carry any current. Fig. 5.4.2 shows the waveforms. The transistor T_1 and T_2 conduct from 0 to $\frac{T}{2}$. Equivalent circuit-I in Fig. 5.4.2 shows the current path when T_1 and T_2 conduct. The output voltage and current are positive. Note that the amplitude of load voltage is V_S.

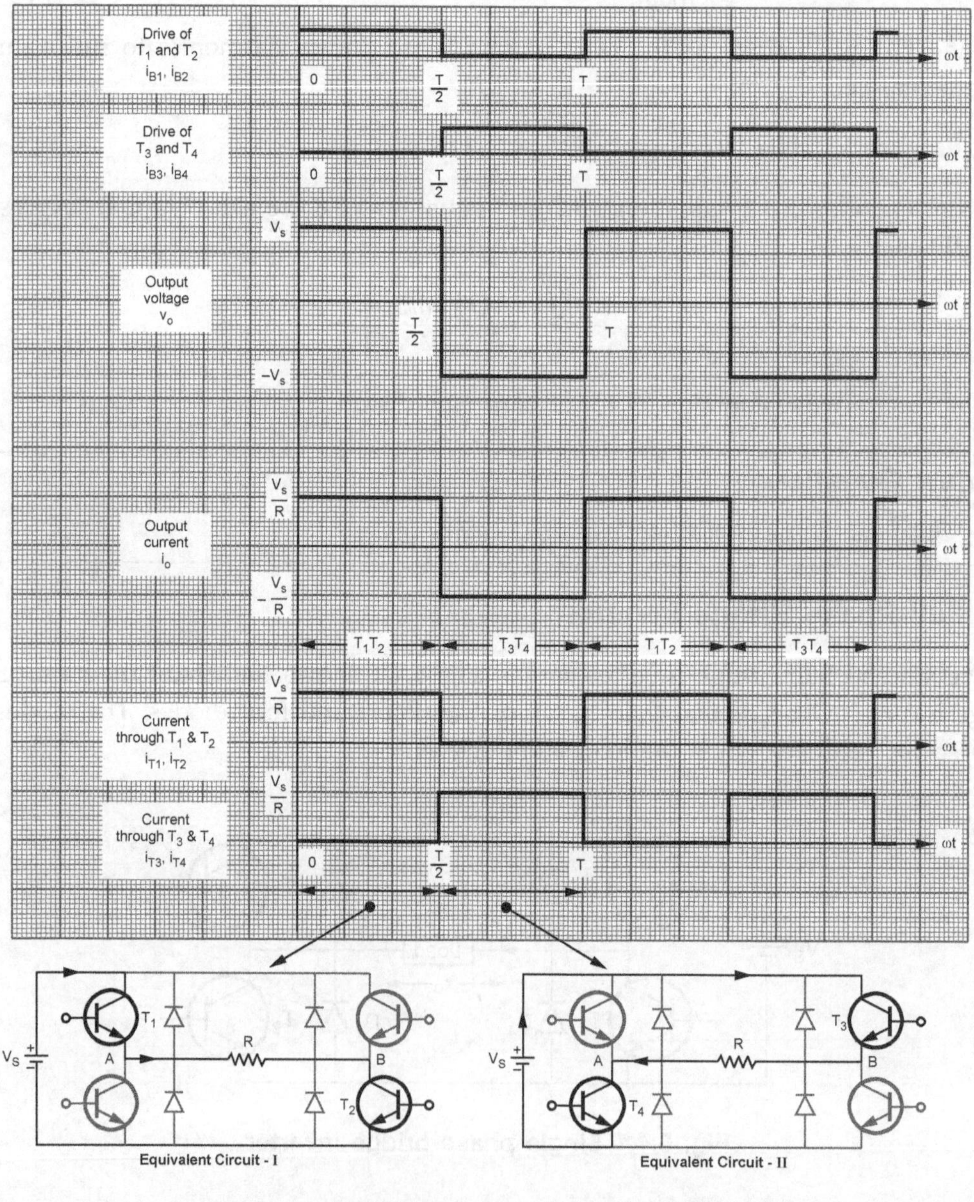

Fig. 5.4.2 Operation and waveforms of 1ϕ bridge inverter for resistive load

At $\frac{T}{2}$, transistors T_1 and T_2 are turned off. Transistors T_3 and T_4 conduct from $\frac{T}{2}$ to T. Equivalent circuit-II shows the current path. Note that the output current is negative. The voltage is also negative. Thus in positive half cycle, T_1 and T_2 conduct. And in negative half cycle, T_3 and T_4 conduct. The amplitude of the output voltage is $\pm V_s$. The output is the square wave. The currents through transistors are also shown in the Fig. 5.4.2.

5.4.2 Operation with Inductive Load

Now let us consider the operation of 1ϕ bridge inverter with inductive (R - L) load. Fig. 5.4.3 shows the circuit diagram of bridge inverter having RL load. The waveforms of this circuit are shown in Fig. 5.4.4.

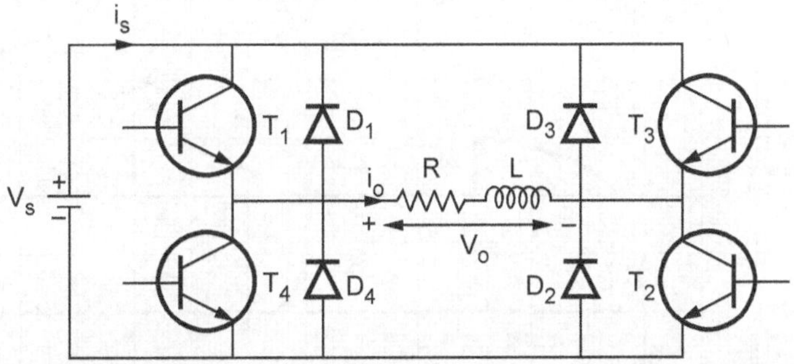

Fig. 5.4.3 1ϕ bridge inverter having inductive load

Please refer Fig. 5.4.4 on next page.

The operation of this circuit can be explained in four modes as follows :

Mode - I (T_1 , T_2 conducts)

T_1 and T_2 are applied the drive at $t = 0$. But they does not conduct till t_1. Diodes D_1 and D_2 conduct from 0 to t_1.

Hence T_1 and T_2 are reverse biased and they do not conduct. From t_1 to $\frac{T}{2}$, T_1 and T_2 conduct. The equivalent circuit is shown in Fig. 5.4.5. The load current is positive and it increases from zero to $+ I_{max}$. The output voltage is also positive.

Please refer Fig. 5.4.5 on next page.

Mode - II (D_3 and D_4 conducts)

At $\frac{T}{2}$, transistors T_1 and T_2 are turned off and T_3 , T_4 are applied drives. The load inductance generates the large voltage $L \frac{di_o}{dt}$ with polarities shown in Fig. 5.4.6. The diodes D_3 and D_4 are forward biased due to inductance voltage. These diodes conduct

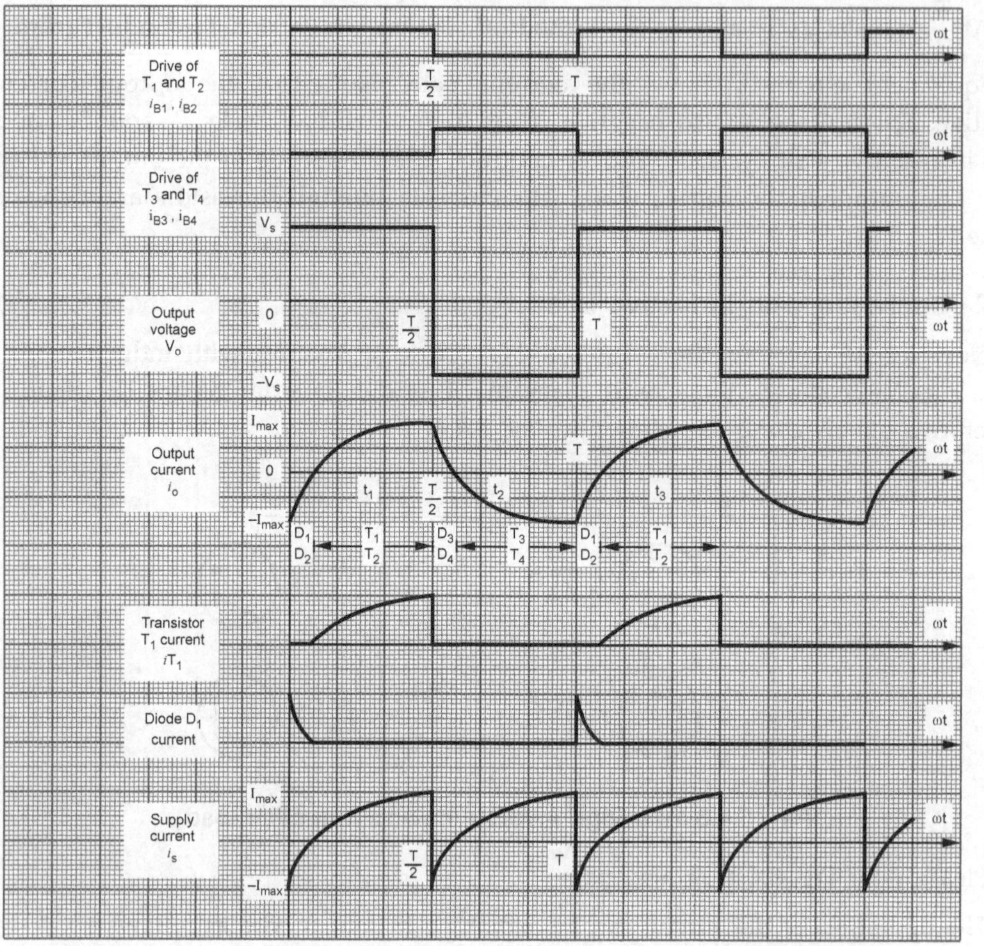

Fig. 5.4.4 Waveforms of bridge inverter for inductive load

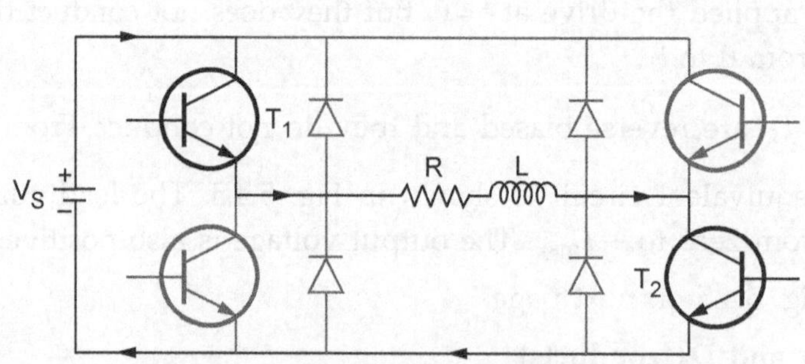

Fig. 5.4.5 T_1 and T_2 conduct from t_1 to $\frac{T}{2}$

and output current flows through DC supply. The energy stored in the load inductance is supplied to the DC supply. This operation is called *feedback operation.* Note that the supply current i_s is negative when diodes D_3, D_4 are conducting. Due

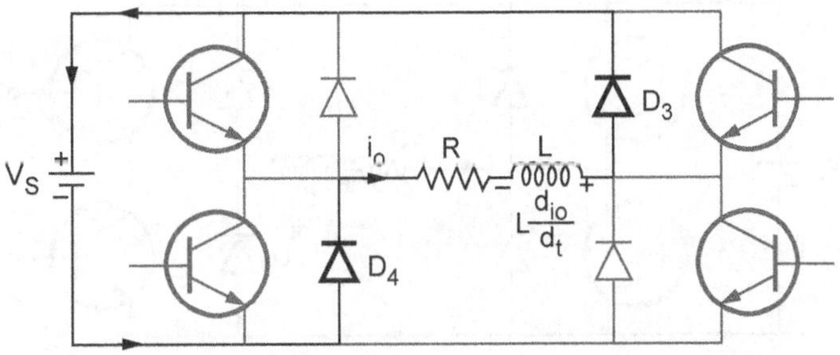

Fig. 5.4.6 D_3 and D_4 conduct from $\frac{T}{2}$ to t_2

to conduction of D_3 and D_4, transistors T_3 and T_4 are reverse biased. Hence they do not conduct, even though base drives are applied. At t_2, the load current becomes zero. Hence transistors T_3 and T_4 start conducting.

Mode - III (T_3 and T_4 conduct)

At t_2 the transistors T_3 and T_4 start conducting. Fig. 5.4.7 shows the equivalent circuit for this operation. The output current is negative and increases towards $-I_{max}$. The supply current i_s is positive. The output voltage is negative during this period.

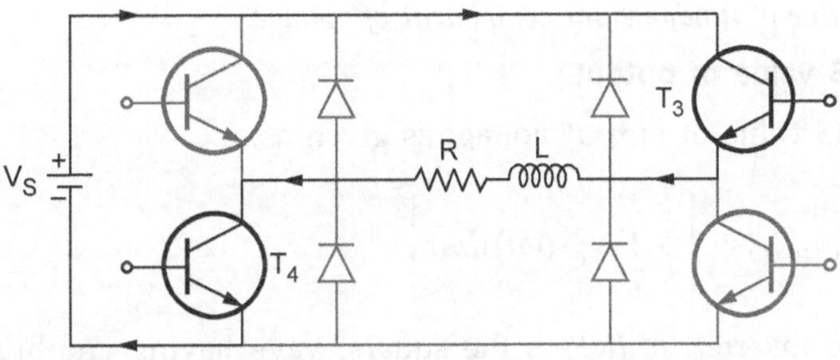

Fig. 5.4.7 T_3 and T_4 conduct from t_2 to T

Mode - IV (D_1 and D_2 conduct from T to t_3 i.e. from 0 to t_1)

At T, transistors T_3 and T_4 are turned off and T_1, T_2 are applied the drive. The output current is at $-I_{max}$. Hence load inductance generates the large voltage $L\,\frac{di_o}{dt}$ with polarities as shown in Fig. 5.4.8. Due to this voltage the diodes D_1 and D_2 are forward biased. Hence they start conducting. The output current flows through the DC supply and it goes on decreasing. The energy stored in the load inductance is supplied to the DC supply during this period. This is called feedback operation. Transistors T_1 and T_2 are reverse biased due to conduction of D_1 and D_2. Hence T_1 and T_2 do not conduct even if their base drive is applied. At t_3 (i.e. at t_1), the output current becomes zero. Hence T_1 and T_2 start conducting. The output current becomes positive. This is beginning of mode - I discussed earlier. Then the cycle repeats.

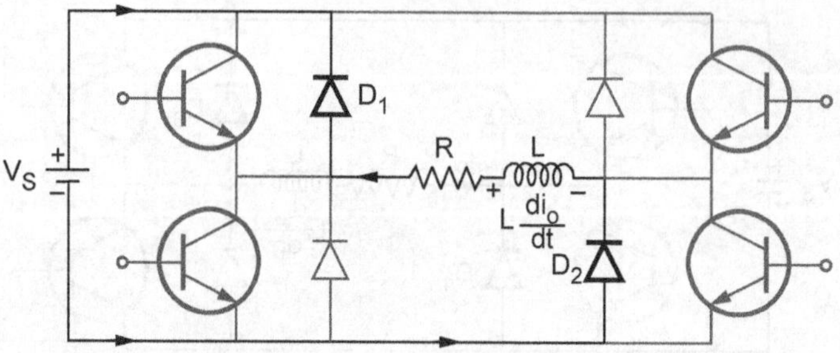

Fig. 5.4.8 D_1 and D_2 conduct from 0 to t_1 i.e. from T to t_3

The load energy is feedback to DC supply whenever diodes conduct. The supply current is negative. And energy flows from supply to the load whenever transistors conduct. The supply current is positive when transistors conduct. The output voltage waveform is square wave having amplitudes of $\pm V_s$.

Example 5.4.1 : *Derive the following for the single phase bridge inverter having square wave output :*

(i) RMS value of output

(ii) Fourier series for output voltage

(iii) RMS value of fundamental component of voltage.

Solution : (i) RMS value of output

The RMS value of output voltage is given as,

$$V_{o\,(rms)} = \left[\frac{1}{T}\int_0^T v_0^2\,(\omega t)\,d\,\omega t\right]^{\frac{1}{2}}$$

For the bridge inverter, $v_o\,(\omega t)$ is the square wave having amplitudes of $+V_s$ and $-V_s$. Hence above equation can be written as,

$$V_{o\,(rms)} = \left\{\frac{1}{T}\left[\int_0^{\frac{T}{2}} V_s^2\,dt + \int_{\frac{T}{2}}^{T} (-V_s)^2\,dt\right]\right\}^{\frac{1}{2}}$$

$$= \left\{\frac{V_s^2}{T}\left[\int_0^{\frac{T}{2}} dt + \int_{\frac{T}{2}}^{T} dt\right]\right\}^{\frac{1}{2}}$$

$$= V_s \qquad \text{... (5.4.1)}$$

Thus the rms value of the output voltage is same as the DC supply.

ii) Fourier series for output voltage

We have derived an expression for fourier series of output voltage for half bridge inverter. The output waveforms of half bridge inverter and full bridge inverter are same, i.e. square wave. There is only difference of amplitudes. The amplitude is $\pm V_s$ in full bridge inverter. Where as, it is $\pm \frac{V_s}{2}$ in half bridge inverter. Let us rearrange equation 5.2.6 as,

$$v_o(\omega t) = \sum_{n=1,3,5,......}^{\infty} \frac{4\left(\frac{V_s}{2}\right)}{n\pi} \sin n\omega t$$

This is Fourier series for half bridge inverter. For full bridge inverter we have to replace $\frac{V_s}{2}$ by V_s i.e.

$$v_o(\omega t) = \sum_{n=1,3,5,......}^{\infty} \frac{4V_s}{n\pi} \sin n\omega t \qquad ... (5.4.2)$$

This is Fourier series for full bridge inverter. The output voltage waveform remain same for resistive and inductive loads. Hence above expression is applicable resistive as well as inductive loads. The above equation shows that output voltage contains only odd harmonics.

iii) rms value of fundamental components

With n = 1 in equation 5.4.2 we get equation of fundamental component of output voltage i.e.,

$$v_1(\omega t) = \frac{4V_s}{\pi} \sin \omega t$$

This has the same frequency (ωt) as that of square wave. The peak value of the fundamental is,

$$c_1 = \frac{4V_s}{\pi}$$

rms value of the fundamental is,

$$V_{1(rms)} = \frac{c_1}{\sqrt{2}}$$

$$= \frac{4V_s}{\sqrt{2}\,\pi}$$

$$= \frac{2\sqrt{2}\,V_s}{\pi}$$

$$= 0.9\ V_s \qquad ... (5.4.3)$$

This is the rms value of fundamental. Similarly rms value of nth harmonic can be obtained as,

$$V_{n(rms)} = \frac{4V_s}{\sqrt{2}\, n\pi} = \frac{0.9\, V_s}{n} \quad \text{... (5.4.4)}$$

Example 5.4.2 : *For a full bridge inverter feeding RL load, derive an expression for output current. Determine maximum and minimum values of load current.*

Solution : Fig. 5.4.9 shows the output current waveform for full bridge inverter.

Note that this waveform is exactly same as that of 1ϕ half-bridge inverter shown in Fig. 5.2.6 earlier. But for full bridge inverter, the load voltage is V_s. In example 5.2.3 we have derived various expressions for currents. If $\frac{V_s}{2}$ is replaced by V_s, we get those expressions for full bridge inverter. i.e. from example 5.2.3 we have following results for 1 ϕ full bridge inverter.

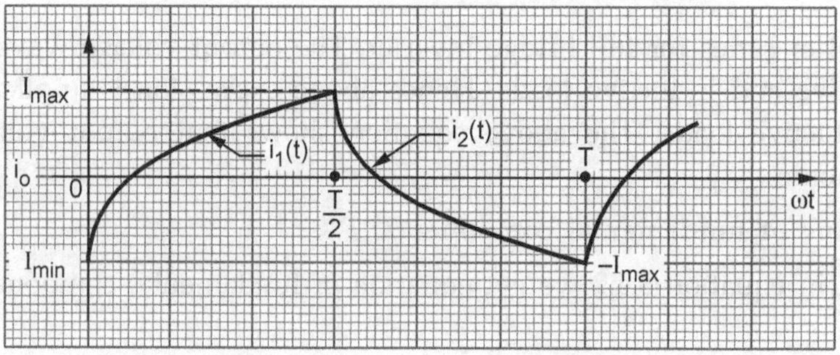

Fig. 5.4.9 Output current waveform

$$i_1(t) = \frac{V_s}{R}(1 - e^{-t/\tau}) - I_{max} e^{-t/\tau} \quad \text{from equation 5.2.12} \quad \text{...(5.4.5)}$$

$$i_2(t) = -\frac{V_s}{R}\left(1 - e^{-\left(t - \frac{T}{2}\right)/\tau}\right) + I_{max} e^{-\left(t - \frac{T}{2}\right)/\tau} \quad \text{from equation 5.2.15} \quad \text{...(5.4.6)}$$

$$I_{max} = \frac{V_s}{R} \frac{1 - e^{-T/2\tau}}{1 + e^{-T/2\tau}} \quad \text{from equation 5.2.13} \quad \text{...(5.4.7)}$$

$$I_{min} = -\frac{V_s}{R} \frac{1 - e^{-T/2\tau}}{1 + e^{-T/2\tau}} \quad \text{from equation 5.2.14} \quad \text{...(5.4.8)}$$

- In all the above equations $\frac{V_s}{2}$ is replaced by V_s.

Example 5.4.3 : *A single phase bridge inverter having square wave output has the DC supply of 48 V and output resistance of 4.8 Ω. Determine (i) rms value of output and (ii) rms value of fundamental.*

Solution : Given

$$V_s = 48$$

$$R = 4.8\ \Omega$$

(i) To determine rms value of output

From equation 5.4.1 we know that,

$$v_{o(rms)} = V_s$$

$$= 48$$

Thus the rms value of output is same as the DC supply.

(ii) To determine rms value of fundamental

From equation 5.4.3 we know that,

$$V_1 = 0.9\ V_s$$

$$V_1 = 0.9 \times 48 = 43.2$$

This is the rms value of fundamental component.

Example 5.4.4 : *The single phase full bridge inverter has a resistive load of R = 2.4 Ω, and the DC input voltage of V_s = 48 volts. Determine,*

i) rms output voltage at the fundamental frequency.

ii) the output power

iii) the peak and average currents of each transistor.

Solution : Given

$$V_s = 48\ \text{V}$$

$$\text{R} = 2.4\ \Omega$$

i) To determine rms value of the fundamental

From equation 5.4.3, rms value of the fundamental component of output is given as,

$$V_1 = 0.9\ V_s$$

$$= 0.9 \times 48 = 43.2\ \text{volts}$$

ii) To determine output power

From equation 5.4.1, the rms output voltage is,

$$V_{o(rms)} = V_s = 48 \text{ volts}$$

The output power can be calculated as,

$$P_o = \frac{V_{o(rms)}^2}{R} = \frac{(48)^2}{2.4}$$

$$= 960 \text{ W}.$$

iii) To determine transistor currents

Fig. 5.4.2 shows the transistor current waveforms for resistive load. These waveforms are reproduced below for convenience.

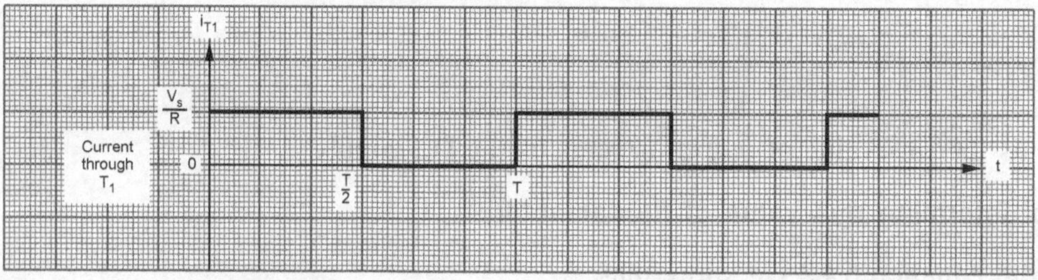

Fig. 5.4.10 Transistor current for 1ϕ bridge inverter

In the above figure, observe that the transistor current is a square wave of amplitude $\frac{V_s}{R}$. Hence peak transistor current is,

$$i_{T(peak)} = \frac{V_s}{R}$$

$$= \frac{48}{2.4} = 20 \text{ A}$$

Each transistor current is a square wave of duty cycle 0.5. Hence average transistor current will be equal to average value of square wave. i.e.

$$i_{T(av)} = i_{T(peak)} \times \text{duty cycle}$$

$$= 20 \times 0.5 = 10 \text{ A}$$

5.4.3 Shoot through or Cross Conduction in Inverters

Definition of Cross Conduction

When the dc supply in inverters gets shorted through power devices of the same link, it is called cross conduction or shoot through.

Effect :

Due to cross conduction power devices are damaged.

Explanation

Fig. 5.4.3 shows the circuit diagram and Fig. 5.4.4 shows the waveforms of 1ϕ bridge inverter. Observe that at $\frac{T}{2}$, transistor T_1 is turned off and transistor T_4 is turned on. But due to switching delays, transistor T_1 may turn-off little after $\frac{T}{2}$. But by this time T_4 is turned on. Fig. 5.4.11 shows this situation.

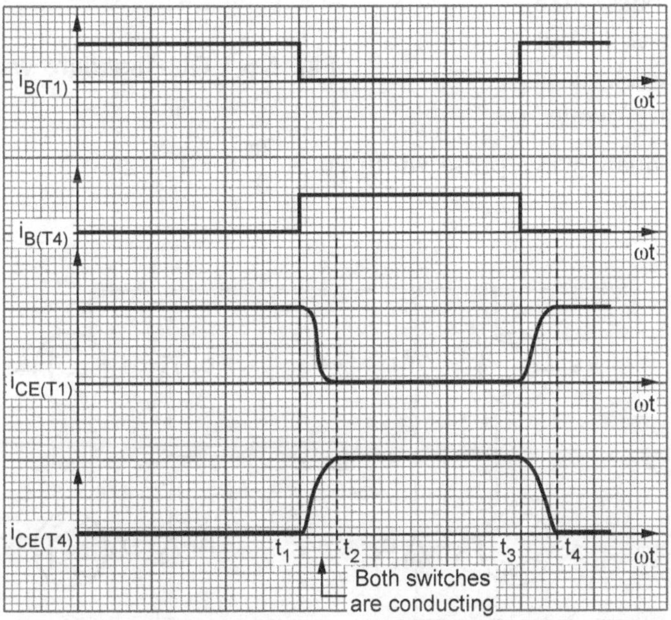

Fig. 5.4.11 Cross conduction in inverters

In this figure observe that in the time interval $t_1 - t_2$, $i_{CE(T1)}$ is reducing and $i_{CE(T4)}$ is increasing. Thus both T_1 and T_4 are in conduction state. This will short the DC supply through T_1-T_4.

How cross conduction is avoided ?

Cross conduction can be avoided by keeping the delay between the drives of switches in same link. For example, for single phase bridge inverter drives of T_1 and T_4 are delayed. Similarly drives of T_2 and T_3 are delayed.

Review Question

1. *With the help of circuit diagram and waveforms, explain the working of 1ϕ bridge inverter.*

5.5 Quasi-Square Wave Inverters

Definition of quasi-square wave : The quasi-square wave is shown in Fig. 5.5.1. In this figure observe that the waveform has zero values near $\frac{T}{2}$ and T. Since this waveform looks like square wave (but it is not exactly square wave), it is called quasi-square wave.

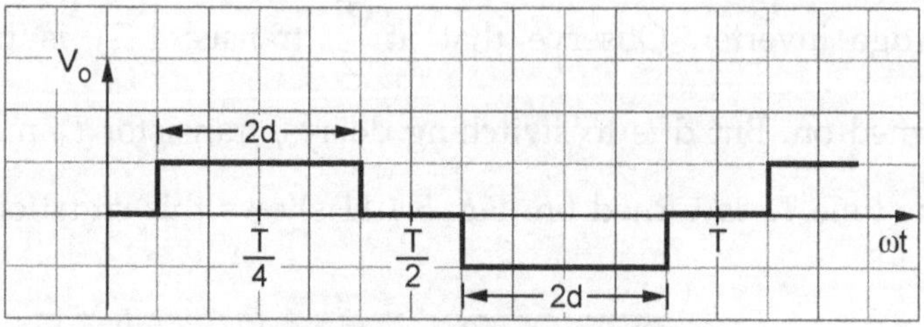

Fig. 5.5.1 Quasi-square wave

For perfect quasi-square wave pulse width 2d is $\frac{2\pi}{3}$ for T = 2π.

Circuit diagram

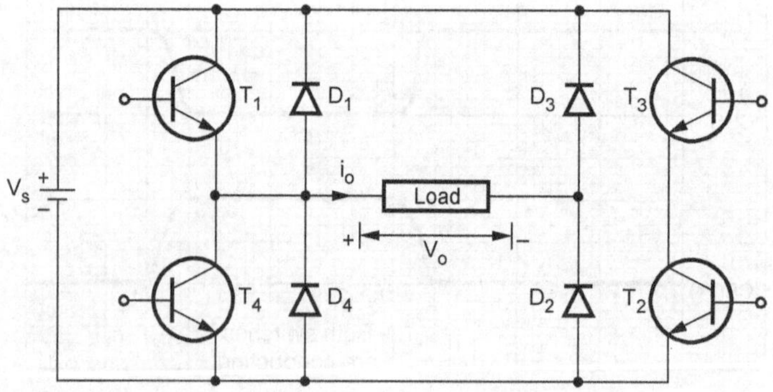

Fig. 5.5.2 Circuit diagram of quasi-square wave inverter

- The above figure shows the circuit diagram of 1ϕ bridge inverter.
- This inverter can operate as square wave, quasi-square wave or PWM inverter depending upon switching used for power transistors.
- Therefore above circuit diagram is same as that shown in Fig. 5.4.1.

5.5.1 Quasi-Square Wave Inverter for Resistive Load

When the load is resistive, the operation of quasi-square wave inverter can be explained in two modes as follows :

Mode-I

In this mode transistors T_1 and T_2 are turned on. As shown in the output voltage waveform of Fig. 5.5.3(c), the output voltage is $+V_s$. Equivalent circuit-I in Fig. 5.5.3 shows the current path.

Mode-II

In this mode transistors T_3 and T_4 are turned-on. The output voltage is $-V_s$ as shown in Fig. 5.5.3(c). Equivalent circuit-II shows the current path.

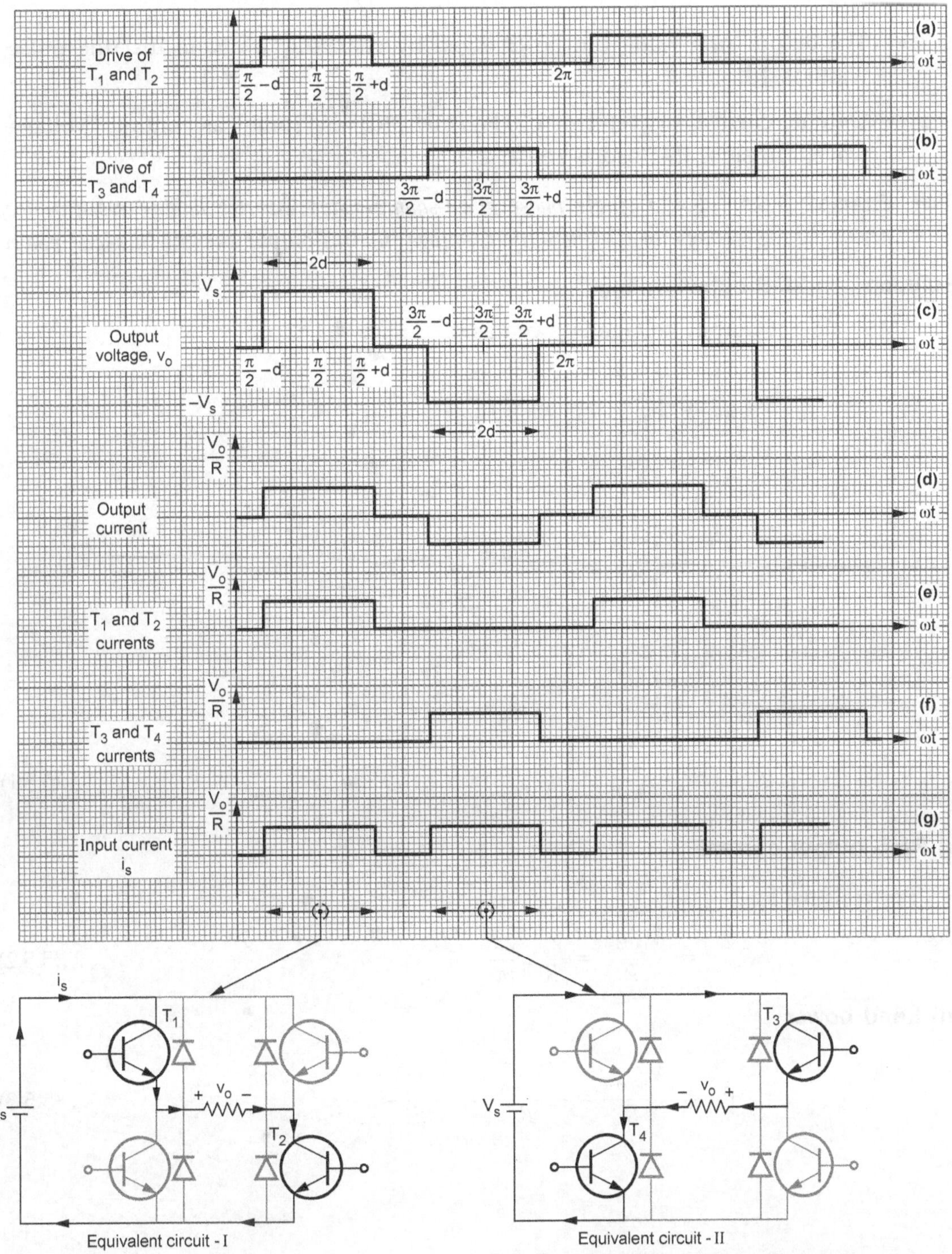

Fig. 5.5.3 Waveforms of quasi-square wave inverter for resistive load

- Since the load is resistive, there is no freewheeling or feedback. Hence the diodes $D_1 D_2 D_3$ and D_4 are not used.
- The pulse width is ±d with respect to $\frac{\pi}{2}$ and $\frac{3\pi}{2}$.
- The output rms value can be changed by changing 'd'. Note that the pulse width is 2d.
- Output current waveform of Fig. 5.5.3(d) is same as output voltage waveform since the load is resistive.

Example 5.5.1 : *For a single phase quasi-square wave inverter having resistive load derive the expressions for (i) rms output voltage (ii) rms output current (iii) load power.*

Solution : i) RMS output voltage

RMS value of output voltage is given as,

$$V_{o(rms)} = \left[\frac{1}{T}\int_0^T v_0^2(\omega t) d\omega t\right]^{\frac{1}{2}}$$

$$= \left\{\frac{1}{2\pi}\left[\int_{\frac{\pi}{2}-d}^{\frac{\pi}{2}+d} V_s^2\, d\omega t + \int_{\frac{3\pi}{2}-d}^{\frac{3\pi}{2}+d} (-V_s)^2\, d\omega t\right]\right\}^{\frac{1}{2}}$$

$$= \left\{\frac{V_s^2}{2\pi}\left[\frac{\pi}{2}+d-\frac{\pi}{2}+d+\frac{3\pi}{2}+d-\frac{3\pi}{2}+d\right]\right\}^{\frac{1}{2}}$$

$$\therefore \quad \boxed{V_{o(rms)} = V_s\sqrt{\frac{2d}{\pi}}} \qquad \text{...(5.5.1)}$$

ii) RMS output current

For resistive load,

$$I_{o(rms)} = \frac{V_{o(rms)}}{R} = \frac{V_s}{R}\sqrt{\frac{2d}{\pi}} \qquad \text{...(5.5.2)}$$

iii) Load power

$$P_o = \frac{V_{o(rms)}^2}{R} = \frac{V_s^2 \cdot \frac{2d}{\pi}}{R} = \frac{2dV_s^2}{\pi R} \qquad \text{...(5.5.3)}$$

5.5.2 Quasi-Square Wave Inverter for Inductive Load

The operation of quasi-square wave inverter can be explained with the help of six modes. The waveforms and equivalent circuits are shown in Fig. 5.5.4.

Mode-I (t_1 - t_2) - Devices conducting (T_1 - T_2)

In this mode transistors T_1 and T_2 conduct. As shown in the waveforms of Fig. 5.5.4(e), output voltage is $+V_s$. The load current increases exponentially from t_1 to t_2. The current path is shown by equivalent circuit-I in Fig. 5.5.4.

Mode-II (t_2 - t_3) - Devices conducting (T_2 - D_4)

At t_2, transistor T_1 is turned-off. But T_2 remains in conduction. Because of inductive load, output current continues to flow in the same direction. Equivalent circuit-II in Fig. 5.5.4 shows that the current flows through T_2 - D_4. It is free wheeling action. Therefore output voltage is zero. The output current (Fig. 5.5.4(e)) reduces exponentially from t_2 to t_3.

Mode-III (t_3 - t_4) - Devices conducting (D_3 - D_4)

At t_3, transistor T_2 is turned-off. Hence the load current now flows through D_3 and D_4. Observe that the load current continues to flow in the same direction due to load inductance. Equivalent circuit III in Fig. 5.5.4. shows the current path. Since D_3 - D_4 are conducting, the output voltage is $-V_s$ as shown in Fig. 5.5.4(e). Note that transistors T_1 and T_2 are applied base drive at t_3. But they donot conduct since diodes D_3 and D_4 are conducting. The conduction of these diodes puts reverse bias on T_3 and T_4. The load current delays to zero at t_4.

Mode-IV (t_4 - t_5) - Devices conducting (T_3 - T_4)

When load current becomes zero at t_4, the transistors T_3 and T_4 start conducting. Equivalent circuit IV in Fig. 5.5.4(e), observe that the load voltage is $-V_s$ and load current is also negative. The load current increases exponentially in negative direction till t_5.

Mode-V (t_5 - t_6) - Devices conducting (T_4 - D_2)

The transistor T_3 is turned-off at t_5. Hence the load current now continues to flow through T_4 and D_2. Equivalent circuit-V in Fig. 5.5.4 shows the current path. It is freewheeling action, hence output voltage is zero. The output current decays exponentially from t_5 to t_6.

Mode-VI - (t_6 - t_7) - Devices conducting (D_1 - D_2)

The transistor T_4 is turned-off at t_6. But due to load inductance the current tries to flow in the same direction. Hence current now flows through D_1 - D_2 as shown in equivalent circuit-VI of Fig. 5.5.4. This is feedback action. The load voltage is positive.

The cycle repeats with mode-I when T_1 - T_2 are turned-on.

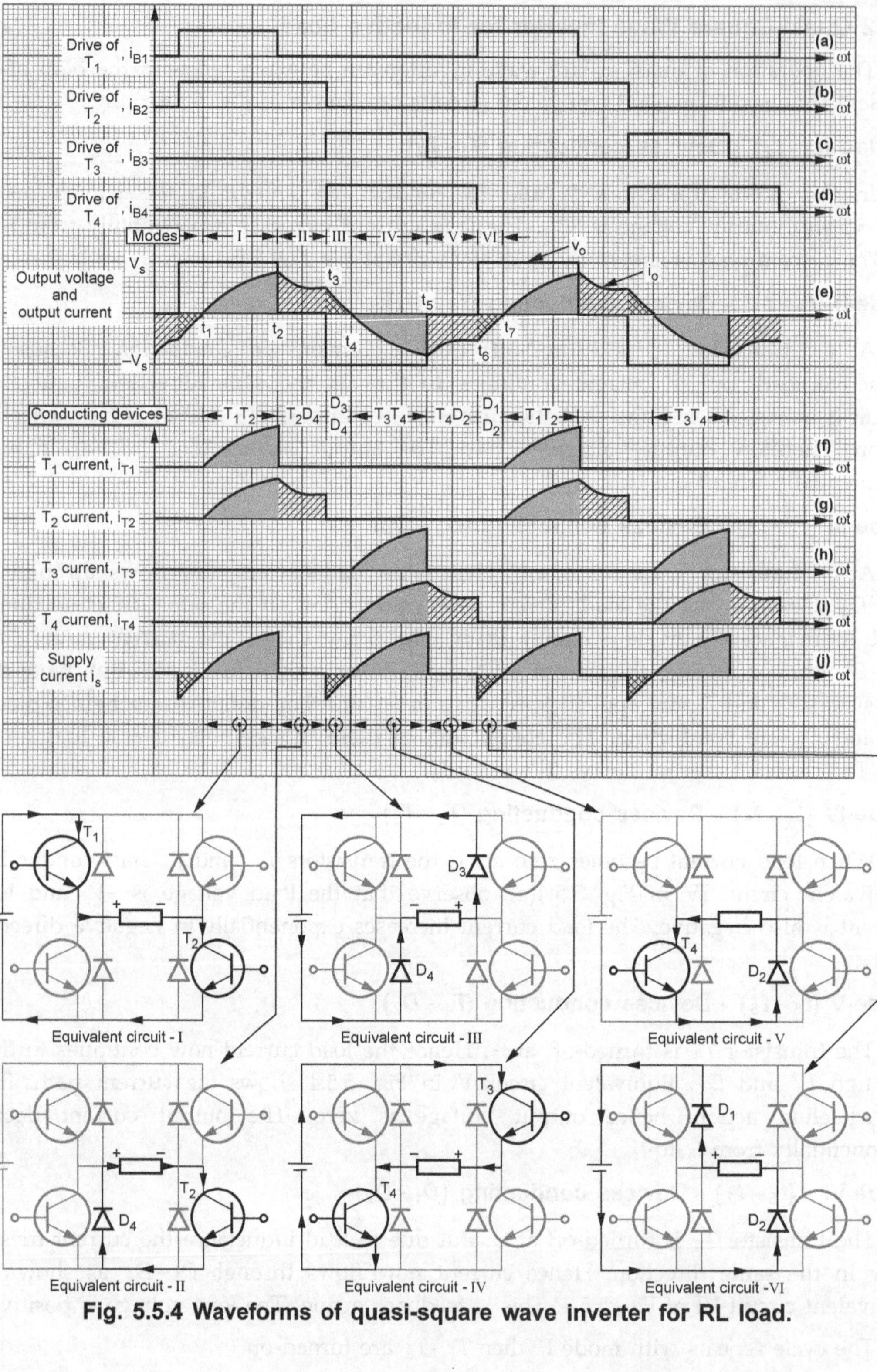

Fig. 5.5.4 Waveforms of quasi-square wave inverter for RL load.

Summary of Operation

Mode	Interval	Conducting devices	Output voltage	Output current	Power flow
I	t_1 - t_2	T_1 T_2	$+V_s$	Positive	Source to load
II	t_2 - t_3	T_2 D_4	zero	Positive	Freewheeling
III	t_3 - t_4	D_3 D_4	$-V_s$	Positive	Load to source i.e. feedback
IV	t_4 - t_5	T_3 T_4	$-V_s$	Negative	Source to load
V	t_5 - t_6	T_4 D_2	zero	Negative	Free wheeling
VI	t_6 - t_7	D_1 D_2	$+V_s$	Negative	Load to source i.e. feedback/

Table 5.5.1 Summary of operation of 1 φ quasi-squarewave inverter

Example 5.5.2 : *For a single phase quasi-square wave inverter feeding inductive load, obtain the expression for rms output voltage.*

Solution : The waveform of quasi-square wave inverter is shown in Fig. 5.5.5. Observe that this waveform is same as that for resistive load shown in Fig. 5.5.3(c). Therefore RMS value of output voltage with inductive load will be same as that with resistive load. Therefore from equation 5.5.1 we can write.

$$V_{o(rms)} = V_s\sqrt{\frac{2d}{\pi}} \quad \text{...(5.5.4)}$$

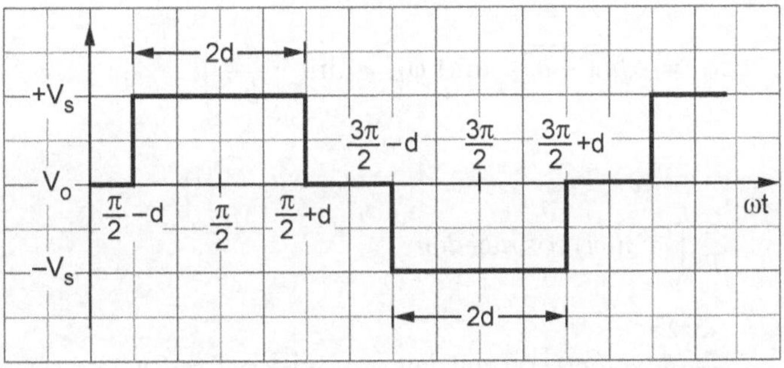

Fig. 5.5.5 Output voltage waveform of quasi-square wave inverter

For perfect quasi-square wave $2d = \frac{2\pi}{3}$ i.e. 120° pulse width. Then above value will be,

$$V_{o(rms)} = V_s\sqrt{\frac{2\pi/3}{\pi}}$$

$$\therefore \quad \boxed{V_{o(rms)} = V_s\sqrt{\frac{2}{3}}} \quad \text{...(5.5.5)}$$

Example 5.5.3 : *Obtain the Fourier series for output voltage waveform of quasi-square wave inverter feeding RL load. Derive an expression for rms value of nth harmonic of output voltage.*

Solution : For the perfect quasi-square wave, pulse width is 120° or $\frac{2\pi}{3}$.

Therefore $2d = \frac{2\pi}{3}$. Hence the waveform of Fig. 5.5.5 can be represented as shown in Fig. 5.5.6.

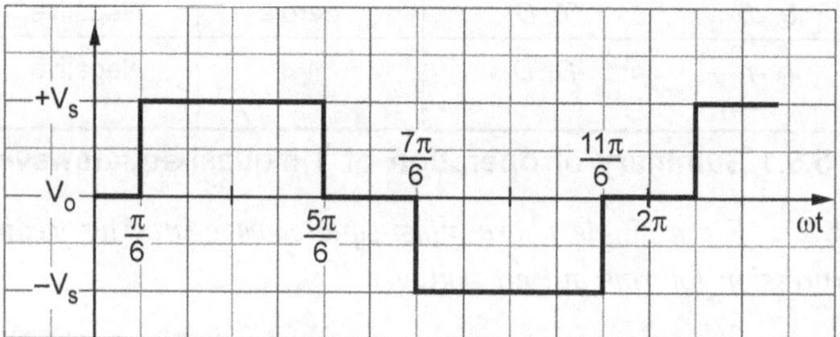

Fig. 5.5.6 Quasi-square wave

The general expression for Fourier series is given as,

$$v_o(\omega t) = V_{o(av)} + \sum_{i=1}^{\infty} C_n \sin(n\omega t + \phi_n) \qquad ...(5.5.6)$$

Here $\qquad C_n = \sqrt{a_n^2 + b_n^2}$ and $\phi_n = \tan^{-1}\left(\frac{a_n}{b_n}\right)$

To obtain a_n

$$a_n = \frac{2}{T}\int_0^T v_o(\omega t)\cos n\omega t \, d\omega t$$

$$= \frac{2}{2\pi}\int_0^{2\pi} v_o(\omega t)\cos n\omega t \, d\omega t \qquad \text{since } T = 2\pi$$

$$= \frac{1}{\pi}\left\{\int_{\frac{\pi}{6}}^{\frac{5\pi}{6}} V_s \cos n\omega t \, d\omega t + \int_{\frac{7\pi}{6}}^{\frac{11\pi}{6}} (-V_s)\cos n\omega t \, d\omega t\right\}$$

$$= \frac{V_s}{\pi}\left\{\left[\frac{\sin n\omega t}{n}\right]_{\frac{\pi}{6}}^{\frac{5\pi}{6}} - \left[\frac{\sin n\omega t}{n}\right]_{\frac{7\pi}{6}}^{\frac{11\pi}{6}}\right\}$$

$$= \frac{V_s}{n\pi}\left\{\sin\frac{5\pi}{6}n - \sin\frac{\pi}{6}n - \sin\frac{11\pi}{6}n + \sin\frac{7\pi}{6}n\right\}$$

$$= \frac{V_s}{n\pi}\left\{\sin\left(\pi - \frac{\pi}{6}\right)n - \sin\frac{\pi}{6}n - \sin\left(2\pi - \frac{\pi}{6}\right)n + \sin\left(\pi + \frac{\pi}{6}\right)n\right\}$$

$$= \frac{V_s}{n\pi}\left\{\sin\left(n\pi - \frac{n\pi}{6}\right) - \sin\frac{n\pi}{6} - \sin\left(2n\pi - \frac{n\pi}{6}\right) + \sin\left(n\pi + \frac{n\pi}{6}\right)\right\}$$

$$= \frac{V_s}{n\pi}\left\{-\cos n\pi \sin\frac{n\pi}{6} - \frac{\sin n\pi}{6} + \sin\frac{n\pi}{6} + \cos n\pi\ \sin\frac{n\pi}{6}\right\}$$

$$= 0 \quad \text{for all n.}$$

To obtain b_n

$$b_n = \frac{2}{T}\int_0^T v_o(\omega t)\sin n\omega t\ d\omega t$$

$$= \frac{2}{2\pi}\left\{\int_{\frac{\pi}{6}}^{\frac{5\pi}{6}} V_s \sin n\omega t\, d\omega t + \int_{\frac{7\pi}{6}}^{\frac{11\pi}{6}} (-V_s)\sin n\omega t\, d\omega t\right\}$$

$$= \frac{V_s}{\pi}\left\{\left[\frac{-\cos n\omega t}{n}\right]_{\frac{\pi}{6}}^{\frac{5\pi}{6}} - \left[\frac{-\cos n\omega t}{n}\right]_{\frac{7\pi}{6}}^{\frac{11\pi}{6}}\right\}$$

$$= \frac{V_s}{n\pi}\left\{-\cos\frac{5n\pi}{6} + \cos\frac{n\pi}{6} + \cos\frac{11n\pi}{6} - \cos\frac{7n\pi}{6}\right\}$$

$$= \frac{V_s}{n\pi}\left\{-\cos\left(n\pi - \frac{n\pi}{6}\right) + \cos\frac{n\pi}{6} + \cos\left(2n\pi - \frac{n\pi}{6}\right) - \cos\left(n\pi + \frac{n\pi}{6}\right)\right\}$$

$$= \frac{V_s}{n\pi} 2\cos\frac{n\pi}{6}(1 - \cos n\pi)$$

Here $(1 - \cos n\pi) = \begin{cases} 0\ for\ \ n = 2,4,6,\ldots\ldots\ldots i.e. even\ n. \\ 2\ for\ \ n = 1,3,5,7,\ldots\ldots i.e. odd\ n. \end{cases}$

and $\cos\frac{n\pi}{6} = 0$ for n = 3, 9, 15, ... i.e. multiples of 3

$\therefore \quad b_n = \frac{4V_s}{n\pi}\cos\frac{n\pi}{6}$ for odd values of n except multiples of 3

...(5.5.7)

To obtain Fourier series

$$c_n = \sqrt{a_n^2 + b_n^2} = b_n \quad \text{since } a_n = 0$$

$$\therefore \quad c_n = \frac{4V_s}{n\pi}\cos\frac{n\pi}{6} \qquad ...(5.5.8)$$

$$\phi_n = \tan^{-1}\left(\frac{a_n}{b_n}\right) = 0 \quad \text{since } a_n = 0 \qquad ...(5.5.9)$$

From the waveform of Fig. 5.5.6, it is clear that $V_{o(av)} = 0$. Hence Fourier series will be,

$$v_{o(wt)} = \sum_{i=1}^{\infty} \frac{4V_s}{n\pi}\cos\frac{n\pi}{6}\sin(n\omega t) \qquad ...(5.5.10)$$

for n = odd values except multiples of 3.

Comment :

In the quasi-square wave all even harmonics and all multiples of 3rd harmonics are absent.

To obtain rms value of nth harmonic

RMS value of nth harmonic component of output will be,

$$V_{on(rms)} = \frac{C_n}{\sqrt{2}} = \frac{\frac{4V_s}{n\pi}\cos\frac{n\pi}{6}}{\sqrt{2}}$$

$$\therefore \quad \boxed{V_{on(rms)} = \frac{2\sqrt{2}\,V_s}{n\pi}\cos\frac{n\pi}{6}} \qquad ...(5.5.11)$$

Here n = odd values except multiples of 3.

= 1, 5, 7, 11, 13, 17, 19,.......

Example 5.5.4 : *For the quasi-square inverter feeding inductive load obtain the following :*

i) RMS value of fundamental component of output voltage

ii) Distortion factor of output voltage.

iii) Harmonic factor of output voltage.

Solution :

i) RMS value of fundamental component of output voltage

With n = 1 in equation 5.5.11 we get,

$$V_{01(rms)} = \frac{2\sqrt{2}\,V_s}{\pi}\cos\frac{\pi}{6}$$

$$\therefore \quad \boxed{V_{01(rms)} = 0.78\ V_s} \qquad ...(5.5.12)$$

ii) Distortion factor (DF)

$$DF = \frac{V_{01(rms)}}{V_{0(rms)}}$$

$$= \frac{0.78V_s}{V_s\sqrt{\frac{2}{3}}}$$

$$\therefore \quad DF = 0.955 \quad \text{or } 95.5\ \%$$

iii) Harmonic factor (HF)

$$HF = \sqrt{\frac{1}{DF^2} - 1}$$

$$= \sqrt{\frac{1}{(0.955)^2} - 1}$$

$$\therefore \quad HF = 0.309 \quad \text{or } 30.9\ \%$$

Example 5.5.5 : *For a quasi-square wave inverter feeding inductive load derive expressions maximum and minimum values of output current.*

Solution : The output current waveform is shown in Fig. 5.5.7 for reference. This waveform is given in Fig. 5.5.4 earlier.

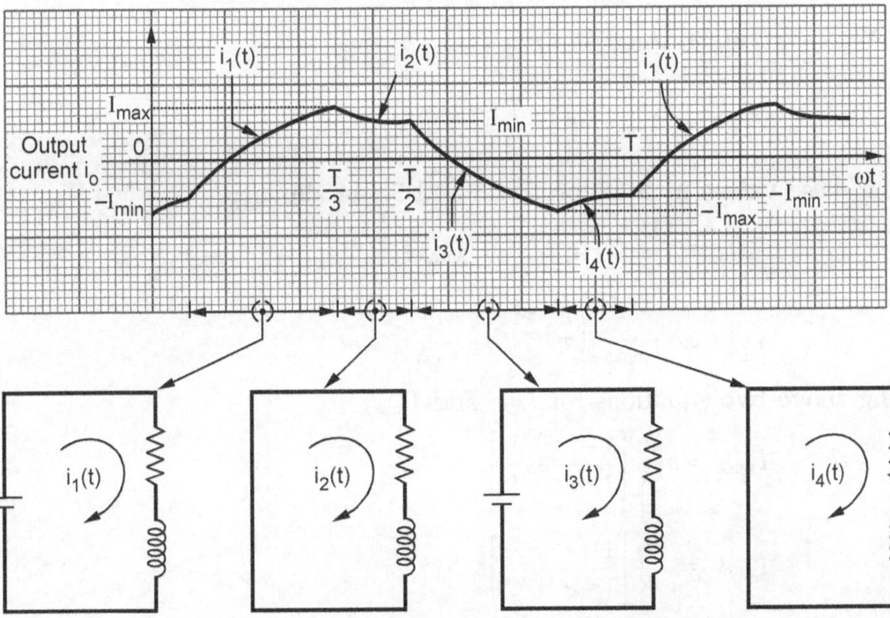

Fig. 5.5.7 Various currents in different modes

By applying KVL to the above equivalent circuits we can write,

$$L\frac{di_1(t)}{dt} + Ri_1(t) = V_s$$

$$L\frac{di_2(t)}{dt} + Ri_2(t) = 0$$

$$L\frac{di_3(t)}{dt} + Ri_3(t) = V_s$$

$$L\frac{di_4(t)}{dt} + Ri_4(t) = 0$$

Let us solve first two equations of $i_1(t)$ and $i_2(t)$ using Laplace transform with initial conditions :

$$i_1(t) = -I_{min} \quad \text{at} \quad t = 0$$

$$i_2(t) = I_{max} \quad \text{at} \quad t = \frac{T}{3}$$

$$\therefore \quad i_1(t) = \frac{V_s}{R}(1-e^{-t/\tau}) - I_{min}e^{-t/\tau} \qquad \text{...(5.5.13)}$$

and $$i_2(t) = I_{max}e^{-(t-T/3)/\tau} \qquad \text{...(5.5.14)}$$

Here $$\tau = \frac{L}{R} \text{ and } T = \frac{1}{f}$$

We have, $$i_1(t) = I_{max} \text{ at } t = \frac{T}{3}$$

and, $$i_2(t) = I_{min} \text{ at } t = \frac{T}{2}$$

Putting these values in equation 5.5.13 and 5.5.14 above,

$$I_{max} = \frac{V_s}{R}(1 - e^{-T/3\tau}) - I_{min}\, e^{-T/3\tau}$$

and $$I_{min} = I_{max}e^{-\left(\frac{T}{2}-\frac{T}{3}\right)/\tau} = I_{max}e^{-T/6\tau}$$

Solving above two equations for I_{max} and I_{min},

$$I_{max} = \frac{V_s}{R}\left[\frac{1-e^{-T/3\tau}}{1+e^{-T/2\tau}}\right] \qquad \text{...(5.5.15)}$$

$$I_{min} = \frac{V_s}{R}\left[\frac{1-e^{-T/3\tau}}{1+e^{-T/2\tau}}\right]e^{-T/6\tau} \qquad \text{...(5.5.16)}$$

5.5.3 Comparison between Square Wave and Quasi-Square Wave Inverter

Following table gives the comparison between the two inverters :

Sr. no.	Parameter	Square wave	Quasi-square wave
1.	Waveform	Complete square wave 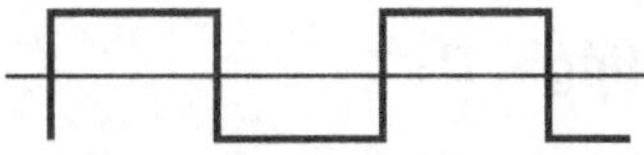	Quasi-square wave 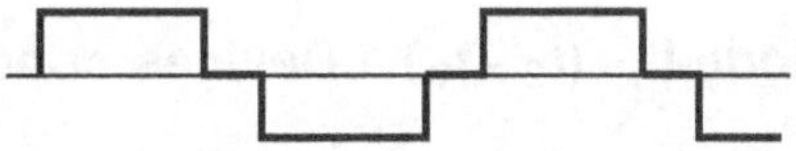
2.	Harmonics	Even harmonics are absent	Even harmonics and multiples of 3rd hamonic are absent
3.	V_{01} (rms)	0.9 V_s	0.78 V_s
4.	HF	48.4%	30.9%
5.	Cross conduction	Should be avoided by providing delays in drives	Totally absent
6.	Operation & control	Simple	Complex

Table 5.5.2 square wave and quasi-square wave inverters

5.6 Push-Pull Inverter

5.6.1 Circuit Diagram

Fig. 5.6.1 shows the circuit diagram of a push-pull inverter. T_1 and T_2 are two switches that generate positive and negative half cycles. D_1 and D_2 are feedback diodes. Note that center tap transformer is necessary for push-pull inverter.

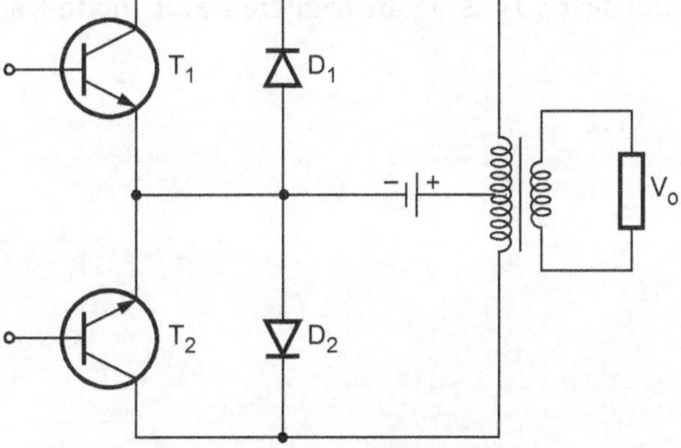

Fig. 5.6.1 Push-pull inverter

5.6.2 Operation with Inductive Load

Fig 5.6.2 shows the circuit diagram and waveforms of 1 ϕ push-pull inverter. The operation of the circuit can be explained with the help of 4 modes starting from t_1.

Mode- I - (t_1 - t_2) - Devices conducting - T_1

In this mode T_1 conducts. Equivalent circuit-I in Fig. 5.6.2 shows the current path. Observe that output voltage is positive and load current increases from zero to positive value.

Mode-II - (t_2 - t_3) - Devices conducting - D_2

When T_2 is turned-off, this mode begins. The load is inductive. Therefore it tries to maintain current in same (i.e. positive) direction. Therefore output voltage polarity is reversed. Observe that output voltage becomes negative at t_2. Equivalent circuit-I in Fig. 5.6.2 shows the current path. Because of this reversal of polarity, diode D_2 is forward biased. Note that this diode D_2 current flows against the DC supply. Thus the energy from load is fed back to DC supply through center-tap transformer. This mode comes to end when output current becomes zero at t_3.

Mode-III - (t_3 - t_4) - Devices conducting - T_2

Note that even though T_2 is given base drive at t_2, it does not conduct, since reverse bias is placed across T_2 due to diode D_2. But as soon as D_2 stops conducting, T_2 starts conducting. Current increases in negative direction. Equivalent circuit - III in Fig. 5.6.2 shows the voltage polarities and current paths. Output voltage is negative when T_2 conducts.

Mode-IV - (t_4 - t_5) - Devices conducting - D_1

This mode starts when T_2 is turn-off. The load inductance tries to maintain current in the same direction. Hence its polarity is reversed. Note that v_0 is positive after t_4. This also generates reverse voltage in center tap transformer primary. Therefore diode D_1 is forward biased and it conducts. This mode comes to end at t_5 when current becomes zero. At t_5, transistor T_1 is again triggered and mode-I starts. Thus the cycle repeats.

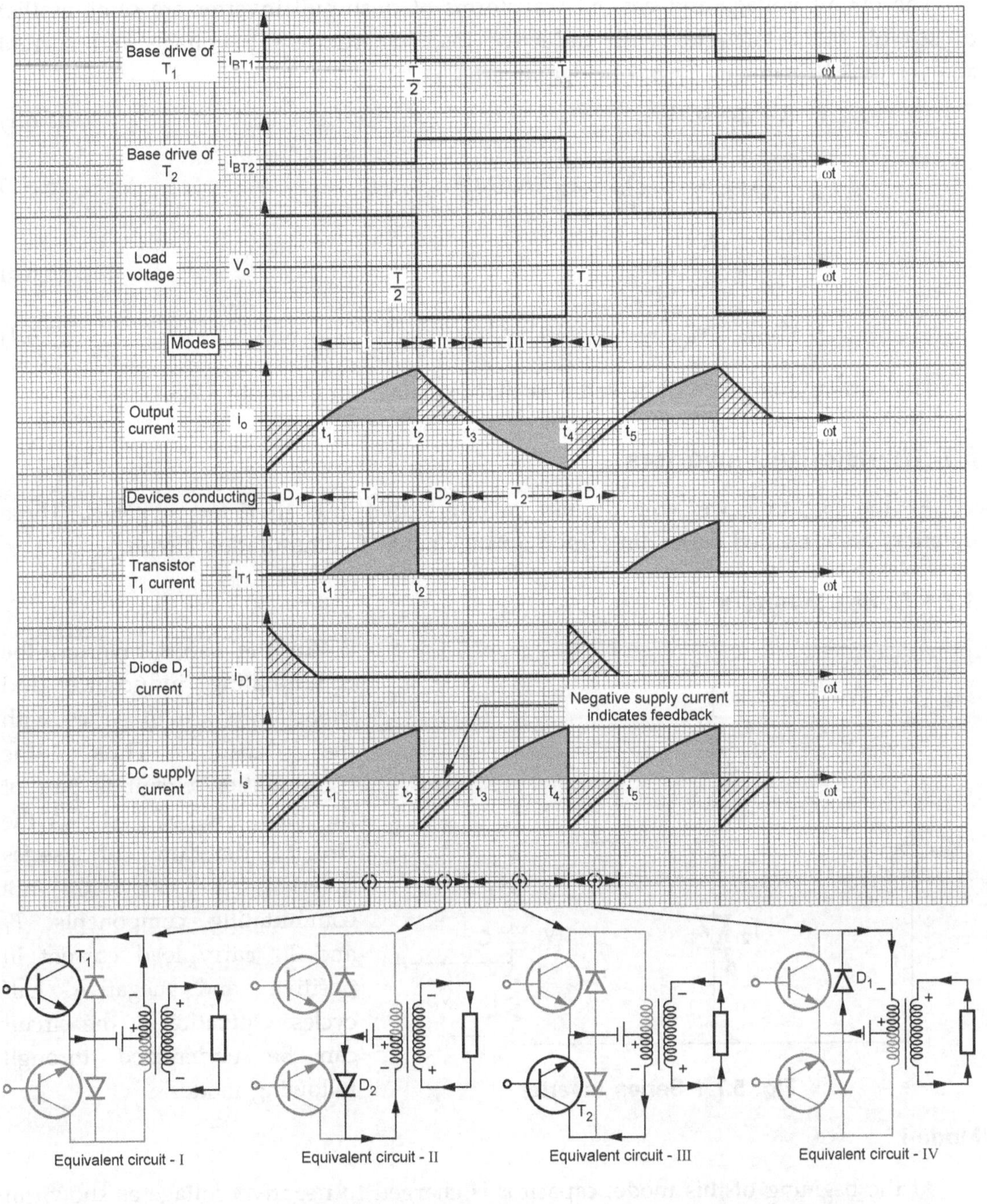

Fig. 5.6.2 Operation of push-pull inverter with inductive load

5.6.3 Mathematical Analysis of Push-Pull Inverter

The output voltage and current waveforms of push-pull inverter are same as that of 1ϕ full bridge inverter with inductive load. Therefore the voltage and current analysis will be same as 1ϕ full bridge inverter i.e.,

$$V_{o(rms)} = V_s \quad ...(5.6.1)$$

$$v_o(\omega t) = \sum_{n=1,3,5,.......}^{\infty} \frac{4V_s}{n\pi} \sin n\omega t \quad ...(5.6.2)$$

$$V_{o1(rms)} = 0.9V_s \quad ...(5.6.3)$$

$$V_{on(rms)} = \frac{0.9V_s}{n} \quad ...(5.6.4)$$

The current analysis will be same as that explained in example 3.2.3.

5.7 Thyristorized Inverters

At the begining we will discuss inverters that use thyristor switches. These inverters are classified depending upon arrangement of commutation circuit.

5.7.1 Series Inverters

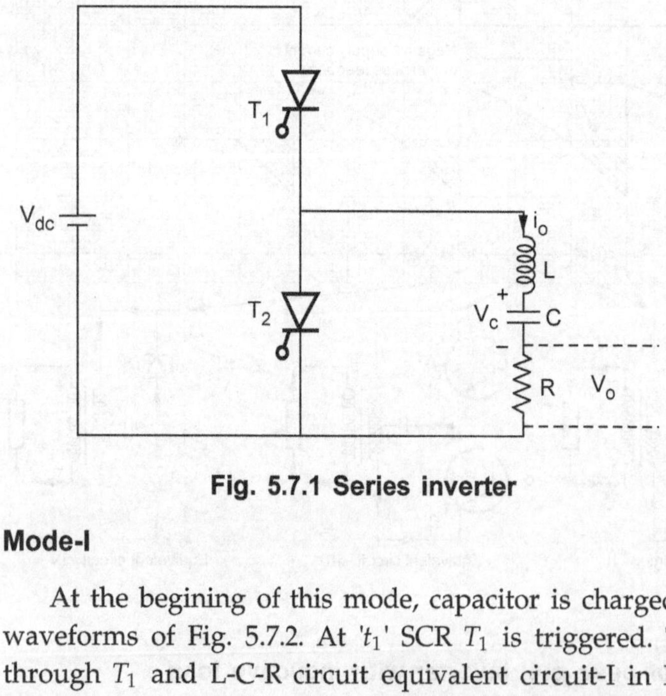

Fig. 5.7.1 Series inverter

In the series inverter, the commutating inductance and capacitance are in series with the load. Thus the commutation circuit is part of the load. Fig. 5.7.1 shows the circuit diagram of series inverter. L and C are commutating components. T_1 and T_2 carry load current in positive and negative half cycles. Operation of the circuit can be understood through following modes.

Mode-I

At the begining of this mode, capacitor is charged to negative voltage as shown in waveforms of Fig. 5.7.2. At 't_1' SCR T_1 is triggered. The output current starts flowing through T_1 and L-C-R circuit equivalent circuit-I in Fig. 5.7.2 shows the current path. Because of the RLC circuit, the current increases sinusoidally. The current becomes

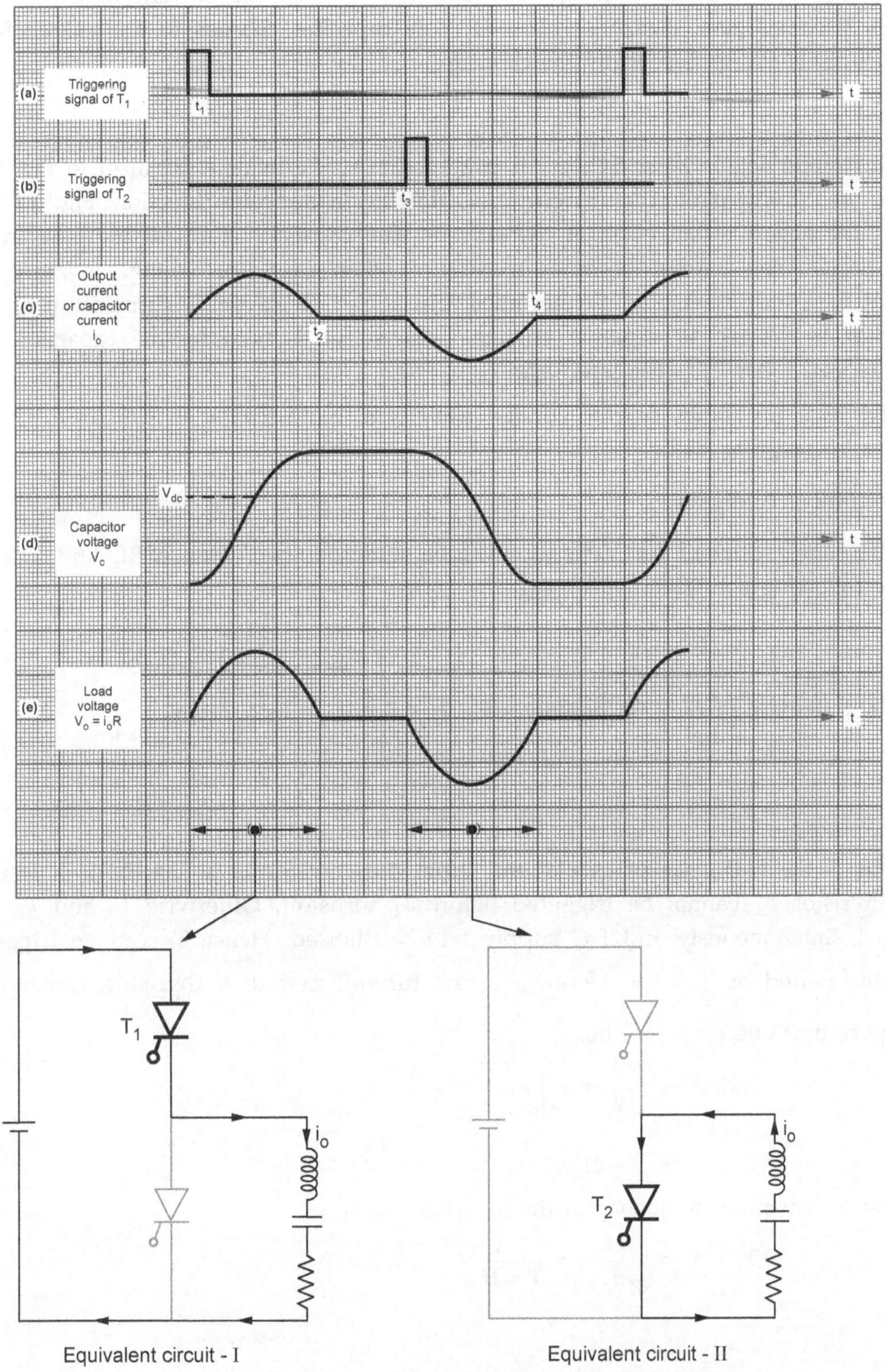

Fig. 5.7.2 Waveforms of series inverter

maximum when capacitor voltage is equal to V_{dc}. Then the current reduces. At t_2 current becomes zero. Hence T_1 turns-off. The capacitor charges to the value higher than V_{dc}. This charge is hold by the capacitor.

Mode-II

This mode begins when SCR T_2 is triggered at t_3. Equivalent circuit-II in Fig. 5.7.2 shows the current path. The current starts flowing in opposite direction. Fig. 5.7.2 (c) shows the negative half cycle of the current. The capacitor starts discharging in the RLC circuit. The current becomes maximum when capacitor voltage is zero. The current then starts reducing and becomes zero at 't_4'. Therefore T_2 turns-off at t_4. The capacitor is charged to negative voltage. This charge is hold by the capacitor. The cycle repeats when T_1 is triggered again.

Fig. 5.7.2 (e) shows output voltage waveform. Note that it is similar to output current for resistive load.

Mathematical Analysis

The series resonant RLC circuit must be underdamped. Hence RL and C must satisfy following condition,

$$R < \sqrt{\frac{4L}{C}}$$

The half cycle of RLC oscillations is $t_2 - t_1 = t_4 - t_3$. It is given as,

$$\frac{T}{2} = t_2 - t_1 = t_4 - t_3 = \frac{\pi}{\sqrt{\frac{1}{LC} - \frac{R^2}{4L^2}}}$$

This is the half cycle of oscillations. From the waveforms of Fig. 5.7.2, it is clear that thyristor T_2 cannot be triggered before T_1 turns-off. Otherwise T_1 and T_2 will conduct simultaneously and DC supply will be shorted. Hence T_2 can be triggered after the period of $\frac{T}{2} + t_{off}$. Here t_{off} is the turn-off period of thyristor. Hence total time period of one cycle will be,

$$\text{Period} = \left(\frac{T}{2} + t_{off}\right) + \left(\frac{T}{2} + t_{off}\right)$$

$$= T + 2t_{off}$$

Hence maximum frequency of the inverter will be,

$$f = \frac{1}{period} = \frac{1}{T + 2t_{off}}$$

Disadvantages of series inverter :

1. Maximum frequency of the inverter is limited to ringing frequency of the RLC circuit.
2. Load current flows through commutating inductance and capacitance. They must have high current rating.
3. Energy is taken from the source only in one half cycle. Hence peak current requirement is more.
4. There is distortion in the output current and voltage waveform.

5.7.2 Modified Series Inverter

Modified series inverter eliminates some limitations of basic series inverter. Fig. 5.7.3 shows the circuit diagram of modified series inverter. In this circuit L_1 and L_2 are two mutually coupled inductors. This circuit operates as follows :

i) T_1 is triggered. Then current starts flowing through T_1-L_1-C-R-V_{dc}. Capacitor charges and current goes on reducing.

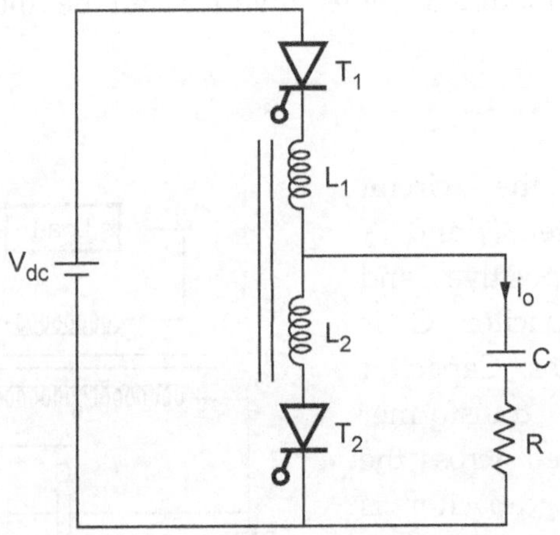

Fig. 5.7.3 Modified series inverter

ii) Let us assume that T_1 is conducting and small current is flowing. At this instant T_2 is triggered. Then capacitor starts discharging through $C - L_2 - T_2 - R$. This means 'i_o' reverses its direction. A large voltage is induced across L_2 with the polarity as shown in Fig. 5.7.4.

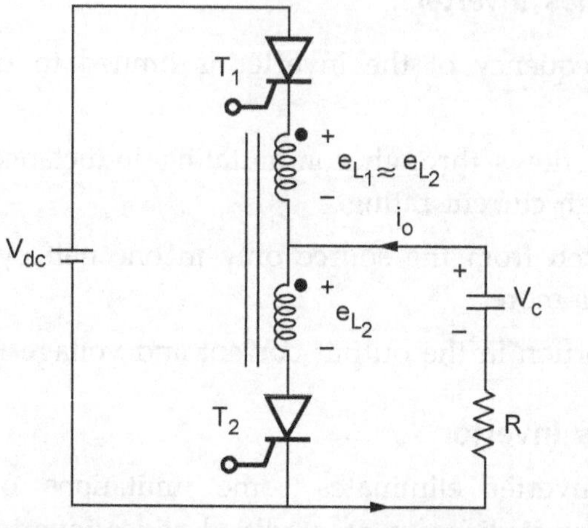

Fig. 5.7.4 Equivalent circuit when T_2 is triggered

Since L_1 and L_2 are coupled, same voltage is induced in L_1 also. Therefore $e_{L1} = e_{L2}$. The polarity of 'e_{L1}' is such that it reverse biases T_1. For reverse biasing of T_1, $e_{L1} + e_{L2}$ must be greater than V_{dc}. Therefore T_1 is turned-off. Thus T_2 can be triggered before T_1 is turned off. T_1 turns-off in this inverter due to coupled inductors. Therefore maximum frequency of modified series inverter can be made higher than ringing frequency of RLC circuit.

5.7.3 Parallel Inverter

Fig. 5.7.5 shows the circuit diagram of parallel inverter. T_1 and T_2 carry the current in positive and negative half cycles. Capacitor 'C' is the commutating capacitor. Capacitor is connected across the transformer primary. Load is connected across the secondary. Observe that capacitor is indirectly connected in parallel to the load. Hence this inverter is called parallel inverter. Operation of this circuit can be explained with following modes :

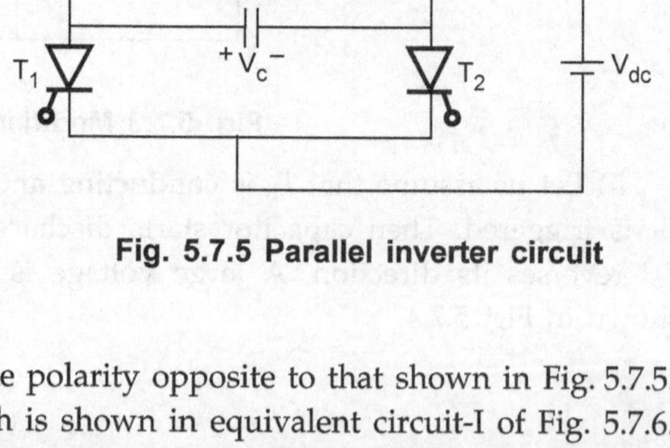

Fig. 5.7.5 Parallel inverter circuit

Mode-I

Initially the capacitor is charged to the polarity opposite to that shown in Fig. 5.7.5. Thyristor T_1 is triggered. The current path is shown in equivalent circuit-I of Fig. 5.7.6. Observe that the supply voltage 'V_{dc}' appears across half of the primary winding.

Hence the voltage of $2V_{dc}$ is induced across the primary by autotransformer action. The capcitor is connected across primary. Hence it also charges to voltage $2V_{dc}$. The capacitor remains charged to this level.

Mode-II

To initiate negative half cycle, thyristor T_2 is triggered. As soon as T_2 is triggered, a capacitor voltage of $-2V_{dc}$ is applied across T_1. Hence T_1 immediately turns-off. The load current starts flowing through other half of the primary winding. This is shown in equivalent circuit-II of Fig. 5.7.6. The load voltage becomes negative as shown in the waveforms.

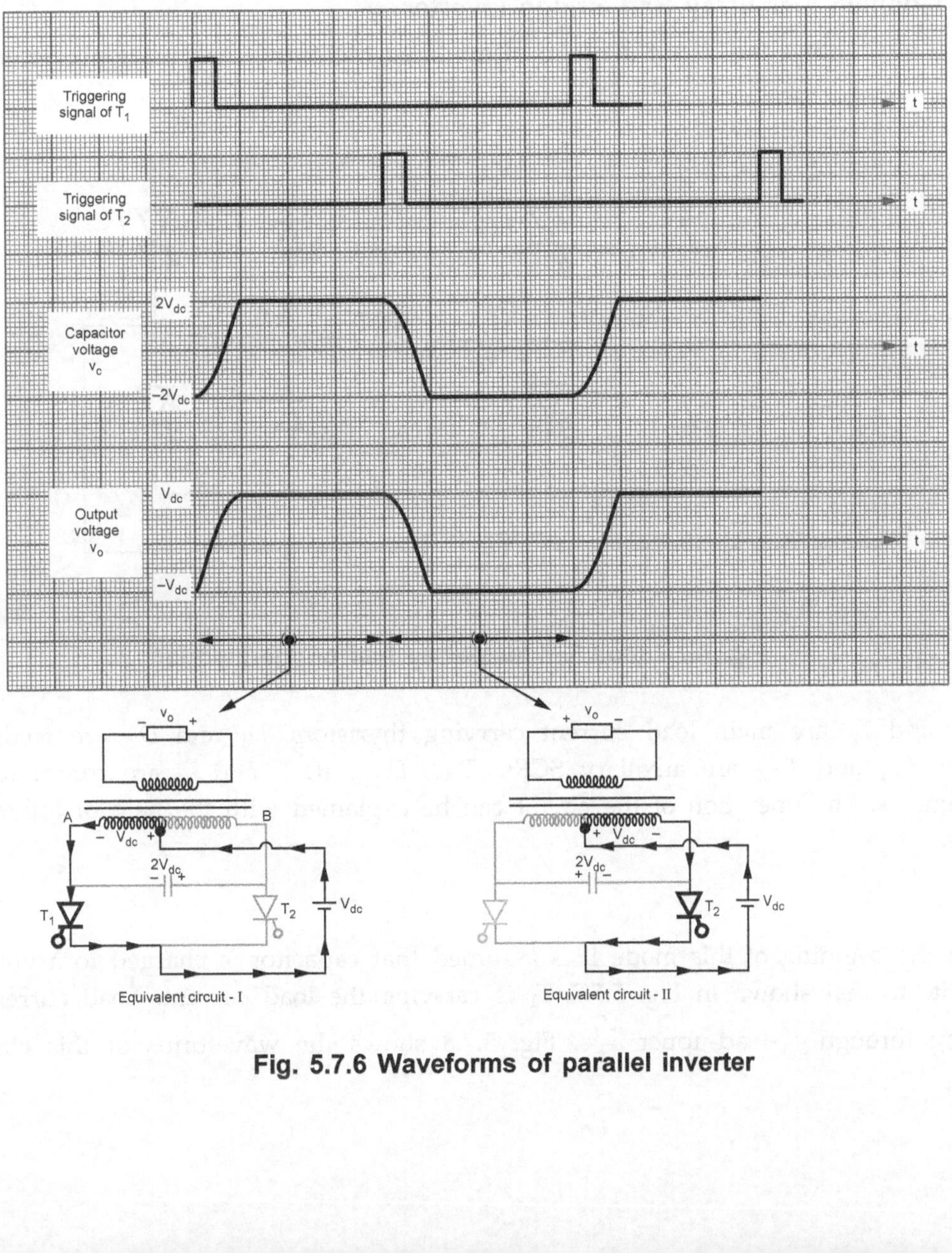

Fig. 5.7.6 Waveforms of parallel inverter

Advantages

1. Output voltage is square wave, which is better than series inverter.
2. Simple commutating circuit.
3. Switching frequency is higher.

Disadvantages

1. Large amount of energy in trapped in commutating capacitor. It is to be removed with the help of additional feedback diodes.
2. Heavy transformer is required to carry load current.

5.7.4 Modified McMurray Half Bridge Inverter

Fig. 5.7.7 shows the circuit diagram of modified McMurray half bridge inverter.

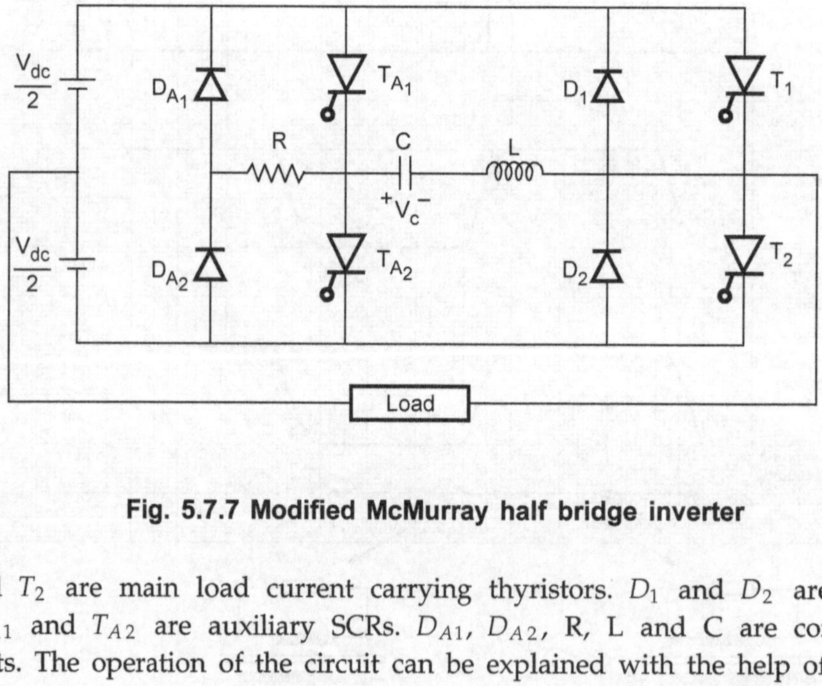

Fig. 5.7.7 Modified McMurray half bridge inverter

T_1 and T_2 are main load current carrying thyristors. D_1 and D_2 are feedback diodes. T_{A1} and T_{A2} are auxiliary SCRs. D_{A1}, D_{A2}, R, L and C are commutating components. The operation of the circuit can be explained with the help of following modes.

Mode-I

At the begining of this mode it is assumed that capacitor is charged to a voltage opposite to that shown in Fig. 5.7.7. T_1 is carrying the load current. Load current is flowing through T_1-load-upper $\frac{V_{dc}}{2}$. Fig. 5.7.8 shows the waveforms of this circuit.

Turn-off of T_1 is initiated by triggering T_{A1}. Hence capacitance starts discharging through $C-L-T_1-T_{A1}$. Equivalent circuit-I in Fig. 5.7.8 shows this current path by dotted line. It is assumed that load current (I_o) is constant during commutation. Hence it continues to flow through T_1 as earlier. The capacitor discharge current goes on increasing. Therefore current through T_1 goes on reducing. At 't_1' $i_c = I_o$, hence current through T_1 becomes zero and it turns-off.

Mode-II

This mode begins at 't_1' when capacitor current becomes equal to load current. Hence T_1 turns-off. Equivalent circuit-II in Fig. 5.7.8 shows the current path. The capacitor discharge current now flows through two paths. One path is $C-L-D_1-T_{A1}$ shown by dotted line. This is an excess current, that is above load current. Diode D_1 starts conducting after T_1 is turned off. Second path is C-L-load- $\frac{V_{dc}}{2}$ -T_{A1}. Current is fixed (I_o) in this path. The capacitor current reaches to maximum when capacitor voltage is zero. The current then starts reducing and capacitor starts charging in positive direction. This mode ends when $i_c = I_o$ again at t_2.

Mode-III

The capacitor current now falls below I_o. The capacitor current flows through T_{A1}-C-L-load-upper $\frac{V_{dc}}{2}$. This path is shown in equivalent circuit-III in Fig. 5.7.8. This current flow charges the capacitor above $\frac{V_{dc}}{2}$. The load inductance tries to maintain load current fixed. Hence remaining load current starts flowing through D_2-load-lower $\frac{V_{dc}}{2}$. Thus capacitor current and diode D_2 current combinely make load current. D_2 current goes on increasing as capacitor current reduces. Fig. 5.7.8 (d) shows the waveform of D_2 current. This mode ends at t_3 when capacitor current is completely zero and load current is fully carried by D_2. Therefore T_{A1} turns-off.

Mode-IV

The capacitor current now reverses its direction. It passes through C-R-D_{A1}-upper $\frac{V_{dc}}{2}$ – lower $\frac{V_{dc}}{2}$ – D_2-L. Equivalent circuit-IV in Fig. 5.7.8 shows this current path. And the load current flows through D_2-load-lower $\frac{V_{dc}}{2}$. Note that the load current is now flowing against the supply voltage, hence it starts reducing after t_3. It is shown in Fig. 5.7.8 (b). The capacitor current is negative and its energy is fedback to the source through diode D_{A1} and D_2. The capacitor voltage also reduces. This mode ends at t_4 when capacitor current again goes to zero.

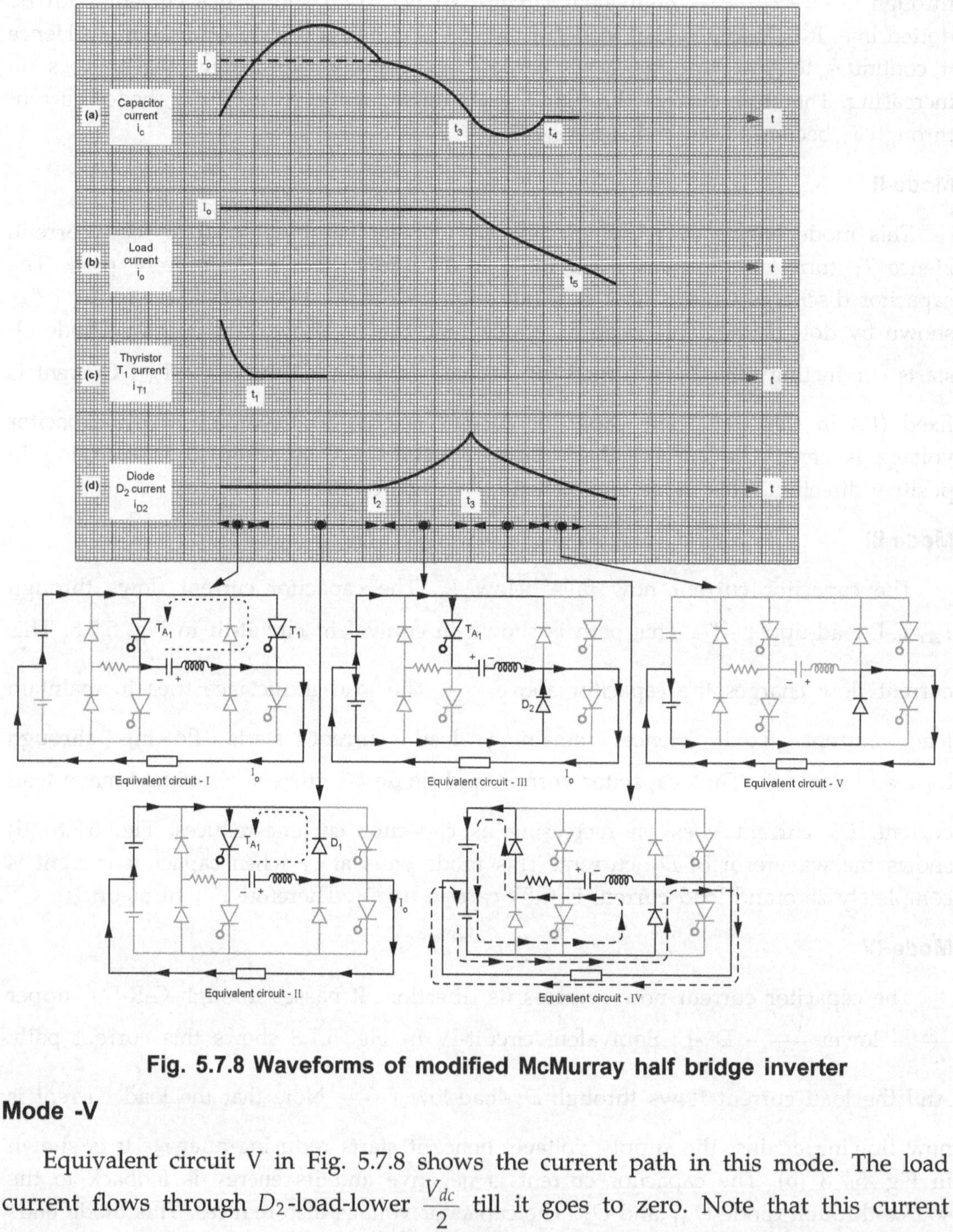

Fig. 5.7.8 Waveforms of modified McMurray half bridge inverter

Mode -V

Equivalent circuit V in Fig. 5.7.8 shows the current path in this mode. The load current flows through D_2-load-lower $\frac{V_{dc}}{2}$ till it goes to zero. Note that this current flows due to load inductance. At t_5 the load current is zero and this mode ends. When load current becomes zero, D_2 turns-off and T_2 is forward biased. Hence T_2 can be triggered to initiate negative half cycle.

Design of commutation components

A reverse bias is placed across T_1 as long as D_1 conducts from t_1 to t_2. Hence t_2-t_1 is the commutation period. i.e.,

$$t_c = t_2 - t_1$$

Then capacitance and inductance values are given as,

$$C = 0.893 \frac{I_{o(max)}\, t_c}{V_{min}}$$

and
$$L = 0.397 \frac{V_{min}\, t_c}{I_{o(max)}}$$

Here V_{min} is the minimum value of supply voltage.

$I_{o(max)}$ is the maximum value of load current.

5.7.5 Modified McMurray Bedford Half Bridge Inverter

Fig. 5.7.9 shows the circuit diagram of modified McMurray Bedford half bridge inverter. In this circuit, T_1 and T_2 are the main load current carrying thyristors. L_1 and L_2 are two magnetically coupled inductors. L_1, L_2, C_1 and C_2 form a commutation circuit. L_{e1}, L_{e2} and four diodes form an energy recovery circuit. The commutation energy is recovered and fed back to the source. This circuit has only two SCRs. There are no auxiliary thyristors.

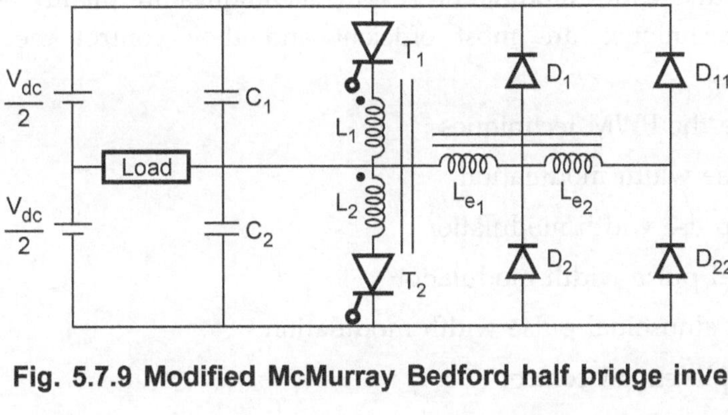

Fig. 5.7.9 Modified McMurray Bedford half bridge inverter

Let us assume that T_1 is conducting and carrying the load current. Then the load current flows through $T_1 - L_1-$ load–upper $\frac{V_{dc}}{2}$. The inductance L_1 is saturated and there is no voltage drop across it. Hence there is no charge on capacitor C_1. Therefore capacitor C_2 is charged to $\frac{V_{dc}}{2}+\frac{V_{dc}}{2}=V_{dc}$. Now T_2 is triggered to initiate turn-off of T_1 and begin negative half cycle of load current. As soon as T_2 is triggered, the complete capacitor voltage appears across L_2. This voltage is V_{dc} with polarity as shown in Fig. 5.7.10. This voltage is also induced in L_1 as shown in Fig. 5.7.10. Therefore T_1 is reverse biased and it turns-off. Thyristor T_2 now starts carrying the load current. The capacitor C_2 discharges through $C_2 - L_{e1} - D_1 - upper\ \frac{V_{dc}}{2} - lower\ \frac{V_{dc}}{2}$. Since L_{e1} and L_{e2} are coupled, current also flows through L_{e2} and D_{11}. Complete energy in C_2 and L_2 is recovered and fed back to DC source.

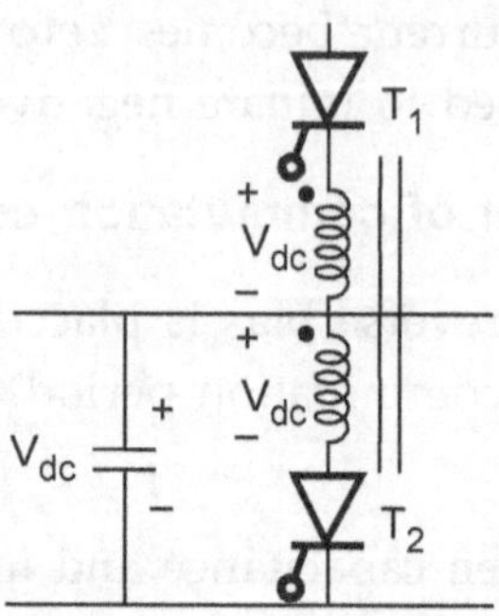

Fig. 5.7.10 Equivalent circuit showing commutation of T_1

5.8 Switching Techniques (Voltage Control and Harmonic Reduction)

The output voltage of the inverter needs to be varied as per load requirement. Whenever the input DC varies, the output voltage can change. Hence these variations needs to be compensated. In case of motor drives the ratio of voltage to frequency $\left(\frac{V}{f}\right)$ is maintained constant. The output voltage and frequency of the inverter is adjusted to keep $\frac{v}{f}$ constant. Similarly, in UPS the output voltage of inverter is to be regulated. These all the reasons indicate that the output voltage of inverter is to be controlled. The pulse width modulation (PWM) techniques are mainly used for voltage control. These techniques are most efficient and they control the drives of the switching devices.

Following are the PWM techniques :

(i) Single pulse width modulation

(ii) Multiple pulse width modulation

(iii) Sinusoidal pulse width modulation

(iv) Modified sinusoidal pulse width modulation

(v) Phase displacement control

Out of the above techniques, sinusoidal PWM techniques are most widely used. They control the output voltage as well as reduce the harmonics.

5.8.1 Single Pulse Modulation (SM)

This modulation gives quasi-square wave output.

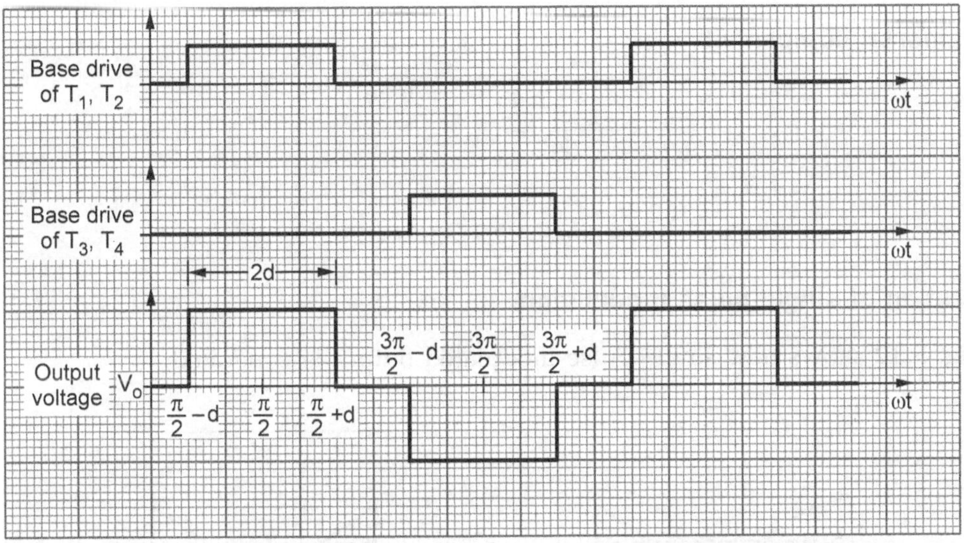

Fig. 5.8.1 Output voltage control using single pulse modulation

- Observe that there is single pulse of output voltage during each half cycle.
- The width of the pulse is '2d'. RMS value of output voltage can be controlled by varying the pulse width.

Mathematical analysis

We have done the complete mathematical analysis of quasi-square wave inverter. The same analysis is applicable to single pulse modulation. Following are the relations in terms of 'd'.

$$a_n = 0 \qquad \text{for all 'n'}$$

$$b_n = \frac{4V_s}{n\pi} \sin\frac{n\pi}{2} \sin nd$$

$$= \pm\frac{4V_s}{n\pi} \sin nd \quad \text{for odd n.}$$

$$\therefore \quad c_n = \sqrt{a_n^2 + b_n^2} = \frac{4V_s}{n\pi} \sin \text{nd}$$

$$\phi_n = \tan^{-1}\frac{a_n}{b_n} = 0$$

Fourier series will be,

$$v(\omega t) = \sum_{n=1,3,5,.......}^{\infty} \frac{4V_s}{n\pi} \sin nd \sin n\omega t$$

Fundamental component

$$V_{o1(rms)} = \frac{c_1}{\sqrt{2}} = \frac{\frac{4V_s}{\pi} \sin d}{\sqrt{2}}$$

$$= \frac{2\sqrt{2}\, V_s}{\pi} \sin d = 0.9\, V_s \sin d.$$

Generation of single pulse modulation

Fig. 5.8.2 shows how single pulse modulated output is generated.

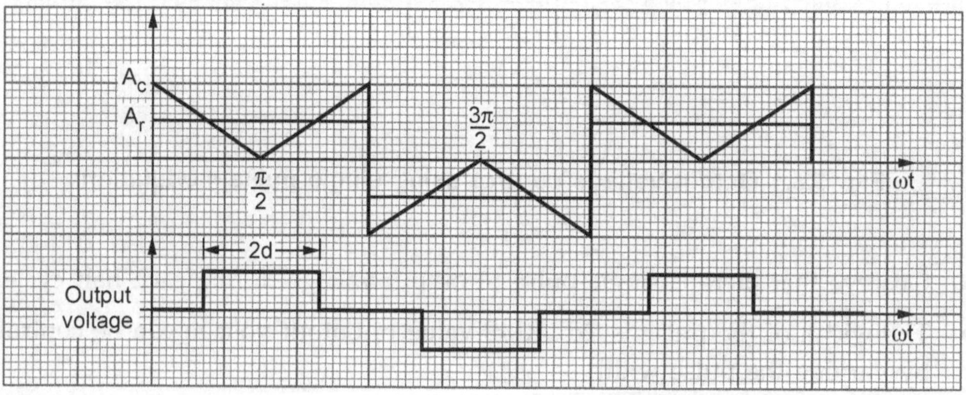

Fig. 5.8.2 Generation of SM output

The ratio of triangular signal amplitude (A_c) and square wave signal amplitude (A_r) is called modulation index. i.e.

$$m = \frac{A_r}{A_c}$$

The width of the pulse can be changed by varying the modulation index.

When m = 1, square wave output is obtained.

5.8.2 Multiple Pulse Width Modulation Technique

Principle : Multiple pulses are used to reduce the harmonic content. The width of all the pulses is same.

Waveforms and explanation

Fig. 5.8.3 shows the waveforms of multiple pulse width modulation.

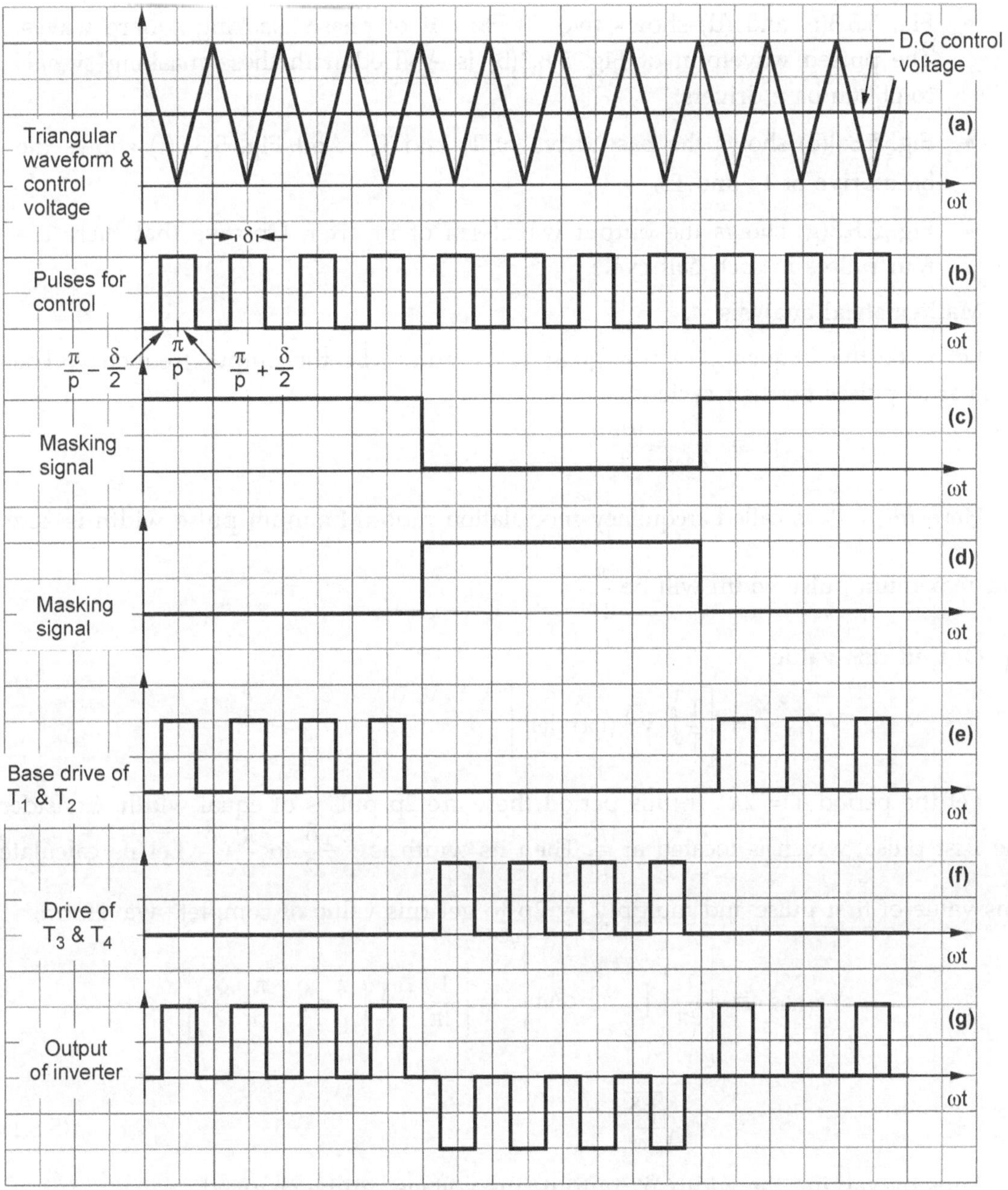

Fig. 5.8.3 Waveforms of multiple pulse width modulation

- Fig. 5.8.3 (a) shows the carrier signal and the reference signal. The carrier signal is the triangular waveform and reference signal is the DC voltage.

- They are compared and the pulsed waveform of Fig. 5.8.3 (b) is obtained. The width of the pulse can be varied by changing the amplitude of DC reference signal.
- Fig. 5.8.3(c) and (d) shows two 50 Hz out of phase masking square waves. The pulsed waveform of Fig. 5.8.3(b) is ANDed with these masking signals to obtain base drivers.
- Fig. 5.8.3(e) shows the base drive of T_1 and T_2. And Fig. 5.8.3(f) shows the base drive of T_3 and T_4.
- Fig. 5.8.3(g) shows the output waveform of inverter. Observe that there are four pulses in each half cycle.

Mathematical analysis

Let f_c be the frequency of triangular wave and f_0 be the output frequency. Then number of pulses per half cycle is,

$$p = \frac{f_c}{2 f_0} = \frac{m_f}{2}$$

Here $m_f = \frac{f_c}{f_0}$ is called frequency modulation ratio. Minimum pulse width is zero, and maximum pulse width will be $\frac{\pi}{p}$.

Output rms value

$$V_{0(rms)} = \left[\frac{1}{T}\int_0^T v^2(\omega t)\, d\omega t\right]^{1/2}$$

Let the period $T = 2\pi$. In this period there are 2p pulses of equal width. Consider the first pulse, which is located at $\frac{\pi}{p}$. Then its width is $\frac{\pi}{p}-\frac{\delta}{2}$ to $\frac{\pi}{p}+\frac{\delta}{2}$. Let us calculate rms value of first pulse and multiply by 2p to get rms value of complete waveform,

$$V_{0(rms)} = \left[\frac{1}{2\pi}\int_{\frac{\pi}{p}-\frac{\delta}{2}}^{\frac{\pi}{p}+\frac{\delta}{2}} V_s^2\, d\omega t\right]^{1/2} = \left[\frac{1}{2\pi}\cdot V_s^2 \cdot \left(\frac{\pi}{p}+\frac{\delta}{2}-\frac{\pi}{p}+\frac{\delta}{2}\right)\right]^{1/2}$$

$$= V_s\sqrt{\frac{p\delta}{\pi}} \qquad \text{... (5.8.1)}$$

Thus by varying the width 'δ', output rms voltage can be changed.

Advantages

i) Distortion factor is reduced compared to single pulse modulation.

ii) As value of 'p' increases amplitudes of lower harmonics reduces.

Disadvantages

i) With increased number of pulses, switching losses are increased.

ii) Control scheme is complex.

5.8.3 Sinusoidal PWM (SPWM)

The sinusoidal pulse width modulation (PWM) is most widely used method of voltage control in inverters. The width of each pulse is weighted by the amplitude of sine wave at that instant. Fig. 5.8.4 shows the scheme of generating sinusoidal PWM.

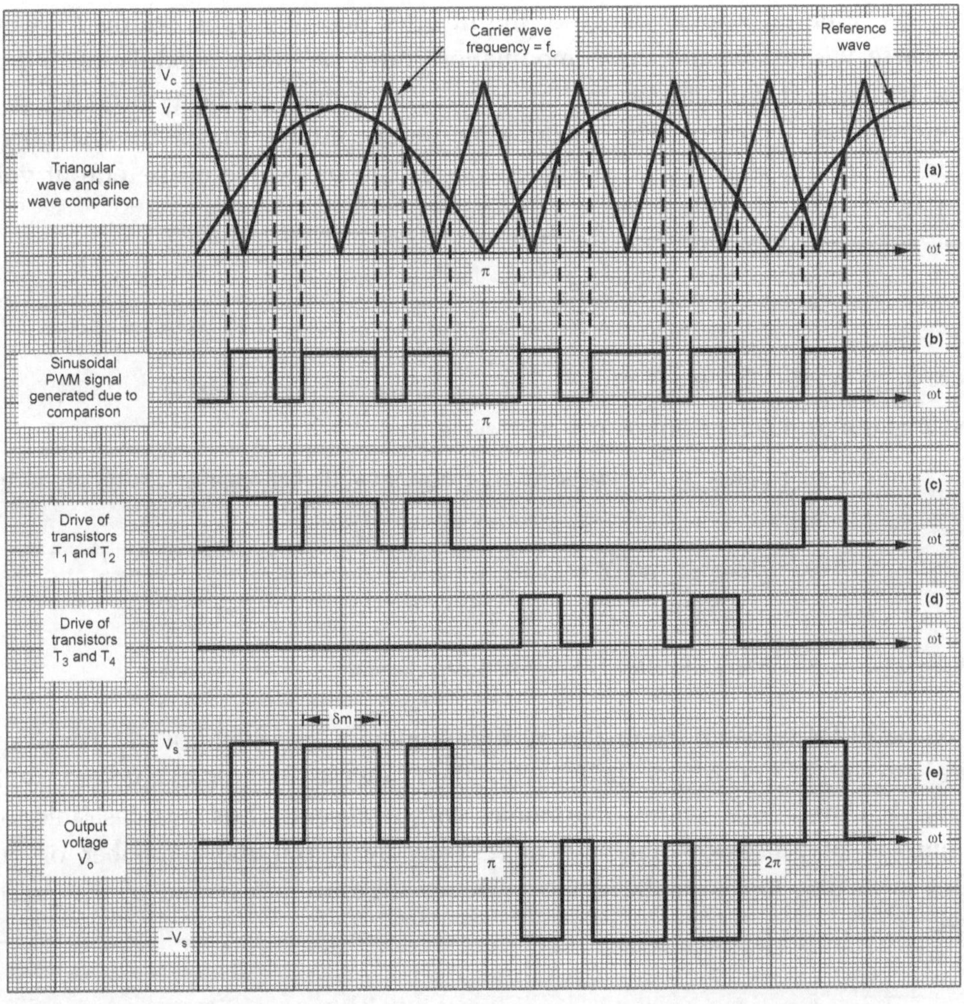

Fig. 5.8.4 Scheme of generating sinusoidal PWM and corresponding inverter output

The carrier signal is a triangular wave of frequency 'f_c' and amplitude 'V_c'. The control signal is a sine wave of frequency 'f' and amplitude 'V_r'. Here 'f' becomes the frequency of the output of inverter. The sine wave and triangular waves are compared and the PWM signal is prepared as shown in Fig. 5.8.4. From this PWM, the drives for the transistors in the inverter are prepared. Fig. 5.8.4 (e) shows the output waveform of the bridge inverter when sine PWM drives are applied. Note that the widths of the pulses are sine weighted. In the waveform of Fig. 5.8.4, there are three pulses per half cycle.

The widths of the pulses depend upon the amplitude (V_r) of the sine wave. If amplitude is increased, widths increase. The ratio of V_r to V_c is called modulation index. i.e.,

$$\text{Modulation index,} \quad m = \frac{V_r}{V_c} \qquad \text{... (5.8.2)}$$

The rms value of output voltage of the inverter depends upon widths (δ_m) of the pulses. These widths depend upon modulation index 'm'. Thus modulation index 'm' controls the output voltage of the inverter. The output voltage increases as the modulation index increases. The modulation index can be varied by changing the amplitude of the reference signal (i.e. sine wave). The rms value of the output voltage is given as,

$$V_{o\,(rms)} = V_s \left[\sum_{m=1}^{p} \frac{\delta_m}{\pi} \right]^{\frac{1}{2}} \qquad \text{... (5.8.3)}$$

Here δ_m is width of the mth pulse

and p is the number of pulses per half cycle.

The number of pulses 'p' depends upon frequency of the carrier i.e. f_c. It is given as,

$$p = \frac{f_c}{2f} = \frac{m_f}{2} \qquad \text{... (5.8.4)}$$

Here $m_f = \frac{f_c}{f}$ is frequency modulation ratio.

The sine PWM (SPWM) eliminates all the harmonics less than $(2p-1)$. For example if there are three pulses per half cycle, then lowest harmonics present will be $(2 \times 3 - 1) = 5$. Thus 2nd, 3rd and 4th harmonics will be absent. Thus it is desirable to use more number of pulses to eliminate more harmonics. But more number of pulses increases switching losses in the devices. Hence the optimum value of the pulses is selected.

5.8.4 Modified Sinusoidal PWM (MSPWM)

The widths of the pulses near peak of the sine wave do not change much when modulation index is changed. Hence carrier is suppressed at ± 30° in the neighbourhood of peak of sine wave. Such scheme is shown in Fig. 5.8.5. Observe that the triangular wave is present for the period of first 60° and last 60° of the half cycle of sine wave. The middle 60° of the sine wave do not have the triangular wave. Hence the generated PWM have less number of pulses. The rms value is more for the same modulation index. The harmonic content is also reduced. This control scheme also reduces switching losses. The implementation of this scheme is relatively complex than sine PWM.

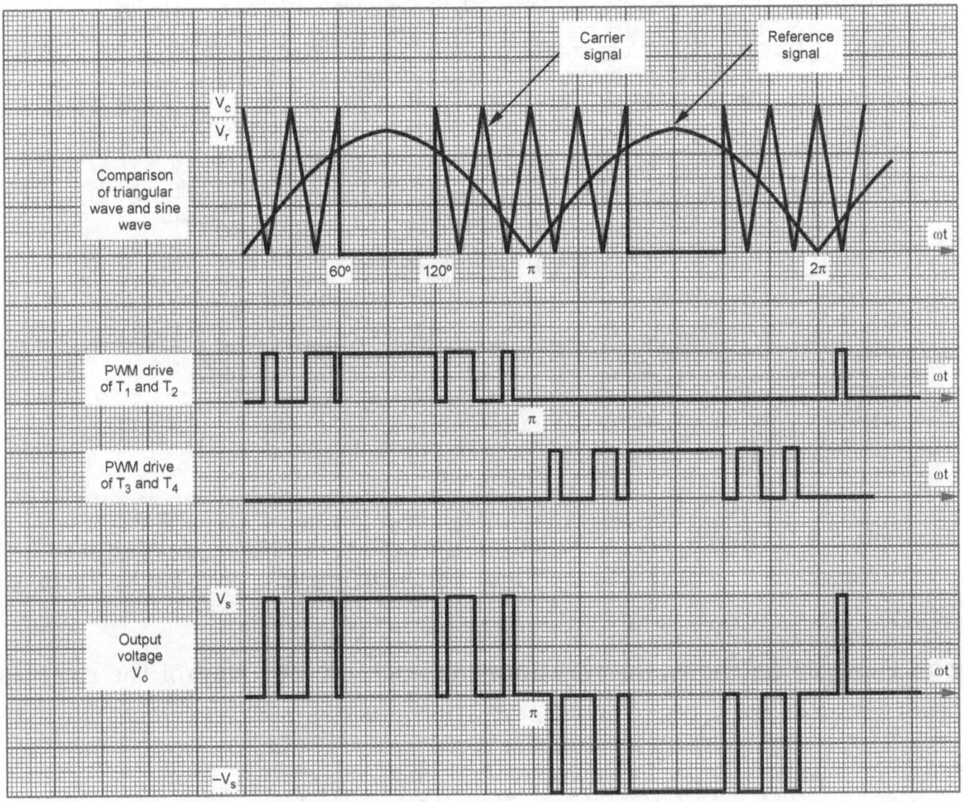

Fig. 5.8.5 Waveforms of modified sinusoidal PWM

5.8.5 Phase Displacement Control

Principle

The two inverters are connected in parallel. Output of one inverter is phase shifted with respect to another inverter. The net output depends upon phase shift between the two inverter outputs.

Waveforms

Fig. 5.8.6 shows the waveforms of this scheme. Fig. 5.8.6(a) and (b) shows the output voltage waveforms of inverter – 1 and inverter – 2. The voltage levels of output are $\frac{V_s}{2}$ in both the inverter outputs. Output of second inverter is phase shifted with respect to first inverter by 'β'. The final output is the difference of two inverter outputs. i.e.,

$$v_o = v_{o1} - v_{o2}$$

Note that $v_o = v_s$ when the two inverter outputs are out of phase. Output voltage can be controlled by varying the phase shift 'β'.

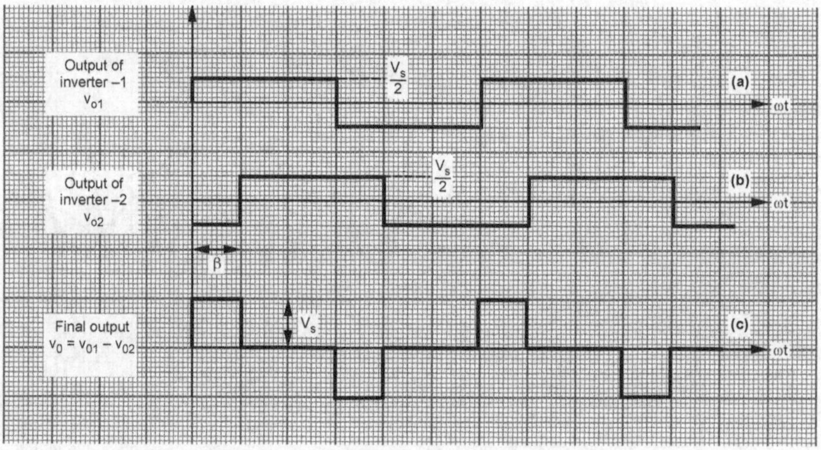

Fig. 5.8.6

Example 5.8.1: *Derive an expression for rms value of output of the inverter having phase displacement control. Assume that the two inverters are supplied with voltage $V_s / 2$.*

Solution : Fig. 5.8.6 (c) shows the waveforms of such inverter control.The rms value is given as,

$$V_{o\,(rms)} = \left[\frac{1}{T}\int_0^T v_o^2(\omega t)\, d\omega t\right]^{1/2} = \left[\frac{1}{\pi}\int_0^{\beta} V_s^2\, d\omega t\right]^{1/2}$$

$$= \left\{\frac{1}{\pi} V_s^2 [\omega t]_o^{\beta}\right\}^{1/2}$$

$$= V_s \sqrt{\frac{\beta}{\pi}} \quad \ldots (5.8.5)$$

And the rms value of fundamental output voltage is given as,

$$V_{o1(rms)} = 2\sqrt{2}V_s \sin\frac{\beta}{2} \quad \text{... (5.8.6)}$$

Example 5.8.2 : *Considering a single phase bridge inverter, explain the phase displacement method of output voltage control. If the dc voltage is 200 V and the required rms fundamental output voltage is 90 V, determine the delay angle β.*

Solution : Here $V_s = 200$ V and $V_{o1\ (rms)} = 90$ V

From equation 5.8.6,

$$V_{o1\ (rms)} = 2\sqrt{2}\, V_s \sin\frac{\beta}{2}$$

Putting the values,

$$90 = 2\sqrt{2} \times 200 \sin\frac{\beta}{2}$$

$$\therefore \quad \beta = 18.3^\circ$$

5.9 PWM IC LM 3524

- LM 3524 is PWM IC which is commonly used for inverter and chopper control.
- It generates the two drives which can be used to control push-pull inverter, half bridge, full bridge inverter or chopper.
- The PWM controller uses output current and output voltage as a feedback to control the width of the drives.

5.9.1 Features of LM 3524

i) ± 1% precision 5 V reference with thermal shut down.

ii) Output drive current upto 200 mA DC.

iii) 60 V output capability.

iv) Wide common mode input range for error amplifier.

v) Improved maximum duty cycle at high frequencies.

vi) Double pulse suppression.

5.9.2 Internal Block Diagram

Fig. 5.9.1 shows an internal block diagram of PWM control IC.

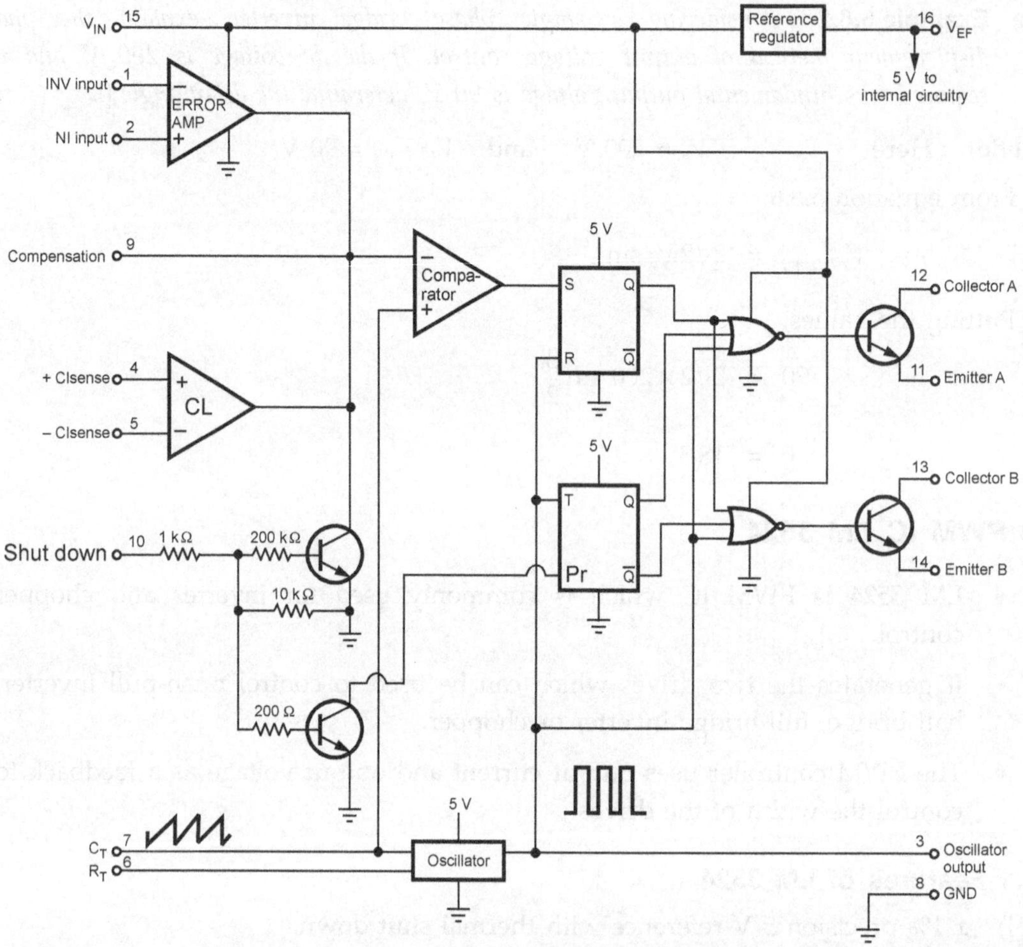

Fig. 5.9.1 Block diagram of PWM controller 3524

- The reference voltage V_{REF} at pin no. 5 is used by non-inverting input (pin2) of error amplifier.
- The inverting input of error amplifier (pin 1) is fed from output voltage feedback.
- Pin no. 4 and 5 are used for current limit sensing.
- Pin no. 10 is used for shut-down in case of fault.
- The combined output of shut-down, error amplifier and current sensing is given to inverting input of comparator.
- The other input of comparator (noninverting input pin no. 7) is fed from sawtooth waveform of external capacitor.

- The comparator produces the drive depending upon the amplitudes of sawtooth signal and reference voltage at inverting input.
- The oscillator work on the external R and C connected at pin no. 6 and 7.
- The output of oscillator is given to two flip-flops.
- The two flip flops generate two drives which are out of phase.
- The two open collector drives are available at pin no. 11-12 and pin no. 13-14.

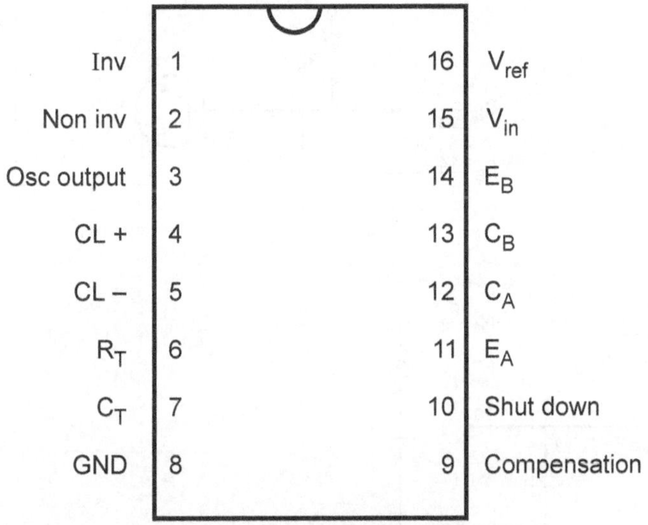

Fig. 5.9.2 Pin diagram LM 3524

5.9.3 Push-Pull Inverter Control using LM 3524

Fig. 5.9.3 shows the circuit diagram of control circuit for push-pull inverter using LM 3524.

- In the above circuit diagram observe that reference voltage is fed to non-inverting input through divider arrangement of R_1 an R_2.
- The feedback from output is taken through transformer and fed to inverting input through R_3 and R_4.
- The CL+, CL– inputs are grounded.
- The R_T and C_T values are selected depending upon oscillator frequency. Normally $f = \frac{1}{R_T C_T}$. Therefore output frequency will be $f_o = \frac{f}{2} = \frac{1}{2R_T C_T}$.
- The compensation pin (pin no. 9) is provided suitable RC compensation.
- The emitters E_A and E_B are connected to base of T_1 and T_2.
- When internal transistor turns-on, the base drives are applied through C_A and C_B.

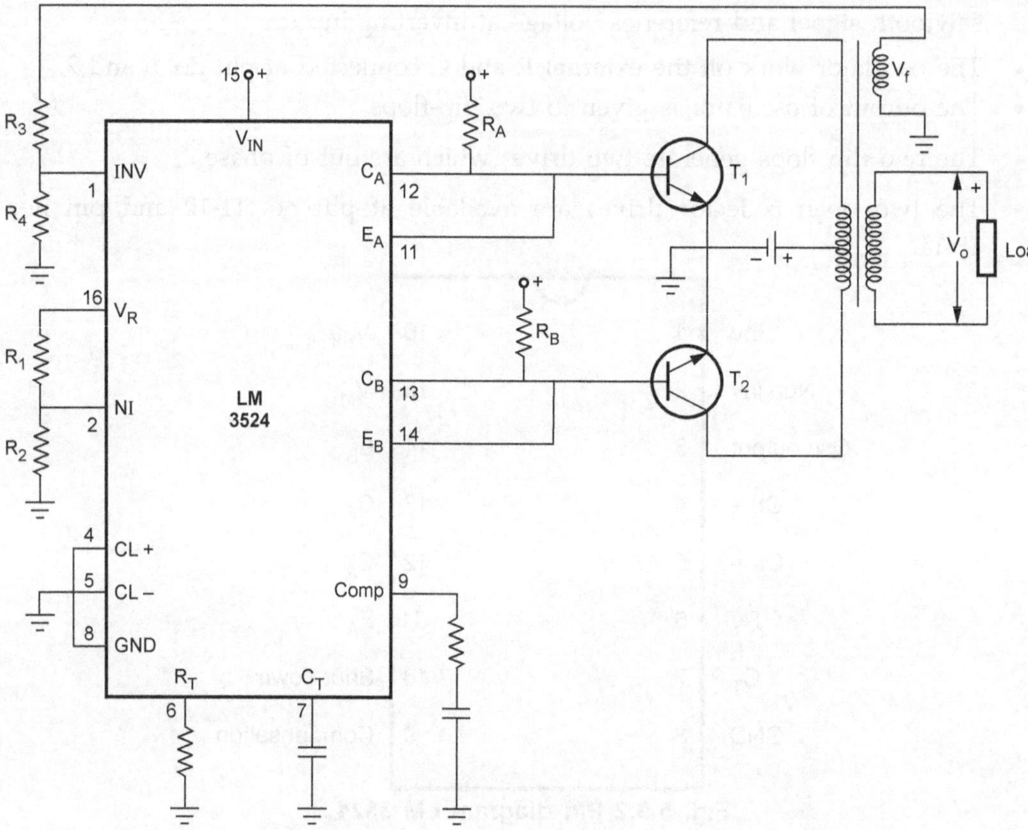

Fig. 5.9.3 Push-pull inverter control using LM 3524

- Output of the push-pull inverter is quasi-square wave depending upon voltages at inverting and non-inverting inputs of LM 3524.

Review Questions

1. *State the various techniques of voltage control in inverters. Explain the sinusoidal PWM technique.*
2. *Explain the selective elimination of harmonics. How it is done ?*
3. *Write short notes on*

   (i) Modified sinusoidal PWM

   (ii) Feedback operation in inverter

   (iii) Harmonic distortion in inverter.

Summary

1. Half bridge inverter

$$V_{o(rms)} = \frac{V_s}{2}$$

$$V_{1(rms)} = 0.45\ V_s$$

$$V_{n(rms)} = \frac{0.45\, V_s}{n}$$

$$v_o(\omega t) = \sum_{n=1,3,5,\ldots}^{\infty} \frac{2V_s}{n\pi} \sin n\omega t$$

2. Full bridge inverter

$$V_{o(rms)} = V_s$$

$$v_o(\omega t) = \sum_{n=1,3,5,\ldots}^{\infty} \frac{4V_s}{n\pi} \sin n\omega t$$

$$V_{1(rms)} = 0.9\ V_s$$

$$V_{n(rms)} = \frac{0.9\, V_s}{n}$$

3. 180° mode of 3 φ inverter

$$V_{line(rms)} = V_s\sqrt{\frac{2}{3}}$$

$$V_{ph(rms)} = \frac{\sqrt{2}}{3} V_s$$

$$V_{RY}(\omega t) = \sum_{n=1,3,5,\ldots}^{\infty} \frac{4V_s}{n\pi} \cos\frac{n\pi}{6} \sin n\left(\omega t + \frac{\pi}{6}\right)$$

$$V_1 = 0.7797\ V_s$$

4. 120° mode of 3 φ inverter

$$V_{line(rms)} = \frac{V_s}{\sqrt{2}}$$

$$V_{ph(rms)} = \frac{V_s}{2}\sqrt{\frac{2}{3}}$$

$$v_R(\omega t) = \sum_{n=1,3,5,\ldots}^{\infty} \frac{2V_s}{n\pi} \cos\frac{n\pi}{6} \sin n\left(\omega t + \frac{\pi}{6}\right)$$

$$V_{1(rms)} = 0.6752\ V_s$$

5. Voltage control in inverters

Sinusoidal RWM,

$$V_{o(rms)} = V_s \left[\sum_{m=1}^{p} \frac{\delta_m}{\pi} \right]^{\frac{1}{2}}, \qquad p = \frac{mf}{2}$$

6. Current source inverter

$$i_o(\omega t) = \sum_{n=1,3,5,...}^{\infty} \frac{4I}{n\pi} \sin \frac{n\delta}{2} \sin n\omega t$$

❑❑❑

A

Important Mathematical Relations

Trigonometric Relations :

$$\sin(x) = \frac{e^{jx} - e^{-jx}}{2j} \tag{A-1}$$

$$\cos(x) = \frac{e^{jx} + e^{-jx}}{2} \tag{A-2}$$

$$\tan(x) = \frac{\sin(x)}{\cos(x)} = \frac{e^{jx} - e^{-jx}}{j(e^{jx} + e^{-jx})} \tag{A-3}$$

$$e^{\pm jx} = \cos(x) \pm j\sin(x) \text{ (Euler's theorem)} \tag{A-4}$$

$$\cos(x \pm y) = \cos(x)\cos(y) \mp \sin(x)\sin(y) \tag{A-5}$$

$$\sin(x \pm y) = \sin(x)\cos(y) \pm \cos(x)\sin(y) \tag{A-6}$$

$$\cos\left(x \pm \frac{\pi}{2}\right) = \mp \sin(x) \tag{A-7}$$

$$\sin\left(x \pm \frac{\pi}{2}\right) = \pm \cos(x) \tag{A-8}$$

$$\cos(2x) = \cos^2(x) - \sin^2(x) \tag{A-9}$$

$$\sin(2x) = 2\sin(x)\cos(x) \tag{A-10}$$

$$2\cos(x)\cos(y) = \cos(x-y) + \cos(x+y) \tag{A-11}$$

$$2\sin(x)\sin(y) = \cos(x-y) - \cos(x+y) \tag{A-12}$$

$$2\sin(x)\cos(y) = \sin(x-y) + \sin(x+y) \tag{A-13}$$

$$2\cos^2(x) = 1 + \cos(2x) \tag{A-14}$$

$$2\sin^2(x) = 1 - \cos(2x) \tag{A-15}$$

$$4\cos^3(x) = 3\cos(x) + \cos(3x) \tag{A-16}$$

$$4\sin^3(x) = 3\sin(x) - \sin(3x) \tag{A-17}$$

$$8\cos^4(x) = 3 + 4\cos(2x) + \cos(4x) \tag{A-18}$$

$$8 \sin^4(x) = 3 - 4\cos(2x) + \cos(4x) \quad \text{(A-19)}$$

$$A\cos(x) - B\sin(x) = R\cos(x + \theta) \quad \text{(A-20a)}$$

$$\text{where} \quad R = \sqrt{A^2 + B^2} \quad \text{(A-20b)}$$

$$\theta = \tan^{-1}(B / A) \quad \text{(A-20c)}$$

$$A = R\cos(\theta) \quad \text{(A-20d)}$$

$$B = R\sin(\theta) \quad \text{(A-20e)}$$

Derivatives :

Definition :

$$\frac{df(x)}{dx} = \lim_{\Delta x \to 0} \frac{f\left(x + \frac{\Delta x}{2}\right) - f\left(x - \frac{\Delta x}{2}\right)}{\Delta x} \quad \text{(B-1)}$$

Differentiation Rules :

$$\frac{du(x)v(x)}{dx} = u(x)\frac{dv(x)}{dx} + v(x)\frac{du(x)}{dx} \quad \text{(products)} \quad \text{(B-2)}$$

$$\frac{d\left(\frac{u(x)}{v(x)}\right)}{dx} = \frac{v(x)\frac{du(x)}{dx} - u(x)\frac{dv(x)}{dx}}{v^2(x)} \quad \text{(quotient)} \quad \text{(B-3)}$$

$$\frac{du[v(x)]}{dx} = \frac{du}{dv}\frac{dv}{dx} \quad \text{(chain rule)} \quad \text{(B-4)}$$

Derivative Table :

$$\frac{d[x^n]}{dx} = nx^{n-1} \quad \text{(B-5)}$$

$$\frac{d\sin(ax)}{dx} = a\cos(ax) \quad \text{(B-6)}$$

$$\frac{d\cos(ax)}{dx} = -a\sin(ax) \quad \text{(B-7)}$$

$$\frac{d\tan(ax)}{dx} = \frac{a}{\cos^2(ax)} \quad \text{(B-8)}$$

$$\frac{d\sin^{-1}(ax)}{dx} = \frac{a}{\sqrt{1-(ax)^2}} \quad \text{(B-9)}$$

$$\frac{d\cos^{-1}(ax)}{dx} = -\frac{a}{\sqrt{1-(ax)^2}} \quad \text{(B-10)}$$

$$\frac{d\tan^{-1}(ax)}{dx} = \frac{a}{1+(ax)^2} \quad \text{(B-11)}$$

$$\frac{d[e^{ax}]}{dx} = ae^{ax} \quad \text{(B-12)}$$

$$\frac{d[a^x]}{dx} = a^x \ln a \quad \text{(B-13)}$$

$$\frac{d(\ln x)}{dx} = \frac{1}{x} \quad \text{(B-14)}$$

$$\frac{d(\log_a x)}{dx} = \frac{1}{x}\log a \quad \text{(B-15)}$$

$$\frac{d\left[\int\limits_{a(x)}^{b(x)} f(\lambda, x)\, d\lambda\right]}{dx} = f(b(x), x)\frac{db(x)}{dx} - f(a(x), x)\frac{da(x)}{dx} + \int\limits_{d(x)}^{b(x)} \frac{\delta f(\lambda, x)}{dx}\, d\lambda \quad \text{(Leibniz's rule)} \tag{B-16}$$

Integration :

Definition :

$$\int f(x)\, dx = \lim_{\Delta x \to 0} \left\{\sum_n [f(n\,\Delta\, x)]\,\Delta\, x\right\} \tag{C-1}$$

Integration Techniques :

1. Change in variable. Let $v = u(x)$:

$$\int\limits_a^b f(x)\, dx = \int\limits_{u(a)}^{u(b)} \left(\frac{f(x)}{dv / dx}\bigg|_{x = u^{-1}(v)}\right) dv \tag{C-2}$$

2. Integration by parts

$$\int u\, dv = uv - \int v\, du \tag{C-3}$$

Integral Tables :

Indefinite Integrals :

$$\int (a + bx)^n\, dx = \frac{(a + bx)^{n+1}}{b(n + 1)}, 0 < n \tag{C-4}$$

$$\int \frac{dx}{a + bx} = \frac{1}{b} ln\,|a + bx| \tag{C-5}$$

$$\int \frac{dx}{(a + bx)^n} = \frac{-1}{(n-1)\, b\, (a + bx)^{n-1}}, 1 < n \tag{C-6}$$

$$\int \frac{dx}{c + bx + ax^2} = \begin{cases} \dfrac{2}{\sqrt{4ac - b^2}}\, tan^{-1}\left(\dfrac{2ax + b}{\sqrt{4ac - b^2}}\right), b^2 < 4ac \\[2ex] \dfrac{1}{\sqrt{b^2 - 4ac}}\, ln\left(\dfrac{2ax + b - \sqrt{b^2 - 4ac}}{2ax + b + \sqrt{b^2 - 4ac}}\right), b^2 > 4ac \\[2ex] \dfrac{-2}{2ax + b}, b^2 = 4ac \end{cases} \tag{C-7}$$

$$\int \frac{x\, dx}{c + bx + ax^2} = \frac{1}{2a} ln\,|ax^2 + bx + c| - \frac{b}{2a}\int \frac{dx}{c + bx + ax^2} \tag{C-8}$$

$$\int \frac{dx}{a^2 + b^2 x^2} = \frac{1}{ab}\, tan^{-1}\left(\frac{bx}{a}\right) \tag{C-9}$$

$$\int \frac{x\, dx}{a^2 + x^2} = \frac{1}{2}\, ln\,(a^2 + x^2) \tag{C-10}$$

$$\int \frac{x^2\,dx}{a^2 + x^2} = x - a\,tan^{-1}\left(\frac{x}{a}\right) \tag{C-11}$$

$$\int \frac{dx}{(a^2 + x^2)^2} = \frac{x}{2a^2\,(a^2 + x^2)} + \frac{1}{2a^3}\,tan^{-1}\left(\frac{x}{a}\right) \tag{C-12}$$

$$\int \frac{x\,dx}{(a^2 + x^2)^2} = \frac{-1}{2\,(a^2 + x^2)} \tag{C-13}$$

$$\int \frac{x^2\,dx}{(a^2 + x^2)^2} = \frac{-x}{2\,(a^2 + x^2)} + \frac{1}{2a}\,tan^{-1}\left(\frac{x}{a}\right) \tag{C-14}$$

$$\int \frac{dx}{(a^2 + x^2)^3} = \frac{x}{4a^2\,(a^2 + x^2)^2} + \frac{3x}{8a^4\,(a^2 + x^2)} + \frac{3}{8a^5}\,tan^{-1}\left(\frac{x}{a}\right) \tag{C-15}$$

$$\int \frac{x^2\,dx}{(a^2 + x^2)^3} = \frac{-x}{4\,(a^2 + x^2)^2} + \frac{x}{8a^2\,(a^2 + x^2)} + \frac{1}{8a^3}\,tan^{-1}\left(\frac{x}{a}\right) \tag{C-16}$$

$$\int \frac{x^4\,dx}{(a^2 + x^2)^3} = \frac{a^2\,x}{4\,(a^2 + x^2)^2} - \frac{5x}{8\,(a^2 + x^2)} + \frac{3}{8a}\,tan^{-1}\left(\frac{x}{a}\right) \tag{C-17}$$

$$\int \frac{dx}{(a^2 + x^2)^4} = \frac{x}{6a^2\,(a^2 + x^2)^3} + \frac{5x}{24a^4\,(a^2 + x^2)^2} + \frac{5x}{16a^6\,(a^2 + x^2)} + \frac{5}{16a^7}\,tan^{-1}\left(\frac{x}{a}\right) \tag{C-18}$$

$$\int \frac{x^2\,dx}{(a^2 + x^2)^4} = \frac{-x}{6(a^2 + x^2)^3} + \frac{x}{24a^2(a^2 + x^2)^2} + \frac{x}{16a^4(a^2 + x^2)} + \frac{1}{16a^5}\,tan^{-1}\left(\frac{x}{a}\right) \tag{C-19}$$

$$\int \frac{x^4\,dx}{(a^2 + x^2)^4} = \frac{a^2x}{6(a^2 + x^2)^3} - \frac{7x}{24(a^2 + x^2)^2} + \frac{x}{16a^2(a^2 + x^2)} + \frac{1}{16a^3}\,tan^{-1}\left(\frac{x}{a}\right) \tag{C-20}$$

$$\int \frac{dx}{a^4 + x^4} = \frac{1}{4a^3\sqrt{2}}\,ln\left(\frac{x^2 + ax\sqrt{2} + a^2}{x^2 - ax\sqrt{2} + a^2}\right) + \frac{1}{2a^3\sqrt{2}}\,tan^{-1}\left(\frac{ax\sqrt{2}}{a^2 - x^2}\right) \tag{C-21}$$

$$\int \frac{x^2\,dx}{a^4 + x^4} = -\frac{1}{4a\sqrt{2}}\,ln\left(\frac{x^2 + ax\sqrt{2} + a^2}{x^2 - ax\sqrt{2} + a^2}\right) + \frac{1}{2a\sqrt{2}}\,tan^{-1}\left(\frac{ax\sqrt{2}}{a^2 - x^2}\right) \tag{C-22}$$

$$\int cos\,(x)\,dx = sin\,(x) \tag{C-23}$$

$$\int x\,cos\,(x)\,dx = cos\,(x) + x\,sin\,(x) \tag{C-24}$$

$$\int x^2\,cos\,(x)\,dx = 2x\,cos\,(x) + (x^2 - 2)\,sin\,(x) \tag{C-25}$$

$$\int sin\,(x)\,dx = -\,cos\,(x) \tag{C-26}$$

$$\int x\,sin\,(x)\,dx = sin\,(x) - x\,cos\,(x) \tag{C-27}$$

$$\int x^2\,sin\,(x)\,dx = 2x\,sin\,(x) - (x^2 - 2)\,cos\,(x) \tag{C-28}$$

$$\int e^{ax}\,dx = \frac{e^{ax}}{a},\ \text{a real or complex} \tag{C-29}$$

$$\int xe^{ax}\,dx = e^{ax}\left(\frac{x}{a} - \frac{1}{a^2}\right),\ \text{a real or complex} \tag{C-30}$$

$$\int x^2 e^{ax}\, dx = e^{ax}\left(\frac{x^2}{a} - \frac{2x}{a^2} + \frac{2}{a^3}\right), \text{ a real or complex} \quad \text{(C-31)}$$

$$\int x^3 e^{ax}\, dx = e^{ax}\left(\frac{x^3}{a} - \frac{3x^2}{a^2} + \frac{6x}{a^3} - \frac{6}{a^4}\right), \text{ a real or complex} \quad \text{(C-32)}$$

$$\int e^{ax} \sin(x)\, dx = \frac{e^{ax}}{a^2+1}[a \sin(x) - \cos(x)] \quad \text{(C-33)}$$

$$\int e^{ax} \cos(x)\, dx = \frac{e^{ax}}{a^2+1}[a \cos(x) + \sin(x)] \quad \text{(C-34)}$$

Definite Integrals :

$$\int_0^\infty \frac{x^{m-1}}{1+x^n}\, dx = \frac{\pi / n}{\sin(m\pi / n)}, n > m > 0 \quad \text{(C-35)}$$

$$\int_0^\infty x^{\alpha-1} e^{-x}\, dx = \Gamma(\alpha), \alpha > 0 \quad \text{(C-36a)}$$

where

$$\Gamma(\alpha+1) = \alpha\Gamma(\alpha) \quad \text{(C-36b)}$$

$$\Gamma(1) = 1; \Gamma\left(\frac{1}{2}\right) = \sqrt{\pi} \quad \text{(C-36c)}$$

$$\Gamma(n) = (n-1)! \text{ if n is an integer} \quad \text{(C-36d)}$$

$$\int_0^\infty x^{2n} e^{-ax^2}\, dx = \frac{1 \cdot 3 \cdot 5 \ldots (2n-1)}{2^{n+1} a^n}\sqrt{\frac{\pi}{a}} \quad \text{(C-37)}$$

$$\int_{-\infty}^\infty e^{-a^2x^2+bx}\, dx = \frac{\sqrt{\pi}}{a} e^{b^2/(4a^2)}, a > 0 \quad \text{(C-38)}$$

$$\int_0^\infty e^{-ax} \cos(bx)\, dx = \frac{a}{a^2+b^2}, a > 0 \quad \text{(C-39)}$$

$$\int_0^\infty e^{-ax} \sin(bx)\, dx = \frac{b}{a^2+b^2}, a > 0 \quad \text{(C-40)}$$

$$\int_0^\infty e^{-a^2x^2} \cos(bx)\, dx = \frac{\sqrt{\pi}\, e^{-b^2/4a^2}}{2a}, a > 0 \quad \text{(C-41)}$$

$$\int_0^\infty x^{\alpha-1} \cos(bx)\, dx = \frac{\Gamma(\alpha)}{b^\alpha} \cos\frac{1}{2}\pi\alpha, 0 < \alpha < 1, b > 0 \quad \text{(C-42)}$$

$$\int_0^\infty x^{\alpha-1} \sin(bx)\, dx = \frac{\Gamma(\alpha)}{b^\alpha} \sin\frac{1}{2}\pi\alpha, 0 < |\alpha| < 1, b > 0 \quad \text{(C-43)}$$

$$\int_0^\infty x e^{-ax^2} Ik(bx)\, dx = \frac{1}{2a} e^{b^2/4a} \quad \text{(C-44a)}$$

where $$I_k(bx) = \frac{1}{\pi}\int_0^{\pi} e^{bx\cos(\theta)}\cos(k\theta)\,d\theta \tag{C-44b}$$

$$\int_0^{\infty}\frac{\sin x}{x}\,dx = \int_0^{\infty} Si(x)\,dx = \frac{\pi}{2} \tag{C-45}$$

$$\int_0^{\infty}\left(\frac{\sin x}{x}\right)^2 dx = \int_0^{\infty} Si^2(x)\,dx = \frac{\pi}{2} \tag{C-46}$$

$$\int_0^{\infty}\frac{\cos(ax)}{b^2+x^2}\,dx = \frac{\pi}{2b}e^{-ab},\ a>0,\ b>0 \tag{C-47}$$

❑❑❑

B

Important Laplace Transform Relations

The laplace transform of a function f(t) is given as,

$$F(s) = \int_{0-}^{\infty} f(t) e^{-st} dt$$

Table B.1 : Properties of Laplace Transform

Sr.No.	Name	Property
1.	Shifting theorem	$\mathcal{L}\,[f(t-t_0)] = e^{-st_0} f(s)$
2.	Differentiation theorem	$\mathcal{L}\,\frac{d}{dt}f(t) = sF(s) - f(0^-)$
3.	Integration theorem	$\mathcal{L}\int_0^t f(t)\,dt = \frac{F(s)}{s}$
4.	Convolution therom	$\mathcal{L}\,[f_1(t) * f_2(t)] = F_1(s) \cdot F_2(s)$
5.	Complex translation	$\mathcal{L}\left[e^{at} f(t)\right] = F(s-a)$
6.	Differentiation by s	$\mathcal{L}\,[t\,f(t)] = -\frac{d}{ds}F(s)$
7.	Initial value therom	$f(0^+) = \lim_{t \to 0^+} f(t) = \lim_{s \to \infty} [sF(s)]$
8.	Final value therom	$\lim_{t \to \infty} f(t) = \lim_{s \to 0} [sF(s)]$

Table B.2 : Important Laplace Transform Pairs

Sr. No	Name of the function	Laplace transform pair
1.	Exponential function	$e^{at} \xleftrightarrow{\mathcal{L}} \dfrac{1}{s-a}$
2.	Unit step function	$Au(t) \xleftrightarrow{\mathcal{L}} \dfrac{A}{s}$
3.	Ramp function	$tu(t) \xleftrightarrow{\mathcal{L}} \dfrac{1}{s^2}$
4.	Impulse function	$\delta(t) \xleftrightarrow{\mathcal{L}} 1$
5.	Sine wave	$A \sin \omega_0 t \xleftrightarrow{\mathcal{L}} \dfrac{A\omega_0}{s^2 + \omega_0^2}$
6.	Cosine wave	$A \cos \omega_0 t \xleftrightarrow{\mathcal{L}} \dfrac{A \cdot s}{s^2 + \omega_0^2}$
7.	Damped sine wave	$e^{-at} \sin \omega t \xleftrightarrow{\mathcal{L}} \dfrac{\omega}{(s+a)^2 + \omega^2}$
8.	Damped cosine wave	$e^{-at} \cos \omega t \xleftrightarrow{\mathcal{L}} \dfrac{s+a}{(s+a)^2 + \omega^2}$
9.	Hyperbolic sine wave	$e^{-at} \sin hbt \xleftrightarrow{\mathcal{L}} \dfrac{b}{(s+a)^2 - b^2}$
10.	Hyperbolic cosine wave	$e^{-at} \cos hbt \xleftrightarrow{\mathcal{L}} \dfrac{s+a}{(s+a)^2 - b^2}$

Table B.3 : Laplace Transformations of Basic Elements

Name of element	Element in time domain	Impedance - voltage source representation in 's' domain	Admittance current source representation in 's' domain
Resistance	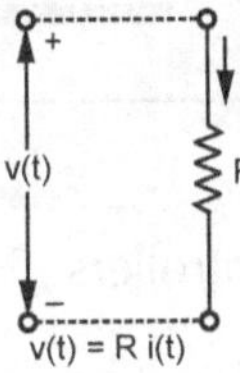 $v(t) = R\, i(t)$	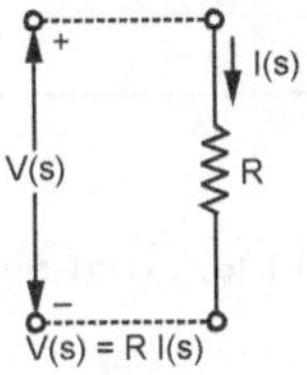 $V(s) = R\, I(s)$	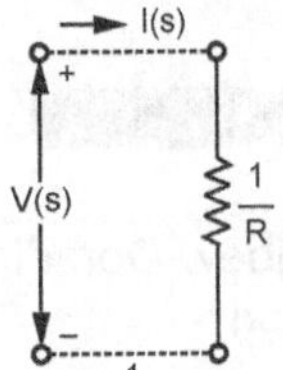 $I(s) = \frac{1}{R} . V(s)$
Inductance	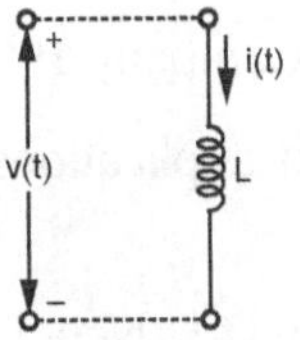 $v(t) = L \frac{d\, i(t)}{dt}$	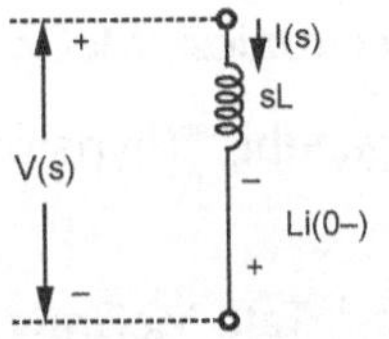 $V(s) = sL\, I(s) - Li(0-)$	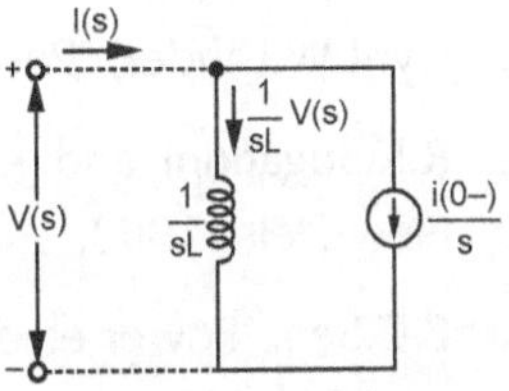 $I(s) = \frac{1}{sL} V(s) + \frac{i(0-)}{s}$
Capacitance	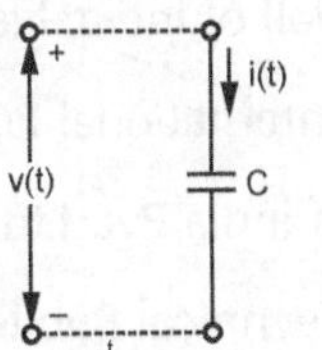 $v(t) = \frac{1}{C} \int^{t} i(t)\, dt + v(0-)$	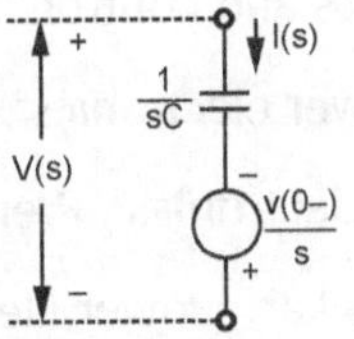 $V(s) = \frac{1}{sC} (s) + \frac{v(0-)}{s}$	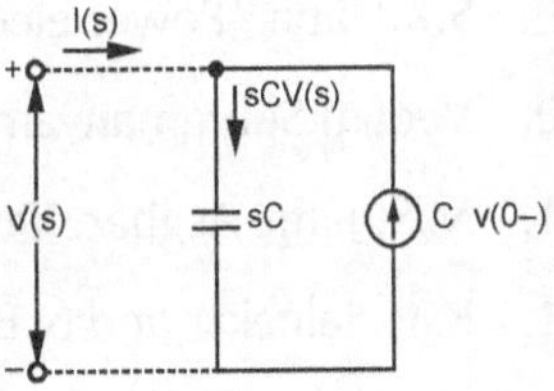 $I(s) = sC\, V(s) - C\, v(0-)$

❑❑❑

References

1. Dubey, Doradla, Joshi and Sinha, "Thyristorised power controllers", New Age International, 1999.
2. M.H.Rashid, "Power electronics", Prentice Hall of India pvt. Ltd., New Delhi 2004.
3. M.Ramamoorty, "Thyristors and their applications", East-west Press Ltd., New Delhi 1991.
4. Cyril W.Lander, "Power electronics", McGraw Hill Book Company 1993.
5. R.K.Sugandhi and K.K.Sugandhi, "Thyristors theory and applications,", Wiley Eastern Ltd., New Delhi 1993.
6. P.C.Sen, "Power electronics", Tata McGraw Hill Company Ltd., New Delhi 1991.
7. P.C.Sen, "Thyristor DC drives", John Wiley and Sons, 1981.
8. Mohan, Undeland and Robbins, "Power electronics", John Wiley and Sons 1989.
9. S.K.Datta, "Power electronics and controls", Prentice Hall of India, New Delhi 1989.
10. Vedam Subramanyam, "Power electronics", New Age International Publishers 1996.
11. M.S. Jamil Asghar, "Power Electronics", Prentice-Hall of India Pvt. Ltd., New Delhi 2005.
12. R.M. Jalnekar and N.B. Pasalkar, "Power Electronics", Technical Publications, Pune 2001.
13. M.D. Singh and K.B. Khanchandani, "Power Electronics", Tata McGraw Hill Publishing Company Ltd., New Delhi 2003.
14. Joseph Vithayathil, "Power Electronics Principles and Applications", McGraw Hill, NY 1995.
15. David C. Griffith, "Uninterruptible Power Supplies", Marcel Dekker Inc., NY 1994.
16. Dubey G.K., "Power Semiconductor Controlled Drives", Prentice-Hall Englewood Cliffs, 1989.
17. "GE SCR Manual", 6th edition, General electric, NY.
18. Chrysis G.C., "High Frequency Switching Power Supplies : Theory and Design", Mc Graw Hill NY 1988.

❑❑❑